Circular Saws

Circular Saws

Their Manufacture, Maintenance and Application in the Woodworking Industries

Eric Stephenson

Stobart Davies
Hertford

Stobart Davies Limited,
Priory House, Priory Street,
Hertford SG14 1RN UK.

British Library Cataloguing in Publication Data
A catalogue record for this book is available from the British Library
ISBN 0 85442 091 6
Published 2002

By the same author:

The Spindle Moulder Handbook
The Shaper Handbook
The Autolathe

To Eunice, my dear Wife. Writing technical books and articles, and preparing and presenting lectures over the past 45 years would have been impossible without her understanding and tolerance.

Acknowledgments

During my fifty-odd years in the woodworking trade I have had the good fortune to meet people from all parts of the world who manufacture or use woodworking machinery in all its forms, and many of these became my personal friends. My thanks are due to them all - far too numerous to mention individually by name - the help, encouragement and comprehensive information they supplied was invaluable when composing this and other woodworking books.

I pay particular thanks to my co-author of our first 1972 book on Circular Saws, Dave Plank, who regrettably is no longer with us. Without his enthusiastic encouragement this first book may never have been published.

Individuals and Companies who checked sections of the book and made helpful suggestions are listed below.

David Aspinwall, Head of Metal Cutting/Material Removal Research, University of Birmingham, UK
Fred Andrianoff, Portland, Oregon, USA
Keith Dobson, MIM Wood T, of **Yew Lodge Health & Safety Training**, Henlow, Bedfordshire, UK.
Fred Duckworth, Pendleton, Lancs, UK (nomograms)
Chandler Jones, Woodworking Writer, Seattle, Washington, USA
Geoff Woods, AIW Sc, MIM Wood T, Woodworking Consultant, Cleveleys, Lancs. UK
Chris Walker of **Atkinson Walker (Saws) Ltd** , Sheffield, UK
Jean-Piere Baron of **J. Chaland & Fils** SA, Cedex, France
DeBeers Industrial Diamond Division, Shannon, Co. Clare, Ireland
Deloro Stellite Inc. Belleville, Ontario, Canada
Deloro Stellite Ltd., Swindon, Wiltshire UK
Alasdair Boag of **Dymet Alloys Ltd.**, Aldershot, Hampshire. UK
Jack Sigrist of **Equipment Ltd**., Hickory, NC, USA, and **Saturn**, Rafz, Switzerland
Metcalfe & Tattersall Ltd., Oswaldtwistle, Lancs. UK
Gary Metzger of **North American Products Corporation**, Jasper, Indiana, USA
John Dexter of **Skellingthorpe Saw Services**, Skellingthorpe, Lincs UK
Spear & Jackson International Ltd., Sheffield, UK
Richard Lee of **Supreme Saws Ltd.**, Romford, Essex, UK
Vollmer UK Ltd., Sandiacre, Nottingham UK
Walter GB Ltd., Reddich, Worcestershire UK

Companies who were more than usually helpful are:-

Ake Knebel GmbH & Co, Balingen, Germany
AnThon GmbH & Co, Flensburg, Germany
Armstrong Manufacturing Co. Portland, Oregon, USA
Autool Grinders, Sabden, Lancs. UK
SCM GB Ltd, Bulwell, Nottingham, UK
Felder Maschinenbau, Hall in Tirol, Austria
Format-4, Loreta 42, Hall in Tirol, Austrai
GreCon Dimter Ltd, (Weinig), Tauberbischofsheim, Germany
Leicester Wood Technique Ltd., Lutterworth, Leicestershire UK.
Leitz Tooling, Harlow, Essex UK
Leuco Oertli, Horb am Neckar, Germany
Michael Weinig (UK) Ltd, Abingdon, Oxfordshire UK

Book References

Armstrong Saw Filers Handbook - PS Quelch, *Armstrong Manufacturing Co, Portland, Oregon. USA*
Bandsaws, Wide and Narrow Blade Types, also **Cutterhead & Knives for Machining Wood** - *Chandler Jones, Seattle, USA*
Forest Products Research Laboratory, various publications.
Hanchett Saw & Knife Manual, *Hanchett Manufacturing Co, Big Rapids, Michigan, USA*
Story of the Saw - *P d'A Jones & EN Simonds, Spear & Jackson Ltd., Sheffield, UK*
The Story of the Mushets - *Fred M Osborn, Thomas Nelson & Sons Ltd., London, UK*
Two hundred years of History and Evolution of Woodworking Machinery - *William L Sims, Leicestershire, UK*
Woodworking Machinery 1800-1880 - *M Powis Bale, London* (Reprinted by Glen Moor Press, Lakewood, CO 80215 USA)

All other sources of material, text, drawing and illustrations, are listed in the final pages.
To all of them I say a very big thank you.

See the Inside Rear Cover for details of Advertisers.

Contents

Chapter 1
Introduction

The saw doctor, or filer, was always considered an important person in all early log breakdown and sawmills, without him fast and economic breakdown of timber would virtually have been impossible.

Although the quality of tooling has advanced dramatically in recent years the tool room remains just as essential now as it was in then. Whilst machines have become highly efficient they are even more costly to buy and run. This makes it more important than ever to keep them in constant production - and the whole operation hinges on the sawblades and tools being kept in the best possible condition.

As with the machines themselves tool room equipment has improved almost beyond recognition, but using the equipment correctly is up to the individual skills, and there still remains that air of mystique where circular saws are concerned.

Fig. 1.1

Most saw doctors problems do not include cutoff saws of this size, but the same dedication to the art is just as important today as when this photograph was taken in the 1970's.

Source - The Armstrong Saw Engineer

Learning the art of saw doctoring cannot be taught from a book, the individual has to find out for himself with hands-on practice on the bench; using hammers and noting their effect; checking how the setting of grinders effects the quality and life of the sawblades - and so on.

Hopefully this book will explain the whys and wherefore in a simply-understood way. Greater than usual detail is given where background information would help to better explain the behavior of materials the saw doctor uses in his everyday work. But whilst the book lays out the principals of maintaining sawblades, it is practice, practice, practice that makes perfect.

Although most books saw doctoring begin with a chapter on the saw shop, such as why it should be insulated from the mill and provided with good natural light and heat and/or cooling, the subject is not covered in this book except in part. Much of it is plain common sense anyway, and a quick visit to the tool room of any big-name company would show users what is needed. In any event, variation in the numbers and types of sawblades and other tools means that each tool room is different.

The Saw in history

Originally logs and flitches were converted by placing the timber over a pit and sawing vertically using a straight, two-handed saw. The sawblade was kept in tension by both sawyers pulling the it in turn, one straddling the timber and the other in the pit - but keeping the sawblade in tension was the problem. This could be solved by fitting the sawblade in a frame (a frame saw) and stretching it using screws or wedges, but the frame had to be made wide enough to straddle the log or flitch - and one for large logs was too unwieldy. There also was difficulty in making a frame strong enough to tension the sawblade but light enough for manual use.

Fig. 1.2

A hand-cranked frame saw as used in the 15th. & 16th. Centuries

Source - Story of the Saw

The origins of woodworking machinery can be traced to mechanical frame saws used in Augsburg, Germany, in the 15th. Century, using a type of saw frame which was strong and heavy. It was moved up and down on pivots or in guides, and driven mechanically by man, animal, wind or water power. The feed was usually linked in some way to the saw frame movement to make this automatic.

The first machines had only single sawblades, but later more sawblades were stretched within the single frame to make multiple cuts - now called gang saws. By the 17th. Century single frame and gang frame saws were in wide use on the European continent and in the British colonies of New England.

But frame saws were slow, and the rapidly increasing demand for goods following the industrial revolution of the 18th century needed a faster, continuous cutting action.

This was met by the circular saw which most attribute to Samuel Miller of Southampton, England. He used circular saws and lathes from 1777 onwards to manufacture pulley blocks for the British Navy. By 1781 circular saws were in wide use in England and their use quickly spread around the world.

Sheffield, where swords had been made for generations,

quickly became the leading British steel city. In Sheffield there were several major makers and exporters of circular saws (and other tools) - at first using the closely-guarded Sheffield crucible cast steel process. The later inventions of the Bessemer converter (1856) and high speed steel (1868) further confirmed Sheffield as the leading name for sawblades and tools of all types, and Sheffield became a byword for quality tooling.

In America quite innovative circular sawing machines were developed for converting timber, but these all initially used British sawblades. Henry Disston was an early supplier of sawblades made from crucible steel imported from England, but he began manufacturing crucible steel in his own plant from 1855 and in doing so doing laid the foundations of the American tool industry.

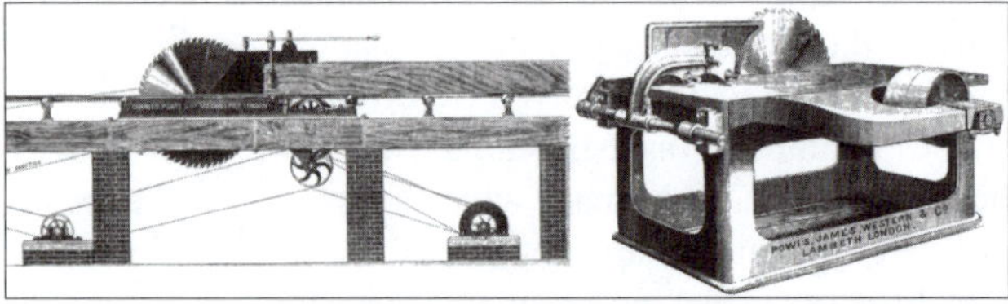

Fig. 1.3

This Powis rack-feed saw bench was typical of the designs of the early 1870's, using a mixture of timber and iron in the construction (left). Eventually they were made with an iron frame (right) but little guarding was supplied.

Source - Powis-Bale

The first sawblades was simply rough and untensioned ground-flat plates - plate saws - with teeth cut into the rim and which ran at about half today's speeds. Hammer tensioning of circular saws dated from about 1814 at Bentonville in New York, and this made sawblades more stable and so able to run at higher speeds. Saw hammering eventually became one of the black arts of the developing industry - which some millmen could master and others could not.

The conflicting needs of a one-piece (plate) saw - to provide both a stable saw body and keen and long lasting teeth - meant that the saw steel had to be a compromise - ideally suited to neither purpose.

Fig. 1.4

This British circular logmill used a single sawblade which often was an American-designed insert-tooth saw, inset view. The insert-tooth saw proved invaluable in log and flitch conversion in the early days - it was tough and stood a lot of abuse, but it had an excessive and very wasteful saw kerf width.

Source - Robinson/Wadkin.

From early times the link between hardness of a saw tooth and its lasting qualities was well understood - the harder the points and the longer the saw ran before needing filing or re-grinding. Unfortunately, in those days, steel hard enough to give a long life in sawblades made the saw body prone to cracking - a problem that haunted the first saw makers for many years.

The ideal was to find a suitable body with a tip so hard that re-sharpening would never be needed. This was an impossible dream then, but modern tooth tips are now much closer to this ideal than earlier users ever imagined.

To meet these conflicting needs many designs using mechanically-held loose teeth were attempted from about 1824, and in 1859 Spaulding of Sacramento invented the successful curved socket type inserted tooth that still remains in use today. Often known as Yankees in England, these saws are still used for converting small diameter logs and for resawing timber reclaimed from derelict buildings - where regular saws are quickly wrecked.

Another pattern, this time using a straight tooth held in position by a rivet or stud, was used for the large diameter crosscut saws in log breakdown mills. These have largely been replaced by mechanically-operated chainsaw machines.

Very much later other hard tips were brazed or welded on circular saw teeth, as later described, but the body for all these sawblades is steel - and understanding the properties of steel is important to the saw doctor and filer.

Steel Manufacture

Steel is mostly iron, which was likely first produced by damping down a fire with soil rich in iron ore. The iron ore smelted down because the hot and enclosed conditions drove off oxygen to leave a clump of poor quality iron amongst the ashes.

Although these clumps (or blooms) still contained natural impurities that made it of little use in this state, early man found that by reheating the bloom in hot ashes and folding and working it with crude hammers it could be made stronger and ideally suited for domestic use and weapons.

From these first crude fires the bloomery evolved - an enclosed furnace with a restricted air supply - a process which was widely adopted. Charcoal was first used as fuel because coal, although capable of raising enough heat for the process, added impurities that made the iron brittle and unusable. In the fifteenth century production was vastly improved by adding a forced draft, then called a blast furnace, to create a fiercer heat and so produce a liquid iron. Limestone was added to combine with any impurities to form a floating slag which could then be run off before the iron was tapped.

Charcoal, produced by heating timber in an enclosed oven, was also used to make the quicklime needed for building and as a soil conditioner in farming. As a result charcoal production became a major industry to such an extent that it created a shortage of timber in England - where timber was also widely used in house and ship building.

The fuel crisis was solved in 1709 when Abraham Darby successfully converted a specific type of coal into coke, and this made a better (basic) pig iron (so-called because liquid iron was run off into a series of side-by side depressions in the foundry floor which resembled pigs feeding). Quality cast iron could then be made by remelting both pig iron and scrap iron together in a copula - a smaller blast furnace which allowed iron to be refined by adding other materials. In the early years, though, pig iron was mainly converted into wrought iron in a puddling furnace as blooms - the first major iron-based construction material. In a puddling furnace long poles stirred the molten iron manually to collect pure iron on them as blooms, with all the natural impurities burnt-out by the fierce heat. These blooms were mixed with floating slag and then folded and hammered or rolled repeatedly to produce a fibrous material strong in tension.

By repeated working wrought iron it was possible to produce swords and cutlery with excellent cutting edges. The material actually became steel - but manufacture in this way was slow, labour intensive and very costly.

Steel is basically iron with most of the natural carbon and impurities removed, and with the remaining carbon bonded to the iron as an iron carbide. It is now no longer produced from wrought iron, but from pig iron, scrap and the alloying elements needed to change its characteristics.

Steel is now produced by continuous casting, or cast as an ingot and either hot-rolled into sheets or billets, or remelted and mixed to produce special steels. Hot rolling leaves steel coated with oxide which needs grinding off before use. Cold rolling of cleaned slabs produces a shiny steel which can be used without further surface treatment.

Fig. 1.5

The modern blast furnace converts iron ore into liquid iron, with coke as the fuel and using a hot air blast (left). Limestone is added to separate impurities from the pure iron by forming a floating slag. A slabbing mill reduces hot billets to slabs of smaller section in preparation for further work (right, top). This multistage cold-rolling line produces flat steel strip which is coiled before shipping (right, bottom). The strip can be sheared and shaped for conventional saw manufacture or laser-cut straight from the strip.

Source - British Steel (now Corus)

One important element in steel is carbon, the main hardening agent, and the amount of carbon in steel classifies all basic steel as low, medium or high-carbon.

Low carbon or mild steels, which are relatively soft and unaffected by heat treatment, are widely used for manufacturing machinery. Medium and high carbon steels are further alloyed to make tool steels which can be hardened by heat treatment.

High carbon steels can be hardened because of the carbon in them. When heated to above the HCP (higher critical point) a change takes place - the vibration that heat causes so shakes the bond between atoms that they break up and switch to a more rigid structure in a complex crystalline form.

If the steel cools slowly the atoms switch back to their original bonding into a soft or annealed condition - something that happens very quickly and shows physically for a few seconds as an apparent reheating of the steel. If instead steel is immediately quenched, in a water, oil or water and oil mix, the hot-bond structure remains frozen to create a hard and brittle steel with large and coarse iron carbide crystals. Steel in this form is quite useless for saw bodies as it readily cracks.

To make a usable steel, hardening is followed by tempering, where it is reheated to a very precise point between the higher and lower critical points, and then again cooled. This makes the crystals smaller, more numerous and more evenly distributed. It also slightly reduces the hardness of the steel - but makes it both hard and tough enough for commercial use. Normally only single tempering is needed, but further improvement is possible by double or triple tempering, or cryogenically - using very low temperatures.

Sawblade manufacture

The first crude circular saws were made from Sheffield crucible cast steel, where carbon and wrought iron, as blister steel, was heated in clay crucibles, then poured into an ingot and finally reworked by hammers and rollers.

Modern circular saws are now made from alloyed steel, where various metals and nonmetals are added as alloying elements to the basic steel to make it more flexible and tough without sacrificing hardness. In present-day manufacture the analysis of the steel mix, its quality and purity, is very tightly and very carefully controlled.

The main element which adds strength and toughness to steel is carbon. Other elements modify the steel by combining with iron or other elements to make crystals more malleable and less likely to break down. In greater detail, manganese hardens the hardest part of steel, the iron carbides, whilst nickel both hardens and toughens it.

Fig. 1.6

Casting steel ingots using the Sheffield Crucible Cast Steel process invented by Benjamin Huntsman in 1742 (left). The process can still be seen at Abbeydale in Sheffield.

Source - The Story of the Mushets.

Large plates are hot-rolled individually as squares (right).

Source - Spear & Jackson

Most sawblades are made from chrome-vanadium steel. Manganese and chromium increase steel hardness and help to distribute the carbides more evenly. Tungsten, vanadium, chromium and molybdenum bond with carbon to form complex carbides which enhance the existing iron carbides.

To get the best results from these complex and precise mixes all saw plates need very careful hardening and tempering, with temperatures and quenching fluids varying according to the specific mix of alloyed steel.

Fig. 1.7

Sixty year ago large-diameter saws were a regular product in Sheffield, an example of which is this 78in (2M) diameter plate saw destined probably for Africa. The shark's tooth form was used both there and S.E. Asia.

Source - Atkinson Walker

Ingots are cog-rolled to slabs, cross-rolled to plates, or cut to size in the steel mill and usually annealed so they can be machined more easily. The plates for the larger saws may be squares or bought as circles.

The blanks are then bored out for the centre hole and toothed around this centre. Blanks are hardened and tempered in an electric or gas-heated oven, or by salt bath, and gripped between open frames whilst quenching to keep them flat. A flattening process is needed for all saw plates, especially hot-rolled plates, by smithing them using special hammers and anvils similar to those used by saw doctors or filers.

Fig. 1.8

Saw production showing quenching of a large diameter saw (left), and standard smaller and large diameter saw plates in process (right).

Source - Atkinson Walker

Following this the sawblade is surface ground to thickness, using a plain grinding wheel mounted on an arbor parallel to the saw plate to give concentric grinding marks, or by a grinding wheel mounted on an arbor at right angles to the saw plate to give part-circular grind marks

The sawblade is then blocked to remove lumps and twists and to even out tension by using blocking hammers - these are similar to but less effective than regular smithing hammers, but do not mark the surface of the sawblade.

The saw teeth are then ground and set (or in the case of tipped saws tipped and ground) before the final smithing and roller tensioning. To remove the tip brazing residues on tipped saws the outer section is usually shot-blasted. The last process is checking and correcting the balance before inspection, marking and packing.

Many sawblades are now manufactured from cold rolled strip which can be sheared into squares for circular trimming before punching-out the teeth. Alternatively they are laser-cut - complete with the saw tooth form, tip seating and center hole - straight from the strip or from squares. This gives a distortion and stress-free cut, and allows makers to very simply add complex expansion and silencing slots.

The better quality saws made from steel strip are normally surface ground on both sides - even though cold-rolling gives an acceptable finish. For lower quality saws the bodies are more often simply polished or spray-treated.

Although most circular plate saws are manufactured from a chrome-vanadium steel, some makers offer more than one class of body for tipped sawblades - according to the use they are intended for - and will advise on the specification once cutting details are given.

Alloy Steel Sawblades

Parallel plate saws

As its name implies, the sides on plate saws are parallel to one to the other. The teeth are spring or swage-set to give clearance at the body of the saw. Saw plate thickness increases in step with sawblade diameter to keep the saw rigid enough for fast-feed deep sawing, but alternative gauges are offered in most sawblade diameters.

Exceptionally thin sawblades are best avoided unless both sawing practice and saw doctoring are of a very high standard - a thin sawblade can be very quickly ruined if badly treated. Thicker-than-normal sawblades are needed when dealing with rough timber or with fast feed rates.

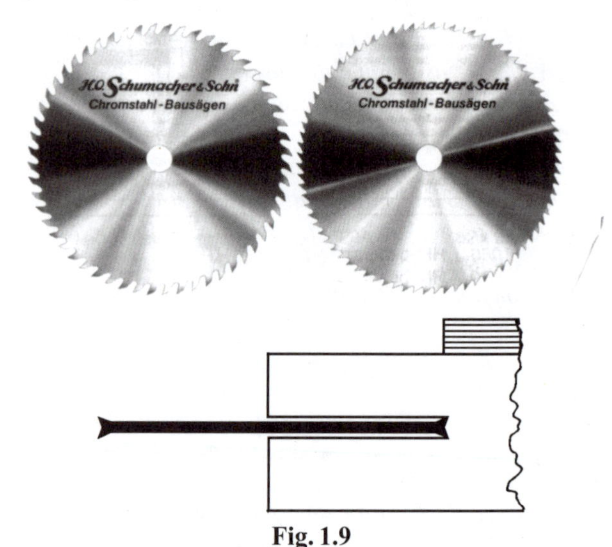

Fig. 1.9

Above, typical chrome-steel plate saws - regular tooth left, peg tooth right. Below, teeth are spring or swage set to give working clearance.

Source - Schumacher.

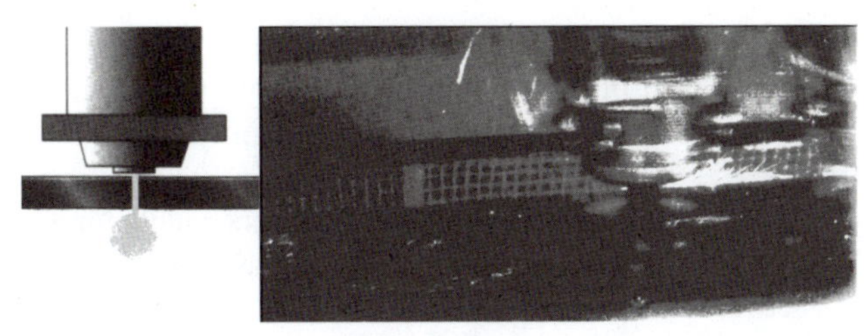

Fig. 1.10

Plate saws can be laser-cut direct from a steel strip, complete with gullets, tip seating recesses, silencing and expansion slots, etc. Unlike die punching, laser cutting does not stress the saw body and can be successfuly used on the harder bodies needed for thin-kerf sawblades.

Source - Freud

Narrow kerf saws

Swage, ground-off and taper saws are ground to a narrow saw kerf width, perhaps only 0.75mm (1/32 in) or thereabouts, but retain a thick body in the collar area to withstand the unusual cutting forces. Because the sawn timber contacts the sides of the sawblade and has to be deflected by it wedge-fashion, they can be used when deep-cutting thin timber only.

Sawblades of this type must only be used on a power-bench machine because the wedge shaped body forces the timber up. Because the teeth should barely break through the top surface of the timber, these narrow-kerf sawblades must also be used on machines with a rise and fall spindle, or in different diameters to match the depth of cut. Teeth are usually spring-set, and tooth pitch is generally smaller than on a plate saws.

They produce an exceptionally fine finish, but require very careful treatment in preparation and use. They are now hardly used, having been replaced by the more versatile and less troublesome wide bandsaws. For the record, though, there are three basic types; swage, taper-ground and ground-off.

Fig. 1.11

This 1870's roller-feed sawbench is typical of the machines using narrow-kerf swage saws for board conversion during this period, such as that shown alongside.

Source - Powis-Bale.

Swage Saws

The swage saw has a thick body which is taper ground on one side only from just beyond the collars to the saw teeth points.

The taper-ground side is always on the fence side of the machine, normally the right-hand side (though benches were also made left-hand at one period).

The gauge-sawn thickness is limited to a maximum of 16-19mm ($^5/_8$ in or $^3/_4$ in) - depending on type and condition of the timber. Swage saws are ideal for through and through conversion of box boards where economy of the saw kerf and a fine finish are important.

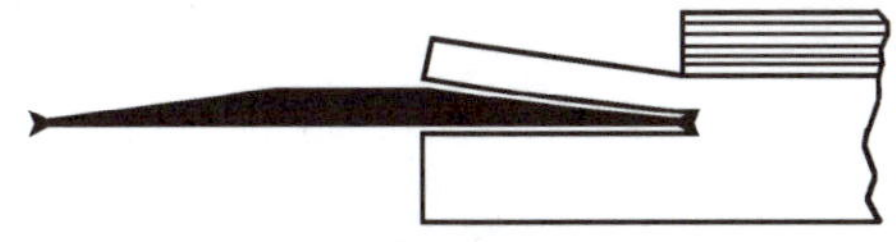

Fig. 1.12

The swage saw is usually made as shown, but can be either hand.

Swage saws are normally spring-set unevenly - with a little more set on the bevelled than the flat side of the sawblade to slightly offset the effect of the bevel. It is not practical to swage-set teeth as the side dies of the swager would need re-grinding.

Taper-ground saws

Fig. 1.13

The taper-ground or double swage saw is suitable only for deep-cut splitting of thin boards.

The taper ground saw is bevelled on both sides by an equal amount from just outside the saw collars to the teeth, and is sometimes known as a double swage saw. This type is suitable only for splitting timber of up to 37mm ($1^1/_2$in) thick maximum.

Ground-off Saws

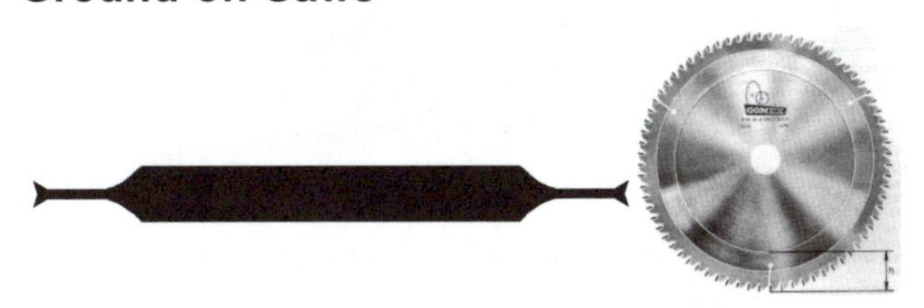

Fig. 1.14

The ground-off saw is for special applications where an extra-narrow kerf is needed for a limited depth of cut. Right, A modern tungsten-carbide tipped ground-off saw. The maximum depth of cut is indicated by 'h'.

Source - Gomex

The ground-off saw is ground parallel at the collar - and for the greater part of its diameter - to a much thicker gauge than a regular plate saw. From here there is a short concave grind from both sides from where it is ground parallel and to a much thinner gauge outwards to the teeth.

Ultra-thin kerf saws are now made for economic production of such as lamellas and louvre slats, either as a parallel plate supported by special spacers, or as a bossed-body sawblade. See page 131 for details.

Hollow-ground saws

With hollow-ground saws the collar area only is ground parallel, but from there it is taper-ground, equally on both sides, increasing in thickness gradually from the collar to the teeth. The teeth are usually the same gauge as the collar area, and the thinnest section is just beyond the collar area - also its weakest point.

The saw kerf has to be greater than with a plate saw to retain enough body strength at its weakest part, which limits hollow-ground saws to a maximum diameter of 600mm (24in) or thereabouts.

Hollow grinding gives side clearance to the teeth without need of spring or swage setting. As tooth alignment is guaranteed by machine-grinding, the tips are always perfectly aligned - giving good quality results.

Unfortunately hollow grinding does not always give enough side clearance - this readily shows up when the saw motor overloads. To cure this the teeth can be spring set - not recommended but quite a regular practice. Because a smaller amount of set is needed with a hollow-ground saw than with a plate saw, setting is usually more precise.

Fig. 1.15

Hollow-grinding of these sawblades gives clearance to the teeth without need of setting, and so produces a fine cut.

These sawblades are not for rough and tumble of conversion, but are suitable for crosscutting, mitring, general dimension trimming of timber and sizing of plywood and plastic panels, etc. They have, of course, virtually been replaced by tc tipped saws.

Insert-tooth saws

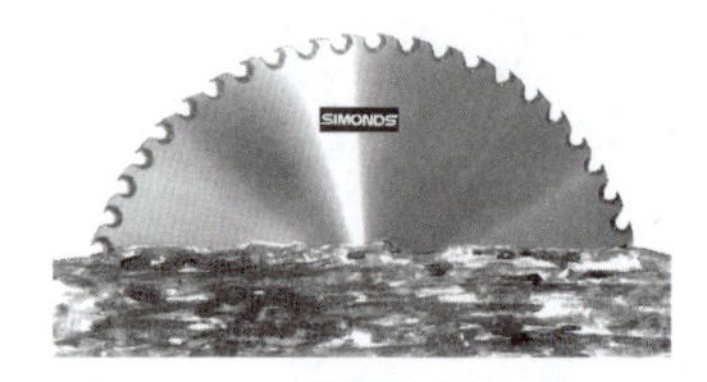

Fig. 1.16

A typical insert-tooth saw.

Source - Simonds

Insert tooth saws have loose teeth inserted in the rim which are replaced when worn or damaged. As a result these sawblade hardly reduce in diameter and so the arbor speed of machines they are used on never needs to alter.

(With plate saws the diameter reduces as they are ground or filed down so the spindle speed should, theoretically, be increased to compensate for the resulting loss of rim speed.)

The teeth and body of insert-tooth saws are of different metals, each of which is ideal for the job. As a result teeth retain their edge longer than on plate saws, and the body keeps its tension well. The teeth are formed slightly wider than the body to give the necessary clearance.

The method of fixing the teeth needs a thick body, so the saw kerf is particularly wide and kerf wastage is rather excessive. To keep kerf wastage to a minimum when converting logs these sawblades were often used as an upper and lower in-line pair, both sawblades having a thinner kerf than would a single sawblade to complete the same cut depth.

See Chapter 8.

Insert-tooth cutoff saws

The other form of insert tooth is the straight type used on cutoff saws in sawmills, shingle and pulp mills for crosscutting logs.

They were made in very large diameters mainly for the large Western-American logs, but generally have been replaced by mechanical chainsaws.

Saw plating

Chrome plating is sometimes applied on both plate and tungsten carbide saw bodies. This reduces friction to make the sawblade run cooler and reduce power consumption. In the case of plate saws it can also increase saw life, but makes filing difficult or impracticable. Another type is a black or coloured, low-friction coating similar to that used in cooker pans - and for a similar reason - to prevent buildup. It also reduces power consumption and extends the saw run-time.

Tipped saws

Fig. 1.17

Typical tungsten-carbide tipped sawblades. These are tipped with Freuds own micro-grain tungsten carbide tips - with added titanium.

Source - Freud

Most saws in common use are now tipped with Stellite, tungsten carbide or polycrystalline diamond. All these tips keep sharper considerably longer than alloy saw teeth, especially on difficult hardwoods and manufactured materials.

Fig. 1.18

This is just a small selection from the wide range of Guhdo alloy, carbide-tipped and diamond-tipped sawblades

Source - Guhdo

All these saws are dealt with separately later in the book.

Chapter 2
Circular Saw Technology

A circular saw is basically a tensioned steel disc with a toothed rim formed to bite into timber and divide it by removing waste as sawdust. Any disc with gashes around the rim will cut if pushed hard enough, but the further the "tooth" shape is from the ideal tooth shape and the less efficient it becomes. Circular saw tooth shapes have evolved over the years to relatively few patterns. Some freak shapes still persist which individual sawyers claim to have unique properties, but the shapes shown later are the ones in most regular use, and from which most sawing needs can be met.

Direction of Saw Rotation

Resistance Sawing

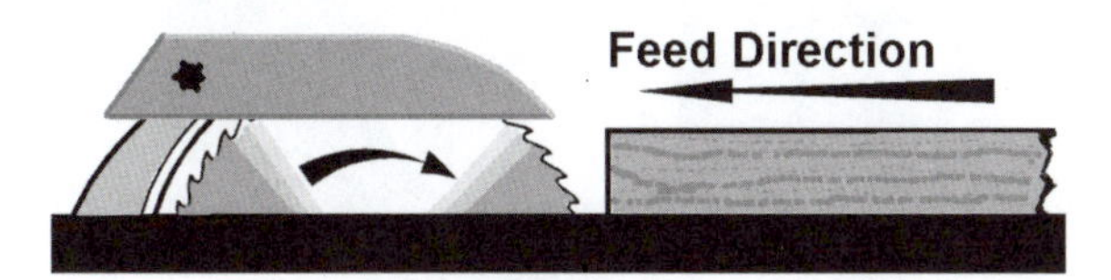

Fig. 2.1
Resistance, or rip sawing, on a regular table saw. The sawblade rotates clockwise and the timber feeds right to left.

Fig. 2.2
This typical small table saw, a Delta 36-830 has a 250mm (10in) rise, fall and tilt sawblade for cutting up to 80mm (3 1/8in) deep when vertical. Shown with a rear extension table.
Source - Delta

In all manual sawing, and with mechanically-fed single sawblades, it is essential for the sawblade to rotate against the feed. The correct term is resistance sawing, but commonly it is called ripping or ripsawing. Ripsawing is the regular way of converting both solid timber and composite panels.

Power-fed straight-line ripsaws have an accurately guided endless feed chain in the bed which traverses timber in a perfectly straight line and so cuts it straight and true. In order to cut clean through the timber the saw teeth project into a groove machined into the top surface of the feed chain.

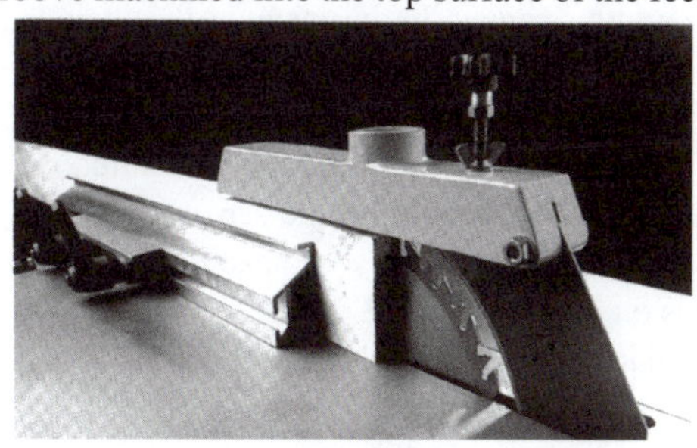

Fig. 2.3
Deep cutting on a Sedgwick TA 400. In addition to the regular rip fence a sliding table is available for crosscutting and mitring.
Source - Sedgwick

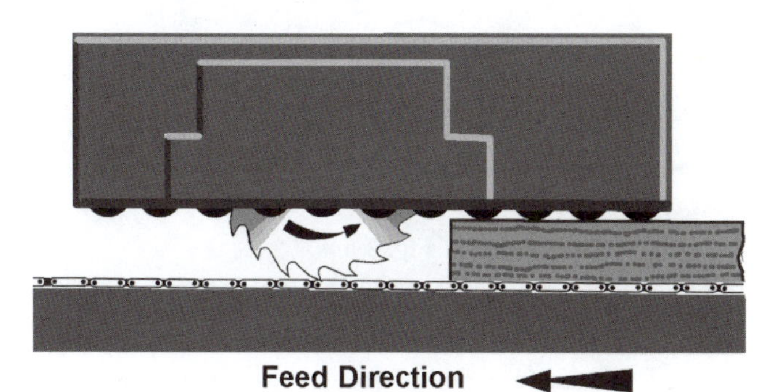

Fig. 2.4
Power-fed ripsawing with a single overcut sawblade.

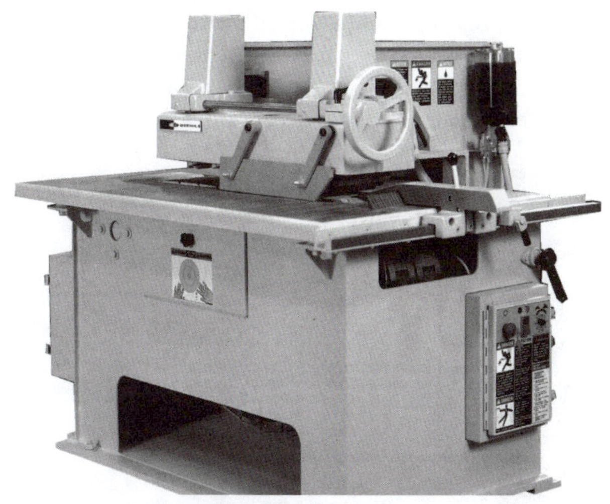

Fig. 2.5
This Diehl midsize overcutting Straight-line Rip Saw ESL-25 can have a sawblade diameter of either 360mm (14in) or 405mm (16in), and feeds at speeds of between 15 and 42M/min (50-140 fpm).
Source - Diehl

Another version of the straight line rip saw (straight line edger) has the sawblade below table level passing between two sets of chains running side by side. Wear in the chain links of this type can slew the timber in an irregular way, so older 2-chain machines tend to be less accurate than the single chain, overcutting types.

In addition to straight line ripsaws, overcutting sawblades are also used in gang ripsaws. With these machines several sawblades are mounted on the same arbor to make multiple cuts, being separated by precision spacers according to the rip widths wanted. In this instance the chain has guides to dip it immediately under the saws, so allowing the teeth to project below the top surface of the chain.

In most cases the sawblades are mounted on a single arbor, but some feature two in-line arbors each carrying one or more sawblade. The allows the spacing between the outer sawblades on each arbor to be instantly altered to suit changing conversion requirements.

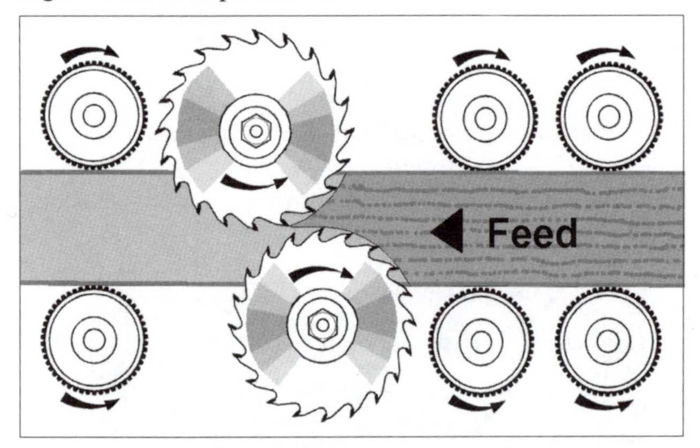

Fig. 2.6
Over and undercutting rip sawblades used as a matched pair.

Twin-arbor machines are regularly used when deep cutting - to allow the cut to be split between two sawblades. Such sawblades can be much thinner than a single sawblade would need to be for the same depth of cut, so both saw kerf wastage and saw motor loading are both reduced. The sawblades are matched in thickness and mounted in line with one another, but slightly staggered in line of feed so the teeth can marginally overlap In some cases the top sawblade climb-cuts to reduce feed loading.

Fig. 2.7
This Multiple Double Arbor Saw Unit VSV 80 is for ripping-off sideboards from round logs, and producing centre yield from four and two-sided cants.
Source - Linck

The same twin-arbor arrangement is also used on circular gang saws with multiple sawblades. Often these machines have twin feed chains, one in front and the second following the sawblades, plus pressure rollers above both.

This is also common on log conversion machines using a combination of vertical and horizontal feed rollers. These machines may have twin arbors each carrying paired sets of sawblades, or two separate in-line arbors above the cut and two opposing in-line arbors below the cut, each one mounting a single or multiple set of matching sawblades

This arrangement allows the opposing arbors, together with their paired arbors, to be moved together or apart to give instant change of centre-sawn widths. Often these machines also have hoggers or profile heads to convert logs to sawn boards at a single pass.

Fig. 2.8
View from the outfeed showing the twin and in-line arbor circular gang-saw arrangement. As is clearly seen, the left and right-hand saws are mounted on separate opposing arbors allowing instant change of the centre width according to the size of timber being converted.
Source - Linck

Climb Sawing

In a few special cases the sawblade rotates with the feed - termed climb-sawing. Examples of climb sawing are the top sawblade of twin-arbor saws for log conversion, and the scoring saw used on board conversion machines.

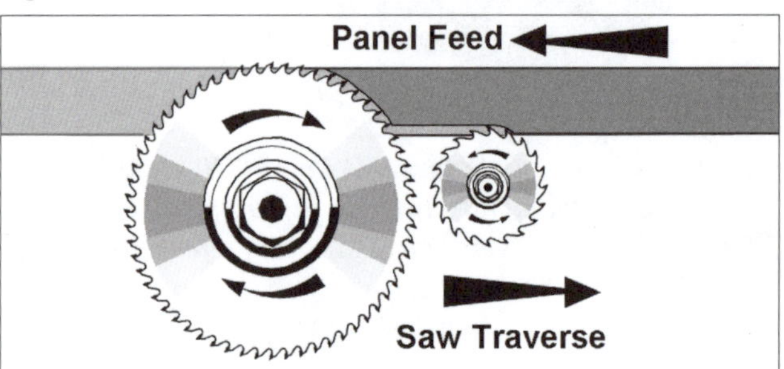

Fig. 2.9
This in-line sawblade combination is ideal for sizing faced panels. The main sawblade rotates against the feed, whilst the scoring saw immediately in front climb-cuts - rotates with the feed. This gives perfectly clean-cut top and under faces

Because climb-cutting sawblades rotate with the feed they can snatch the timber or board and throw it rearwards with great force, and maybe wreck the sawblade and perhaps part of the machine in the process. For this reason both the feed rate and the timber have to be rigidly controlled using a power feed and some reliable form of mechanical hold-down.

The one exception is when using a climb-cutting scoring saw on a panel saw. As this projects only a small amount into the underside of the material being sawn there is little tendency to throw it rearward. Because of this a climb-cutting scoring saw can be used safely even with a manual feed - provided all other normal precautions are taken.

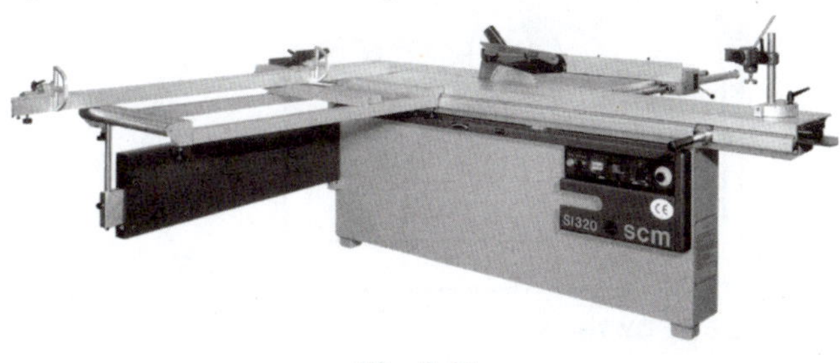

Fig. 2.10
Main and scoring saws are regularly incorporated into panel-sizing and dimension saws such as this SCM SI 320. The machine has a rise, fall and tilting saw unit, and can have extension tables added.
Source - SCM

Fig. 2.11
This close-up shows clearly the main and scoring saws on the SCM SI 320. The sawblades, riving knife and guard move as a single unit when tilted, as shown here.
Source - SCM

Panel cutting saws

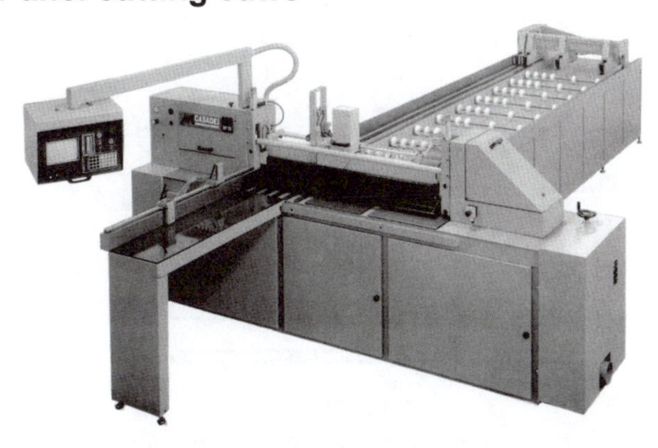

Fig. 2.12
On beam-type panel saws of the type shown here, the board is fixed and held down by an overhead pressure. The saw unit carrying the main and scoring sawblades travels underneath the board to cut through on the forward movement only, returning well clear of the board below table level. This narrow machine, the SV15, is intended for crosscutting strips of boards previously ripsawn on a wider companion machine.
Source - Casadei

On most board and panel machines the main sawblade is of the undercutting type and rotates against the feed - resistance sawing. If the main sawblade is used alone this will give a perfectly clean top face, but will break-out on the underface.

To avoid this the scoring sawblade cuts a shallow groove in the underside of the board directly in line with, but in front of, the main sawblade. Because the scoring sawblade climb-cuts upwards into the underside, the slot it forms is perfectly clean. The main sawblade then cuts precisely within this slot giving absolutely clean edges on both top and bottom surfaces. Of course the process is not as simple as this, more details are given later. Machines for converting panels are of two main types:-

Sliding-table panel saws. These have fixed main and scoring sawblades. The right-hand table is used for ripsawing material to width, and the sliding table is used for crosscutting. These machines were developed from dimension or variety saws, but the modern version has greatly increased capacity.

Beam Panel Saws. These are built specifically for board conversion and have a level or vertical fixed table and possibly a material clamp. The main and scoring sawblades are traversed manually or by power.

Crosscutting

When crosscutting timber using pull-out, radial cut-off and similar crosscut machines with over-cutting sawblades, the cutting conditions are quite different to those when ripsawing.

Resistance sawing when crosscutting with these machines is impractical because sawdust would be thrown at the user, with the added danger of timber being lifted into the sawblade during the cut and possibly thrown at the operator.

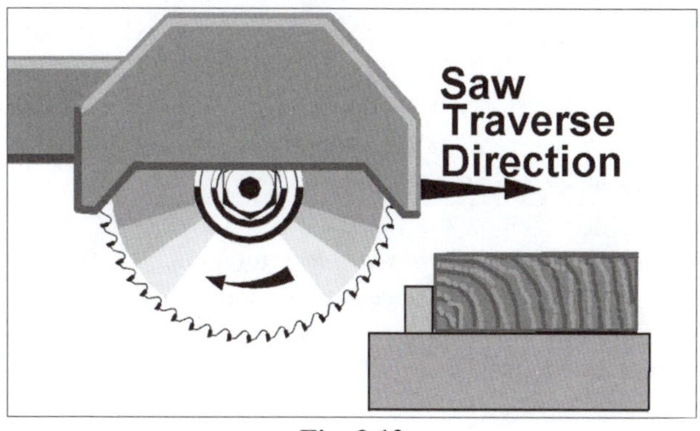

Fig. 2.13
A regular pull-out crosscut saw. This type has and over-cutting sawblade climb-sawing, i.e, rotating with the feed, which must have a negative hook angle.

For these reasons climb-cutting *(rotating with the feed)* is the norm when using over-cutting cross-cut saws of these types - but with this arrangement sawblades tend to run forward in the cut. They are given a negative hook angle to reduce this tendency, but running-out can still be a problem.

The most dangerous combination is when cutting thin but tough material - this is because the sawblade then cuts only with the lowest part of its cutting arc where teeth are at their most aggressive.

There is more resistance with thicker material so the "feel" can actually vary quite a lot. To cater for this a manual cross-cut operator will start the sawblade in the cut and either gently ease it out or hold it back according to how it then behaves. *An automatic cross-cut saw does not have this problem, of course, and there are also other types of cross-cut saws which do not behave in this way.*

Fig. 2.14
This Omga Radial 700 2V is typical of the universal type of cross-cut saw in wide use throughout the trade.
Source - Omga

Saw tooth Geometry

This chapter deals with the general principles of saw tooth geometry. For specific angles and measurements, etc., see Chapters 5 & 16.

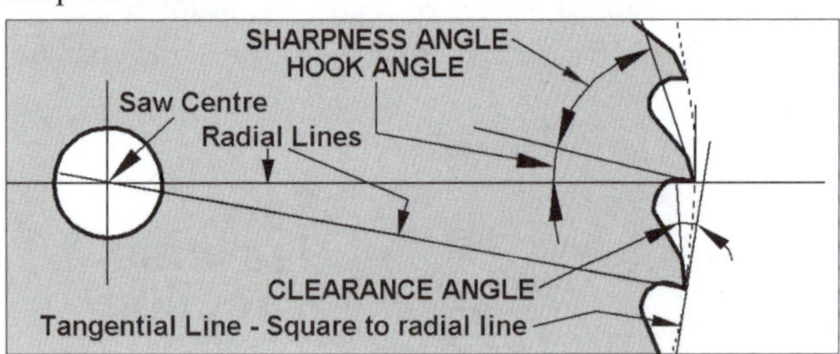

Fig. 2.15
The hook, clearance and sharpness angles of circular saw teeth are measured as above.

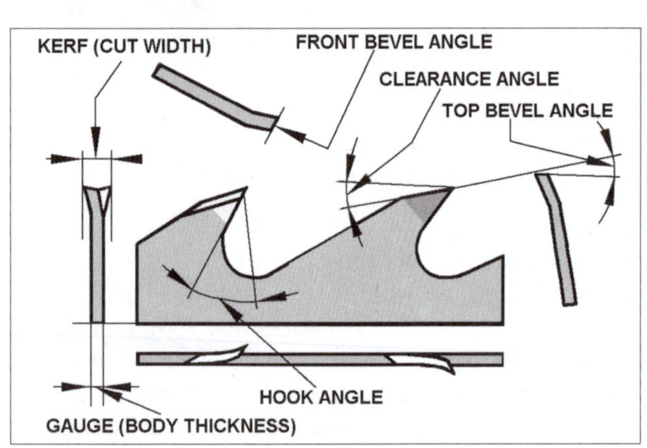

Fig. 2.16
Typical softwood plate ripsaw, with narrow, spring-set teeth and a large hook angle. As with all ripsaws the front bevel angle is zero.

Several factors affect how a sawblade performs and, regardless of the type of sawblade in use, there is a certain basic geometry common to them all - alloy, insert and tipped. To get the best out of a sawblade the shape and geometry of the saw teeth must be correct for the type and condition of the material to be sawn.

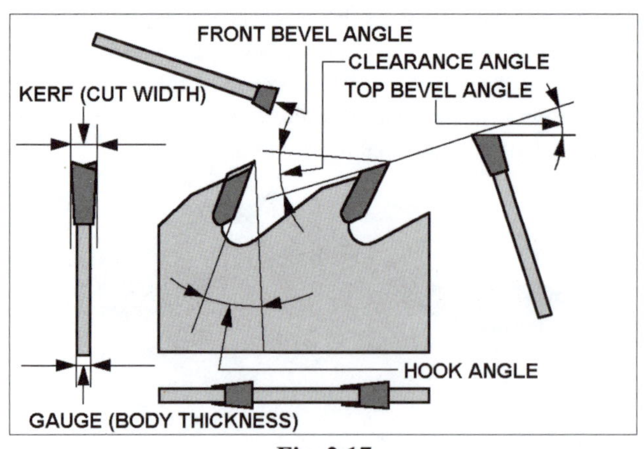

Fig. 2.17
Typical tct ripsaw for softwood, with large hook angle and alternate top bevel teeth.

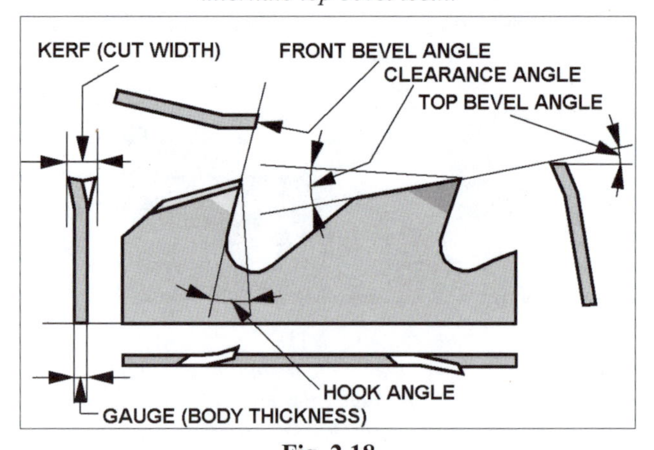

Fig. 2.18
Typical hardwood plate ripsaw, with a smaller hook angle and stubbier teeth.

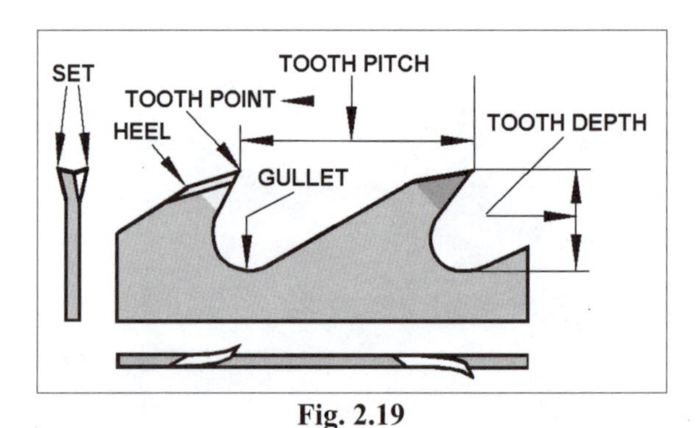

Fig. 2.19
Other important features of saw teeth are shown here.

Saw tooth geometry becomes much more critical when sawing hard, interlocked-grain and abrasive hardwoods, and when converting manufactured composites such as plastic sheets and plastic-faced chipboard, medium density fibreboard, and so on.

The saw doctor needs to take considerably more care with sawblades for these applications, as even a slight change in saw tooth geometry can have a marked effect on the performance of a sawblade both in quality and operating time.

Also, because manufactured materials are reasonably consistent in quality, most sawing problems can be traced to the sawblade itself - blaming these is not the get-out option it remains with natural timber.

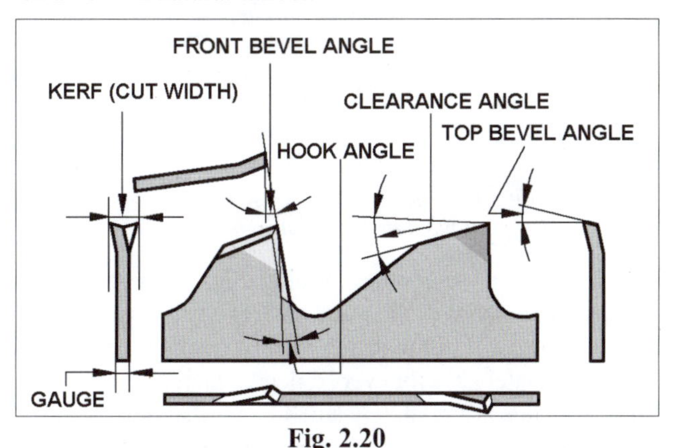

Fig. 2.20
Typical crosscut plate saw teeth, with negative hook angle and a front bevel angle.

Although quality manufactured materials are now reasonably consistent, there still can be differences between batches of inferior composites. Certain changes, for example a higher trash or resin content in chipboard, can play havoc with sawblades and other woodworking tools. So it pays to deal with an accountable supplier who can guarantee the quality of his products and so eliminate problems from this source.

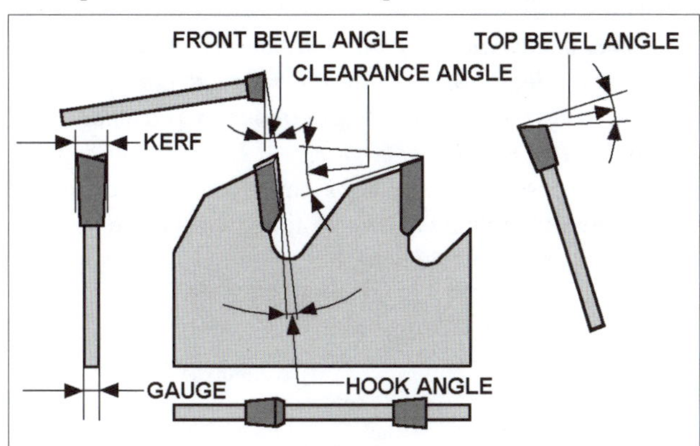

Fig. 2.21
Typical tct crosscut teeth, with negative hook angle and front bevel angle.

The writer well remembers an unexpected series of problems with tipped sawblades when working as a product specialist for a saw maker. Several customers in turn complained of inferior quality sawblades which I could find no fault with.

The problem was finally traced to a rogue batch of chipboard. This was manufactured from trees taken from a forest where fierce fighting took place in the early 1940s. The trees were riddled with bullets and shrapnel, and some of these were shredded and included in the board mix. It was this that caused the rapid breakdown of sawblades - not an inherent fault in the sawblades themselves.

When the reason finally dawned on the current board users they sold on the remaining stock at a "good" price as "surplus to requirement" - so another manufacturer inherited the problem. These rogue boards passed through several hands, creating havoc wherever they went, until they were finally used up many years ago.

Hook angle - ripsaw teeth

The hook angle is measured between the front face of the tooth and a line joining the centre of the saw and the tooth point. A saw tooth for ripping leans forward, having a positive hook angle, and a saw tooth for cross-cutting leans back, having a negative hook angle. In both cases the amount of hook angle determines the "feel" of the cut, the quality of the finish and the power requirement to drive the sawblade.

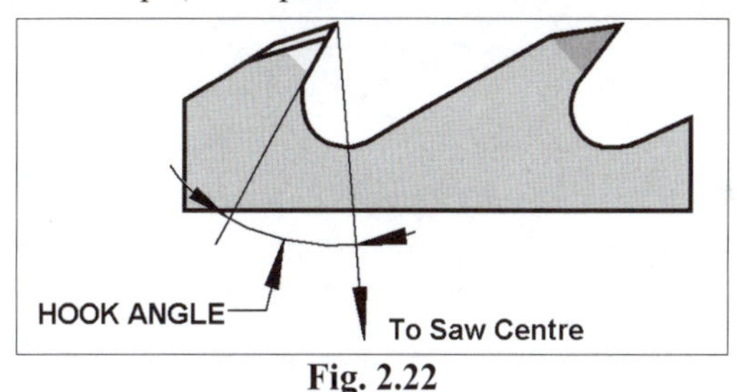

Fig. 2.22
Measuring the hook angle on a softwood plate ripsaw

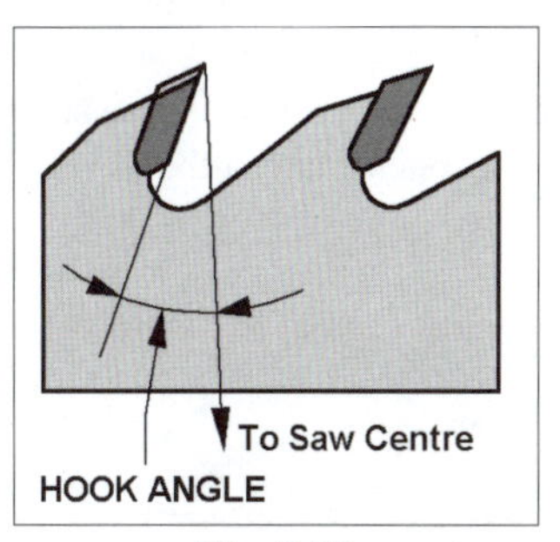

Fig. 2.23
Measuring the hook angle on a tct ripsaw.

Unlike wide bandsaw teeth which have a constant approach angle (the angle of the front face of the tooth to horizontal), the approach angle of circular saw teeth changes dramatically during the cutting action - Figs. 2.24 & 2.25 clearly show how the cutting action of the two saws differ.

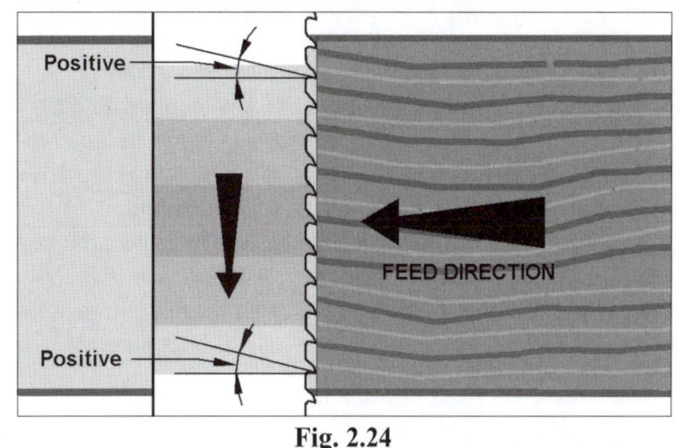

Fig. 2.24
Because bandsaws travel in a straight line the hook angle is also the approach angle, shown as positive, and does not alter throughout the depth of cut.

In conventional or rip (resistance) sawing the first point of contact is with the top face of the timber. Here, and during this first arc of contact, the front face of the tooth first bends and compresses the grain in a scraping action before the tooth point severs the forming chip.

Because this forms a negative approach angle (between the tooth front and the grain of the timber) this creates a resistance to the feed.

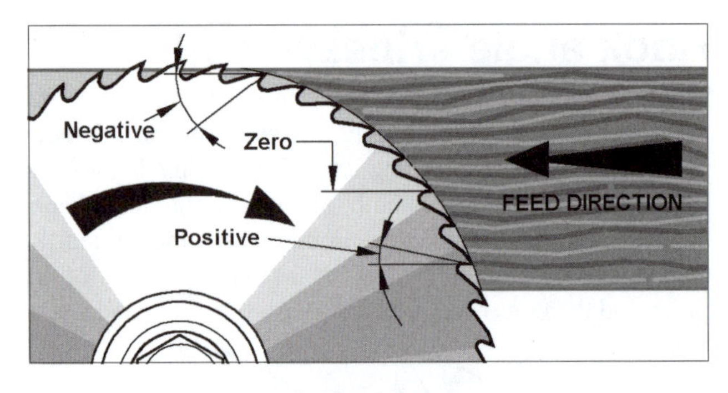

Fig. 2.25
Because circular saw teeth travel in an arc the approach angle varies from start to finish of the cutting action. For most of the time the approach angle is negative and chips are compressed before being severed by the tooth point.

Further down the cutting arc a point is (usually) reached when the tooth face is parallel to the top and bottom surfaces of the timber, zero approach angle. Beyond this the tooth point makes contact in advance of the front face so the timber is severed before being compressed and feed resistance is slightly reduced. Here the approach angle, although positive, is still less than the hook angle.

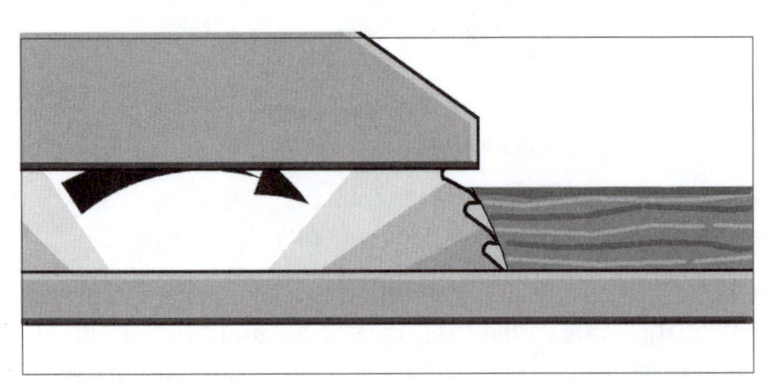

Fig. 2.26
When cutting thin material with the sawblade at its maximum height the teeth cut almost vertically through the timber

When cutting thin materials, with the saw teeth set to barely break through the top surface, sawing is always in the compression-before-cut condition. This takes more power to drive the sawblade and feed resistance is at its maximum.

If the saw teeth cut almost vertically (by raising the saw arbor to its maximum height) feed resistance is considerably lower and less power is needed. The arbor height setting, however, also affects the quality of the cut - as later detailed.

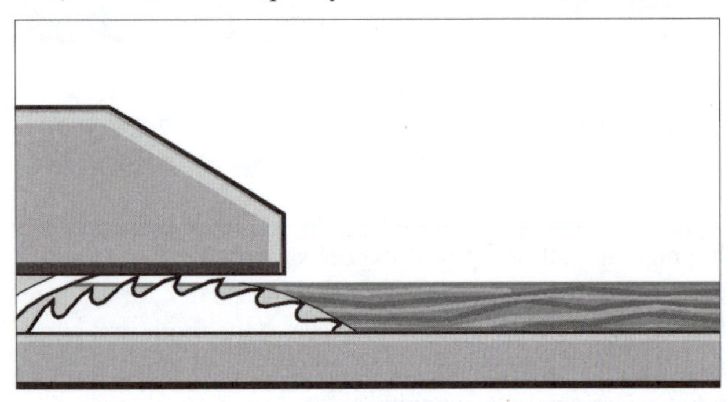

Fig. 2.27
When cutting thin material with the sawblade barely breaking through the top face the teeth cut in a long arc and at a smaller approach angle to the grain.

Large Hook Angles

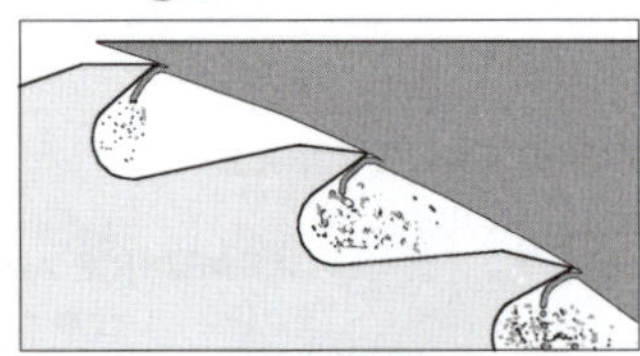

Fig. 2.28
This shows the bending of the forming chips at the start of a deep cut when using a large hook angle.

With a large hook angle the forming chip is bent through a smaller angle - so the cut is free and easy and least power is needed to drive the sawblade. Power consumption reduces by approximately 2% for every 5 degree increase in the hook angle - but this is a relatively small factor that only becomes important when deep sawing at fast feed speeds with multiple sawblades.

But there are disadvantages. A lower quality of sawn finish will always result from using a large hook angle, and the "feel" of the cut is also affected. When ripsawing in some conditions, saw teeth with a large hook angle tend drive into the timber to drag it forward - or create unacceptable vibration.

This unwanted affect increases as the hook angle increases - especially with dense timbers and when the cut is near vertical.

This makes the sawblade cut harshly in dense timber and when cutting through softwood knots and pitch pockets - it may even rip out softwood knots - and there is a tendency to tear timber and rag the underside.

An excessively large hook angle can make teeth less rigid to possible vibrate and score the sawn surface in an irregular and unacceptable pattern.

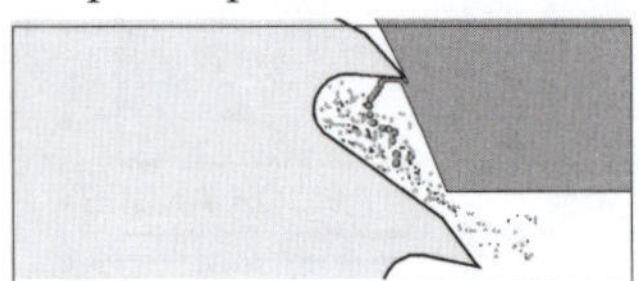

Fig. 2.29
This shows the forming chip being bent less at the exit of a deep cut when using a large hook angle.

Small Hook Angles

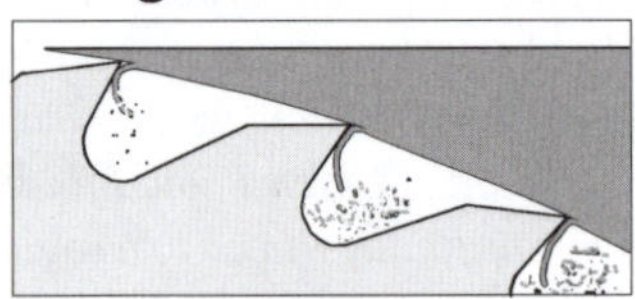

Fig. 2.30
This is similar to Fig. 2.28, but with a smaller hook angle giving severe bending of the chips at the start of a deep cut

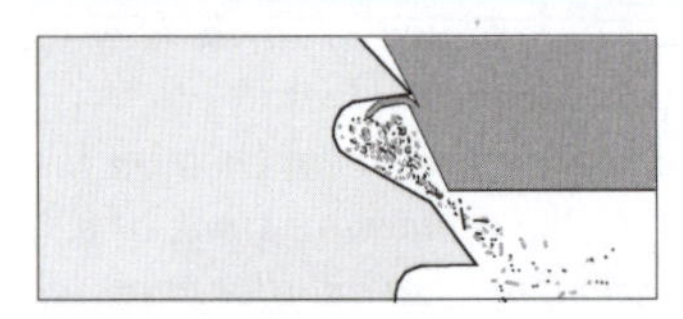

Fig. 2.31
This is similar to Fig. 2.29, but with a smaller hook angle bending the chips more - even at the exit point.

With a small hook angle the forming chips are bent more prior to being severed, giving even more of a scraping action. The sawblade then cuts much more cleanly and with a better "feel" - but takes more power to drive and feed it. Too small a hook angle results in an unacceptable feed resistance and makes manual sawing much more difficult.

Regular Hook Angles

To take account of all these conflicting factors the hook angle has to be a working compromise - based on the type and state of the timber being sawn and the working conditions. A sawblade intended for less dense and wet timber, and for sawing at fast feed speeds, needs more hook to "bite" cleanly and cut easily. A saw intended for dry and dense timber needs less hook to avoid snatch and vibration - but should be used at lower feed speeds.

The best advice is to use the biggest hook angle that gives an acceptable finish - provided this allows safe operation. Recommended hook angles are given later.

Hook angle - crosscut teeth

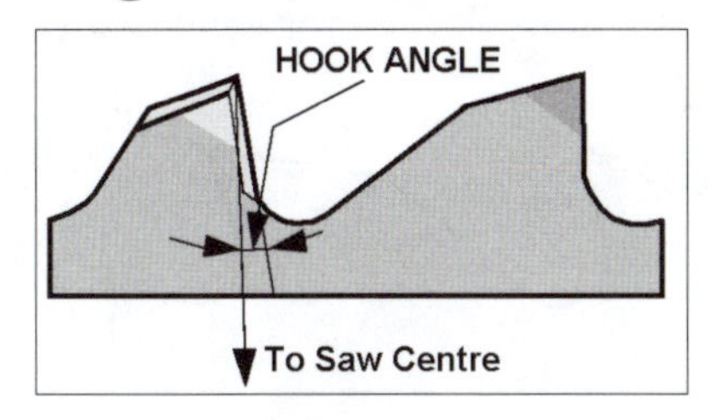

Fig. 2.32
Typical plate cross-cut tooth. The two different features when compared to a regular rip tooth is the negative, or backward leaning hook angle and a front bevel angle.

A negative hook angle is essential on over-cutting cross-cut sawblades to create a balanced resistance to the tendency of these to run out - this allows the manual crosscut operator to more safely control the cutting action. A negative hook angle of about 5 degrees is the norm, but with hardwoods this could be increased slightly to give even better control.

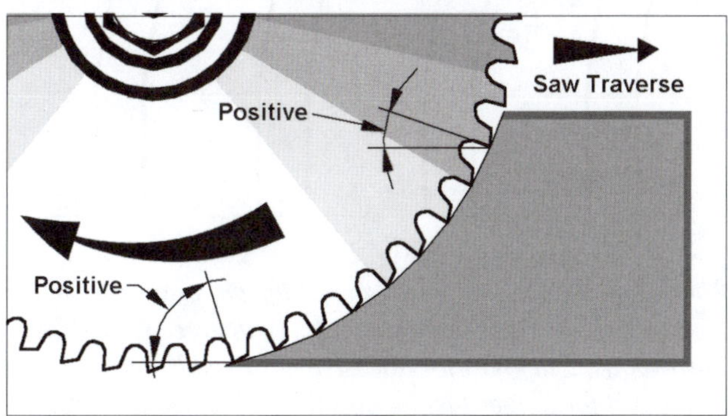

Fig. 2.33
Even with a negative hook angle of 5 degrees, crosscut saw teeth still have a positive approach angle to the timber - much more than a softwood tooth when ripsawing.

Front bevel angle

The front bevel angle is measured between a line parallel to the saw arbor and the front face of the tooth, looking down the tooth along the hook angle.

The front of a rip tooth for natural timber is should be formed square across. This reduces tooth vibration and deflection sideways - and keeps sawdust within the gullet.

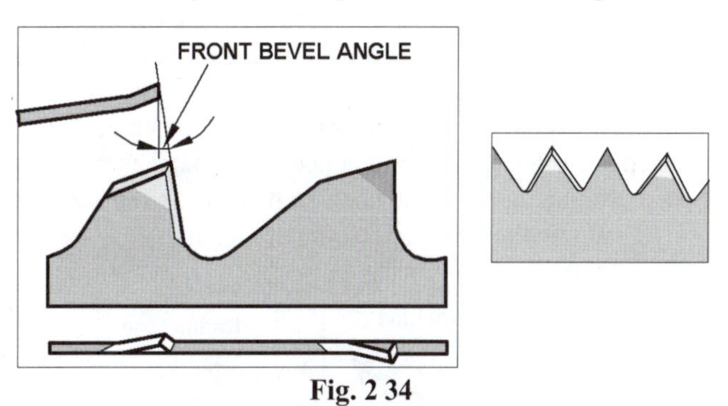

Fig. 2 34
A front bevel angle as shown on a crosscut plate saw (left) and on a peg tooth saw (right).

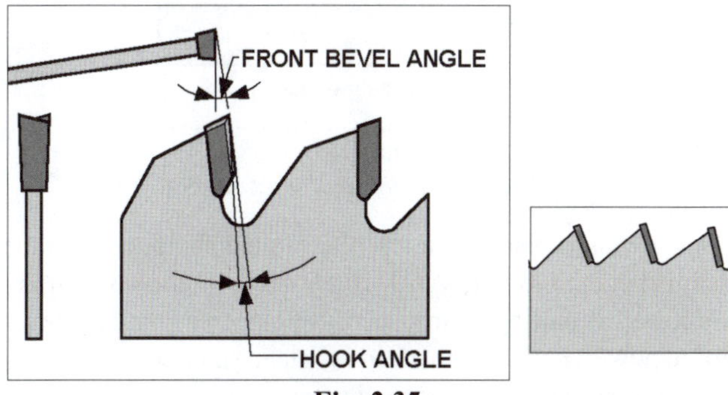

Fig. 2.35
Typical tct crosscut tooth (left). These have a negative hook angle and a front bevel angle. The tct peg-tooth form (right) is used on twin mitre saws for moulded sections.

Different conditions apply on cut-off and crosscut saws because the saw cuts at right angles to the grain of the timber. The teeth should have a front bevel to form sharp outer points in order to cut across the walls of the wood in advance of the main cutting edge. This makes a cleaner cut and reduces the tendency to rag or spelch on the underside and rear faces of the timber.

Clearance angle

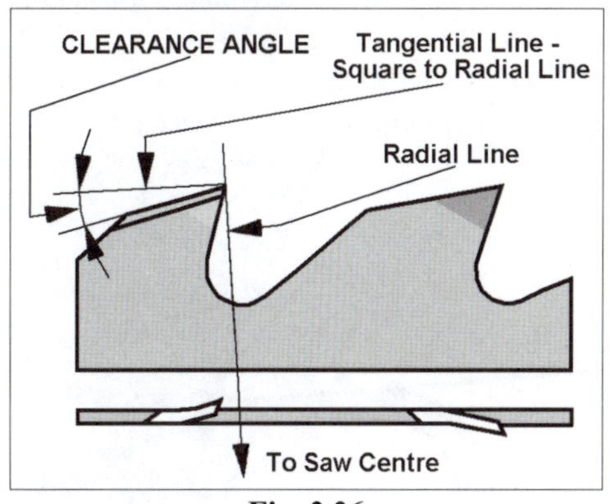

Fig. 2.36
Clearance angle of a hardwood plate saw tooth

The clearance angle is measured between the top of the tooth and a line normal to the cutting circle at the tooth point *(at right angles to a line joining the tooth point and the saw center)*. Only the front face of the tooth actually cuts, the tooth top merely backs-up and gives support to this.

The clearance angle must be big enough for the heel to clear the timber during the cut. Due to the continuous feeding action, however, the timber moves towards the sawblade between the time the point and heel of each tooth passes the same spot. For this reason the heel must be fractionally lower than the point to clear the timber at all. Only a degree or two is needed in theory, but in practice much more is required because timber is a resilient material that tends to spring outwards and into the sawing line after cutting.

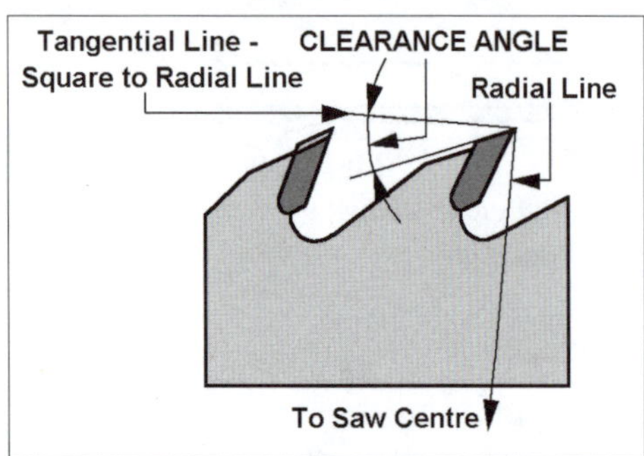

Fig. 3.37
Clearance angle of a tct saw.

As wear takes place teeth lose their initial sharpness and the points begin to round over. This tends to compress the timber more before parting it - instead of cutting it cleanly. This takes more power, increases the feed resistance and gives the sawn face a woolly and more ragged appearance

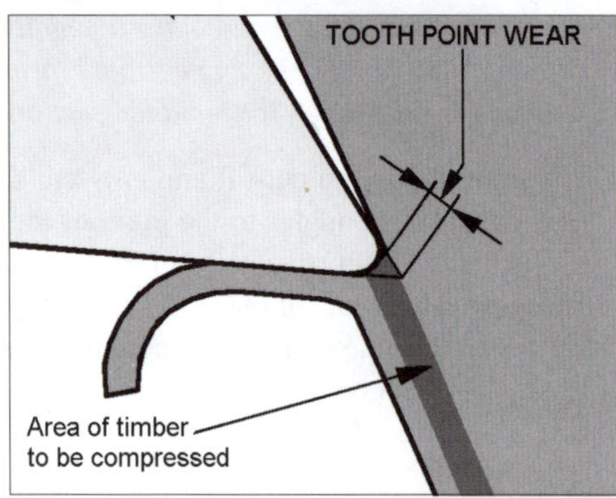

Fig. 2.38
When the tooth point wears the round formed at the tip compresses the timber instead of severing it cleanly. This rags the surface and also takes more power. This shows the effect of too-small a clearance angle.

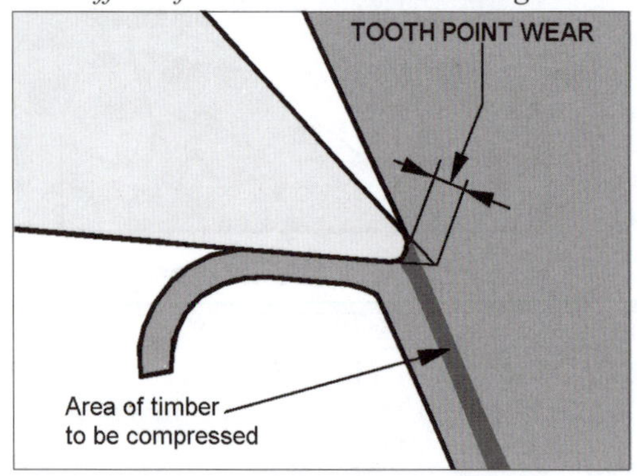

Fig. 2.39
The correct clearance angle compresses less for the same amount of tooth point wear - to take less power and give a cleaner cut.

The amount of wear is roughly proportional to the distance each tooth travels through the timber, but the amount of rounding this creates on the points is not consistent.

For a given amount of wear, rounding-over and subsequent timber compression is greater with teeth having a small clearance angle and less with teeth having a large clearance angle. Note the dark bands in Figs. 2.38 & 2.39. These show how compression differs from two sets of tooth points which, although having worn down equally, have different clearance angles.

It might be a temptation to use a very large clearance angle to reduce the compression effect still further, but using too large a clearance angle gives a weak tooth point which can quickly break down.

The clearance angle is always a compromise. Use a bigger clearance angle for soft, wet and stringy timbers as these require more clearance, and a smaller clearance angle for hard and dry timbers to provide a stronger cutting edge.

Top bevel angle

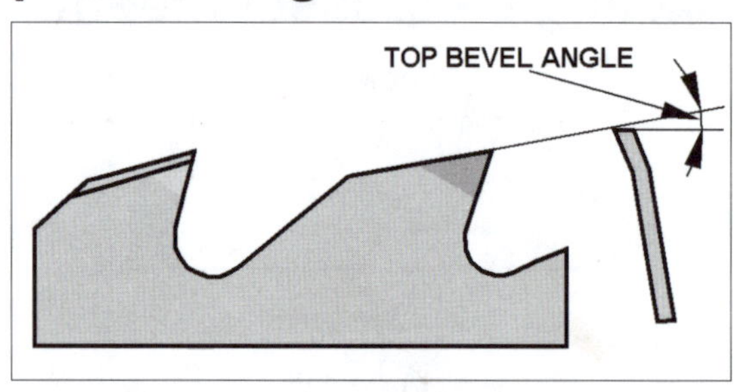

Fig. 2.40
Measuring the top bevel angle on a plate saw.

The top bevel angle is measured between the top face of the tooth and a line parallel with the saw arbor, looking along the clearance angle. The top bevel angle, to some extent, determines how cleanly the saw cuts.

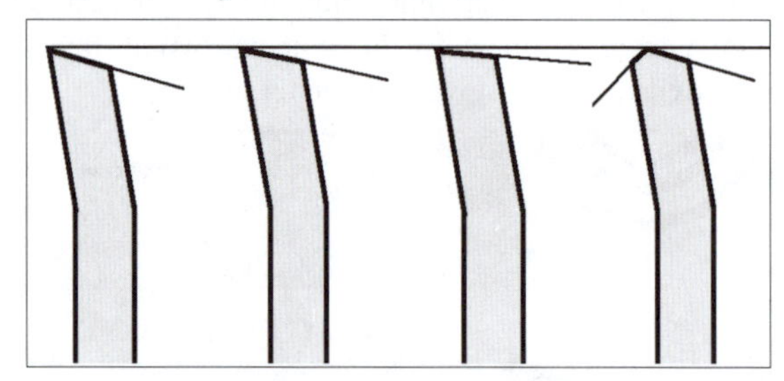

Fig. 2.41
This shows typical top bevel angles for plate saws; from left to right, 15 degrees for softwoods, 10 degrees for medium hardwoods, 5 degrees for abrasive hardwoods and a double bevel for longer lasting points on abrasive timbers

Dense hardwoods naturally cut more cleanly than stringy softwoods, so use the smallest practical top bevel angle to give a point strong enough to resist the tougher cutting requirements.

Soft and stringy timbers need a sharper point to give a clean cut. Tests show that, when the tooth bite nears maximum in feeding softwood at a high feed rate, a large top bevel angle on a plate ripsaw allows a faster feed than a small top bevel angle.

With too small a top bevel angle the inside of the tooth

can actually rub on the timber and greatly increase the power consumption of the sawblade - but only on spring-set plate saws. Swage-set plate saws are always formed with a flat top, and tct saws have a full width tip - so neither suffer from this particular defect.

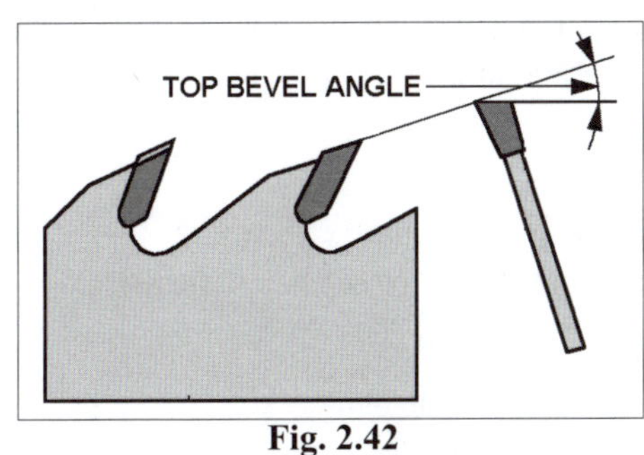

Fig. 2.42
Measuring the top bevel angle on a tct saw

Where particularly hard and abrasive timber is being cut the fine point formed by a regular top bevel angle on plate rip-saws can be very quickly destroyed, leading to a short saw life.

In this particular case saw life can be increased by forming a secondary bevel *(on the points only)* at about 40 degrees or so and to a width of roughly 1/4 of the saw kerf. The much stronger point this gives stands up better to hard and abrasive timbers than a regular point, but resharpening entails much more work on older grinding machines.

Plate saw teeth are formed to these patterns only, but tct saws can be ground to a flat top, alternate bevel or any one of several of the other tooth forms shown later.

Saw tooth set

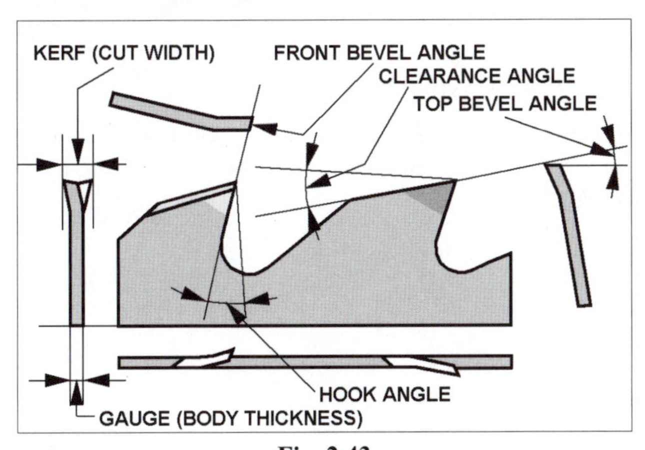

Fig. 2.43
Spring set, where plate saw teeth are bent alternately to left and right.

The teeth of all circular saws must be wider than the saw body in order to cut efficiently. "Set" is the term used for the amount the tooth point projects beyond the saw body.

With too little set the saw body binds against the sawn timber sides - to possibly overheat and perhaps form a blister.

With too much set there is unnecessary saw kerf wastage, more power is consumed, sawing becomes rough, vibration or chattering of the teeth may occur, and there is more chance of sawdust spilling from the gullets.

The amount of set has to be more if the timber is soft, wet or woolly, and should be less if the timber is hard and dry. More set is needed for larger saws and for saws of above average thickness.

The teeth of plate saws can be spring or swage set to give the necessary body clearance in the cut. In spring setting tooth points are bent alternately to one side and then the other. In all cases the sharp point is on the same side as the set. Because setting is a simple and practical operation most plate saws are spring set.

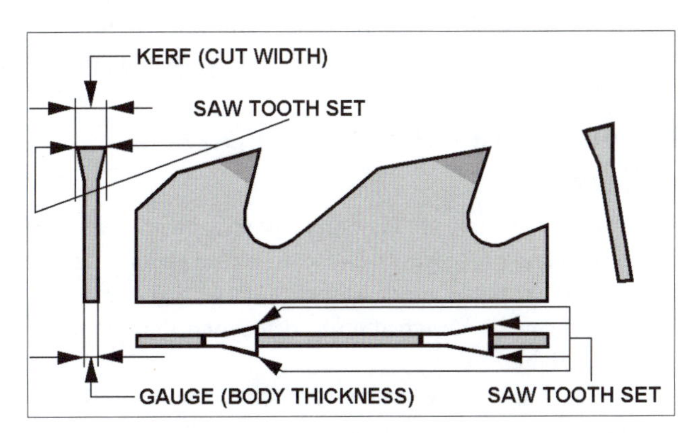

Fig. 2.44
Swage set, where plate saw teeth are first mechanically squeezed from the front to widen them, then squeezed back by shaped dies to a pre-set cutting width.

In swage setting, the tooth front is first squeezed to spread it outwards using a hardened, eccentric die, then squeezed back at a second operation with hardened side dies to a definite shape and specific kerf width.

Amount of set

The amount of set required varies with sawblade diameter and the timber to be sawn as noted above. For more detail see Chapter 5, saw setting.

With tungsten carbide tipped saws the tips themselves project beyond the saw body - each normally by an equal amount on either side - to give the necessary working clearance. Side projection details are given later.

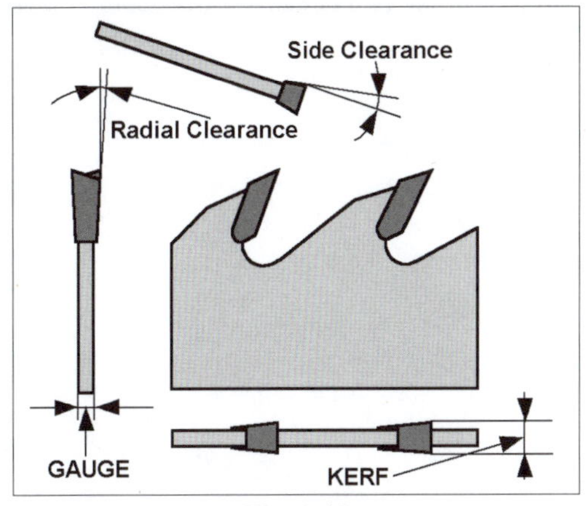

Fig. 2.45
With tct saws the brazed-on tips are initially slightly wider than actually required, and are then ground back to give the correct kerf width and the required side and radial clearances.

New tips are extra wide than so that, after brazing to the sawblade, they can be ground on both sides to the required saw kerf width. To avoid rubbing behind the cutting point or cutting edge the tips are ground to two distinct angles, termed side and radial clearance.

Side clearance is formed to an angle of about 5 degrees, as measured looking down the hook angle, so that the trailing heel of the tip is narrower than the front face.

Radial clearance is formed with an angle of about 1.5 degrees, as measured looking at the front of the tooth, so that the tip gradually narrows towards the saw centre. In this way the kerf width is measured between the two leading points.

Power requirement and power efficiency

Power requirement is not normally a problem with saws used in the finishing processes but, in timber conversion, log size and the timber types are so highly variable that converting machinery can be stretched up to or beyond the maximum power available.

Where this can be troublesome, for example with multi-sawblade machines, there are certain factors that can be changed to reduce power consumption. *(These mainly apply to older and sometimes under-powered saws - modern machinery using sawblades from reputable makers provide a combination that is equal to any likely requirement.)*

Each saw tooth absorbs a certain amount of power during its pass through the timber. The amount of power absorbed is in direct relationship to the saw kerf thickness and the length of travel through the timber - but in a non-proportional relationship* to the chip thickness produced. The amount of power absorbed is also in direct relationship to the specific gravity of the timber**.

**A basic amount of power is absorbed by the cutting action regardless of the chip thickness produced. This then increases as the chip thickness increases - but not linearly, i.e., for double the chip thickness the power consumption rises by between 60 and 75%. The original chip thickness is the minimum bite figure as given later. A scraping cut resulting from too-small a bite can actually absorb more power than the recommended minimum bite in all species.*

***Specific gravities for some timbers is given in Chapter 5 under sawblade types.*

Shown below is a guide giving power consumption figures for some Canadian species based on a 1M22 (48in) insert-tooth saw running at 700 revs/min (45M/sec, 9000ft/min)) and cutting 150mm (6in) deep timber at 80M/min (170ft./min.).

Species	H.P.
Balsam fir, Spruce	34
Basswood, White & Red pine, Eastern hemlock	37
Western hemlock, Jack pine	45
Douglas fir	50
White birch, Cherry, white elm	59
Black Walnut, Yellow birch	65
Beech, Sugar maple	92
Rock elm, White oak	99

The several ways of reducing power consumption, and/or increasing power efficiency are detailed in Chapter 3, but in brief these are :-

1. Use the minimum saw kerf practical.
2. Use the smallest practical sawblade diameter.
3. Reduce the tooth travel by adjusting the saw arbor to give a more square cut.
4. Use the widest tooth pitch practical.
5. Increase the bite to the maximum practical, either by reducing the saw arbor speed but retaining the feed speed (when nearing maximum saw motor loading), or by increasing the feed speed and retaining the saw arbor speed.
6. Maintain the correct hook angle.

Of course, the practicability of all the above depends upon many other factors - which must be taken into account. These are dealt with later

Chapter 3

Efficient use of Saws

To use sawblades efficiently in converting either natural timber or manufactured boards it is essential to know what each is capable of. Failure to do so results in loss of efficiency - lower production rate, rapid dulling and/or poor quality sawing.

The two factors which control cutting efficiency - assuming all other factors are correct - are the tooth bite and the gullet capacity. Between them they determine the maximum practical feed speed.

The tooth bite, sometimes called the feed-per-tooth, is the distance moved forward by the material being cut between one tooth and the next passing the same point. (In this context the "tooth" means a single flat tooth, or a tooth group - two or more teeth which between them complete the cut.)

The tooth bite is determined by the number and type of teeth in the sawblade, the saw arbor speed and the feed speed. For them to operate efficiently, the bite should be within the maximum and minimum values for the class and condition of material being sawn.

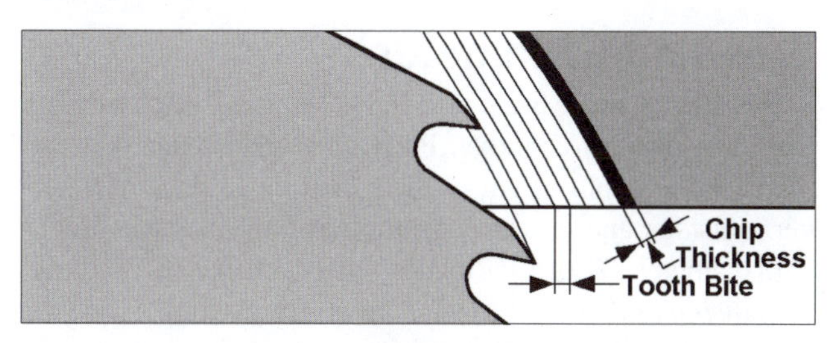

Fig. 3.1
The tooth bite is the feed per tooth or tooth group. The chip thickness is the thickness of the sliver formed by the cutting action.

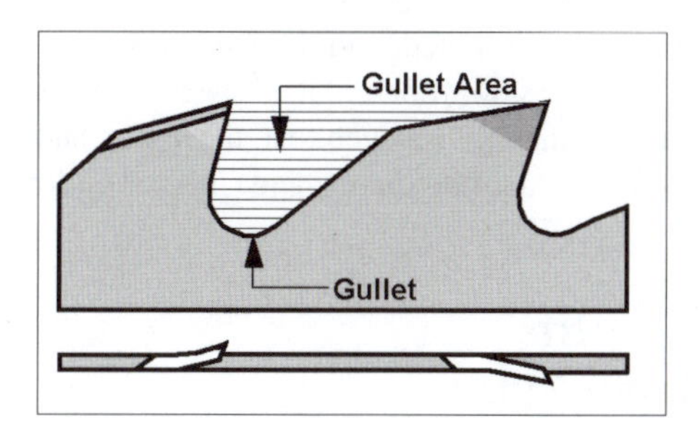

Fig. 3.2
The gullet contains the waste generated during the cutting action. The gullet capacity is the amount of waste the gullet can contain, shown as a shaded area.

The gullet removes the waste generated by the cutting action. The sawblade diameter and the type and number of teeth determine the gullet capacity - the amount of waste the gullet holds during the cut - which in turn limits the feed speed. Calculations need take account of the waste formed by a single pass only of each tooth - and all waste must fully discharge as the tooth clears the cut.

This chapter hopefully explains what effects these factors have on the quality of cut and cutting efficiency of the sawblades used. Chapter 16 deals with the calculations.

Tooth Bite

Natural timber needs big bites, the biggest for primary conversion of wet timber, and the smallest for finish-dimension sawing of seasoned timber. Manufactured materials need much smaller bites, the least for hard and brittle plastic facings and veneers. Within the ranges of bites recommended for different materials, extreme bites give the following effects:-

Small bites give a good finish but a shorter run-time - sometimes called the saw life - the time period a sawblade can be run before it becomes dull. Small bites produce small particles - which tend to escape between the saw plate and the sawn surface to build-up on the sides of the sawblade - especially if the material is gummy. This increases the power consumption, and the friction created can overheat and blister the saw body - a particular problem with multi-saws.

Large bites give a poor finish - but a longer run-time and higher productivity. Large bites also form coarse waste which compacts more and escapes less.

For these reasons the bite is always a compromise between quality of finish and productivity.

Relationship of Tooth Bite to Chip Thickness

Although the tooth bite is the most influential factor, chip thickness must also be taken into account. The forming chip grows in thickness from the entry to the exit point - when resistance sawing - but when climb sawing it reduces during the cut.

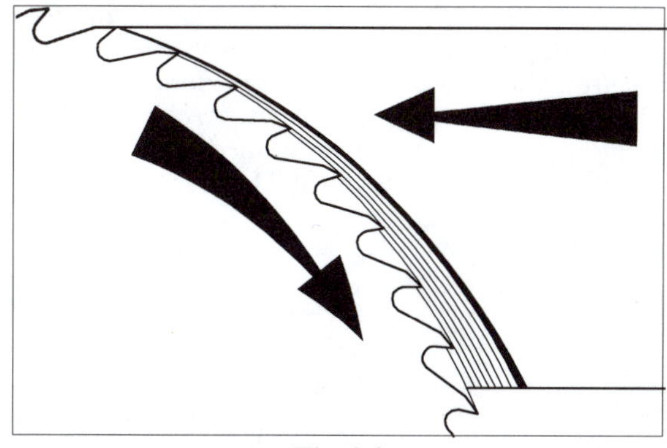

Fig. 3.3
This shows the curved path followed by each tooth. The black band is the chip thickness - the waste to be removed by the tooth just entering the cut.

Saw Arbor Setting

The angle of the curved cutting path can be varied by increasing or decreasing the distance from the material to the saw arbor. By decreasing it the angle becomes steeper and the chip thickness increases - for the same bite. By increasing it the angle grows less and the chip thickness reduces.

The chip thickness is least when ripsawing with teeth barely breaking through the top surface of the material, but the teeth may then rub and dull more quickly than would otherwise be the case. This setting, however, gives the best possible finish, and is used when converting manufactured boards, material that is thinner than the tooth pitch, and when dimension sawing natural timber.

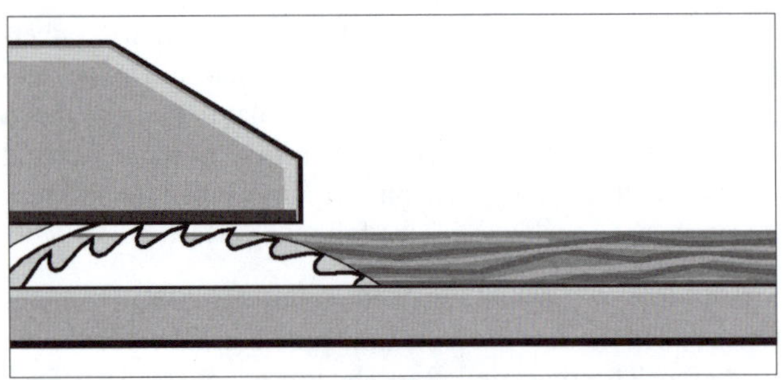

Fig. 3.4
Barely breaking through the top surface is not the most efficient way to convert timber, but it gives the best results where quality of finish is important.

These cutting conditions also apply to straight-line edgers and single-arbor multi-ripsaws. Their sawblades barely project through the timber being converted to give a starting chip thickness of almost zero when resistance sawing. With twin-arbor multi-ripsaws the first sawblade is buried in the cut and the following one barely overlaps, so again the starting chip thickness is virtually zero.

Fig. 3.5
On both over and under-cutting power ripsaws the sawblade projects through the timber by only a few millimetres. The saws on this M3 multi-ripsaw have typical ripsaw teeth. The extra timber pressures allow short lengths to be handled.
Source - SCM

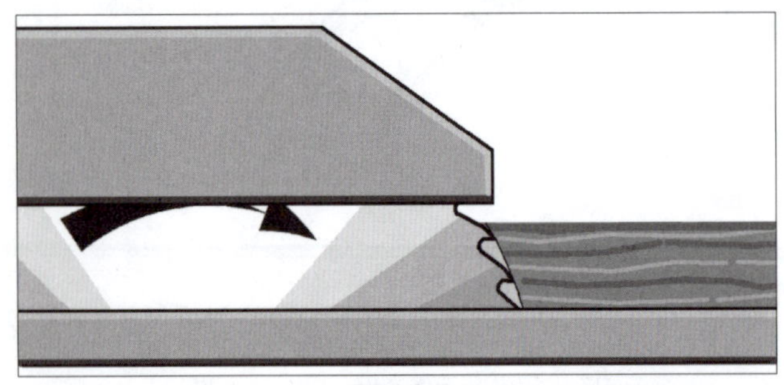

Fig. 3.6
When resawing natural timber on a power-feed saw, set the saw arbor to its maximum height for maximum sawing efficiency.

On machines for converting natural timber, the production rate is more important than the quality of finish, so their sawblades have a slightly larger pitch and bite. This partially offsets the dulling effect that near-zero chip thickness creates. The sawblade diameter should be matched to the depth of cut needed - and set to cut as near vertically as practical - this is a more efficient way of sawing.

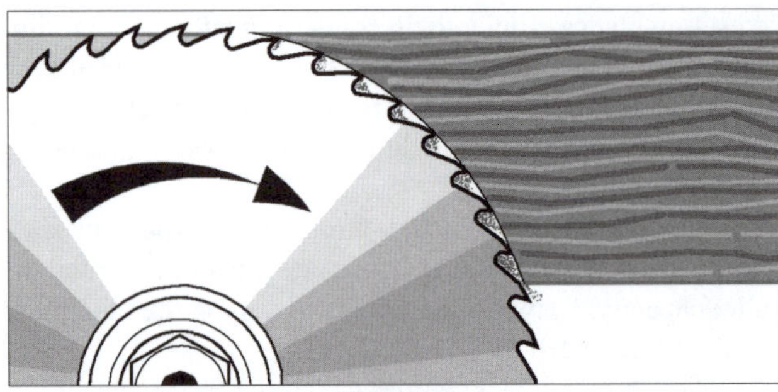

Fig. 3.7
The sawblade should be near-square to the timber for much of the cut when converting natural timber.

General-purpose table saws used for manually converting natural timber may have a rise and fall saw arbor. If so set the arbor to the highest position - perhaps using a smaller diameter of sawblade if projection is excessive, and use a power feed. Lower arbor settings are used when sawing panels and when dimension-sawing.

Sawblade Diameter

Sawblades are made in a limited range of sizes, which can vary from one maker to another. Preferably choose one that best matches the depth of cut needed, bearing in mind that saw kerf wastage and power consumption both increase in step with diameter. *There is not an absolutely free choice, of course, the sawblade diameter a machine is designed for - and the saw arbor speeds available on it - restrict what can actually be used.*

To avoid mistakes record details of all machines, including the maximum and alternative sawblade diameters and their maximum depths of cut. Also note the saw arbor speed, if more than one is provided, and the feed speeds on powered machines. Where saw arbor and feed speeds are not given, or where the information has been lost, these can easily be calculated, see Chapter 4 on drives, and Chapter 16 for calculations.

Cut Depth

Cut Depth Guide

TYPE OF MACHINE	TYPICAL CUT DEPTH CAPACITY		
	Percentage of diam.	Maximum saw diam mm	Cut depth mm
Saw benches	34%	300	104
Dimension Saws	28%	350	100
Panel saws	25%	430	110
Power ripsaws	32%	400	125
Deep power ripsaws	37%	810	300

Makers specifications for single-arbor fast-feed saws usually allow a projection of the sawblade through the cut of perhaps 5 to 15 mm, but slightly less with over-cutting straight-line-edgers and multi-ripsaws. With twin-arbor saws there is usually an overlap of about 5mm, or so.

As a proportion of the sawblade diameter, the maximum cut depth varies slightly from one machine to another depending upon the diameter of the saw collars fitted and the type and construction of the machine. The correct information will be given by the machine maker, but as a general guide use the table above.

Fig. 3.8
With twin-arbor paired saws the lower sawblade is buried in the cut and the upper sawblade is offset to overlap by a few millimetres. This Multibi 18 twin-arbor multi ripsaw has feed chains at both infeed and outfeed, and powered pressure rolls above. Feed speed is controlled automatically according to the cutting conditions. The top arbor is raised in this particular view.
Source - Storti

The cut depth capacity is always a smaller percentage of the sawblade diameter when using sawblades of smaller diameter than the maximum.

For example, a deep ripsaw normally taking an 810mm saw to cut 300mm deep would need a 610mm diam. to cut 200mm deep. A smaller sawbench would cut 200mm deep using a saw of only 540mm diameter.

The simple calculation to find a suitable saw diameter for a smaller depth of cut is:- the maximum sawblade diameter for the saw bench, minus double the difference between the maximum and required depths of cut.

Saw Arbor Speed

The saw arbor speed is stated in revolutions per minute *(revs/min)*. This, together with the sawblade diameter, determines the saw rim speed - which should match the type of sawblade and the material to be cut, see Chapter 16 for details.

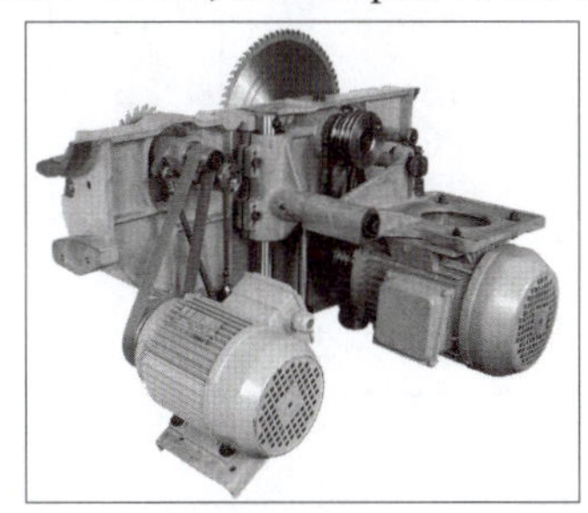

Fig. 3.9
This view of the saw unit of the SCM Hydro panel saw shows the multispeed drive to the main sawblade and a single-speed drive to the scoring saw.
Source - SCM

With few exceptions sawblades are driven by regular two-pole, induction-type mains-frequency SCR motors, either direct-on the saw arbor or via a belt drive. Most direct-drive saw motors have a single operating speed fixed by the mains frequency - and this restricts their sawblades to the maximum diameters given under 'Notes'.

Fig. 3.10
On some machines the sawblade is mounted directly on an extension of the motor arbor, as on this Bauerle Cross-cut Saw.
Source - Bauerle

Sawblades directly-driven by high-frequency motors *(via a frequency changer)* operate at much higher speeds than are possible from a regular direct-drive mains motor. Also modern controls fitted on some machines allow a regular SCR motor to be driven direct from the mains supply at surprisingly wide range of speeds. See notes in Chapter 4.

Most machines, though, have a short-centre drive fitted with pulleys of different diameter - to allow the saw arbor to be driven at the most appropriate speed for each application. (*Short-centre drives are so-called because old belt drives, powered by a single power source for the whole mill, had long driving centres.)*

Rim Speed

The rim speed is the speed of the saw teeth - sometimes called the saw speed - and is stated in either metres per second (M/sec) or feet per minute (fpm.). The rim speed is determined by the sawblade diameter and saw arbor speed, and should correspond with the information given in Chapter 16.

Notes:-

The traditional rim speed of alloy plate saws is around 50 M/sec (10,000 fpm.), but there is a general trend to run sawblades faster. This has been made possible through improved materials and higher standards of engineering - better bearings, more precise balancing of high speed rotating parts and closer machining tolerances. Higher speeds give the best finish, so are mainly used when finish-sawing - especially with carbide-tipped saws.

On some sawing machines a higher sawblade speed than 50 M/sec is quite normal. This is because sawblades are often fitted direct-on the motor arbor, normally a two-pole SCR type running direct off the mains supply. With a mains frequency speed of just below 3,000 revs/min on 50Hz and 3,600 revs/min on 60Hz, these give rim speeds of around 66 & 71 M/sec (13 000 & 14 000 ft./min) for the regular sawblade diameters of 450 & 350mm (18in.& 14in). respectively.

Whatever saw arbor speed is used, it is essential that

this also matches the speed for which the sawblade is tensioned - see under 'Tensioning Saws'. Whilst some sawblades state on them the maximum recommended speed, most do not.

Most sawblades are tensioned to operate within the rim speeds listed in Chapter 16 - unless ordered otherwise - but there is always some latitude in the speed a sawblade can be run at. However, running a sawblade outside the range for which it is tensioned only becomes critical when dealing with very thin kerfs and large diameter sawblades.

Number of Teeth in Saws

The relationship of pitch to sawblade diameter and number of teeth in a sawblade are shown in a nomogram in Chapter 16. From this the tooth pitch can be determined from the sawblade diameter and number of teeth, or the number of teeth for any sawblade diameter can be found from the tooth pitch.

Number of teeth in the cut

One important factor is that there should always be at least two teeth in the cut at any time when feeding manually. This gives safer operation, a better "feel" and avoids the dangerous vibration that occurs when the tooth pitch exceeds the material thickness.

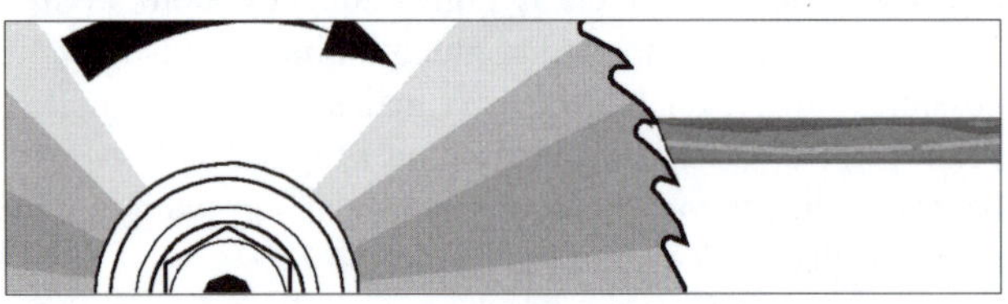

Fig. 3.11
When cutting thin material with a regular sawblade and the saw arbor set at its highest position there may be less than two teeth in the cut.

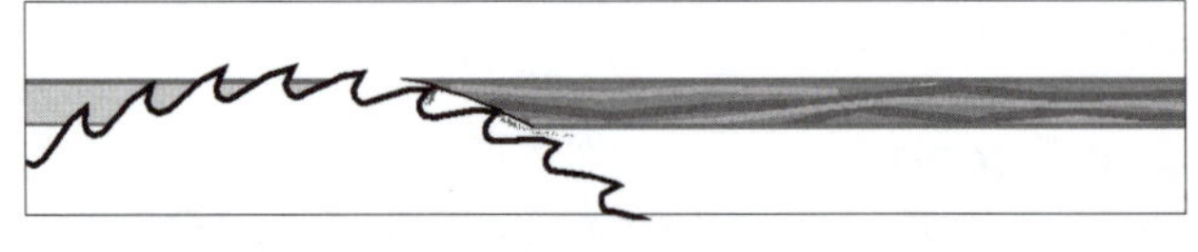

Fig. 3.12
Simply by lowering the saw arbor more teeth can be brought into the cut - without making any other change.

This is important when cutting thin timber or boards singly. Vibration is set up if each tooth strikes after the previous tooth clears the material. On many machines this can be avoided by substituting a sawblade with a finer pitch, or simply by lowering the saw arbor.

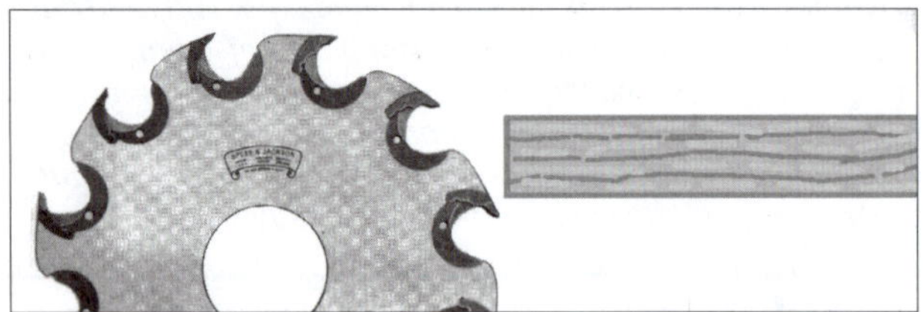

Fig. 3.13
Insert-tooth edging saws often have a greater tooth pitch than the flitch thickness. This may result in vibration - caused by each tooth striking the timber after the previous tooth has already cleared.

Edgers using insert-tooth saws have a fixed-height saw arbor - and often the smallest tooth pitch is greater than the flitch thickness. Vibration is at its worst when the rate at which teeth strike the timber is in harmony with the resonance (natural vibration rate) of the timber. As this is more likely at the regular rim speed of 40M/sec, increasing the rim speed to 60M/sec can force the tooth strike rate out of sequence with the timber resonance - and so greatly reduce vibration.

Alloy saws

The numbers of teeth in alloy sawblades for natural timber relate only to the class of timber to be sawn - not to the sawblade diameter.

The logic behind this is quite simple. Large sawblades have a large pitch for feeding deep timber at a medium feed rate, while small sawblades with the same number of teeth have a small pitch for feeding thinner timber at a higher feed rate.

When fed at a suitable feed speed for the conditions, all sawblades have more or less the same bite and a gullet capacity to match the rate of sawdust generation.

Numbers of teeth in typical alloy saws

Softwood ripsaws, less than 560 kg/mm^3. *(Hemlock, Obeche & Scots Pine)*	48
Hardwood ripsaws, 560-800 kg/mm^3. *(Ash, Larch, Sapele)*	54
Hardwood ripsaws 800-1040 kg/mm^3. *(Ebony, Jarrach & Rosewood)*	60
Hardwood ripsaws over 1040 kg/mm^3. *(Ekki, Greenheart & Wamara)*	80
Peg teeth crosscut saws	84
Regular crosscut saws	64

Most requirements in sawmilling can be met by sawblades within the ranges given above, but sawblades with a different number of teeth to standard can usually be supplied by a specialist maker.

Insert-tooth saws

Maximum Number of teeth in typical insert-tooth saws

Diam (in.)	Pattern				
	B	F	$2\frac{1}{2}$	3	$4\frac{1}{2}$
12	8	10	12	-	-
16	12	14	16	12	-
20	16	18	22	16	
24	18	22	26	18	16
32	26	28	34	26	22
36	26	30	38	30	26
48	-	44	48	-	-
54	44	50	-	-	-
60	52	56	-	-	-

The maximum number of teeth in insert-tooth saws depends upon the pattern number and the sawblade diameter, although sawblades with fewer teeth can be provided. The original tooth patterns are designated by numbers from $2^1/_2$ to $4^1/_2$, with later patterns given the letters B & F. Insert-tooth saws are made only in imperial sizes.

Carbide Tipped Saws

With tungsten carbide tipped saws the tooth pitch remains fixed according to the application - so the number of teeth in carbide saws vary according to diameter and application. As a general rule, sawblades for ripping natural timber have a large pitch and a large bite, and those for converting difficult board materials have a short pitch and a small bite.

Fig. 3.14
Close-up of the main and scoring sawblades in a Format-4 Sliding Table Panel Saw. The scoring sawblade on the automatic version has electrical height adjustment - which also lowers the sawblade below table level on switch-off, then raises it to the pre-set height on switch-on.
Source - Format -4

For practical reasons carbide saws are not made in every possible combination of diameter, number and style of teeth - and makers vary as to what they offer.

When ordering a new tct saw the precise operating conditions should be given to a supplier, who can then advise on the most suitable type from manufacturers listings.

Some can offer nonstandard tct saws to suit special conditions, but these are usually on extended delivery times - and can cost more.

Fig. 3.15
Stacked boards being sawn on a Holzma HPP 81 panel saw. Note: the pressure beam has been raised for illustration purposes
Source - Holzma

As a general guide in specifying a suitable tct saw, however, note the calculations given in Chapter 16 relating to the saw arbor speed, tooth pitch, bite and feed speed.

Note, though, the stack height of manufactured boards being converted. TCT saws for converting boards singly or two up generally have a smaller pitch and allow a faster feed speed than similar sawblades for converting the same boards in multiple (or deep timber). In this one instance tungsten carbide saws have a similar logic to alloy saws.

Tooth Bite Recommendations

Whatever working conditions are chosen they must give a bite that is within the recommended range for the material being sawn, see Chapter 16 for details.

Too small a bite gives an excellent finish, but may cause the teeth to rub instead of cutting cleanly and dull more quickly. The minimum bite below which teeth rub varies with the type and condition of the material, and the type of saw teeth and their sharpness. The minimum bite is less for dry, hard and brittle materials, and more for soft, wet, abrasive and stringy materials.

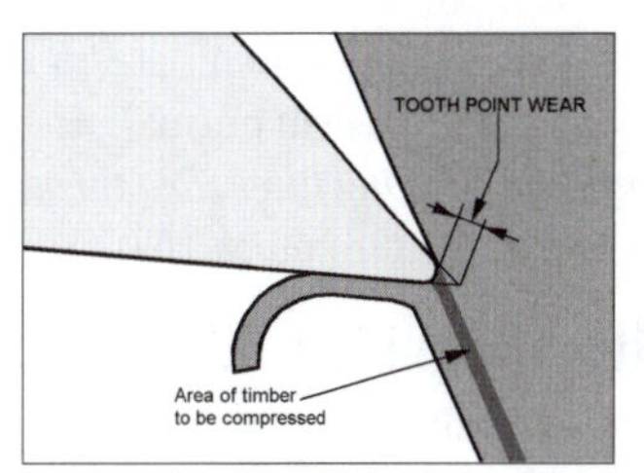

Fig. 3.16
When the tooth points dull they rub and compress timber rather than severing the chips cleanly - so the finish deteriorates, more power is absorbed and the rate of dulling increases.

Sawblades operate at their most efficient at maximum bite - which also means higher feed speeds - but other factors may not allow this. These include problems in physically handling the faster throughput, the saw motor power available, the capability of the sawblades and of the machine to withstand the extra demand, and the lower quality this then produces.

As with other factors the right balance has to be found - a tooth bite within the recommended range, a tooth gullet capable of completely removing the waste as it is generated, an amply-powered saw motor that operates well below overload conditions, and facilities to handle raw and finished material without problem.

Choosing the right sawblade

In theory each different material, feed speed and cutting depth needs a different specification of sawblade.

When dealing with consistent operating conditions it is essential to use a sawblade of the right specification - this is the proper approach which will give maximum profit.

Small custom shops, however, may not stock an ideal sawblade for every one-off job - and it is uneconomic to buy a special sawblade for a limited use.

In practice, though, almost any sawblade will saw almost any material - provided the feed speed chosen gives a suitable bite for the material being sawn, and one is not too picky about the finish and efficiency this gives.

For example, if a sawblade has a pitch or hook angle which gives a poor finish in normal feed conditions, then the quality of cut can be improved either by increasing the saw arbor speed (if practical) or more likely by reducing the feed speed.

On the other hand, if a saw turns waste into fine dust and also dulls the teeth quickly, then this might be improved by either slowing the saw arbor speed (if practical) or by increasing the feed speed.

Converting Natural Timber

Using a sawblade with a large pitch. A sawblade with a large pitch and few teeth produces coarse sawdust, has less tendency to gum up and gives a longer run-time - so more timber is converted before sharpening is needed and power consumption is less. On the down side tooth loading is higher, a poorer finish and rougher cutting will result, and there is less "feel" when feeding manually.

Using a sawblade with a small pitch. A sawblade with a small pitch and more teeth forms finer sawdust, and the finish, smoothness and "feel" of the cut all improve. However more power is needed because more teeth are in the cut, and the teeth will dull at a faster rate than normal. *As the dust of many timbers is harmful, producing finer dust through using a very small bite also increases the health risk.*

Dealing with abrasive materials.

Abrasive hardwoods contain hard mineral deposits within their cells - silica in the case of teak. These timbers themselves may be relatively soft, but the mineral deposits quickly wear saw teeth of all types.

Manufactured boards and plastic veneers may also contain hard and abrasive inclusions, possibly in the bonding, and these also have a similar dulling effect.

To avoid rapid dulling when converting such materials make sure that the bite is somewhere between the recommended maximum and minimum, and preferably towards the maximum. Also the feed must be continuous - never allow it to slow down, nor the sawblade to idle (run in the cut without feeding) - both rapidly dull teeth.

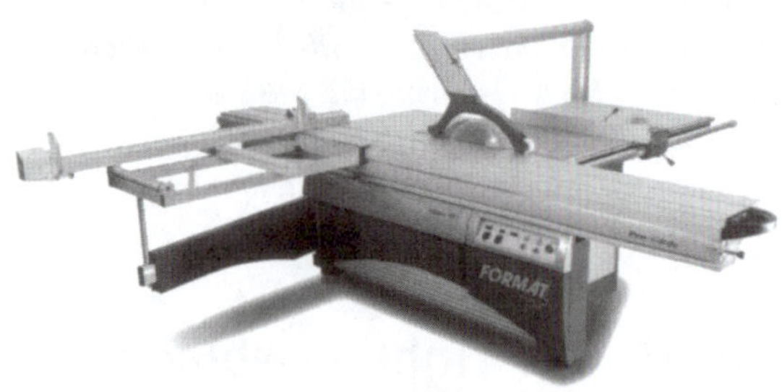

Fig. 3.17
Even large boards and long flitches can be handled on modern sliding table saws. Machines of this type have both a long table traverse movement and a large width-cutting capacity. The version shown is Kappa 450.
Source - Format-4

Maintaining a continuous feed is difficult when sawing manually, so for teak and similar timbers a mechanical feed is much better than a manual feed.

Rapid dulling can also be a problem when crosscutting teak or converting highly abrasive materials on cross-cut or panel saws. Both types can have a high rim speed which actually worsens the problem - abrasive materials have a lesser dulling effect when the rim speed is low.

The most direct way to reduce the rim speed is to lower the saw arbor speed, but on single-speed and direct-driven saws this is not an option. One solution is to use a sawblade with fewer teeth - but this is not practical on pull-out cross-cut saws as this can make the sawblade more prone to run-out. A better alternative is to use the smallest practical sawblade.

A mechanical feed is preferable for board materials - otherwise use a sliding table saw where material can be gripped by clamps to allow a smooth, continuous cutting action. Both avoid the hesitation and associated dulling through changing grip with a regular hand-feed movement.

Fig. 3.18
A sliding table saw also allows flitch edging to be carried out smoothly and accurately and in a single, continuous cutting action, as on this CP Panel saw.
Source - Wadkin

Alloy & Stellite-tipped Saws

Ripsawing Natural Timber

Minimum bite. The bite should be not be less than 0.13mm (0·005in) for hardwoods, not less than 0.20mm (0.008in) for softwoods, and not less than 0.25mm (0.01in) for abrasive hardwoods and other timbers producing powdery sawdust.

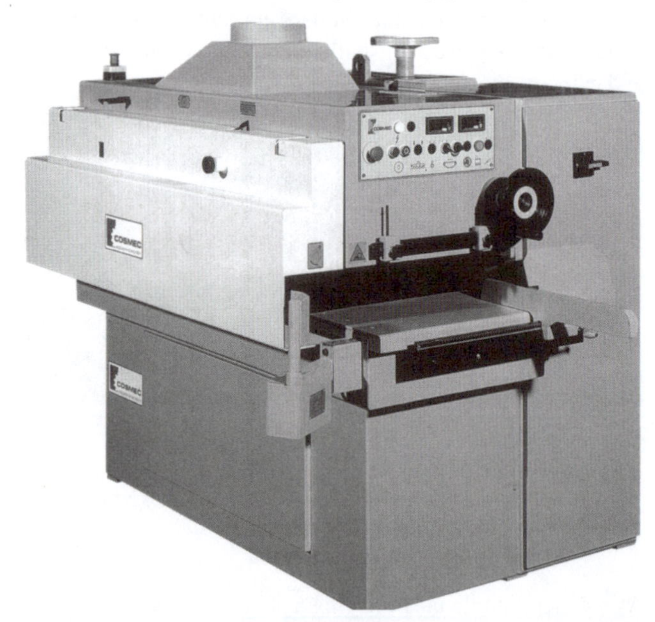

Fig. 3.19
This heavy-duty single arbor, multi-ripsaw converts both softwood and hardwood at high feed rates.
Source - Cosmec

On machines for converting natural timber where teeth just break through the opposite surface of the timber, such as straight-line ripsaws, or where the saw is buried in the cut, such as multi-rip saws, the chip thickness at entry is virtu-

ally zero. To reduce the rubbing effect that this creates, machines of this type often use sawblades with bigger than normal pitches and with slightly more hook.

Maximum bite. There is also a maximum recommended bite which, if exceeded, can vibrate the saw teeth or possibly overload them to rapidly increase power consumption. The figure varies according to the type of sawblade and class of timber. With alloy saws the maximum bite should be no more than 1.00mm (0.04in) for softwoods, and not more than 0.40mm (0.02in) for hardwoods.

Overloading the teeth, attempting a larger bite than recommended, can force alloy spring-set teeth inwards to close-in and narrow the kerf - particularly when teeth begin to dull or when using very thin sawblades. The sawn timber then no longer clears the saw body and this can lead to the timber pinching, i.e., closing-in on the saw body and perhaps being thrown-back, or the saw body becoming burnt and blistered.

Swage-set saws stand up much better to harsh cutting conditions than spring-set saws and, whilst they do not close-in on the cut, overloading the teeth can slow down or possible stall the sawblade.

Fig. 3.20
This M3 multi-ripsaw is fitted with a laser beam to align flitches prior to edge trimming. In this case a single sawblade only is fitted.
Source - SCM

Converting board material

Alloy saws were originally used for converting board materials - there was no other alternative at one time - but also there was much less variety - manufactured boards were mainly plywood, blockboard, low-grade chipboard and some plastic facings. Alloy saws quickly become dull when sawing newer man-made materials as they are much more abrasive, so when carbide saws became widely available they quickly replaced alloy saws in these applications.

Insert-tooth Saws

Ripsawing Natural Timber

Insert-tooth saws are used mainly for wet timber conversion and for resawing used timbers with inclusions which could wreck other sawblades.

Minimum bite. The minimum bite for insert-tooth saws is around 1.00mm (0.03in) on softwood, but users often go for a much bigger bite when these sawblades are used for fast feed conversion. As later mentioned, the limiting factor may, in fact, be the gullet capacity - not the bite.

Maximum bite. A large bite is allowable for insert-tooth saws, with an average of 2mm (0.08in) and up to a maximum of 3.50mm (0.14in). The figure is proportionately less for hardwoods.

Tungsten Carbide saws

Tungsten carbide saws are used for much wider applications than alloy saws. This ranges from wet and dry timber to all types of manufactured materials such as plywood, chipboard, medium density fiberboard, flakeboard, also facing materials such as plastic - both as thin sheets bonded to boards and as veneers.

Many board materials are both abrasive and brittle. For these the bite should be small - the least with the most brittle materials to reduce break-out to the absolute minimum.

Because there is a such a great variety of materials converted by tungsten carbide saws there is a correspondingly wide range of bite recommendations - from as little as 0.02mm for plastic laminated board to 0.90mm for softwood conversion. Maximum and minimum recommended bites for most materials are given in Chapter 16.

Type of Tooth

The type of tooth must be taken into account when deciding a suitable bite, also the scratch pattern formed by the tooth points on the sides of the material - this determines the surface finish.

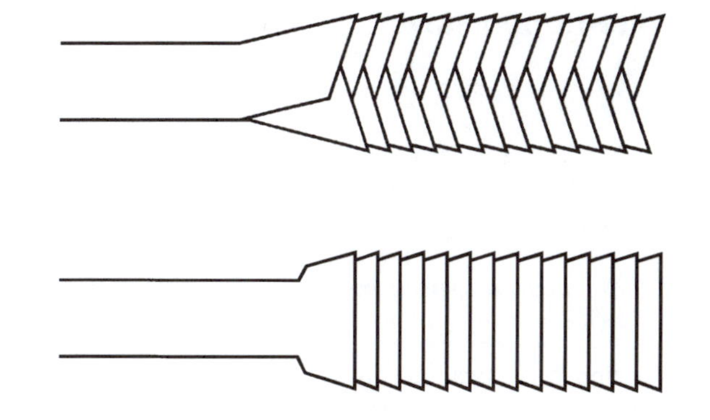

Fig. 3.21
Size and shape of chip removed in equal bite conditions by square-topped teeth (below) and alternate bevel (above).

Insert-tooth saws

The teeth of insert-tooth saws are of the swage-set type, the tip itself projects sideways beyond the saw body to give working clearance, so the bite and surface finish recommendations relate to all teeth.

Alloy Saws

For alloy saws there are two basic tooth types, swage set and spring set. As swage-set saws cut equally at both sides with all teeth, all teeth count in forming the surface finish and the bite recommendations apply to them all.

Spring-set teeth have a top bevel angle and cut alternately, first to one side and then to the other. The actual bite of each tooth is calculated as between its own path and that of the preceding tooth of the same hand, and it is this that must not exceed the maximum recommended figure. The simple way of accounting for this, as far as the bite and surface finish is concerned, is to count two following teeth as a group.

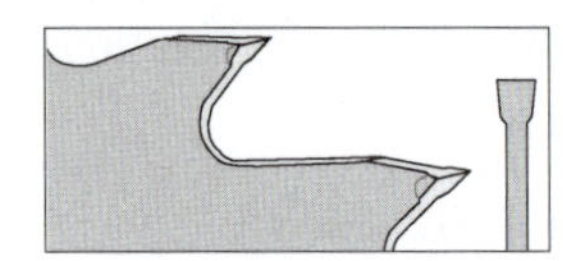

Fig. 3.22
Swage-set teeth cut at both sides, so all teeth count both as regards the surface finish and in respect of the bite.

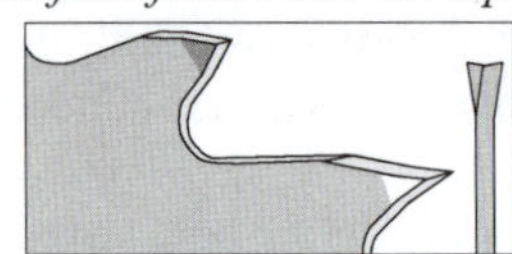

Fig. 3.23
Spring-set teeth cut alternately left and right, so two following teeth count as a single tooth, both as regards the surface finish and in respect of bite.

In practice this means that a swage-set saw will give the same quality of finish when fed at double the feed rate of a spring-set saw having the same number of teeth.

Usually, though, a swage-set saw intended for the same feed speed and bite of an equivalent spring-set saw will be made with only half the number of teeth. This then has double the tooth pitch and a correspondingly much larger gullet capacity - so a greater depth of cut is possible. This is why alloy swage-set saws, or any flat-top Stellite or carbide-tipped saws, are the preferred types for fast-feed conversion.

Stellite-tipped saws

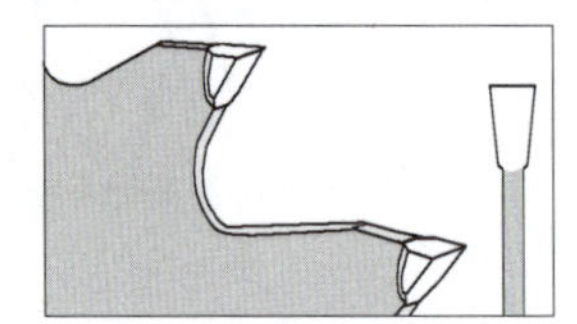

Fig. 3.24
Stellite-tipped saws, like swage-set teeth, cut both sides, and all teeth count as regards surface finish.

Stellite-tipped saws traditionally have flat tops, and each tooth projects equally at each side of the saw body. Technically they operate in a similar way to swage-set alloy saws, but they are far superior - and have virtually replaced swage-set alloy saws for fast conversion of natural timber. Modern Stellite-tipped saws are also more suitable for timbers containing fluids that attack the bonding of tungsten-tipped saws.

Tungsten carbide Saws

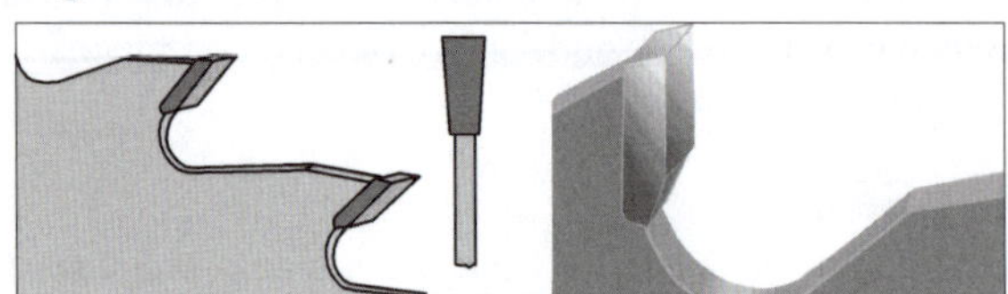

Fig. 3.25
With flat-topped carbide tipped teeth, as used mainly for fast feed conversion, all teeth count.
Source for illustration - Guhdo

Tungsten carbide teeth also project at each side of the saw body to give working clearance. They can be flat-topped, in which case the surface finish recommendations relate to all teeth - but if they have some other profile, then the bite and surface finish relates to whatever number of teeth there are to form a tooth group needed to complete the cut.

Usually there are two teeth per group, for example with alternate bevel or flat/triple chip tooth forms. There can, however, be more, for example with novelty saws having a cleaner tooth followed by two pairs of alternate-bevel teeth - an effective tooth-group of four.

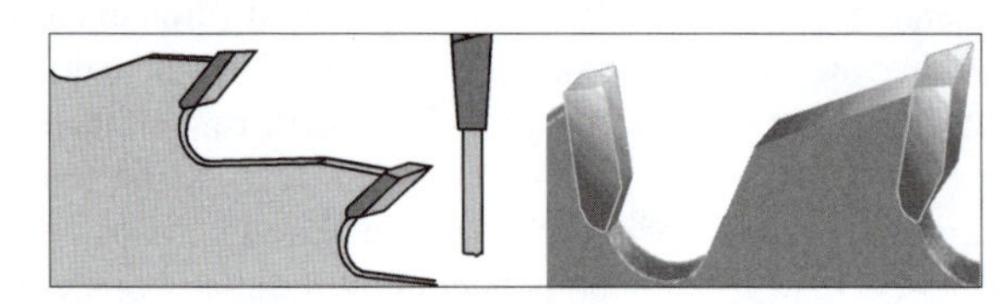

Fig. 3.26
Alternate-bevel carbide teeth cut alternately and count in the same way as spring-set alloy teeth. On some saws the non-cutting side is ground flush with the saw body.
Source for illustration - Guhdo.

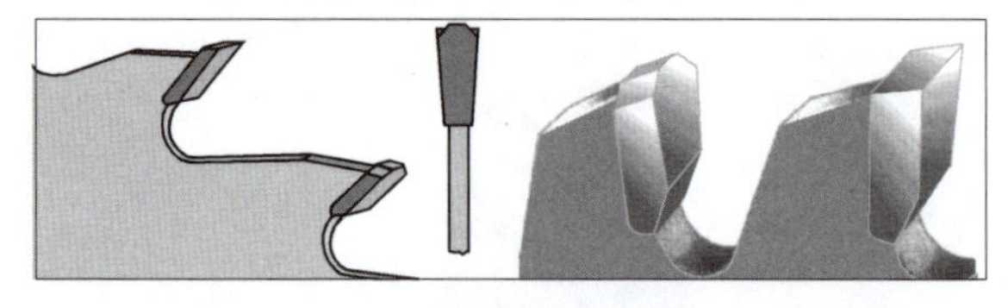

Fig. 3.27
Triple-chip teeth have normally one flat-topped tooth followed by a tooth ground to a higher flat top and two bevels. This gives an excellent finish when cutting decorative facings - and keeps sawblades running true.
Source for illustration - Guhdo.

Like most tables and charts giving this information, the Nomogram in Chapter 16 uses the bite only as a basis for calculations. The difference between bite and chip thickness is only important in certain operating conditions - and this factor has already been taken into account in any event.

Gullet capacity

The gullet, the area immediately in front of each tooth, collects waste as it is generated, then discharges it fully when the tooth is clear of the material being sawn.

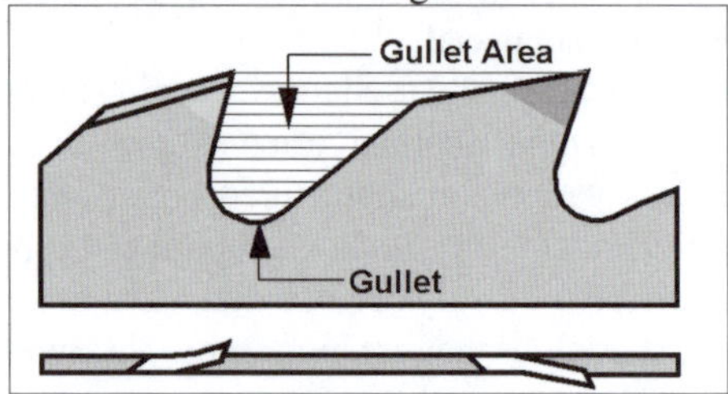

Fig. 3.28
The shaded area of the gullet holds sawdust during the cutting action and is the measure of the gullet capacity.

The maximum cutting conditions in terms of feed speed and cut depth are reached when the gullets are entirely filled with waste at each pass of each tooth. For each specific tooth type and size the point at which this happens is determined by the bite, the depth of cut, and the material being sawn.

Gullets are not always completely filled, in fact this happens only when a fast feed speed is combined with a deep cut. These are conditions usually found when converting natural timber, and when cutting stacked composite material. Most circular saws operate with the gullets only partially filled - this gives no problem - a problem only arises when conditions are such that the gullets become packed - see 'The cost of over-filled gullets'.

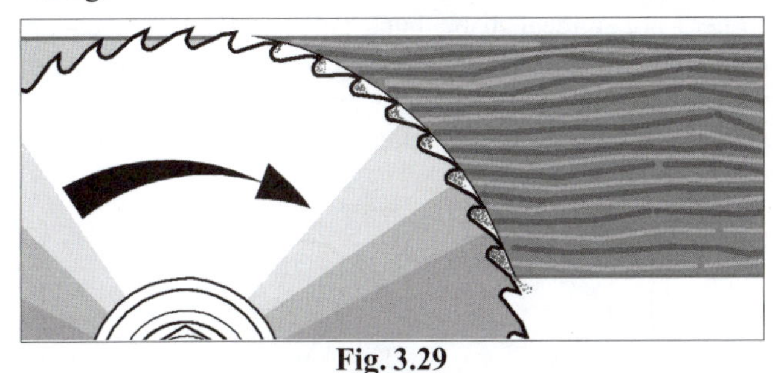

Fig. 3.29
In maximum-feed conditions the gullets are completely filled during the cut and completely discharged as the saw clears the underside of the material

Although gullet capacity is actually a volume measurement, in practice it is measured as the area enclosed by the gullet profile. Kerf width does not actually affect the calculation - a wider kerf generates more waste - but also increases the gullet capacity by the same amount.

Waste is held mostly in the gullet and, although some may escape down the sides of the sawblade, calculations must assume that the gullet alone must hold all the waste generated by the cutting action. Most gullets are roughly triangular in shape, or part-circular in the case of insert-tooth saws, so in either case the gullet capacity can be calculated quite easily. Chapter 16 shows how.

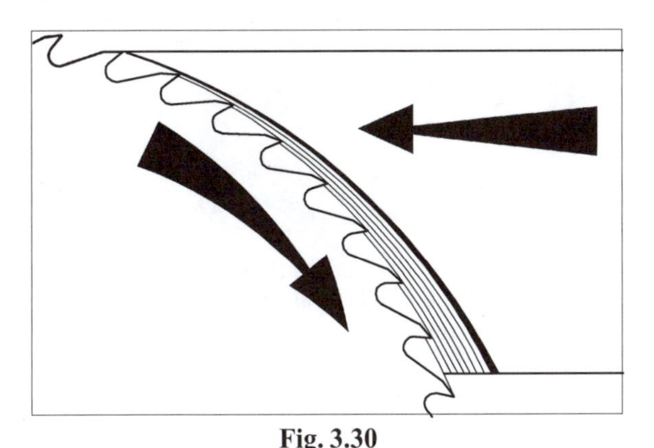

Fig. 3.30
This shows tooth tracks of square-form teeth already in the cut, with the solid waste to be generated by the tooth just entering the cut shown as an enclosed black area.

Measuring the Waste Generated

The waste generated at each tooth pass - the area enclosed by two following tooth tracks - must fit into the gullet. These tracks are almost true arcs of the cutting circle, but technically they are curtate trochoids - a circle drawn with a moving centre. Calculating such an area is both complicated and quite unnecessary.

For square-form teeth the area is the effective depth of cut multiplied by the bite. *(The actual chip thickness is less than the bite, but it is also longer due to the curvature - so the two balance one another out).*

For alternate-bevel teeth the equivalent area is the effective depth of cut multiplied by half the bite. *(This is because alternate-bevel teeth cut only half the kerf width, so the waste removed is half that of a square tooth having the same bite.)*

Effective Depth of Cut

Because the gullets start discharging waste almost immediately the tooth breaks through the underside, this short distance need not count in the calculation. The actual point where waste starts to discharge varies with the tooth shape and the angle of the cutting arc. For practical purposes, however, it can be taken as one quarter of the average tooth pitch, so the effective depth of cut is the material thickness less this amount.

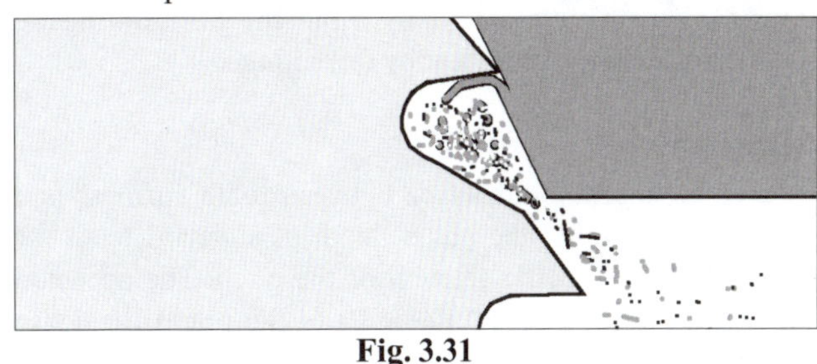

Fig. 3.31
Waste begins discharging from the gullets shortly after the tooth breaks through the underside, so the effective depth of cut is less than the depth being sawn.

Accounting for waste expansion

When the waste is detached and breaks up into particles or dust it occupies a larger area than the solid material from which it is formed. The amount of expansion varies with the dust particle size and the moisture, gum content and the general structure of the raw material. Also, dust does not pack lightly and evenly in the gullet - it tends to concentrate mostly at the tooth front and the base of the gullet.

With a fine or medium tooth bite it can be assumed that the gullet must hold up to three times the area of waste produced by each tooth in a single pass.

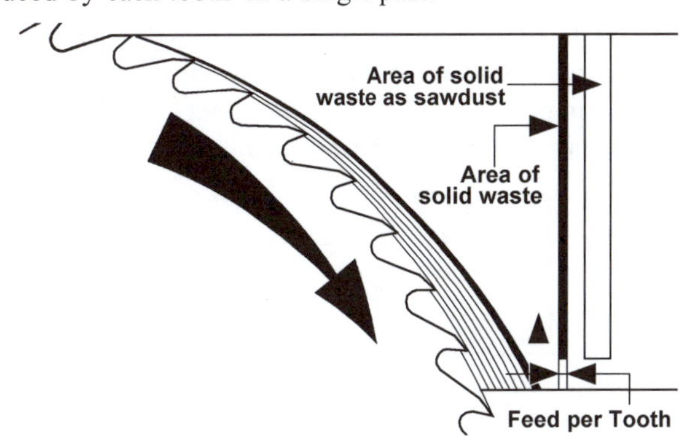

Fig. 3.32
The material removed by square teeth is enclosed within the arc-shaped black area. As the curved shape can be ignored, the area for square teeth is the bite (feed-per-tooth) multiplied by the effective depth of cut. With a small and medium bite waste expands by roughly three times into loosely-packed dust, shown enclosed by the blank rectangle, and this must fit into the gullet area. (In the sketch it will not because the feed-per-tooth has been exaggerated to show the paths of the teeth more clearly).

When converting natural timber using insert-tooth and other

fast-feed sawblades a much smaller figure can be used for sawdust expansion - because of the larger particle size the waste compresses more effectively to give an expansion rate of roughly $1^1/_2$.

Gullet Profiles

The gullet capacity of any sawblade varies with the pitch and type of the teeth. Because there are many different tooth types there is no general rule that applies to all. A later nomogram in Chapter 16 takes account of different pitches and tooth profiles.

Gullets should never contain sharp internal corners - these stress the saw body and provide a starting point for cracks. Also the gullet base should be smooth, regular, square-across and well-rounded - to ensure that sawdust is swirled around to spread out more evenly within the gullet during the cut and to discharge more readily on exiting.

Alloy saws

To give an acceptable balance between tooth stiffness and gullet area capacity, the gullet depth of standard teeth for plate saws is a specific fraction of the pitch. The accepted guide is 0·45P or 0·5P ($^{45}/_{100}$ or $^1/_2$ of the pitch) for dense timbers, and 0·6P ($^6/_{10}$ of the pitch) for light timbers.

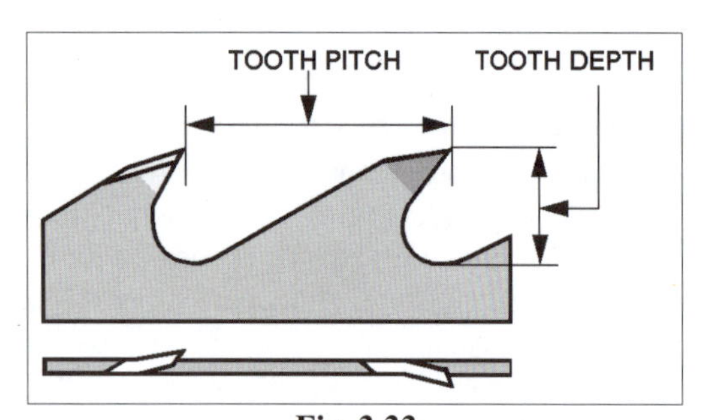

Fig. 3.33
Measuring the tooth pitch and depth. The correct proportions give a gullet capable of holding enough waste in the form of sawdust whilst retaining a rigid tooth form.

Gullet and tooth back

There are two basic tooth shapes for alloy circular saws. One is the continuous tooth shape, or round-back tooth, and the other the file-sharpened or flat-back tooth.

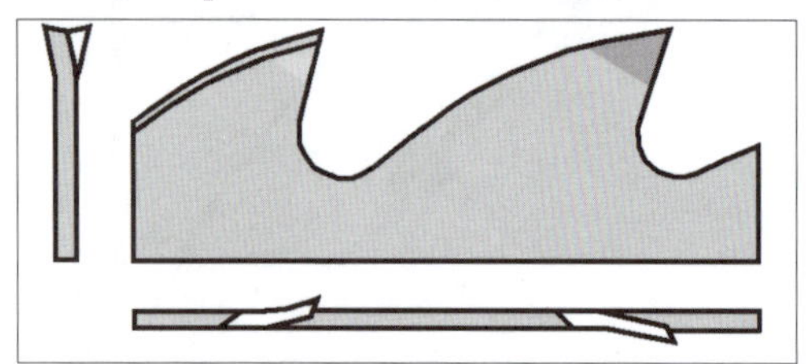

Fig. 3.34
Profile of a machine-sharpened plate ripsaw.

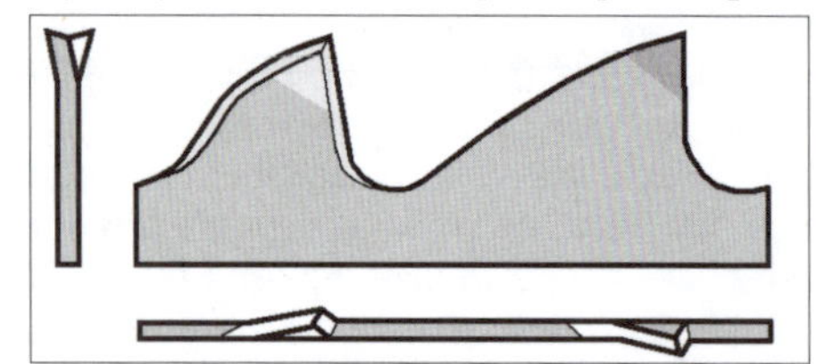

Fig. 3.35
Profile of a machine-sharpened cross-cut plate saw.

The round-back tooth profile is formed by an automatic sharpener, where the grinding wheel grinds down the front of the tooth, around the gullet, up the back and then across the tooth top. With an automatic grinder the correct gullet depth and tooth profile is maintained consistently to always maintain the correct proportions.

Hand-filed teeth have a similar face and gullet, but with a straight back from the gullet to meet the flat tooth top at a corner - not an ideal shape, but highly practical. As teeth are filed down the gullet depth reduces. With this the capacity of the sawblade to feed fast and deep is also progressively reduced - so ideally the gullets should be re-ground periodically.

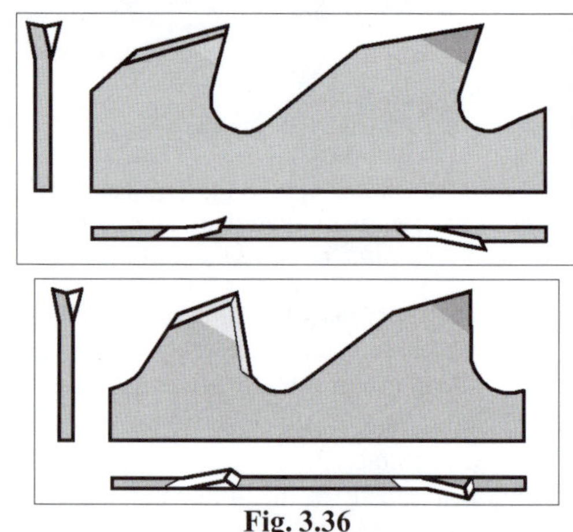

Fig. 3.36
Top - Profile of a file-sharpened plate ripsaw.
Bottom - Profile of a file-sharpened cross-cut plate saw.

File-sharpened sawblades are used mainly for manual feed only, so a reducing gullet depth doesn't matter too much as they are rarely operated to maximum gullet capacity - but watch this factor when power-feeding hand-filed sawblades.

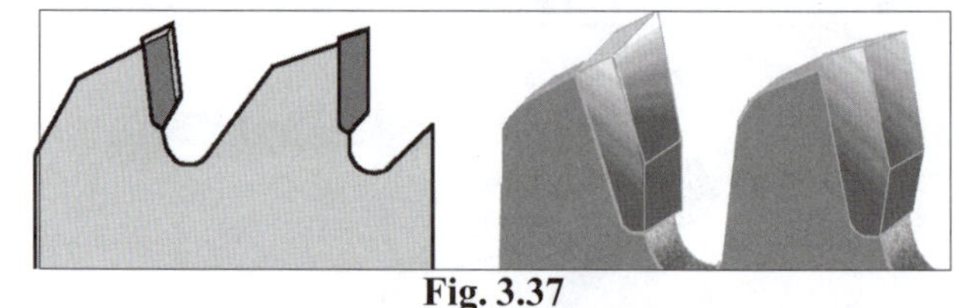

Fig. 3.37
Negative-angle carbide teeth for cross-cutting
Source for illustration - Guhdo

Saw Teeth for Frozen Timber. There is one case for making a tooth gullet other than fully rounded. When a regular rounded gullet shape is used for sawing frozen timber the waste does not readily break-apart into fine sawdust.

Instead the frozen chips may lodge-in and block the gullet so the teeth are no longer able to cut. To prevent this happening the gullet is formed with an angle in the base to smash the forming chips into particles small to be held and discharged efficiently.

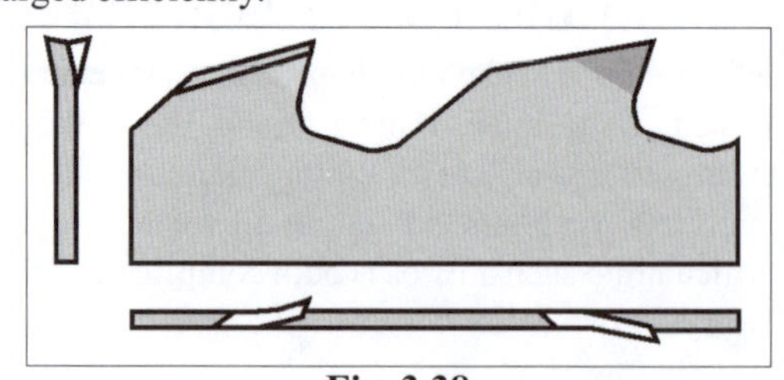

Fig. 3.38
Frozen timber tooth shape for a plate saw.

Insert-tooth Saws

With these saws worn tips are replaced by new tips, so the original sawblade diameter and gullet capacity are maintained consistently. There are alternative holders which slightly change the gullet capacity, but these are used for other reasons. The most common problem is in well-used holders rounding-over, so allowing waste to escape between the sawn timber and saw plate. For frozen timber special holders can be fitted to prevent frozen chips blocking the gullet.

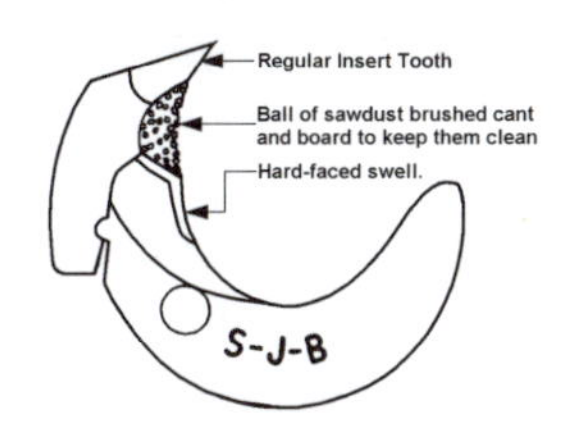

Fig. 3.39
When using an insert tooth saw for frozen timber the regular holder can be replaced by a special frost holder which keeps the saw free-running in cold weather.
Source - Spear & Jackson.

Carbide-tipped Saws.

The gullet shapes in carbide saws vary in depth-to-pitch far more than the equivalent alloy saw teeth.

The teeth of carbide saws for fast feed conversion of natural timber usually have a similar proportions to teeth of machine-sharpened alloy saws, with well-rounded gullets for efficient ejection of sawdust. Others types for a slower mechanical feed have a regular gullet shape but a much longer pitch, sometimes with the tooth backing formed concentric to the saw centre.

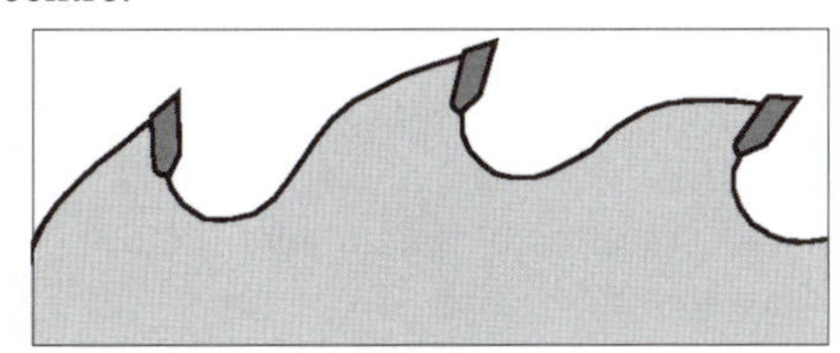

Fig. 3.40
Typical carbide saw for fast-feed conversion of timber.

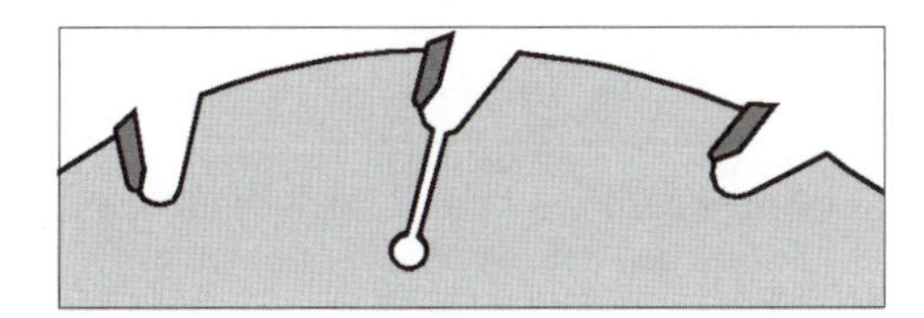

Fig. 3.41
Long-pitch carbide saw for slower feed speeds on timber.

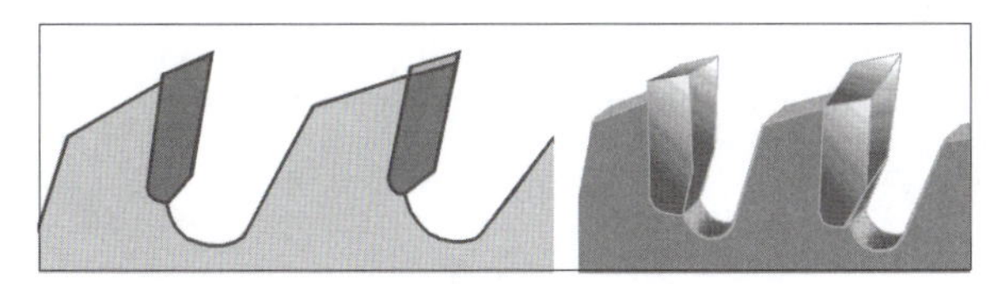

Fig. 3.42
Carbide teeth for panel conversion. The pitch is less but the gullet depth remains roughly the same
Source for illustration - Guhdo

Teeth for converting panels have similar profiles to file-sharpened alloy saws, but with a shorter pitches and deeper gullets.

Large-pitch sawblades for manual feed have anti-kick-back, or chip-limiting, shoulders for safer operation. In their cases the gullet area to hold waste is that immediately in front of the tooth and is relatively small - but hand-fed sawblades do not need a large gullet capacity in any event.

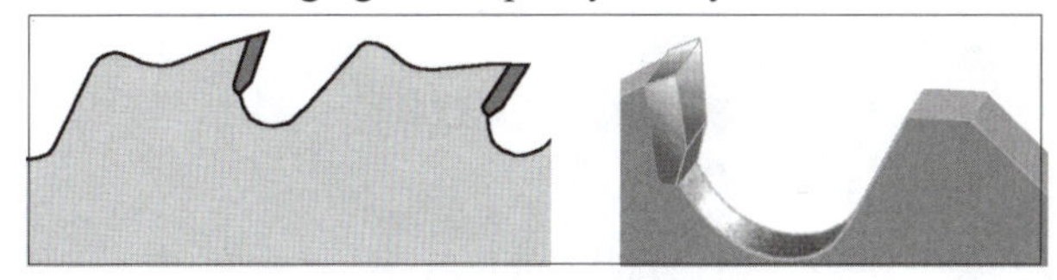

Fig. 3.43
Carbide teeth with anti-kick-back shoulders.
Source for illustration - Guhdo

The gullet shape for new carbide saws is determined by the saw makers - and is correctly proportioned for the intended working conditions.

As tooth tops are ground down, though, a common practice is also to grind down the tooth backing behind the tip to maintain the step-down. When tips are eventually replaced a shorter tip could be used, or the tip seating could be deepened to accept a regular tooth tip. In either case the gullet gradually grows shallower and capacity is progressively reduced. Except for fast-feed timber conversion, though, this loss is not usually critical.

For fast-feed conversion the original gullet proportions must be maintained. The only solution is to fully re-tip the sawblade when the feed speed or cut depth begin to be affected. This involves considerable and costly work, so it is probably more economical simply to replace a worn-out sawblade with a new one - and get a guaranteed performance into the bargain. See 'Carbide Saws' for more details.

Fig. 3.44
View of the SCM M3 single-arbor multirip-saw
Source - SCM

The Cost of Overfilled Gullets

When gullets are overfilled - on any sawblade and for any reason - dust packs so tightly in the gullet during the cut that it does not fully discharge when clear. This is mainly a problem on multirip-saws cutting resinous and stringy timbers.

Packed gullets physically prevent the teeth biting further into the material - and so trigger a rapid rise in power consumption. This may possibly slow-down the saw arbor or even blow the overloads to instantly stop the machine. Stalling in the cut must be avoided as this can cause a jam to hold up production and loose profit - and could damage sawblades extensively,

Fig. 3.45
General view of a Linck Profiling line for processing round logs using a combination of chippers, profiling heads and circular saws to produce square-edge boards without further treatment. Waste in the form of good quality chips and sawdust is extracted and separated automatically.
Source - Linck

Fig. 3.46
On production lines such as that shown above, the whole process is controlled from an enclosed cabin overlooking the equipment. Machinery is fully monitored from the various control displays alongside the operator.
Source - Linck

On multi-ripsaw machines for converting natural timber, one or two dials continuously show the power consumption of the saw motor (or motors). By checking these regularly, also listening to the cutting action, the operator can immediately reduce the feed speed should power loading approach danger point.

In most cases reducing the feed speed immediately reduces the power demand, and the machine can then operate normally. If the gullets are overloaded with gummy sawdust, though, they may first need cleaning-out before production can resume. Overfilling the gullets is not the only possible cause of overloading the saw motor, of course, other factors could cause this.

Many machines now have some form of automatic feed control that reduces the feed speed should the power consumption rise dangerously - by monitoring either the power consumption or the noise of sawing.

The Way Teeth Dull

In ideal conditions teeth bite into timber and other materials without problem, with each tooth immediately forming a chip from the point of entry - which then breaks up into waste.

If the feed speed is gradually slowed, however, a point is reached when the teeth can no longer cut effectively - because the bite is too small to properly form a chip. Instead the teeth rub on and compress the material rather than cutting it cleanly, with larger bites being taken irregularly on the compressed fibres when enough pressure builds-up. This rubbing action dulls teeth at a faster rate than normal - particularly with abrasive materials.

The conditions which cause teeth to rub rather than cut cleanly vary with the type and condition of the material being sawn - and by the sharpness of the teeth. To avoid this problem keep to the recommended bites - and keep the sawblade sharp.

Saw arbor height setting

Converting Natural Timber

The chip thickness on entry is the important factor when resawing natural timber. The average chip thickness might be fine for chip formation but, if it is too thin on entry, rubbing will take place both here and further into the cut where it might otherwise not. (As already mentioned, the thinnest chip is always at the entry point when ripsawing.)

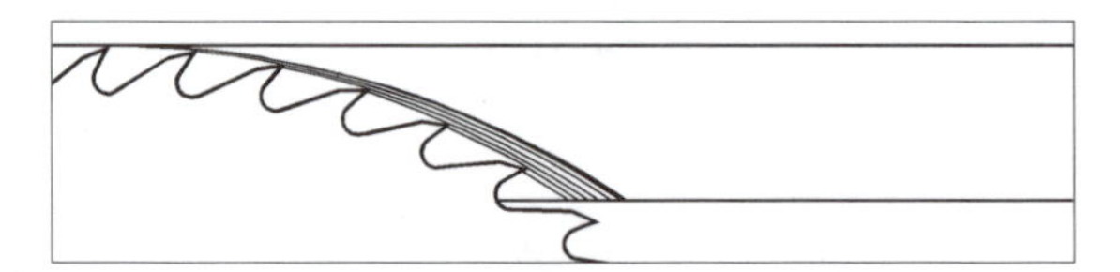

Fig. 3.47
With the saw arbor lowered the chance of the teeth rubbing increases, but the quality improves.

For any given feed condition the chip thickness can be altered by varying the saw arbor height. Lowering the arbor of an undercutting saw reduces chip thickness and increases the chance of the teeth rubbing. Raising it increases chip thickness and reduces the chance of the teeth rubbing.

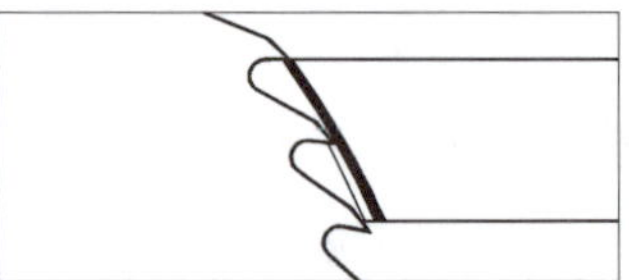

Fig. 3.48
With the saw arbor raised the chips formed are short and thick so the chance of the teeth rubbing is less.

Undercutting sawblades should preferably be at their highest setting when deep-cutting natural timber on power-fed saw benches - the regular setting on slabbers.

When sawing timber depths less than the maximum, there is a choice of using the maximum size of sawblade or a smaller one - see the following notes on 'Quality of finish and run-time' for a guide on this.

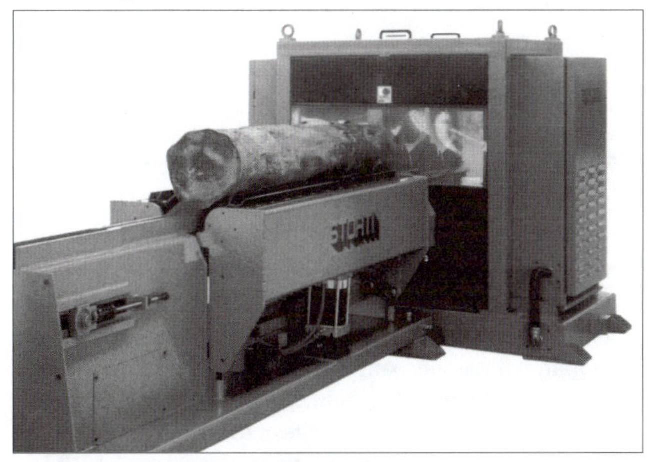

Fig. 3.49
Double and Multiple Slabbers convert round logs into flitches and barkside slabs. This Storti electronic machine can be equipped with four or eight sawblades.
Source - Storti

Altering the saw arbor height to vary the amount of sawblade projection beyond the timber is not an option on over and undercutting straight-line-edgers and multi-ripsaw types. The arbor usually does adjust, but only to take account of variations in the depth of timber being sawn. In these cases thin entry chips are always formed - but the dulling effect this has is more than off-set by the type of sawblades used and the many other advantages these machines offer.

Converting Panel Materials

Fig. 3.50
The panel Saw shown here is the upmarket Giben Prismatic PF equipped with side loading device and front air flotation tables.
Source - Giben

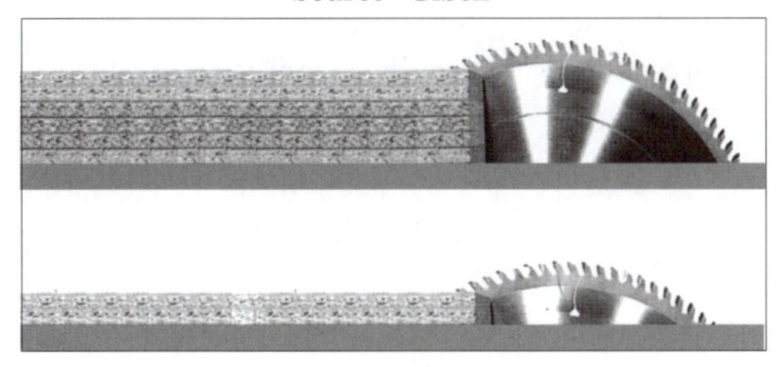

Fig. 3.51
On the Giben panel saw the saw arbor height automatically adjusts to suit the thickness of panel being sawn.
Source - Giben

On all panel converting machines the sawblade gullets should barely break through the top face of the panel as this gives the best possible finish - all else being equal. They have vertically-adjustable arbors for this reason. Manual adjustment of the saw arbor height may be by means of an external lever, but on some panel saws the saw automatically adjusts in height.

Quality of finish & Run-time

All saws leave a scratch pattern on the sides of the material being converted. The closer the marks are the better the surface finish - and there is less likelihood of ragging the underside. Square teeth leave a scratch mark on both sides during the cut, whilst alternate-bevel teeth leave scratch marks first on one side and then the other. To give the same density of scratch pattern as a square-toothed sawblade, an equivalent alternate-bevel sawblade would need double the number of teeth or half the feed speed - the same logic as applied to gullet capacity.

The action of tooth dulling

All teeth wear no matter how hard they are. Alloy saw teeth dull relatively quickly, whilst hard-tipped teeth last much longer before dulling - depending on how hard and tough the tips actually are and the hardness of the material being sawn.

When most teeth dull the points round over to almost a true part-circle, perhaps with an added slight angle to the front and/or top. Tungsten carbide tips tend to break down and loose hard particles unevenly rather than wear progressively - but the following remarks still apply.

The rate at which sawblade teeth dull is in direct relationship to the length of their cutting path - the distance each tooth travels in the cut - the chip thickness actually has little effect on the rate of wear. Because of the curved path circular saw teeth take the cutting path is always more than the depth of timber. This can be reduced by raising the saw arbor, or increased by lowering it.

One exception is when bite is so small that teeth compress the timber rather than cutting it cleanly - so that rubbing takes place to quickly wear the tooth points. The other exception is when the bite is so big that the cutting edge breaks down prematurely. As mentioned previously, the type and condition of the material and the style and sharpness of the saw teeth have a considerable affect on these factors.

When tooth points are fully sharp and work within the recommended bite conditions they form chips from the point of entry. When teeth begin to dull, though, this capability is progressively lost - then tooth wear begins to accelerate.

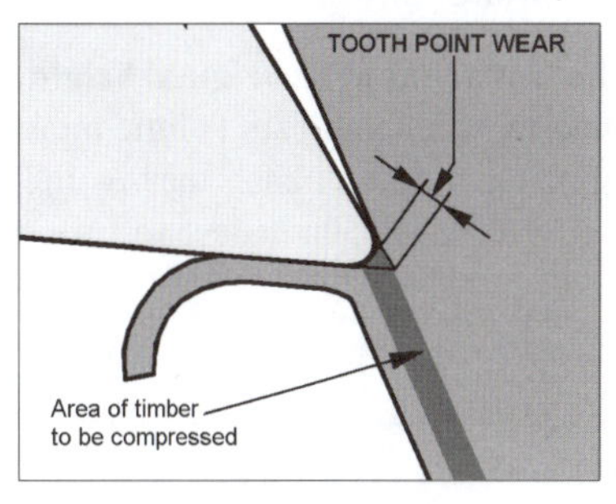

Fig. 3.52
The points of teeth wear almost to a true round, sometimes with a small flat at both top and front of the saw.

The following scratch patterns are based on the appearance of that formed by a swage-set alloy softwood saw fitted on a power-feed bench, having 48 teeth, an arbor speed of 1180 revs/min and a feed speed of approximately 14M/min. This would give the correct rim speed of 50M/sec, an acceptable bite of 0.25mm and an acceptable scratch pattern.

In this and all scratch pattern drawings the bite has been greatly stretched to show the effect more clearly.

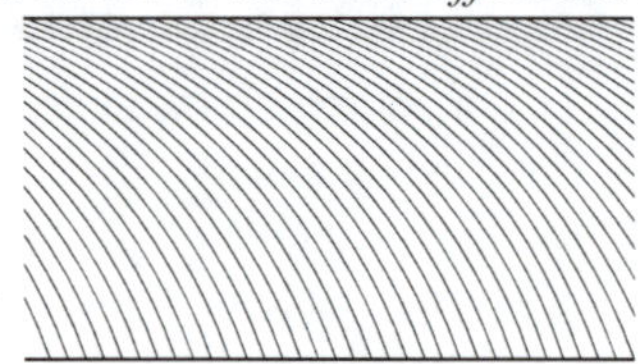

Fig. 3.53
The scratch pattern, or surface finish, formed on the sides of a board when sawing at maximum cut depth.

When deep sawing at the maximum cut depth for the diameter of sawblade used, the actual tooth travel is considerably more than the cut depth. If sawing thinner timber there is a choice - provided the saw arbor height can be varied.

Lowering the saw arbor* increases the tooth travel to dull teeth at a faster rate - but also gives a denser pattern of scratch marks and an apparently better finish. By doing this when sawing thin timber at maximum bite, the chip thickness effectively reduces - so the feed speed can actually be raised to still produce the same quality of finish - without exceeding tooth loading. Provided this does not then overfill the gullets it is an acceptable practice.

Raising the saw arbor* reduces the tooth travel to dull the teeth at a slower rate - but gives a less dense pattern of scratch marks.

**On under-cutting saws.*

Both the run-time and the quality of the surface finish is affected by the bite, the spacing of the scratch marks (chip thickness), and by the saw arbor height setting. The effect all these have on the run-time and surface finish is best illustrated as follows:-

As a basis for comparison, a sawblade diameter of about 810mm would cut 300mm deep, and have a tooth path of roughly 440mm (nearly half as much again as the actual cut depth). If the saw worked efficiently for 4 hours before re-sharpening was necessary, it would convert 3400M of 300mm deep softwood in this time.

If cutting timber 150mm deep under similar conditions there are several alternative ways of doing this - and these affect both the surface finish and run-time in different ways:-

1. Leaving the saw arbor at its original height setting (Fig. 3.54). By doing this the tooth path is little more than the cut depth at 170mm. On the above basis the saw life would be 10 hours to convert 8400M of 150mm deep softwood at 14M/min.

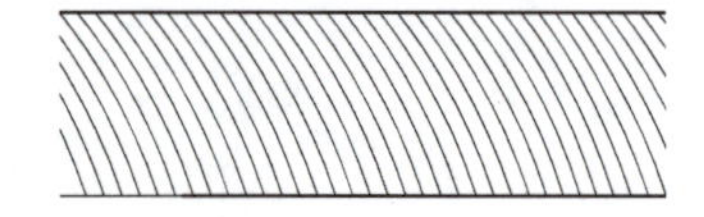

Fig. 3.54
At maximum arbor height the tooth travel is short.

2. Lowering the saw arbor (Fig. 3.55) - so that the sawblade breaks through the timber only by about 15mm (if practical). This increases the tooth path to 270mm, almost double the cut depth - but the feed speed can then be increased to still produce the same quality of finish.

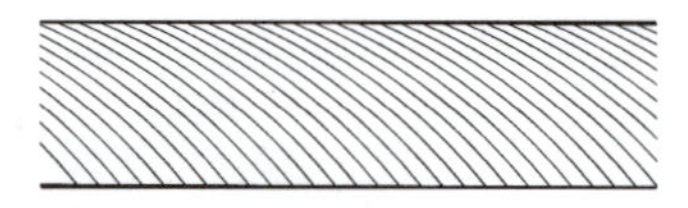

Fig. 3.55
By lowering the saw arbor the scratch pattern is more dense - so the bite can be increased (from 0.25 to 0.3mm.) by raising the feed speed -yet still give an acceptable finish.

An acceptable feed speed increase is a matter of judgement, but a reasonable guide would be to maintain a consistent chip thickness at the exit point. The actual chip thickness here is 0.245mm with an unchanged feed speed (less than average when cutting 300mm deep timber), so the feed speed can be increased to 17M/min to give a bite of roughly 0.3mm. On this basis the saw life would be 6 1/2 hours to convert 6 630M of 150mm deep softwood.

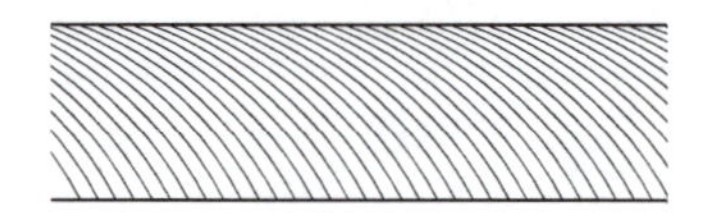

Fig. 3.56
Using a smaller diameter of saw allows a faster feed but still gives an acceptable finish.

3. Using a smaller diameter of saw. (Fig. 3.56) If using a saw of 510mm diameter this should run at a higher arbor speed to keep the same rim speed - and so would allow the feed speed to be raised pro-rata - whilst keeping the bite the same.

When using a 510mm diameter sawblade for a cut depth of 150mm the tooth path is 240mm. If keeping to the same rim speed of 50M/sec, the saw arbor speed should be 1872 revs/min - and the feed speed could then be raised to 22M/min and still give the same bite. On this basis saw life would be 4 1/2 hours to convert 5940M of 150mm deep softwood.

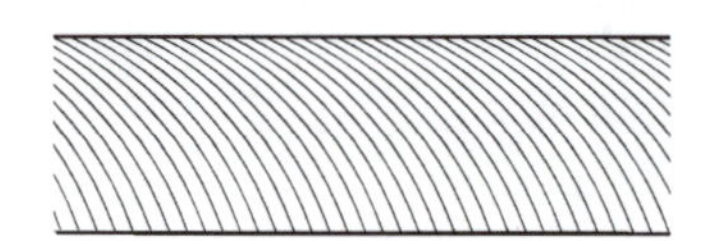

Fig. 3.57
Using the maximum diameter of saw on a smaller saw bench gives the highest efficiency.

4. Using a smaller power-feed sawbench. (Fig. 3.57) A suitable smaller sawbench would have 150mm as the maximum depth of cut for the maximum sawblade diameter of 440mm - so giving a tooth path of 230mm.

The saw arbor speed to give a rim speed of 50M/sec would be 2170 revs/min and the feed speed could then be 26M/min. On this basis the saw life would be 4 1/4 hours to convert 6630M/min of 150mm deep softwood.

NOTE: The above assumes that the correct saw arbor and feed speeds are available. This is not necessarily so - many machines have a limited range of saw arbor and feed speeds, and the running conditions suggested may not be practical. However, they are intended merely as a guide to efficient sawing practice - not specific recommendations.

Sawing Efficiency

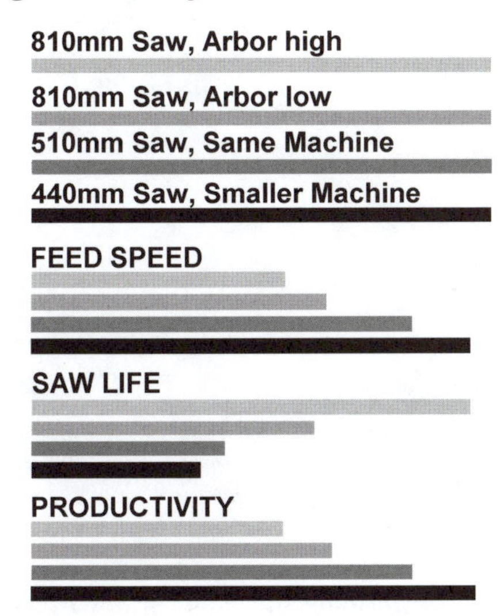

The comparisons detailed above are graphically shown by these bar charts, and illustrate the need for choosing the right sawblade and sawing conditions. Although the shorter sawblade lives mean more saw changes, the actual productivity figures keep more or less in step with the increase in feed speed - even taking sawblade-change downtime into account.

Type of tooth

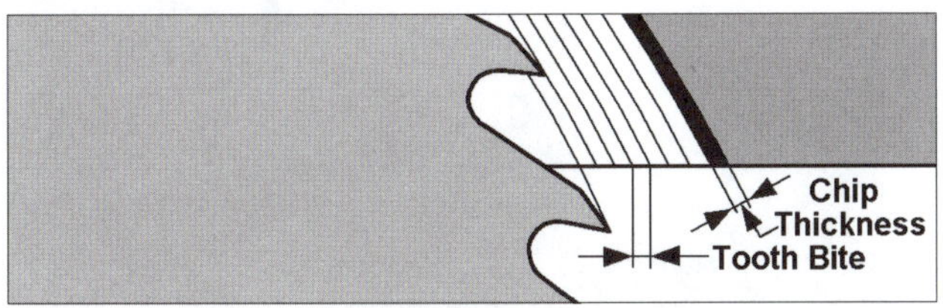

Fig. 3.58
The tooth bite and chip thickness when using square-form, full kerf-width teeth. All teeth cut at both sides.

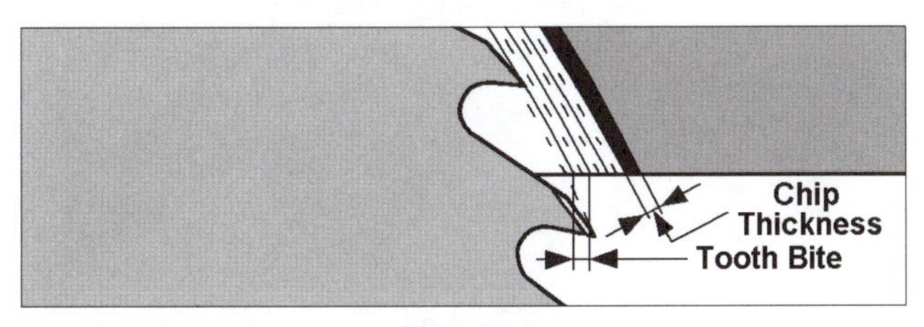

Fig. 3.59
The tooth bite and chip thickness when using spring-set alloy saw teeth, alternate bevelled tct teeth, or any other paired-group teeth.
If the sawblade has the same number of teeth as that shown in Fig. 3.58 and runs at the same arbor speed, then the feed speed must be half the former to give an equivalent bite and scratch pattern.

Teeth conventionally used for timber conversion are of the square or flat-topped type. To produce a better quality of cut, i.e., for finish-sawing and cross-cutting of timber and panels, group-type teeth are used - alternate bevel, triple chip, etc. Unlike square teeth, group teeth make a surface mark every alternate tooth, so the scratch pattern has twice the pitch for the same number of teeth in the same operating conditions.

To keep the tooth bite and surface finish the same, paired group-teeth saws must either have double the number of teeth of square-topped sawblades, or be fed at half the feed speed of the latter. This factor must be borne in mind when making calculations.

Converting timber with group-teeth

In order to give the same surface finish as equivalent sawblades with square teeth, sawblades with two teeth per group need double the number of teeth. As a result their gullets are half the pitch, half the depth, and a quarter* of the gullet capacity of squared-tooth sawblades - for the same operating conditions.

**However, group-teeth remove only half the waste of square teeth for the same tooth bite, so making their gullet capacities half that of square-toothed sawblades.*

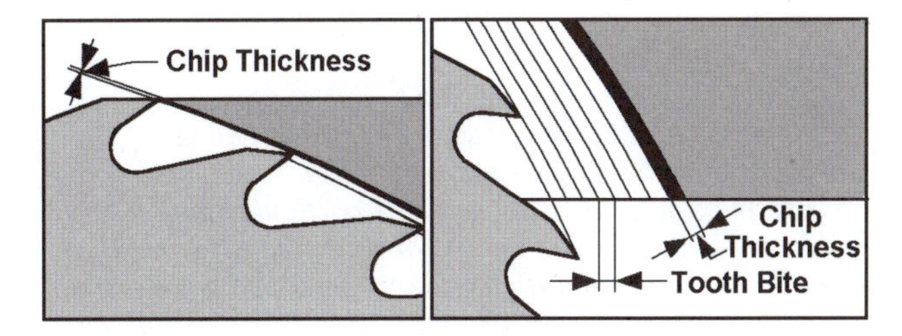

Fig. 3.60
Regardless of the tooth type, the chip thickness varies from the entry point (left) to the exit point (right).

Fig. 3.61
Changing the saws on an SCM single-arbor multi saw. The carbide saws in this multi-ripsaw are typical of their type, long pitch, smooth gullet and square form.
Source - SCM

Finish-sawn timber and panel conversion

For finish-sawing natural timber, and for converting manufactured panels, surface finish is of much greater importance. Sawblades for this purpose have teeth either of the alternate-bevel or a triple-chip type - and a smaller bite. These sawblades mark the sawn surface every second tooth.

For finish-sawing the saw arbor is set so that the tooth gullets barely break through the top face of the material being sawn. When cutting thin material the scratch patterns appears more dense, so it is practical to feed at a slightly faster rate than when the sawblade is set higher.

Quality of Surface Finish

The surface finish produced by Stellite and carbide teeth is infinitely better than that from spring-set alloy teeth. This is due to two factors. Firstly, the amount set on spring-set alloy

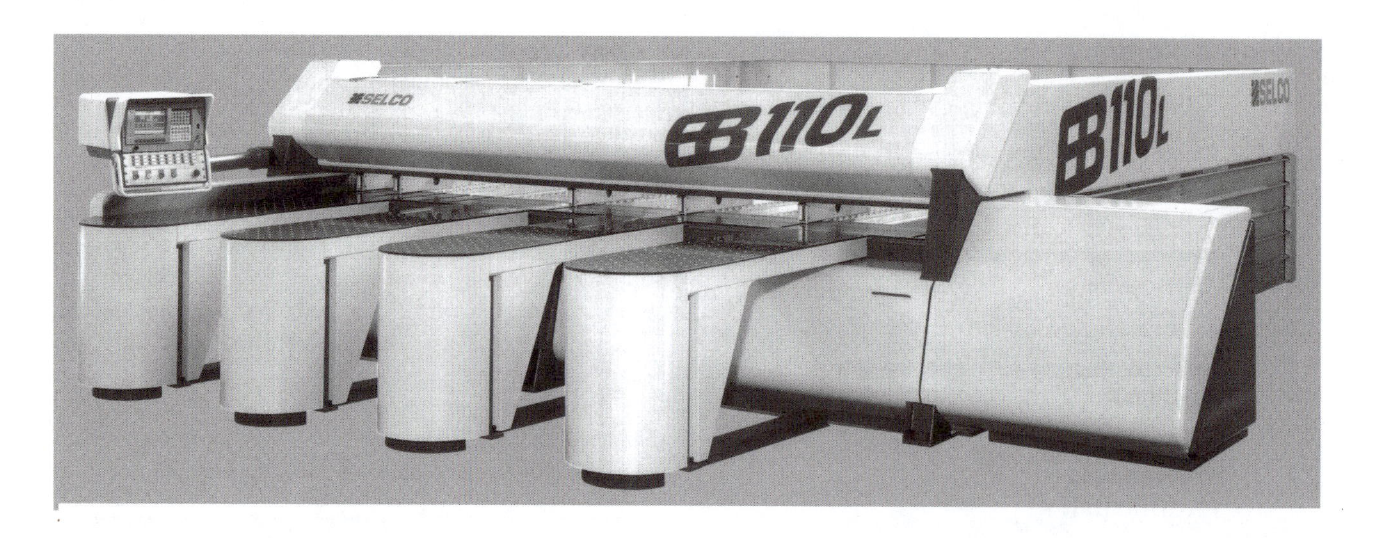

Fig. 3.62
This Selco EB110L panel sizing centre is typical of the modern, purpose-built saws for precision conversion of boards and timber of all types. This particular model has a wide range of optional extras for a limited investment. It has high cutting quality, excellent positioning precision and a powerful user-friendly control. This version has special covers and rounded air cushion tables.
Source - Biesse

saws is not as precise and can and vary from one sharpening to another, whereas carbide and Stellite teeth are precisely ground to size

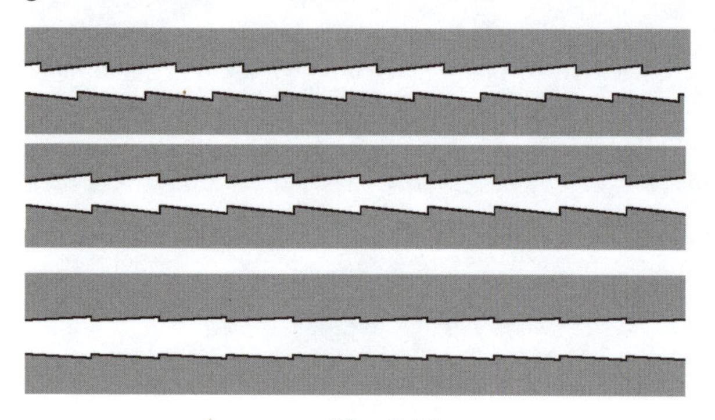

Fig. 3.63
Comparing the surface finish from a spring-set alloy saw teeth (above) and swage-set teeth (centre) with that of a Stellite or carbide saw (below), showing differences in the degree of the zigzag effect.

Secondly, the sharp angle of bend on spring-set alloy saw teeth and swage-set saw teeth makes the points dig-in and give a more pronounced zigzag effect, compared to that from the smaller radial clearance angle ground on Stellite and carbide teeth.

In fact, for finishing sawblades with carbide teeth, a smaller radial clearance angle is used to improve the surface quality, whereas for natural timber conversion a larger radial clearance angle is preferred to give a cleaner cutting action and reduced power demand.

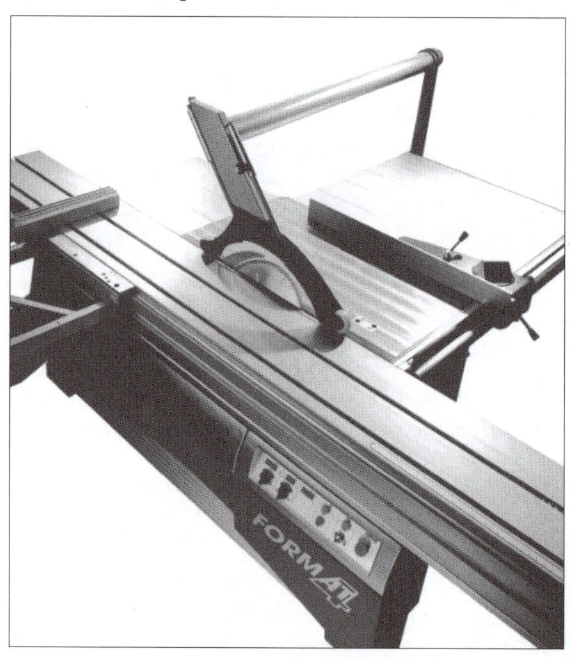

Fig. 3.64
Modern sliding table saws, such as this Kappa 450, have finger-tip controls for start, stop of the saw motor, also for the rise and fall the main sawblade, and the tilt of both the main sawblade and the scoring sawblade.
Ready access is provided for adjusting the scoring sawblade height and alignment.
Source - Format-4

Chapter 4

Motors & Drives

Mains Power Supplies

Electrical power is the flow of electrons through good electrical conductors, such as copper, silver and gold. An electric supply is often compared to a river, where voltage is the pressure of the flow, high in waterfalls and low is sluggish rivers, and amperage is the volume of flow. The energy of electrical current is termed the wattage - or KVA, kilovolt-amps in mains supply terms. This is calculated by multiplying the current flow by the pressure - volts times amps, so there is same energy from a low voltage and high amperage current as from a high voltage and low amperage current.
There are two types of electrical supply:-

DC (direct current)
DC current flows one direction only and is the type of electrical current powering car, torch and portable radio batteries.

AC (alternating current)
AC current reverses its direction of flow many times a second. This is the regular electrical supply for industrial and domestic use.

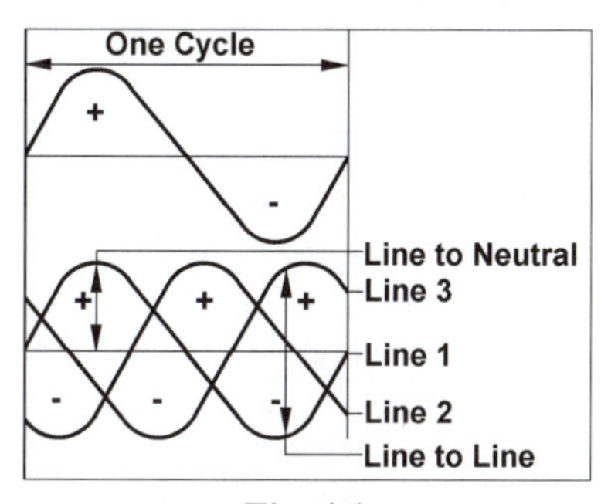

Fig. 4.1
AC supply can be shown as a sine wave - voltage against time. The current goes in one direction for half the cycle, shown as a positive voltage, and in the opposite direction in the second half, shown as a negative voltage. In three phase power supplies the voltage between line and neutral is less than that between lines, 240 and 440 are typical voltages.

A plot of the voltage of an AC supply against time is a sine wave. In one complete cycle the voltage rises from neutral (zero voltage) to a maximum with the current flow in one direction, then reduces to pass neutral and reach a second maximum in the opposite direction with the current flow reversed, and then again returns to neutral.

The cycle is continuously repeated in an AC supply at the mains frequency. In the UK and Europe this is 50 complete cycles per second, written as 50Hz and pronounced 50 hertz. In the USA this is 60 cycles per second - 60Hz.

The reason for using AC rather than DC for mains electrical supplies is because AC can readily be transformed from a high transmission voltage of perhaps 33 000 volts to a lower supply voltage of around 240 or 110 volts - this makes electrical power distribution cheap and reliable.

Three-phase Power

AC is ideal for running SCR induction motors - but three-phase power is required. This needs three live or hot lines, a neutral or return line and an earth or ground line. (Each live line is often carried as a pair of opposing lines on electric pylons, but always as a single line in the final supply)

Why three lines? - three lines is the least number to drive an SCR motor always in the same direction. Two lines would drive an SCR motor at the same speed, but the direction it would rotate in would be a matter of chance. More lines than three could be used - and were experimented with - but more than three is uneconomic.

Each of the three live lines (usually) has the same power - same voltage and amperage capacity. Some supplies have one live or hot line of a different voltage to the two others, but this is not regular practice.

The frequencies of the three lines are precisely the same, but the phasing is different. The second line begins its cycle one third of a cycle behind that of the first line, whilst the third line starts its cycle two-thirds of a cycle behind that of the first line. In technical terms the lines are each 120 degrees out of phase with one another.

If the lines were connected to three light bulbs arranged in a triangle, the rise and fall in brightness of the bulbs due to this phasing would give the effect of a rotating light source. (Actually mains frequency is too fast for this to be seen by the human eye, but the effect is evident when the frequency is slowed down enough.) This same effect provides a rotating magnetic field when three electric coils (stator windings) are placed in a triangular arrangement and separately connected to the three live lines - this is the basis of the SCR motor.

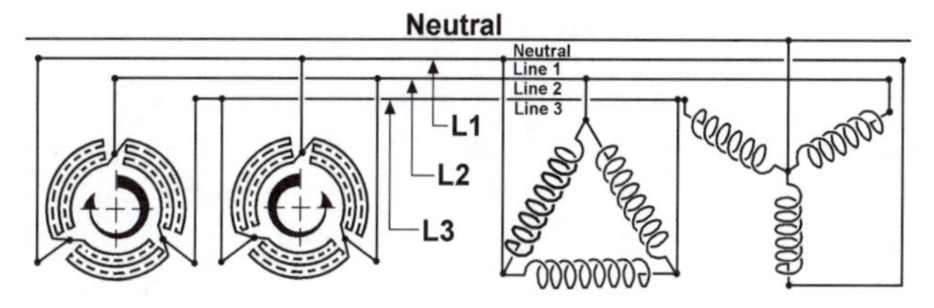

Fig. 4.2
Left - The stator windings form a part-circle in section, and sometimes overlap. The way the windings are connected to the three-phase supply decides the direction of rotation of the magnetic field.
Right - Delta (left) and Star (right) are the two ways in which motors can be connected to a three-phase supply, (shown diagrammatically). Star connection gives a lower voltage to the stator windings and is the starting mode for large motors before switching to Delta - the higher voltage running mode.

The live lines can either be connected via the stator windings to the neutral line (star connection), or to one another via the stator windings (delta connection). Star and delta connections each apply a different voltage to the stator coils - from the same supply - and are used either in sequence to start a motor, or to give the ability to connect to alternative electrical supplies of different voltages.

The earth line or ground line is connected to the motor casing to blow the fuse should any live line accidently contact the machine frame. For safety reasons the machine frame and the motors should be electrically bonded to an earth connection. In some early supplies the earth and neutral lines were a common, single line, but this is no longer recommended.

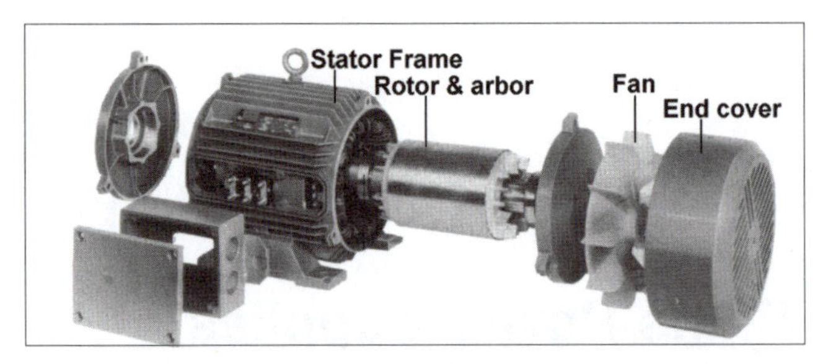

Fig. 4.4
Exploded view of a regular SCR motor
Source - Electrodrives

SCR Motors

Mains motors are induction motors, usually of the SCR (squirrel-cage rotor) type. They are nowadays totally enclosed, with a covered fan blowing air across external fins to keep the motor cool. Some older motors were open to allow air to blow through the windings, but these caused endless burnouts and fires in woodworking mills.

On two-pole motors the motor stator, the stationary outer part, has the three windings connected to the three lines of the mains supply to create the rotating, two-pole magnetic field. The rotor, the heart of the motor, also has closed electrical circuits (actually the squirrel cage) - with no external electrical connections. The bulk of the rotor is made of materials which force it to follow the rotating magnetic field, and it is this that provides the power source. The direction a motor runs in depends on the line connections.

Fig. 4.3
Left, a regular, three-phase SCR Induction motor. Right a single-phase SCR induction motor.
Source - Electrodrives

When the rotor runs slower than the rotating magnetic field, electrical eddy currents induced within the rotor cage speed it up. This also increases the electrical demand of the stator to make SCR motors self-regulating - taking more power under load and less when running idle. Due to this characteristic SCR motors run at an almost constant speed, idling at just below the frequency of the mains supply and with only a slight speed drop when under load.

SCR motors never run at exactly the mains frequency even when idling, because, if they did, the eddy currents induced in the squirrel cage would be zero. The rotor would then slow down until the slip - the difference between the supply frequency and the motor speed - induces enough eddy currents in the rotor to maintain a steady speed.

In the UK, with a mains frequency of 50Hz, a two-pole mains motor has a speed of just below 3 000 revs/min. In the US, with a mains frequency of 60Hz, a two-pole mains motor has a speed of just below 3 600 revs/min. For convenience, the quoted speed of a motor is sometimes the synchronous speed, 3 000 or 3 600. The actual free-running speed (without load) is usually given on the motor plate.

SCR induction motors are also made with four poles to run at 1 500 or 1 800 revs/min, respectively, and with six poles to run at 1 000 or 1 200revs/min, and so on, but these are used for lower-speed applications such as feed drives.

Four-pole motors have six windings arranged in sequence so the rotor turns only half a revolution for one complete phase, and six pole motors have nine windings so the rotor turns only one third of a revolution for one complete cycle - hence the difference in motor speeds from the same supply.

Single-phase SCR Motors.

Some small machines use single rather than three-phase motors, usually fractional horsepower. These run off one hot or live line and the neutral or return line. Single-phase SCR motors have one or two large capacitors which splits the supply into an artificial three-phase supply, so the motors run very much as standard three-phase motors.

Hand-held tools such as drills and sanders use a commutator type motor which gives wild variation in the running speed according to the load - so this type is unsuitable and hardly ever used on mainline woodworking machines.

There is also electronic equipment which allows regular SCR three-phase motors to be run off a single phase supply. The equipment is intended for one or more small motors only, and the rated output of the unit must be balanced against the total horsepower of the motors it runs.

Saw Motors & Drives

Direct Drives

Some machines have a direct drive to the sawblade, either with the sawblade mounted on an extension of the motor arbor, or with a shaftless motor driving an extension of the saw arbor. With such as ripsaw, cross-cut and panel-cutting saws the motors are mains driven but with smaller trimming sawblades these motors are often of the high-frequency type.

Short-centre Drives

The majority of sawblades are driven from a regular SCR mains motor through a short-centre drive using endless belts. On very small machines only one saw arbor speed is provided, driven via single pulleys and with power transmitted by vee, multi-vee or flat belt. Most machines, though, have stepped pulleys, normally on both the motor and the saw arbor, giving perhaps two, three, or more sawblade speeds.

Fig. 4.5
Left, a reduced-diameter SCR motor manufactured specifically for directly driving a circular saw. This type of motor allows a greater depth to be sawn than would a regular SCR motor.
Right, an alternative is a shaftless motor mounted direct-on the extended saw arbor.
Source - Perske

Fig. 4.6
This shows the saw unit of the Kolle panel saw. The main sawblade (right) has a four-speed vee-belt drive, and the scoring saw (left) has a single-speed drive.
Source - Kolle

With multi-speed drives the stepped pulleys are usually so sized as to retain the same drive centre distance whatever set of pulleys are used, so that a simple mechanical lever can be used to quickly re-tension them.

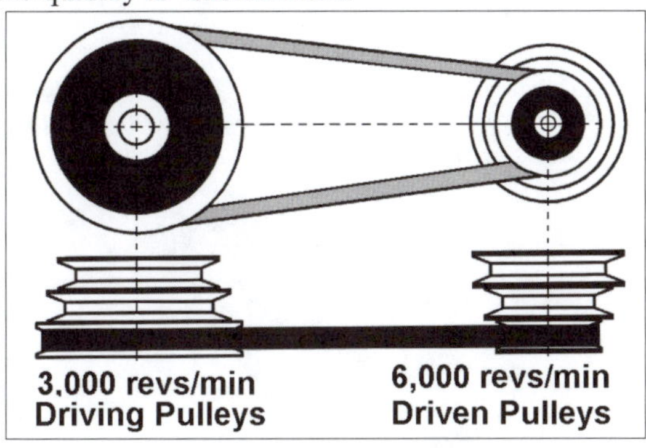

Fig. 4.7
This shows a three-speed drive using stepped pulleys and a single vee belt. The pulleys in use have a drive-up of from 3 000 revs/min on the motor to 6 000 revs/min on the saw arbor. The alternative speeds for the saw arbor are 3 000 & 4 500 revs/min.

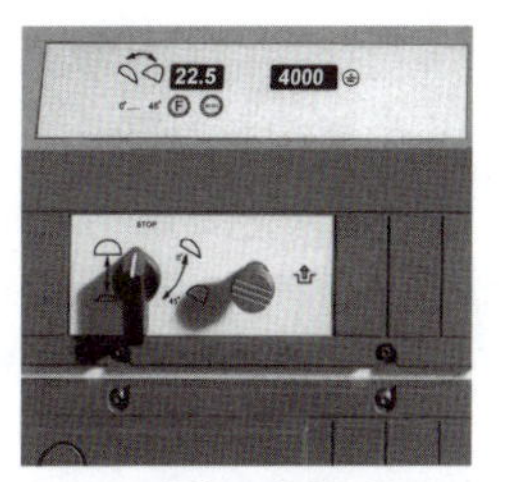

Fig. 4.8
With multi-speed drives it is essential that the user knows the precise running speed of the saw arbor. On the Altendorf standard class of panel saws a digital display shows both the saw tilt angle and the saw arbor speed for which the machine is set.
Source - Altendorf

Vee-belt drives

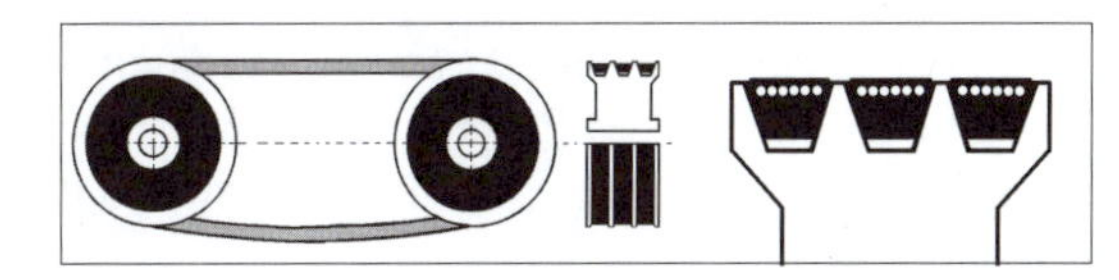

Fig. 4.9
Left, typical vee-belt drive with three vee belts.
Right, Section through vee belts showing the driving ropes. The effective diameter of the drive is measured to the centre of these ropes.

Vee belts are wedge-shaped, with the drive tension taken by ropes within the load-bearing layer just below the top surface. They are encased in rubber and fabric layers which very effectively grip the sides of vee pulleys, and which are flexible enough to readily compress or stretch as the belts repeatedly flatten and wrap around the pulleys. They are used singly, or in multiple sets for more powerful drives.

The effective diameter of the pulleys is where the ropes lie, usually just at or below the outer diameter. Vee belts are very efficient as they wedge further into the vee groove to automatically increase their grip as the load increases, then release their grip partially as the load decreases. Unfortunately they also absorb more power in driving than most other types.

Vee belts drive by contact only with the sides of the vee grooves - should they contact the groove base they are worn and need to be replaced.

There is a wide choice of belt lengths, sizes and types so it is essential that replacements for worn or damaged belts are identical to the originals.

To fit belts first release the tension so that belts can be removed and replaced whilst slack, then retention to makers recommendations. Fitting vee-belts whilst still tensioned can rupture and damage the driving cords so the belts quickly fail.

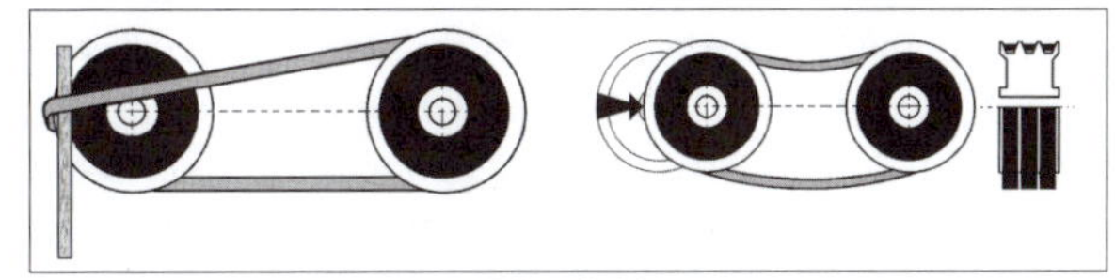

Fig. 4.10
Left, never do this! Forcing vee belts on or off whilst still tensioned can cause early failure.
Right, to remove or change the vee belts first slacken-off the belt tension.

Multiple vee-belt drives.

Where multiple vee-belts are used these should always be a matched set, i.e., all of the same specification - type, size, length and grade.

In manufacturing it is impossible to make all vee-belts of the same nominal length to exactly the same length, so they are carefully stretched and measured individually. Those of absolutely identical length are then given the same identifying letter, number or code - the grade. Oddballs are used for single drives.

When multiple-drive belts wear out, or if a single belt of a multiple set has been damaged, broken or badly worn, all the belts must be discarded and replaced. Replacing a single belt - even of the same grade as the existing belts - is bad practice. The remaining belts will have stretched so that the new belt will take most of the strain and will fail early because of this.

Tensioning vee-belts

After fitting belts the tension must be re-set. Vee-belt manufacturers give specific guides for this, sometimes by stating the amount a belt must deflect when pressed down mid-way between pulley centres.

These guides specify the loading to be used and the correct amount of belt deflection - both of which relates to the type of belt and the distance between driving centres.

Fig. 4.11
The lower marker on the gauge shown here is set to the correct belt deflection. Pressure is applied through spring loading in the gauge until the lower marker is level with the top of the adjacent belt. The applied pressure is then read-off the top marker.

Setting the correct amount of tension is essential to ensure driving efficiency. Belts that are too slack slip and squeal when starting or under load and wear quickly. Belts that are too tight put added load on the bearings of both the motor and the saw arbor. This could wear the belts, the pulleys and the bearings excessively, possibly make the belts fail early, and absorb considerably more power in driving. The golden rule is to tension belts to the least amount needed to drive efficiently.

Pulley and Arbor Alignment

With new machines the vee belt pulleys will be correctly aligned. If the drive motor or pulleys are replaced for any reason it is essential that these are checked for correct alignment. Failure to correctly align vee belt pulleys leads to rapid belt wear and early failure. A small amount of mis-alignment is acceptable with all vee-belts, but check with the makers as to how much.

Adjustment for mis-alignment is normally made by shifting the motor sideways. If this is impractical the pulleys themselves can be shifted on their shafts, but in all cases keep the pulleys as close as possible to their nearest bearing. On machines where the saw arbor has a cross-movement, align the pulleys at the mid-position.

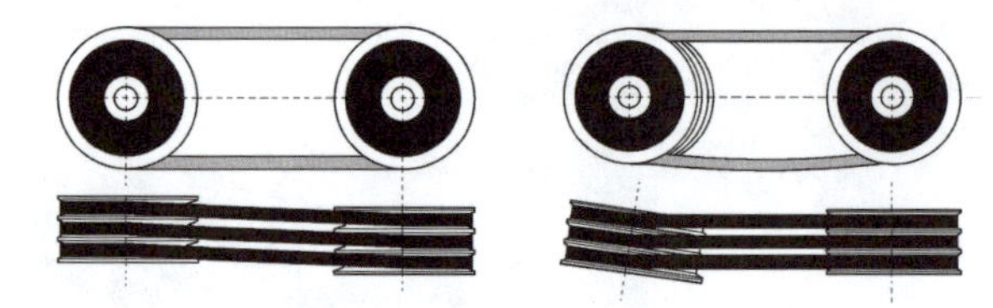

Fig. 4.12
Misaligned drive pulleys, as shown here, wear belts and causes premature failure (left).
Out of parallel (or toe-in) drives wear the belts excessively and gives uneven tension so that the tighter belt fails early Right).

Aligning is normally done by placing a straight edge across the driver and driven pulleys. (Usually the first vee-groove is the same distance in from the outer face of both pulleys - if not, make the necessary compensation.

An error is possible if the motor and saw arbors are at opposite sides of the belt. In this case the vees themselves must be aligned - not the pulley faces.)

It is also essential to ensure the both motor and saw arbor are parallel when finally set. Following alignment always re-check the tension.

Multi-vee belt drives

Some saw arbors are driven by a single, multi-vee belt. These look like regular flat belts from the outside, but inside they have a series of vee ribs which make full contact with the surface of the matching multi-vee pulleys. Because they are a one-piece belt there are no matching problems. Like vee belts they are made in many lengths and sizes.

Both with single and multi-speed drives the saw speed should be changed ONLY after first slackening belt tension. Much less tension is needed with multi-vee belts than with regular vee-belts - check with the makers for details. After any change is made to the drive, multi-vee drives should be aligned in the same way as vee-belt drives.

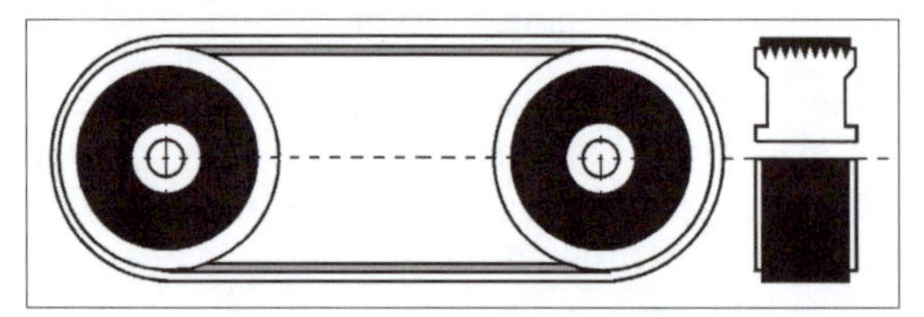

Fig. 4.13
Multi-vee drives are another, highly efficient alternative to vee belts and take less space than equivalent regular vee belts for the same power transmission.

Fig. 4.14
Typical illustration of a multi-vee belts.

Flat belt drives

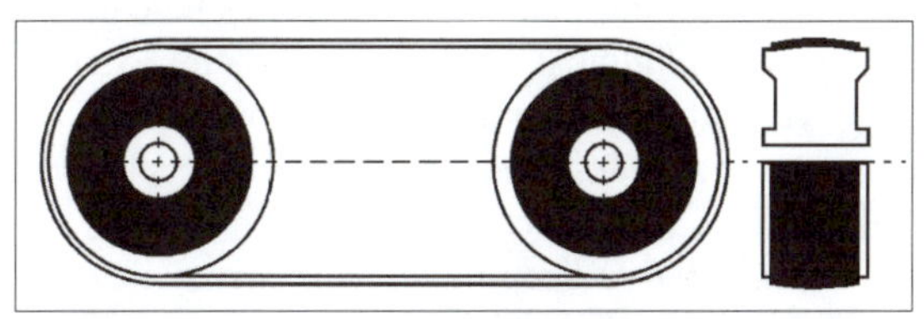

Fig. 4.15
Flat belt drives are becoming increasingly popular, they have very high grip for their width and need less tension than other types.

Flat belts consist of a single or multiple synthetic tension member covered both sides with chrome leather, composite plastic or synthetic rubber. Modern flat belts are produced in long lengths which are then cut, spliced on an angle and glued to produce whatever belt length is required.

It is vital that they are fitted so that the fine end of the splice trails on the pulleys - in this way driving continuously wraps the trailing edge down. Running in the wrong direction could unravel the splice and lead to belt failure. The correct direction of drive is usually shown by an arrow on the outside of the belt.

With multi-speed drives it is not necessary to slacken tension before switching belts. Most modern flat belt drives allow the speed to be readily switched after first shifting the belt off the larger pulley. It is easier then to fit the belt on the smaller pulley of the speed next required and run it onto the corresponding larger pulley.

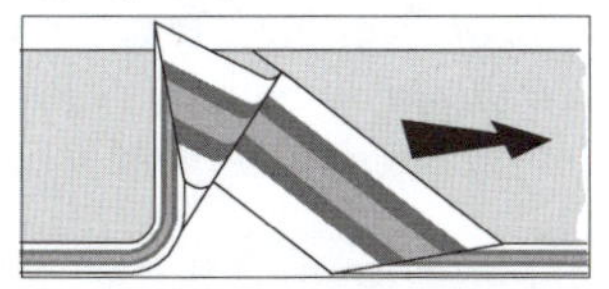

Fig. 4.16
Flat belts are spiced at an angle to give a large bonding surface. The splice must wrap down as the belt rotates. The correct direction is shown by the arrow on the outside of the belt. The shaded areas show the different layers of a modern composite belt.

Flat belt drives must be correctly tensioned in the same way as regular vee belt drives - and for the same reasons. Tension can be measured by deflection, as with a vee belt, but in many cases a specific distance is marked on the belt. When applying tension the belt is stretched until the marked distance increases by a stated percentage, or until it reaches a measurement specified by the makers.

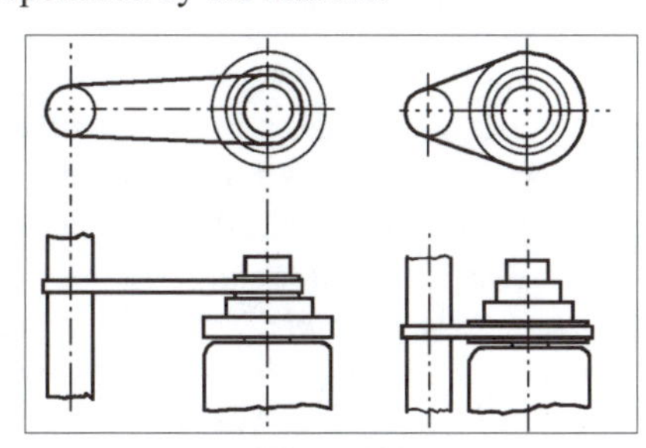

Fig. 4.17
A flat belt drive with stepped pulleys on the motor and a single pulley on the saw arbor.

With flat belts perfect pulley alignment is not absolutely essential - unless the belt and pulleys are the same width. Flat belts run true because one or both pulleys has a crown, ie., the pulley resembles the center section of a beer barrel. If one edge of the belt lies on the crown, the faster speed this is then driven at slews the belt until both edges lie equally on either side of the crown - then the belt remains stable.

In long drives this happens slowly, but in very short drives the belt oscillates rapidly before finally becoming stable. After shifting belts always rotate the drive by hand until the correct belt position is reached.

In a few cases with flat belt drives the saw arbor has a single, wide pulley (to keep this small in size) with stepped pulleys fitted only on the motor - so the drive needs re-tensioning when the belts are switched. With this arrangement each of the stepped driving pulleys will have a crown to keep the belt running true - but the driven pulley will usually be flat.

Variable speed motors.

On a few modern machines the saw arbor speed can be adjusted infinitely by varying the motor speed. This allows the sawblade rim speed to be perfectly matched to the type of material being sawn regardless of sawblade diameter. Speed variation is through the control knob of an electronic speed controller.

Fig. 4.18
Inverters such as this Lenze 8200 gives speed variation to AC motors, so allowing the sawblade rim speed to be perfectly matched to the material being sawn regardless of sawblade diameter.
Source - Lenze

High Frequency Motors

On some machines the sawblade is mounted direct-on the arbor of a high frequency motor. This is a special type of motor which runs faster than a regular motor and which is powered, not direct from the mains supply, but via a frequency changer.

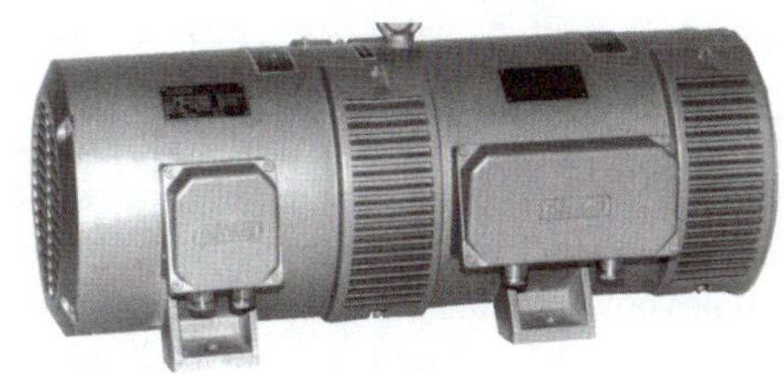

Fig. 4.19
A typical direct-driven frequency changer used for high power motors.
Source - Perske

Frequency changers are themselves powered by the mains supply, but provide an output with a higher frequency than the mains supply. If the output frequency is 100Hz, for example, then the motors run at 6 000 revs/min., if it is 200Hz, then the speed is 12 000 revs/min. The frequency changer may itself be a double or combination motor, or the high frequency supply may be produced electronically.

Using a frequency changer can be a costly option. This is due to the extra electrical equipment needed, high power loss, and problems in replacing obsolete burnt-out motors of older machines. In spite of this there are good reasons for providing a high frequency supply - where several high-speed motors are needed on a single machine, where motor brakes are an essential safety feature and where a short-centre drive would be impractical. The trimming saw motors of edge-banding machines are a good example of this.

Fig. 4.20
12 000 revs/min high frequency saw motors are used for trimming boards on this edge banding machine. The frequency changer in this instance is electronic, and also provides motor brakes.
Source - Brandt

Motor Burnout.

All SCR motors heat up when running - even though the motor fan helps to blow away some heat as it is generated. This is because the windings act partially as heating elements.

Modern motors normally run at a relatively high temperate - hot to the touch - because they are smaller than earlier types and are made from materials that withstand higher temperatures. Even so, if any motor is allowed to heat up excessively it can burn out - and re-winding or replacement is costly and inconvenient.

Most heat is generated when the motor is first started, so most large motors are started slowly on a lower voltage using a star connection, then switched to the higher voltage of delta connection when enough speed has been built up. Delta is then the running mode.

Older machines have a three-way manual switch for this, usually a centre position for off, left for star, than right for delta, often termed start, stop and run. The operator has to judge the correct speed before switching from start to run - and switching too quickly can cause overheating. Fortunately most machines now have automatic star/delta control, engaged by a simple push-button, and give controlled starting that no longer relies on the users judgement.

Overloading the mains supply to machines having two or more large motors is possible if attempting to start them all at the same time. Large motors must be started in sequence, allowing each motor to reach full running speed before starting another. Motors of any type should never be continuously started and stopped - the extra heating this causes can burn them out. A motor which is used irregularly actually runs cooler if left running. Before stopping a machine all material should be cut clear through. Stopping the machine with material in the cut will almost certainly jam when re-started and quickly overheat the motor and possibly burn it out. For the same reason sawblades must be allowed to reach operating speed before the feed is started.

It is essential to keep motors and drives clear of dust and waste - a chip and dust covering on a motor will act as a heat-retaining blanket to prevent the cooling fan from doing its job.

Motor rating

Motors for converting timber and boards are continuously-rated, that is they are capable of running under load for long periods without danger of burnout. Motors for cross-cut saws are intermittently-rated - they can take full load only when actually cutting, then have to run idle for a short time to cool down. Manual cross-cut operation allows this, intensive automatic cross-cutting may not.

Motor Control Gear

Overload protection in modern motor controls is intended to cut-off the power if there is a danger of a burnout for any reason. Many use bi-metal strip relays with heating coils connected to each hot line. The coils heat up in step with the amount of current taken by the motor windings and, if the current flow exceeds the set level, the bi-metal strip heats up enough to bend and trip the connection.

An adjustment often allows the same control gear to be used with a range of motor powers - check that overload settings correspond to the ratings of the motors they protect.

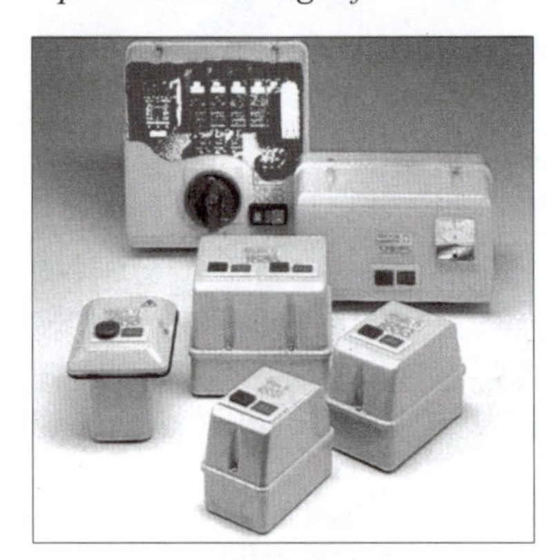

Fig. 4.21
Regular motor control units, some fitted with isolators.
Source - Electrodrives

Following a trip-out the coils have to cool down and reconnect before they allow the motor to be re-started. Modern control gear has automatic re-sets, but with older equipment a re-set button has to be pushed on the relay that tripped.

In some extreme cases bi-metal strip relays do not react quickly enough - for example if there is a total jam - a modern motor could burnout in this circumstance even though the overload trips are correctly set. To prevent this happening some motors have a heat sensor within the windings to give better protection by tripping-out the supply if excessive heat is sensed.

Machine control gear is often interlocked, so that if one motor fails all the remaining motors are immediately stopped.

In some instances machines are fitted with automatic brakes to stop the motors quickly once the electrical supply is broken.

Direct-driven saw motors are very free-running and take a long time to slow down. Apart from the lost time waiting for these to stop, a still-turning sawblade is a danger that may not be so obvious. On these machines a brake is an essential safety feature. Some makers offer add-on electronic braking systems for machines originally supplied without.

Fig. 4.22
Brake motors are fitted when immediate braking is essential. This type has a mechanical DC-triggered brake operated through a built-in rectifier.
Source - Electrodrives

Sophisticated electrical gear starts and stops motors with much more precise control. These provide a ramp-up to full running speed and ramp-down before stopping motors, with timings for both individually adjustable. In addition to the quicker and safer stopping this makes possible, the soft-start is much kinder mechanically to the equipment.

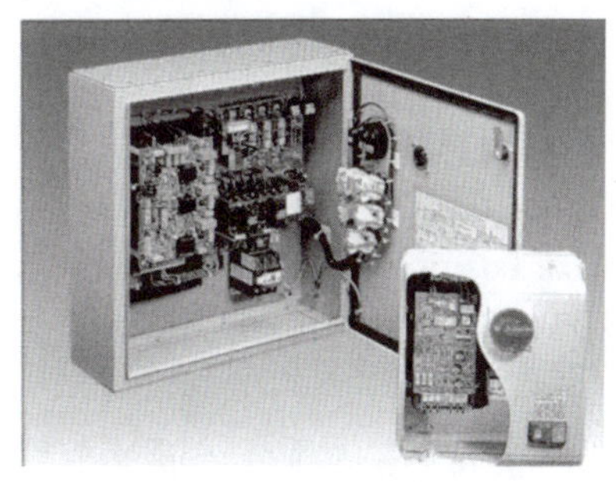

Fig. 4.23
Soft-start control units start motors with a controlled ramp-up and stop them after a controlled ramp-down.
Source - Electrodrives

Electrical control gear is also usually fitted with no-volt protection. These are coils which, once energized, keep motors running only so long as the electrical supply is maintained. They trip-out when the machine is stopped or if the factory supply is cut-off. They prevent machines from re-starting when an interrupted factory supply is restored without warning.

Feed Drives

There are several types of feed drive systems on machines using sawblades. Most are driven electrically by SCR motors that are slower than those driving the sawblades themselves, but are otherwise of the same general type.

All need a speed reduction to match the machine feed speed. This is usually provided by a reduction gearbox, either as part of the motor, or as a separate unit. With the latter it is usual to fit shift levers, rather like a gear-shift car, to engage one of a range of feed speeds. On some a two-speed short-centre drive from the motor doubles the range of speeds the gearbox provides, or the motor itself may be double-wound to give two speeds selected via an electrical switch.

Variable-speed Feed Gear

Mechanical variable speed gear.

Gear-shift drives rarely give a wide enough choice of feed speeds. A small speed change can make a huge difference both to quality and productivity, so many makers now provide infinitely variable-speed feed gear.

Early types used a large power-driven disc driving a roller to power the feed. By shifting the roller across the diameter of the disc the output speed could be varied, and even reversed. This type is rarely used now.

Another type uses special vee-pulleys on the motor and feed arbors. They are made in two halves which move together or apart, with the motor pulley driving the feed pulley via a wide vee belt.

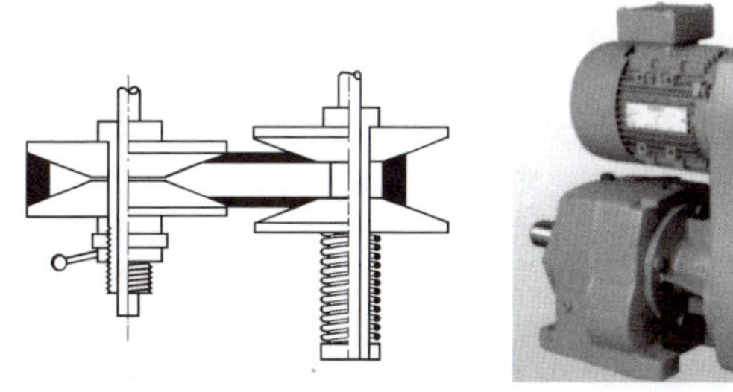

Fig .4.24
The expanding vee-pulley type of mechanical variable speed gear has a speed variation of about 6:1 (left). Complete units are available for driving feed gears which comprise of a motor, mechanical variable speed unit and an output reduction gearbox (right).
Source - Electrodrives.

One vee pulley, usually that on the motor, is screw-adjusted to vary the spacing of the two halves. By shifting the motor pulley halves apart, the belt contacts the pulley sides at their smallest diameter. At the same time spring pressure on the feed pulley arbor forces these two halves together to maintain belt tension and to bring the belt into contact with the pulley sides at their largest diameter.

Fig. 4.25
This Lenze disco variator gives infinitely variable speed output and is ideal from driving grinding wheels at any chosen speed. The unit has a built-on motor drive and a speed indicator on the adjustment dial.
Source - Lenze

This gives the same effect of a small pulley on the motor and a large one on the feed arbor - so the feed is driven at the lowest speed. By moving the motor pulley halves together the ratio of motor pulley to feed pulley is changed, and the feed speed is increased. The fastest feed speed is reached when the belt contacts the motor pulley near the rim and the feed pulley near the arbor.

There are other versions where only one pulley is adjustable, or where a live intermediate arbor mounts two expanding pulleys, one driven by the motor and the other driving the feed - with speed adjustment made by moving the intermediate arbor towards or away from the motor.

Another form of variable speed gear is the disco planetary variator. This has a fixed outer ring with an adjustable outer ring facing it. An inner ring of two facing halves is driven by the input shaft and has a compression spring forcing the halves together. Between the two rings, and in rolling contact with both, are four or more thin, double-coned planetary discs mounted on a carrier to drive the output shaft. As the gap between the outer rings is varied, the planetary discs are forced towards or away from the centre - so driving the output shaft at a steplessly variable speed.

Hydraulic variable speed gear

Hydraulic variable speed gear, as used on woodworking machines, usually has an electric motor powering a hydraulic pump. This, in turn, drives a hydraulic motor connected to the machine feed gear. The pump and motor may be contained within the same unit, or housed in separate units connected by delivery and return pipes.

On many machines the mechanical setting of the pump can be altered to vary the volume of oil delivered. In this case the fixed-setting motor is driven at a speed proportional to this oil volume. Others have a fixed-setting delivery pump and a motor with a variable setting.

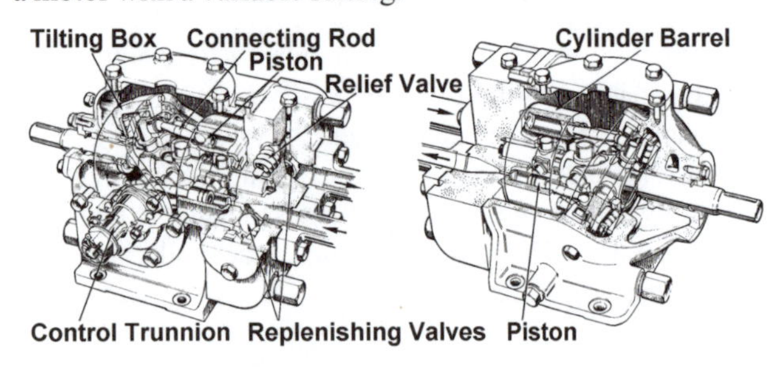

Fig. 4.26
Left, this shows the working details of a Vickers VSG variable-delivery hydraulic pump. The angle of the tilting box is governed by the control trunnion and this, in turn, determines the stroke and volume of oil delivered by the pistons.
Right, the Vickers VSG hydraulic motor has a fixed angle for the tilting box - and an output speed which varies according to the volume of oil delivered.
Source - Vickers

There are different types of variable-speed hydraulic pumps and motors, but they all work in a similar way.

Hydraulic pumps have several oil pistons pumping oil under pressure. The piston movements are powered by an eccentric, swash plate or cam driven by the input shaft (or the pistons themselves are driven by the input shaft and the eccentric, swash plate or cam remains stationary).

The amount of piston movement controls the rate of oil delivery, and this is governed by the setting of the eccentric drive, swash plate or cam, or the angle of the pistons to the input shaft - which can be altered via an external mechanical control.

Hydraulic motors are similar in construction - except that the pistons are driven by oil pressure to rotate the eccentric, swash plate or cam arbor, and so drive the output shaft. With a variable delivery pump the motor has a fixed setting, with a fixed delivery pump the motor has a variable setting.

Most hydraulic systems have a wide range of output speeds, which in some cases can also provide a reverse feed to draw timber out of a jam. On many machines the pump drives a single motor, with the rest of the feed mechanism then driven by connecting chains and flexible couplings. Where the feed arrangement needs extreme feed roller movements it is sometimes more convenient to fit several motors, perhaps even on each individual feed rollers - all connected too and driven at the same speed by a single pump.

Fig. 4.27
This Jonsereds resaw has individual, hydraulically-driven feed rollers - as clearly shown here.
Source - Jonsereds

Power Movements

Many machines now have power movements, to set fences, clamp timber or boards, power saw grinders, etc. Some are quite simple movements such as a plunger or lever moved by a rotating crank, eccentric or cam.

These types were widely used in early machines - and are still used today in some low-cost saw grinders. This type of movement is cheap, but restricts both the speed and the variety of operation. For greater versatility most power movements are controlled by more sophisticated methods.

Many power movements are now fluid-based, powered by air or hydraulic oil. Both air and oil systems use similar equipment, pumps to provide the pressure and flow, pipes to transfer the fluid, pistons to make the movements, and valves to control the direction or speed of fluid flow. Some equipment is suitable for either air or oil operation simply by switching the seals. The main difference between air and oil operation, though, is in the nature of the fluid.

Air power

Air will compress, so it is ideal for such as clamps or pressures as it always maintains the same pressure.

In its simplest form it is not suitable for operating, say, a cross-cut saw because the movement would hesitate if resistance is met, then run forward quickly when resistance is less. This is contrary both to good practice and to safe operation - a precisely controlled speed is essential.

Actually, it is possible to use air power for this type of movement by pressurizing both sides of the cylinder or, more precisely, by connecting a closed-circuit hydraulic cylinder to the air cylinder to regulate its speed. See Fig. 4.34.

Fig. 4.28
Side and top air clamps firmly and safely hold timber whilst being cross-cut on this GreCon Optimizing cross-cut
Source - GreCon

Fig. 4.29
The air-operated snipper saws on this Homag Unit of their edge-bander are for flush end trimming of overhang.
Source - Homag.

Fig. 4.30
Several air cylinders are used to operate this automatic double-end trim, mitre and compound mitre saw.
Source - CTD

Most plants have a central air supply, with air compressors, pressure tanks to maintain a steady pressure, together with connecting air lines. Machine makers often incorporate air-operated movements in a machine - where suitable - as it costs less to use an existing air supply rather than produce equivalent hydraulic power within the machine. Air is a one-way system - the exhaust from cylinders is simply blown-off to atmosphere - so piping is simpler. Leaks are annoying and wasteful, but clean.

Hydraulic oil systems.

Oil does not compress (in practical terms) so it is ideal for making a steady cylinder movement at a speed which is easily variable and precisely controllable. Some engineering plants have central hydraulic systems, but as few woodworking plants do, most woodworking machines operated by oil have their own self-contained hydraulic unit.

Hydraulic systems are totally self-contained - exhaust oil is returned to a tank for filtering and reuse - so they are more complex and costly than comparable air systems. Leaks can be a nuisance as hydraulic oil is an unpleasant fluid - and few hydraulic systems operate without leaks.

Oil can also be used on hydrostatic systems. These are simple systems where a pair of cylinders are connected by one or two pipes. The movement of one cylinder operates the other, possibly with cylinders of different bore so that a long but low-pressure movement of one gives a small high-pressure movement of the other. These operate without a power pump and are ideal for controlling movements remotely where mechanical connection would be difficult.

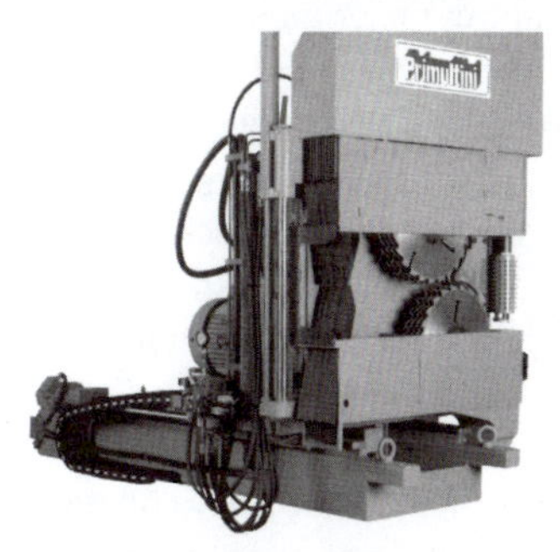

Fig. 4.31
This Primulti two-arbor saw unit is hydraulically controlled and is used in conjunction with a log carriage to precede a bandsaw cut.
Source - Primulti

Most systems, though, are hydro-active, with a central pump running continuously, and one or more cylinders driven by the pump via control valves.

Fluid system operation

Hydraulic pumps and air compressors are made in various types, using cranks and cylinders, intermeshing gears, linear spiral gears or vane-type rotors. Each type has a different characteristic, some produce high volume at low pressure, others low volume at high pressure - it all depends upon the final application as to which type is used.

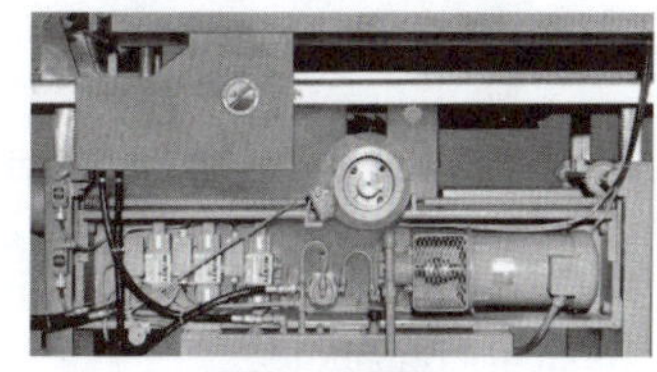

Fig. 4.32
The Northfield Gang Ripsaw has a hydraulic system which provides an infinitely variable feed speed and also powers the table height and upper feed roll adjustments.
Source - Northfield

Some cylinders are pressured on the piston end only and have a spring within the cylinder for the return movement. Others are pressured both at the piston and the rod end alternately to make the movement.

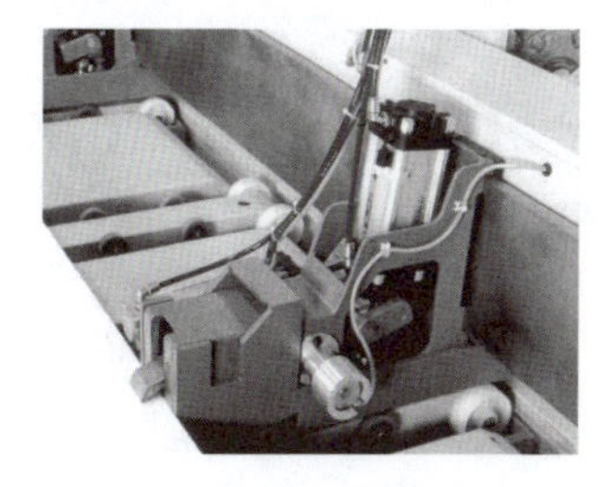

Fig. 4.33
The air-operated clamps shown here are part of the feed system for Gabbiani panel saws. They lift a stack of panels without stopping the work cycle.
Source - Gabbiani.

Control systems.

Fluid movement and direction is controlled by valves of various types. The movement of a cylinder is not necessarily the full movement, the stroke can be changed by a manual adjustment of some sort.

The speed of cylinder movement in hydraulic systems is governed by manually-set restrictor valves - these control the oil flow. Often the speed in one direction has to be faster than in the other, in which case a one-way valve bypasses the restrictor on the return stroke. Similar restrictors are used with air cylinders, but where precise speed control is needed a closed-circuit hydraulic cylinder alongside is needed to control the speed.

In systems which repeat a set pattern of cylinder movements, there may be a series of valves operated via cams on a rotating arbor. This type of system has no feed-back, and operates at the same cycle speed even though the sequence of operation may be changed - and may continue if there is a crash.

Fig. 4.34
In this form the Omga cross-cut saw is fitted with an air-powered traverse and an air clamp - but the saw movement is controlled by a closed-circuit hydraulic cylinder.
Source - Omga

More upmarket systems have feed-back via built-in switches. For example, when a cylinder completes its forward or return stroke it trips a limit switch to start the next sequence. Some use fluid switches which directly control the air or oil to the next valve or cylinder. Most, though, are electrical switches - which can be mechanical switches physically operated by the movement, or light, proximity or magnetic sensors which operate without physical contact.

The lines from the switches can lead directly to the next fluid valve, but usually they connect with a central control box to operate valves through relays. This allows much safer low-voltage switches to be used.

The relays trigger the next sequence - valves or perhaps a timer to delay the next operation, or even a counter. The timers and relays within these control boxes can often be reprogrammed to give a different sequence of operations through external panel switches.

Programmable Logic Controllers

The increasing use of central control boxes lead to the development of the PLC - the Programmable Logic Controller.

These are electrical devices which contain numerous relays, counters, timers and other electrical equipment. Often these are solid state and miniature - and take up far less space than the regular relays and timers they replaced. Input terminals connect to limit switches or counters, etc. on the machine, and output terminals connect to control valves and similar. LED (light emitting diode) displays can show the state of both input and output lines.

The PLC has a memory that stores programs electronically. Usually the program is a ladder program, a series of steps that take place in strict sequence. The program begins when a button is pressed and the PLC then starts the first step. Limit switches on the machine, or a timer within the PLC, signal when this is complete and make the program step to the next sequence, and so, on until it reaches the program end.

In some cases the sequence is started each time by a push button, on others it returns to the program start to repeat the program until stopped - perhaps by an internal saw-tooth counter. The memory can hold several programs which are selected via one or more switches.

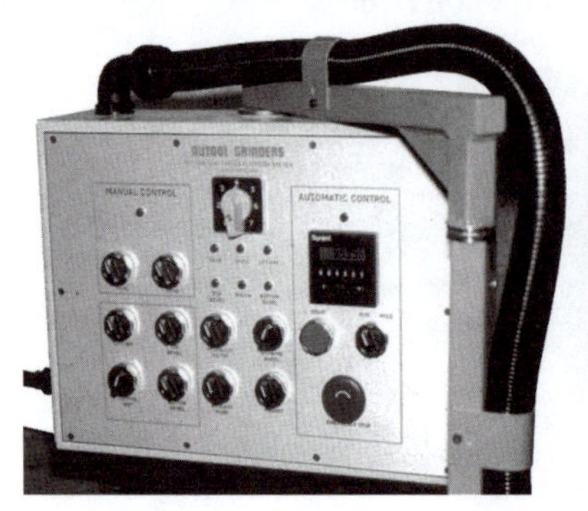

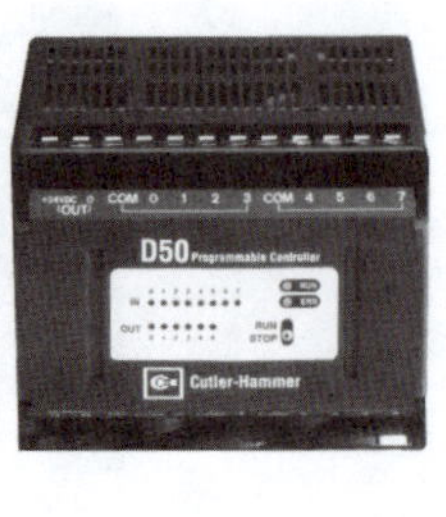

Fig. 4.35
Left, the control panel of this PLC-controlled Autool TCT/25 saw grinder has switches to select any one of several grinding programs, and to trip through any program step by step. A counter switches off the machine on completing a pre-set number of teeth.
Right, a typical Cutler-Hammer PLC unit of the type used in controlling many machines. These contain numerous relays, timers, counters, and connectors etc. in a box.
Source - Autool

As with all electronic memories, the program is retained only as long as the power supply remains on. Whilst this originally had to be the actual mains supply to the machine, most PLC's now have a dry battery to maintain memory.

Some PLCs are operated directly by the memory, but it is also possible to run the PLC program from a PROM (programmable read-only memory) module. A PROM is a separate unit which, when fitted to the PLC, controls the program instead of the PLC's internal memory.

Unlike the latter, the PROM program is retained until altered or erased - without need of a battery back-up. By switching between different PROMs it is possible to completely change the ladder program of the PLC - and with it the operating sequence of the machine. Pre-programmed PROMs can be supplied by the machine makers.

Many PLC's allow the user to write a new program, using either a clip-on unit which connects with the PLC, or by connecting it to a PC (personal computer). Both allow the program to be monitored as it is run. The PC is better as it shows more of the program on the screen - and allows users to add notes to make the program easier to understand.

It is possible to program the PLC memory direct, or to program a EE PROM (electrically erasable PROM). Unlike regular PROMS, these allow the user to remove the original program and replace it with a new one.

PLCs are usually connected to a control panel housing start and stop buttons, selection switches and possibly LEDs or miniature bulbs to show the progress of the program. The control panel can also include a counter.

Digital machine setting

Fig. 4.36
These two digital indicators show the alternative method of number display, either for viewing from the top or from the front. Regular indicators are mechanically-operated, but these are electronic with an LCD (liquid crystal display) read-out.
Source - Siko

Mechanical digital indicators are a simple, manually-operated device used for setting machines very precisely. They are mounted on the adjusting arbor of any suitable screw-adjusted movement - for example a rip fence on a table saw - to indicate its exact position.

Because they are clear, precise and easy to read, digital indicators are now widely used in woodworking machines.

It is essential to take up play by moving always in the same direction to the final setting, usually 'inwards'. Indicators can be simply adjusted to compensate for changes in the saw kerf width, for example, or to correct any error noted in positioning.

Normally digital indicators are mechanically operated by the rotation of the screw, others are electronically-driven - which allows some to be switched between imperial and metric measure.

An alternative to a mechanical digital indicator mounted on the screw arbor is to instead fit an encoder. This puts out electrical signals as the screw is turned to drive a remote read-out. This arrangement is used when the movement is electrically-driven from a remote position - either for convenience or safety - so the read-out is placed near.

Computer Numerical Control (CNC)

Fig. 4.37
This computer-controlled optimizing cross-cut saw is capable of reading and cutting-out crayon-marked defects. As part of the same process it then accurately cross-cuts the remainder to the lengths required - and in the most economical manner.
Source - Paul

The most versatile machine control - providing machine setting to precise measurements rather than to manual-set limits on fluid cylinders - is cnc (computer numerical control).

CNC is used both on power-driven linear (straight-line) and on rotary movements. With both the stopping point is controlled by electrical signals exchanged between the machine and the computer. This is done in one of two ways.

One method uses a precision screw driven by a stepper motor which moves the machine part via the screw nut. The computer controls the stepper motor via series of electric signals using a binary code (strings of ones and zeroes).

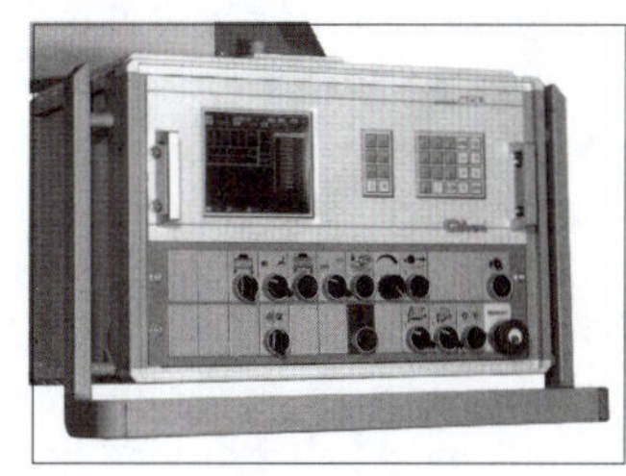

Fig. 4.38
Modern machines are controlled by computer, as on the Giben Fastmatic panel saws. The computer has overall control of the program for cutting patterns, with input via pattern graphics. The cutting pattern is shown on the screen, together with material yield characteristics. Controls permit alternative pattern input and modification, also a step-through of the machine operation.
Source - Giben

Fig. 4.39
Boards are gripped by automatic clamps on pushers which precisely positions the board stack and holds it firmly during the full cutting cycle on this panel saw.
Source - Schelling

Another method uses either an engraved bar alongside the movement scanned by a reader fixed to the traversing machine part, or an encoder driven separately from the main drive. The computer compares the binary codes as they are output with the feed-back from the actual movement - and continues the movement until the correct position is reached.

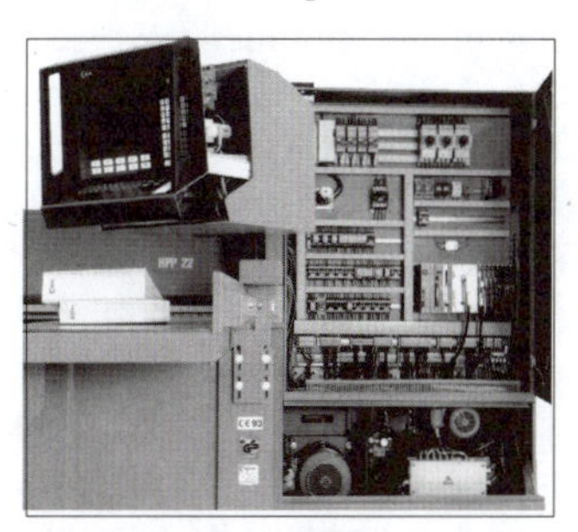

Fig. 4.40
Electrical and pneumatic control gear is neatly housed in readily accessible control boxes as on this computer-controlled panel saw.
Source - Holzma

The bar and reader system is ideal for retro-fitting NC or CNC equipment to a machine originally made for manual setting.

Normally this type separately checks each movement as the distance traversed, not the actual position of the moving part.

For it to operate correctly, avoiding possible error build-up, the machine head or fence is first moved out to the maximum extent, then back to the required setting. This allows the reader to begin its measurement from a precise start point it passes on the inward movement.

On more sophisticated set-ups batteries can be added to allow the last setting to remain in memory even when the machine is switched off - so obviating the need of full outward movement prior to setting.

Precision powered screw movements, as used now on many machines, is not simply power traverse coupled to regular screw adjustments - a much more precise drive is essential.

The traverse ways have to be virtually friction-free, so some form of linear bearings are used rather than plain slides. Some linear bearings have ball bearings held captive by a circular cage. This is placed around a hardened arbor, and inside a hardened outer sleeve on which the movement is both supported and guided. These bearings give a smooth and easy-running rolling support, but the degree of movement is limited.

Fig. 4.41
The board positioner of this panel saw guarantees rigidity of the movements and precise setting. Positioning is numerically-controlled using an encoder (lower left) which operates on an independent track.
Source - Gabbiani

Recirculating linear bearings give the same free-running and precise support, but have unlimited traverse movement. One type has a hardened steel bar around which are several ball bearing units. The units have a circular path which, as movement takes place, feeds the ball bearings first against the bar then around back to the beginning to provide an endless supply of bearings and continuous support. Other types have ground and hardened slides instead of a bar.

The screw and nut have a similar re-circulating ball bearings arrangement to give smooth, fast and precise movement. Any play is taken out of both the nut and the end thrust bearings so that movement in either direction is without backlash and just as precise. Special motors drive the screws and these, or a separate encoder, control the precise positioning of the movement.

The equipment is controlled from a central computer built into the machine. This also has a memory, and it is this that actually runs the program, starting the sequence when a start button is pressed.

On a panel saw the sequence might be: first a fence movement to position the panel, then clamping the panel, starting the saw for the forward cut, followed by the return movement, finally releasing the panel for a second cut, and so on.

Quite complex programs are possible, and these can include any combination of simple movements like clamping, precise fence settings or traverse movements, timing, and number count-down.

Programs are often prepared remotely on a PC or similar, then fed into the machine by a punched tape, magnetic tape, magnetic disc, ROM (read only memory) or direct link to the computer.

Most machines show the operator a visual display of the operating sequence, allowing the program to be stepped-through to check it absolutely prior to start-up. In many cases it is possible to write a program direct on the machine via a built-in keyboard, or to alter an existing program, .

CNC is very versatile indeed, it can, for example, control a cross-cut saw to convert mixed lengths of raw timber into various lengths, each requiring different number counts, and defect-cut at the same time.

The machine measures the length of each piece of timber, perhaps taking into account any defects, calculates what defect-free lengths this would convert into with least waste, then automatically converts the timber, removes the cut pieces and enters the next length, and so on. Modern machines do this at an incredible operating speed.

Similar programs are used on a panel saw for converting panels into a mixed cutting list of quantities and sizes - so as to produce all the cut sizes needed from full-size boards in the most economic way and on a continuous basis - switching cutting programs during the process - if necessary.

Some panel saws are designed to turn partially cut boards during the normal operating cycle and so use a single machine for both rip and cross-cutting - others pass them to a second machine mounted at right angles to the first required.

See Chapter 15 for more details.

Application the grinders

Computer control is now being widely used in sawblade grinding machines. In most cases the positioning and traverse movements of the grinding wheel are computer controlled via the keyboard and screen which forms part of modern equipment.

This permits a whole new way of setting these machines. Instead of adjusting levers and screws in various parts of the machine, and noting the setting on dials or other read-outs, it is now possible to control all these functions and fully set the grinder purely through the keyboard and screen.

The most advanced systems need only the sawblade and grinding wheel data to be entered, from which the computer calculates the machine settings needed - and sets the machine automatically perhaps even to controlling the precise arcuate movement needed for the pawl.

Most use a single-axis movement to traverse the grinding wheel, but it is also possible to use two-axis control when profiling or side grinding tipped teeth, known as continuous-path control. This is fully dealt with in Chapters 10 & 12.

Chapter 5
Saw Tooth Profiles

Alloy Steel Sawblades

The shape of alloy saw teeth evolved through trial and error in the period when only plate saws were made and only natural timber was sawn - so the following remarks apply with this in mind. *(Most tungsten carbide and diamond-tipped saws also follow the same general patterns - but include some forms used only for tc & pcd saws.)* Regardless of the type of sawblade, however, the tooth shape of most is of the rip or crosscut pattern - or a combination or compromise of the two.

Tooth geometry should match the particular type and condition of the timber being sawn - and the method of conversion - in order to get the best possible results. *(This is true for sawblades of all types, not just alloy saws.)* However, saw makers have rationalized alloy saw tooth shapes to a relative few patterns - which between them satisfy most requirements.

Where different timber species are handled, such as softwood and abrasive hardwood in the same mill, then separate sawblades should be used for these - if this is not done then efficiency may suffer.

If existing sawblades are unsuitable for a new species, then ideally an additional sawblade should be purchased. This may be nonstandard - and at a longer delivery and higher cost than a regular sawblade - but this is the right approach if the amount of work justifies it. If the job is a one-off, however, then alter the tooth profiles or number of teeth in an old sawblade - easy with alloy saws - or use the nearest regular sawblade and accept the poorer finish and lower efficiency.

Keeping the shape

The ease by which alloy steel tooth shapes can be changed is a two-edged sword. It can lead to the most common problem with plate saws - slow change of the tooth profile until this hardly resembles the original.

What is not generally appreciated is that these changes can alter the quality of the sawn surface, possibly make sawing less safe, and/or increase the power consumption. Good saw doctors and filers periodically check and correct the tooth shapes of plate saws to make sure they continue to work as they did when new.

The following are fairly widely accepted tooth shapes for alloy steel saws when converting natural timber.

NOTES:

The front bevel angle of all ripsaw teeth is 0^0 - i.e., either filed or ground straight across. Timber densities refer to seasoned timber. Bracketed figures show alternatives. The number of teeth stated apply to regular sawblade diameters. The exceptions are for sawblades of very small diameter where fewer teeth are used. Where two numbers are given the numbers of teeth vary according to diameter - less for small sawblades, more for large sawblades. Gullet depths and tooth top widths are shown as a fraction of the tooth pitch (p).

Sawblades for ripsawing

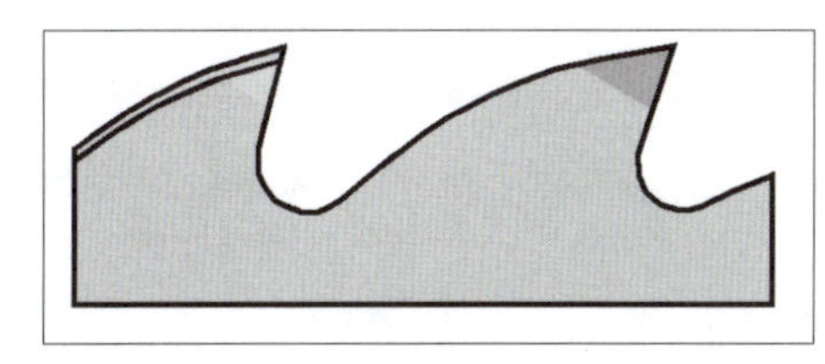

Fig. 5.1

Machine sharpened tooth profiles, as shown above for hardwoods, have the gullet shaped as an S-profile which continues up to the tooth point.
They are identical, as regards number of teeth, hook, clearance and bevel angles, to the drawings following which show file-sharpened tooth profiles with a straight back and flat clearance bevel.

Ripsaw teeth for softwoods

For Timber density	less than 560 Kg/mm³
Typical Species	Hemlock, Obeche, Scots Pine
Number of teeth	48*
Hook angle, Green timber	25^0 (30^0)
Hook angle Seasoned timber	25^0
Clearance angle	20^0
Top bevel angle	15^0
Gullet depth	0·50p
Tooth top	0·25p

* Saws of 355 mm diameter or less normally have 36 teeth.

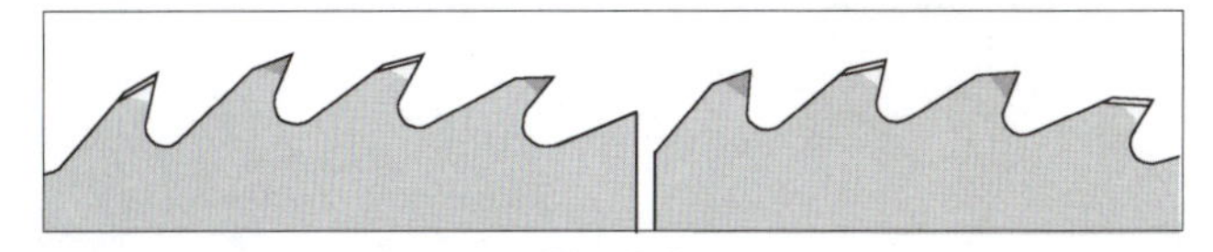

Fig. 5.2

Left, ripsaw teeth for softwoods, right, for hardwoods

Ripsaw teeth for light hardwoods

For Timber density	560 to 800 Kg/mm³
Typical Species	Ash, Larch, Sapele
Number of teeth	54
Hook angle, Green timber	20^0
Hook angle, Seasoned timber	15^0
Clearance angle	15^0
Top bevel angle	15^0
Gullet depth	0·50p
Tooth top	0·30p

Ripsaw teeth for medium hardwoods

For Timber density	800 - 1040 Kg/mm³
Typical Species	Ebony, Jarrah, Rosewood
Number of teeth	60
Hook angle, Green timber	15^0
Hook angle, Seasoned timber	10^0
Clearance angle	15^0
Top bevel angle	12^0
Gullet depth	0·50p
Tooth top	0·30p

Ripsaw teeth for dense hardwoods

For Timber density........... over 1040 Kg/mm^3
Typical Species................ Ekki, Greenheart, Wamara
Number of teeth....................................80
Hook angle, Green timber.................. 10^0
Hook angle, Seasoned timber........... 10^0
Clearance angle................................ 15^0
Top bevel angle.................................. 10^0
Gullet depth..0·45p
Tooth top..0·30p

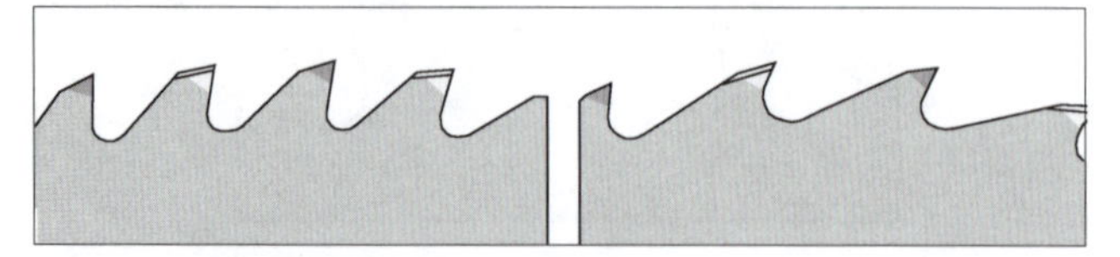

Fig. 5.3
Left, ripsaw teeth for dense hardwood. Right, ripsaw teeth for abrasive timbers

Ripsaw teeth for abrasive timbers

Number of teeth....................................40
Hook angle, Green timber.................. 15^0
Hook angle, Seasoned timber........... 15^0
Clearance angle................................. 15^0
Top bevel angle.................................... 5^0
Gullet depth..0·35p
Tooth top..0·25p

Ripsaw teeth for Straight-line-edgers

Number of teeth....................................36
Hook angle, Green softwood............. 15^0
Hook angle Seasoned softwood........ 10^0
Hook angle Seasoned hardwood....... 10^0
Clearance angle................................. 20^0
Top bevel angle.................................. 10^0
Gullet depth..0·45p
Tooth top..0·35p

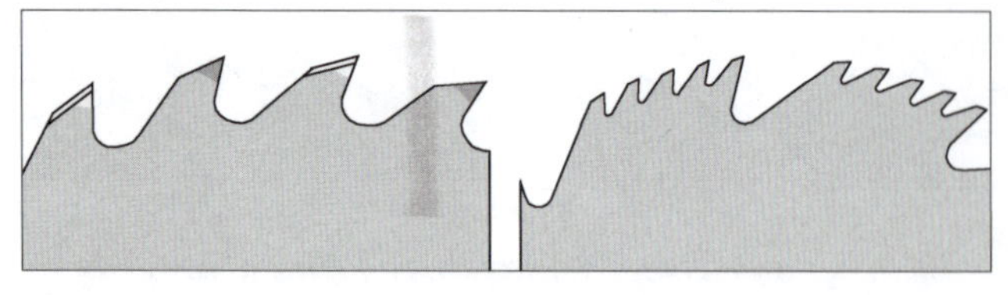

Fig. 5.4
Left, ripsaw teeth for straight-line-edgers and multi-ripsaws. Right, ripsaw teeth for double slabbers, generally called group or cluster saws.

Ripsaw teeth for double slabbers

These are special sawblades made for double slabbing round logs and flitches on machines specifically made for this purpose. *(Slabbing is removing the two outer sideboards from logs.)*

Tooth geometry is generally similar to that of softwood and hardwood ripsaw teeth, as detailed above, but teeth are formed in groups of five separated by a deep, double-tooth-width gullet. The large gap between groups sweeps sawdust away and prevents buildup on the sawblade body. This makes sawblades run cooler and with less chance of blistering than regular-toothed sawblades. File-sharpened sawblades have a flat top on the leading unset tooth, and bevels on the following alternate-set teeth. File-sharpened slabber saws, however, are not very efficient except when converting very low-density timbers. For the best results - as with all fast-feed ripsaws - preferably use saws having swage-set or Stellite-tipped teeth, with all tops normally ground flat.

Sawblades for cross-cutting

Cross-cut teeth for radial and other pull-out cross-cut saws are similar to ripsaw teeth, but have a negative hook angle of 5 degrees or so to stop the saw running-out. To sever the timber fibres ahead of the main cut, the fronts of the teeth are formed to a bevel angle of 5 degrees - unlike rip saw teeth which are filed straight across. The front bevel angle ensures that timber is cleanly cut.

Although regularly of the hollow ground pattern - to ensure the most accurate tooth side alignment - these sawblades usually need to be spring-set to avoid burning in the cut - especially when dealing with softwoods.

NOTE: Other types of cross-cut and mitre saws are now in common use and these use very similar sawblades to those shown previously. Others may use rip & cross-cut teeth similar to those described under dimension sawblades, it all depends upon the way the machine operates.

Cross-cut saw teeth - regular profile

Number of teeth.............................. 56-84
Hook angle.. 5^0 negative
Clearance angle................................ 25^0
Top bevel angle................................. 10^0
Front bevel angle................................. 5^0
Gullet depth.......................................0·55p
Tooth top...0·40p

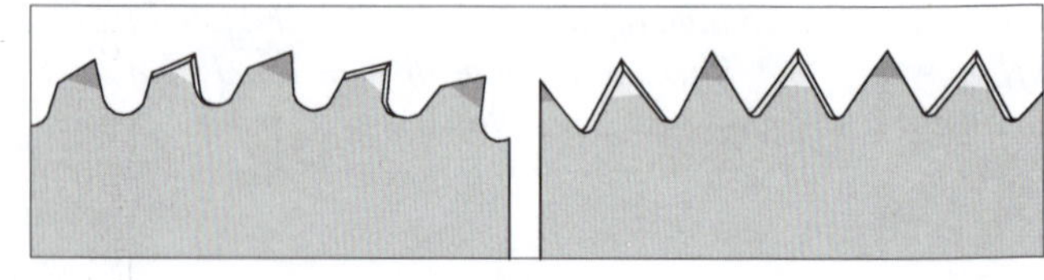

Fig. 5.5
Left, regular cross-cut saw teeth. Right, peg cross-cut teeth have the same hook as the clearance angle. The negative cutting angle on these sawblades prevents them running-out on radial and pull-out crosscuts

Cross-cut saw teeth - peg tooth profile

This is a very old tooth form with many more teeth than any other type. It gives a clean cut, but takes more power and needs considerably more effort in sharpening and setting than regular cross-cut saws.

Number of teeth............................ 76-132
Hook angle... 30^0 negative
Clearance angle................................ 60^0
Back bevel angle................................ 15^0
Front bevel angle............................... 10^0

With having the same hook as the clearance angle, some early types could actually be run in either direction. Many years ago it was claimed that reversing these sawblades periodically actually sharpened the trailing point through the rubbing action whilst cutting. The same claim was also made for odd-looking two-way facing ripsaws - but no-one believes that anymore either.

Dimension Sawblades

Dimension sawblades are used for both for ripping and cross-cutting solid timber, primarily in finishing operations, i.e., sizing cabinet doors and mitring picture frames and similar work. Special machines were built for this purpose, called dimension saws in the UK and variety saws in the USA.

The first types had one-piece tops, sometimes able to be tilted for angle sawing, and a sliding mitre gauge to allow single or compound angle cross-cutting.

The tilting top proved dangerous and was eventually replaced (on other than low-cost hobby machines) by a tilting sawblade. On more upmarket machines the sliding mitre gauge evolved into a fixed angle gauge on a left-hand sliding table.

When alloy saws were the only type in use, a double dimension sawbench was quite popular. This had two quickly switched arbors, one fitted with a cross-cut saw and the other with a ripsaw. This gave excellent results - but the design was costly and complicated and is now no longer produced.

On modern machines only one main saw arbor is fitted - so the sawblades have to be suitable for both rip and cross-cutting*.

So far as manufactured panels (other than plywood) are concerned there is no difference between ripping and cross-cutting - but this can remain a problem with solid timber.

**When alternately ripping and cross-cutting solid timber on custom work this factor needn't be overstated - because the high rim speed of modern machines now gives a far better finish than on the early machines.*

The main difference between regular rip and crosscut teeth is the hook angle, so a sawblade with a zero hook angle is largely acceptable for both ripping and cross-cutting most timbers. *(Although this type of saw tooth is not entirely suited for either operation, simply by using a slower feed speed - not a critical factor for dimension sawing anyway - an acceptable finish is possible. Using a slow feed speed dulls saw teeth more quickly than if using a faster one, but this is a small price to pay for the convenience of using the same sawblade for both rip and cross-cutting solid timber.)*

Dimension saw teeth - regular profile

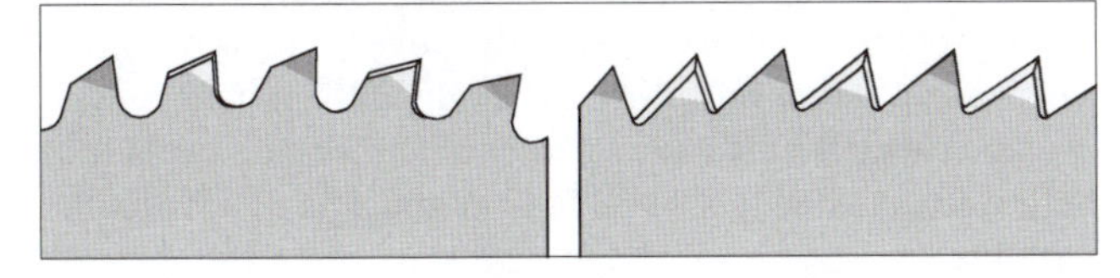

Fig. 5.6

Left, regular dimension saw teeth have a hook angle which is a compromise between rip and crosscutting.
Right, the dimension peg tooth profile is similar to the cross-cut version, but with a more aggressive hook angle.

Number of teeth	56-88
Hook angle	0^0
Clearance angle	25^0
Top bevel angle	10^0
Front bevel angle	5^0
Gullet depth	0·60p
Tooth top	0·35p

The overall capacity of the dimension (variety) saw was increased substantially - and the name changed to panel saw - when manufactured panels came into wide use. These then needed tct tipped main and scoring sawblades to cleanly cut these harder materials. *(The main and scoring alloy sawblades used initially were difficult to maintain in matching kerf widths - and their lifetime was far too short.)*

Dimension saw teeth - peg tooth profile

This is another type of sawblade once widely used on dimension and variety saw benches for smaller and more delicate work such as small picture frame mitring. It is basically a pendulum cross-cut saw, but with a slightly more aggressive hook angle to give acceptable results both when rip and cross-cutting.

Number of teeth	84-120
Hook angle	10^0 negative
Clearance angle	40^0
Back bevel angle	15^0
Front bevel angle	10^0

Dimension saw teeth - novelty profile

This is a combination sawblade having both rip and crosscut teeth in sets of five, with a deepened gullet separating the sets.

The leading tooth of each set is a raker or a cleaner tooth to remove the bulk of the waste in a ripping action. This is filed or ground straight across both front and top to finish at a slightly lower level than the following teeth, and has no set.

The following the raker teeth are two pairs of alternately spring-set teeth. Their points project beyond the raker teeth by 0.8mm for softwoods and 0.4mm for hardwoods - to scribe and sever the fibres ahead of the following raker tooth. The result is the smoothness of a fine-toothed crosscut saw and the "bite" of a ripsaw.

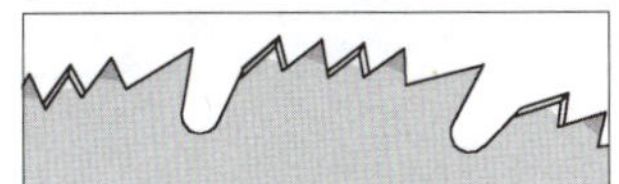

Fig. 5.7

A novelty tooth profile. These sawblades were once widely used on dimension saws for sizing plywood, etc.

Number of tooth groups	12-20
Hook angle, peg teeth	10^0 negative
Clearance angle, peg teeth	40^0
Back bevel angle, peg teeth	15^0
Front bevel angle, peg teeth	10^0
Hook angle, raker teeth	15^0
Clearance angle, raker teeth	25^0
Top bevel angle, raker teeth	0^0
Front bevel angle, raker teeth	0^0

Preparing sawblades for sharpening and setting

To ensure that tooth heights will be correct when filed it may be necessary to *'stone'* a dull sawblade before removal.

Theory of stoning

High-quality, fast-feed sawing is possible only if all the teeth share the work equally, ie. when the cutting points are concentric to the axis of rotation of the saw arbor.

To ensure this an abrasive stone can be used on alloy saws to dress teeth down to a common height prior to removal for sharpening - this leaves a bright 'flat' or witness mark on the points as a guide when filing.

The technique is to file so that the stone witness mark disappears on each tooth at the final filing stroke. By noting the amount by which this reduces at each stroke, the filer can usually pace the process so that the final stroke is always a complete one.

Mounting the sawblade

Before stoning a sawblade check that it is properly fitted. Older machines and sawblades may be worn enough to have developed play between the two. In this case set the driving pin at 12 o'clock and pull the sawblade back against this before tightening the locking nut. If no pin is provided, then file reference marks on the sawblade and on the rear collar, and align these at 12 o'clock each time the sawblade is fitted.

Checking the cut profile

To check the condition of a newly sharpened sawblade, partially cut into a hardwood test piece in a slow, deliberate movement so that the cutting action can be clearly seen.

If the cutting action is steady and even, then the saw teeth are all in the same cutting circle. If the cutting action is irregular or jerky then the saw teeth are not concentric to the arbor centre.

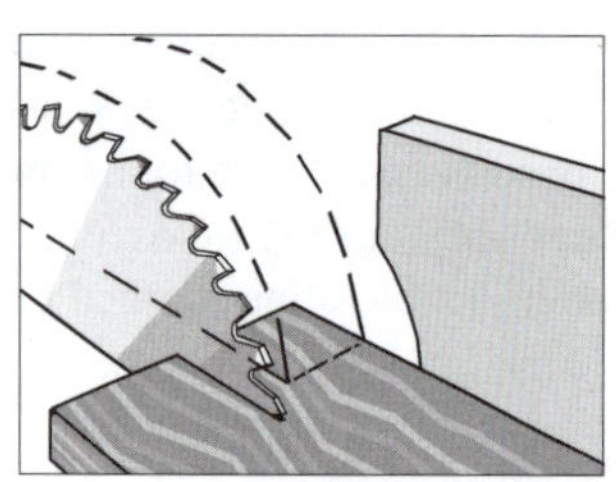

Fig. 5.8
To check the condition of a sawblade make a short cut in a piece of hardwood

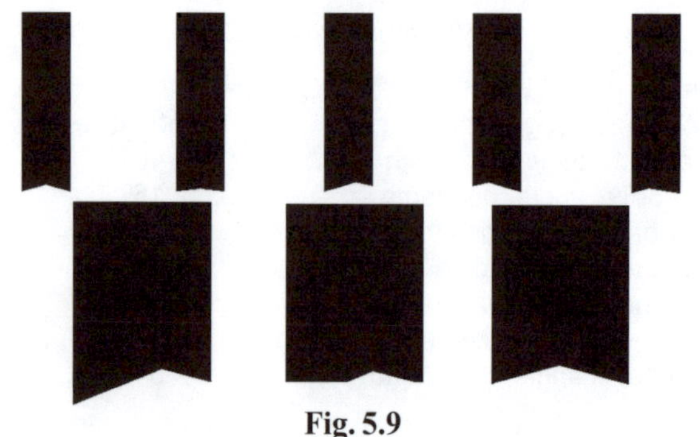

Fig. 5.9
These show typical saw kerf outlines. Top: from left to right, correct softwood profile, correct hardwood profile, left leading tooth, right leading tooth, and unequal bevels. Bottom: from left to right, leading tooth before stoning, after stoning, and after filing.

Carefully draw the timber clear and examine the saw-cut profile. Points should be of equal height and top bevel angle, if they are not then the saw may run in the cut - lead to one side.

If the saw teeth are both concentric to the saw arbor centre and equal in height, then stoning is not needed - simply file each tooth top by an equal amount. If teeth are of unequal height or not concentric then the sawblade needs dressing (also called ranging, stripping or jointing).

Methods of stoning

Caution: whatever method of stoning is used it is essential to wear protective goggles just in case a spark is thrown towards the eyes.

An old method of stoning is to track a section of abrasive wheel across the saw teeth with the saw running. This is not only dangerous, but also gives poor results .

Another way is to first lower the sawblade so that the teeth are fractionally below table level. Using a long abrasive stick, place this at right angles to the sawblade, setting the top guard barely clear. Gripping both the projecting ends of the abrasive stick, carefully and slowly track the stick over the top of the running sawblade. If no contact is made raise the sawblade very slightly and repeat - until all teeth show a bright witness mark indicating contact with the stone.

Stoning jig

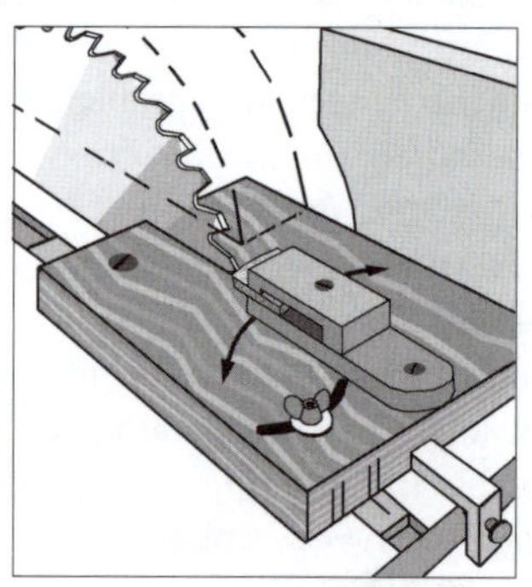

Fig. 5.10
Using a stoning jig allows the abrasive to swing across the saw teeth.

Using a simple stoning jig such as that described is safer and better. It consists of a plywood base with a swinging arm with a small and thin piece of abrasive stone clamped to its end.

With the sawblade raised and running, and with the top guard properly positioned, carefully move the jig towards the sawblade - *control this by the end screw.* When light contact is made swing the abrasive stone clear across the sawblade at both sides to stone it. The pivot point must be exactly in line with the sawblade so that alternate teeth are dressed to the same height.

The jig shown has a strip underneath to fit into the left-hand slot provided on most saw benches. It swivels to allow for precise alignment of the swinging arm pivot to the sawblade when used on different saw benches.

The frequency of stoning depends upon circumstances. Although many users do this as a matter of routine, it is not usually necessary to stone sawblades every time before removal for filing.

After stoning, sawblades are set and then sharpened, or sharpened and then set - which first is a matter of opinion rather than fact. In practice, though, it seems to make little difference provided care is exercised in both.

Spring setting

Spring setting is bending the teeth alternately to right and left to provide the necessary clearance of the saw body - most alloy circular sawblades are spring set.

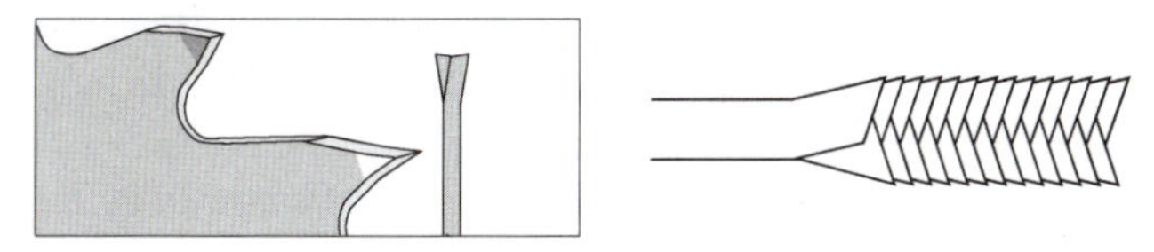

Fig. 5.11
Left, spring-set alloy saw teeth. Right, the chip removal pattern with spring-set teeth

The amount of set varies according to the material and its condition. Hardwoods and dry timber require less set, soft and wet timbers more, see Fig. 5.12. Setting equipment must be capable giving a precise amount of set, or must allow the amount of set to be carefully measured - and then corrected if wrong. Accuracy is of vital importance, high points score the sawn surface of the timber and give a poor finish.

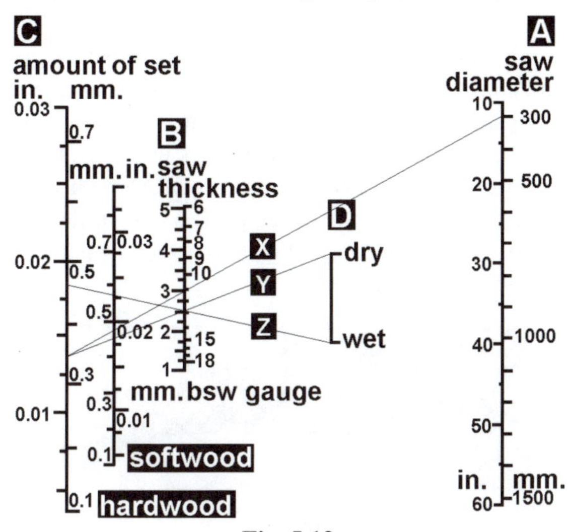

Fig. 5.12
*Use this nomogram to find the amount of set needed for alloy steel sawblades. Draw a straight line from **A** corresponding to the saw diameter, through **B** corresponding to the saw body thickness, to **C** to indicate on either the softwood or hardwood scale the correct amount of set for seasoned timber.*
*The example **X** shows a 300mm. diameter saw with a 3mm. body thickness which needs set of 0.47mm for softwood, or 0.35mm for hardwood. To find the amount of set required for unseasoned timber for the same diameter and thickness of sawblade, draw a straight line **Y** from 'dry' on **D** to where **X** crosses **C** on either the softwood or hardwood scale (this is actually shown for the hardwood scale). Draw a third straight line **Z** from 'wet' on **D** through where **Y** crosses **B** to show the correct amount of set on the appropriate **C** scale. The example shows the correct figure for unseasoned hardwood as 0.47mm - up from 0.35mm for seasoned hardwood.*

The hand saw-set

Conventional hand saw-sets can be single or double handed. Several openings or gates are cut in the set, so use the one that fits exactly.

Take great care that the set does not slip off the tooth point when used. In setting, the tooth point is deliberately bent using the edge of the saw-set as a fulcrum - not simply pushed over from the root. How the saw-set is placed and the way it is held are both critical - but this is a skill soon learned after a little practice.

Do not bend the tooth from the root - this gives a springiness to the tooth which may then chatter in the cut.

Instead give the tooth point a slight twist from a positive line of bend between one half and one third the way down the tooth face, and with a slope angle of about 30 degrees or so upwards towards the heel.

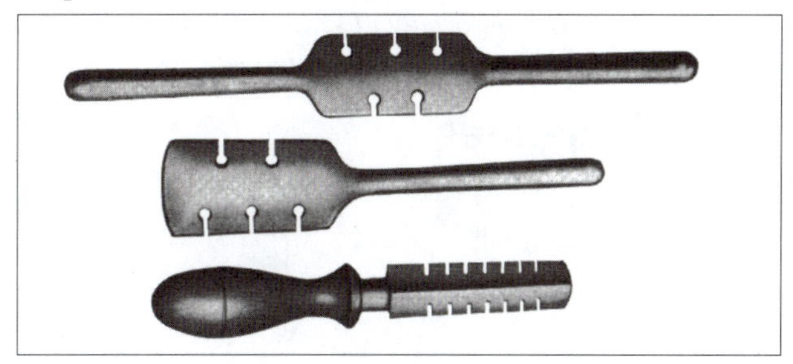

Fig. 5.13
This shows typical hand saw-sets. Top, a two-handed saw-set for thick saws, middle, a single hand saw-set for average saws and, below, a single hand saw-set for small circular saws and handsaws.
Source - Robinson/Wadkin

Using a saw set.

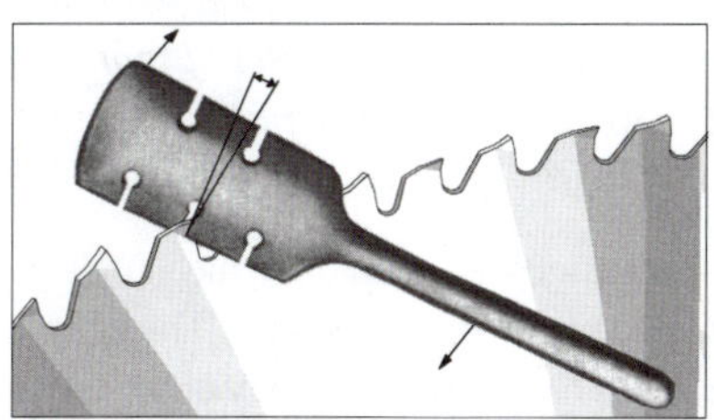

Fig. 514
This shows the twisting movement used with a regular saw-set. This is angled as shown to give the correct line of bend. The set should be placed roughly in line with the tooth face to avoid damaging the points.

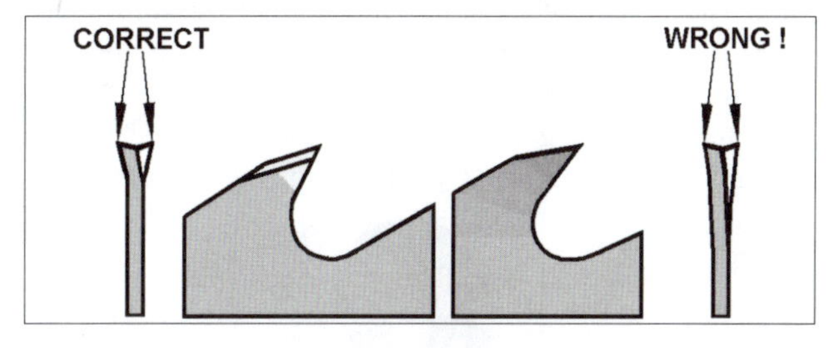

Fig. 5.15
Spring set with a sharp bending line gives a rigid tooth, left. Bending from the root, right, may allow the tooth to spring in the cut and give a poor finish.

Immediately after setting check the amount of set using a gauge - and corrected if necessary. Many filers hold the set in one hand and the gauge in the other, allowing them to set, check and correct the amount of set for each tooth in turn.

When setting the alternate teeth it is better to use the

same hands for both the saw-set and the gauge after reversing the sawblade in the filing stand. Doing this gives more consistency than by switching hands.

Fig. 5.16
The regular practice is to hold the saw-set in one hand and the gauge in the other. Use them in the same hands for both sets of teeth to give uniformity of setting - by reversing the sawblade for the opposite teeth.

Saw-set gauges

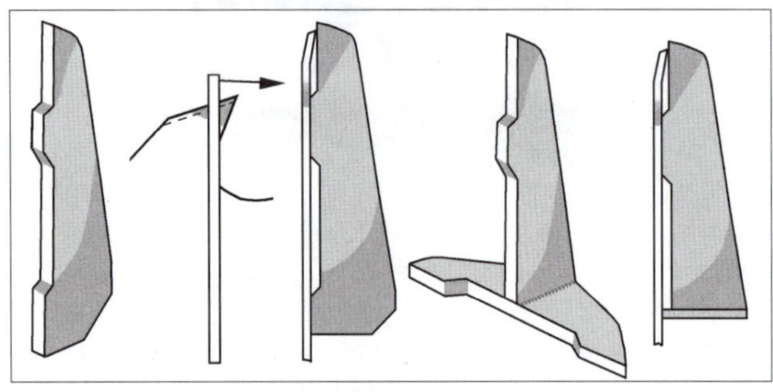

Fig. 5.17
A simple saw-set can be made from a flat piece of steel. Move the set across the teeth to 'click' it in the direction of the arrow, keeping the feet in contact with the saw body. The shaded area shows the section of tooth bent over in setting (left). To keep the saw-set absolutely square to the saw body fit the extra piece shown (right).

Simple saw-set gauges, each made to measure a given amount of set can be made from bandsaw strip. The simplest are made from flat steel, but for greater accuracy make the gauge in a tee form to ensure squareness to the saw body

To check the amount of set place the body of the gauge firmly against the side of the sawblade, with the gauge face actually on the tooth point. If the gauge can be rocked this shows that more set is needed. If the gauge does not rock, then move it across the tooth from the heel and to clear the point. The amount of set needs reducing if a *'clicking'* sound is heard as the gauge face drops off the tooth point. The amount of set is correct if light *rubbing contact* is felt between the tooth and gauge face as it passes. When each tooth point gives the same *'feel'* then the sawblade is correct.

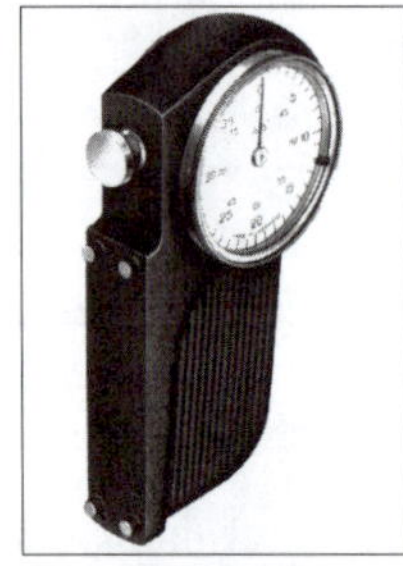

Fig. 5.18
The dial indicator is a very accurate means of measuring the amount of set on spring-set and tc tipped saws.
Source - Wadkin

The best gauge to use is the dial indicator type with a rotating bezel. Pre-set this by placing the gauge on a flat metal surface with a feeler gauge equal to amount of set needed under the measuring disc. Rotate the outer bezel so the indicator is directly opposite the pointer. In use place the indicator's feet firmly on the saw body, and with the measuring disc against the tooth pointer. The amount of set is correct when the indicator is again opposite the pointer.

Great accuracy is possible by using this type of gauge - but only if the saw plate is perfectly flat, free from lumps, and also runs true on the arbor.

Other setting devices

Other types of saw tooth setting devices in general use are the plier types and setting stand types. On both the tooth is bent over against a fixed anvil by a die. The anvil is often a round piece of hardened steel with flat and angled faces ground on it. The die is similar but with a single angle only. To vary the amount of set, the overhang of the tooth beyond the flat of the anvil is adjusted. To vary the line of bend both anvil and die can be partially rotated.

With the simplest type of setting stand the die is struck with a mallet to set the saw tooth. With others a handlever is used instead - but neither can guarantee absolutely accurate setting because the teeth may spring back unevenly. Even so, filers can become so skilled that they can set teeth with acceptable accuracy at the first attempt. The most accurate type has built-in dial indicators that indicate exactly what amount of set is given to each tooth.

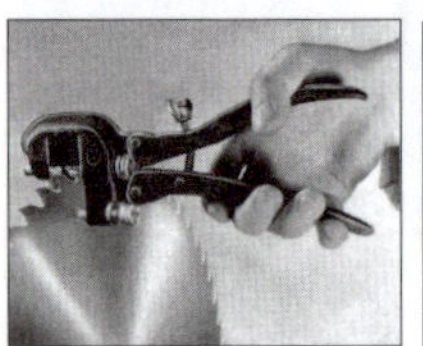

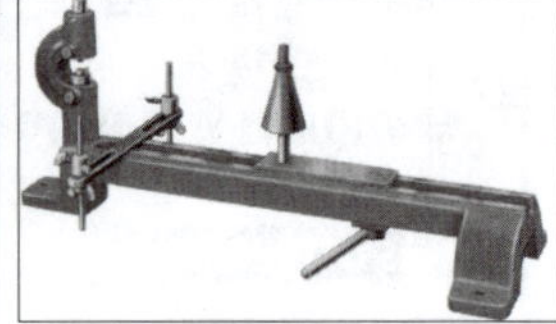

Fig. 5.19
Left: plier type setters are used mainly for handsaws and small circular saws. They set the tooth as the levers are squeezed.
Right: To use this very simple setting stand the tooth face is positioned against a stop and the spring-loaded die struck with a mallet. The centering cone adjusts vertically to set the saw level with both the anvil and the outer supports. The saw is flipped to set teeth on the reverse side.
Source Robinson/Wadkin

Machine Setters

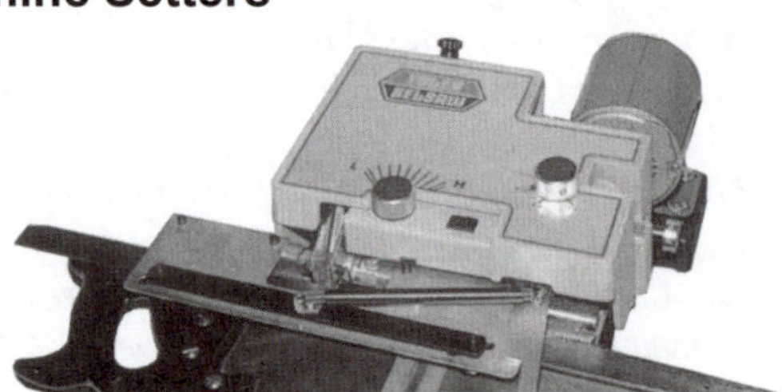

Fig. 5.20
The Foley SS1000 is a fully automatic machine for setting left and right hand teeth at a single pass. Although shown with a handsaw, a circular saw can be substituted.
Source - Foley-Belsaw

Machine setting is not normally used for circular saws - except perhaps for peg-tooth saws with small-pitch teeth. This is possible on saw setting machines primarily intended for handsaws or bandsaws.

Swage Setting

Swage-set teeth have their points spread outwards to give an equal amount of set at each side. Teeth intended for swage-setting should be ground square across both front face and top, and to an included or sharpness angle of 45 degrees.

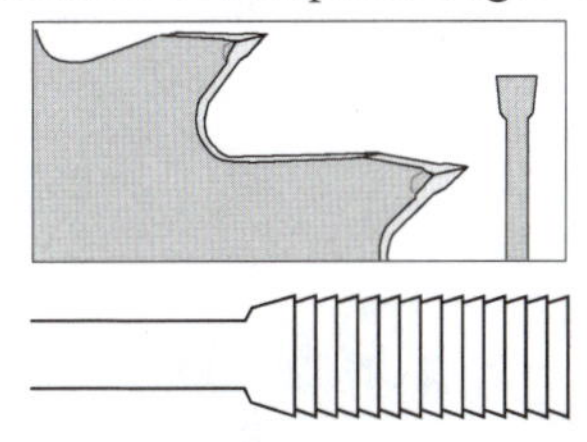

Fig. 5.21
With swage-set teeth each tooth takes a full, kerf-width cut, as shown below.

In swage setting, each tooth is first securely clamped to the swager. Then the tooth face is deformed by rotating an eccentric, hardened die to spread the tip sideways. The die is rotated by a handlever, using a rolling action (and a small amount of oil) to squeeze the tooth face against an anvil set hard against the tooth top.

The tip then deforms and spreads plastically by an amount governed by the pre-set rotary movement of the die. On large teeth two or more swaging movements may be needed to give the depth of swage required and, on the largest teeth, two following swaging passes give better results.

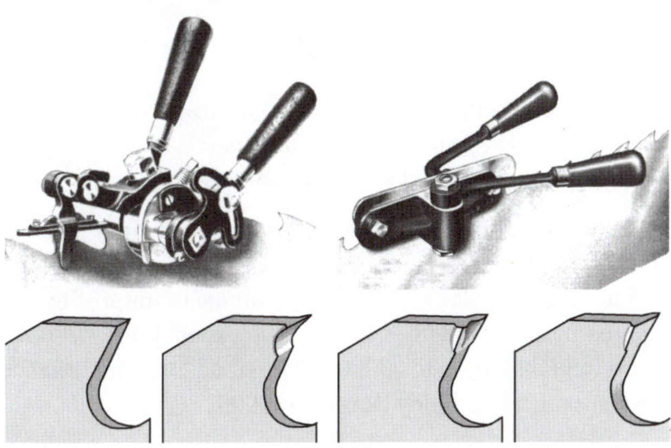

Fig. 5.22
Top left, typical hand-operated saw swager.
Bottom left, the effect of swaging showing the regular tooth form (left) and the tooth after swaging (right).
Top right, typical hand-operated saw dresser (also called a shaper). The rear handle is used to position the dresser, then the front hand lever is used to squeeze the tooth.
Bottom right, shaping the tooth using the side dresser, also the tooth after face filing or grinding (right).
Source Robinson/Wadkin

At a following operation each tooth is finally shaped by a side dresser or shaper. This has a pair of hardened dies which between them squeeze-back the excess tooth spread to a precise kerf width and shape - and give both tangential and radial clearance angles to the tooth point. On some dressers one die is fixed and the second moves, whilst on others both dies move together simultaneously.

By adjusting the setting of the plates which rest on the tooth points, the depth of swage and the kerf width can be altered. The process of swaging and side dressing leaves a somewhat irregular face which is cleaned-up at a second filing or grinding pass.

The same basic style of manual and power-operated swager and side dresser have been in use for many years - though mainly for wide bandsaws. Where swagers and side dressers are used for circular saws, this is mainly to prepare them for manual Stellite tipping, see Chapter 8.

Filing alloy saw teeth

Hand filing of circular saw teeth is relatively easy - provided the art is well practiced and care is taken.

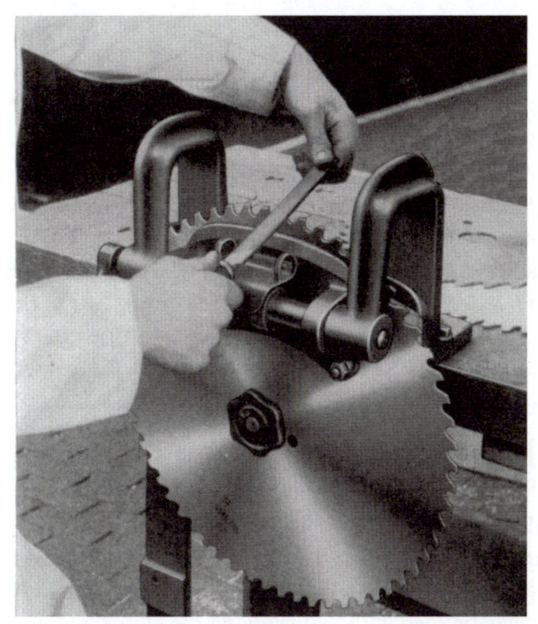

Fig. 5.23
Standing in the correct position and holding the file in the proper way helps to ensure accurate filing. Some companies still offer a filing stand, but most filers make their own wooden filing stand.
Source - Wadkin

The file used for the round gullet type of teeth is a 10in millsaw, second cut, single, 2R.E. *(Ten inches (255mm) is the length of the file, millsaw is the type, second cut is the grade, single indicates that there is one set of teeth only cut diagonally across the surface, and 2 R.E. indicates that the file has two round edges).* This length of file, and a similar 8in (200mm) file, are the ones most commonly used.

NOTE: It is essential to clean sawblades thoroughly before starting to file - otherwise the file teeth quickly become clogged

The sawblade is usually clamped in a filing stand (vice) with the base of the gullets no more than 25mm (1in) above the top of the stand. The tooth to be filed should be uppermost and about elbow height.

Use the smooth portion rather than the body of the file to move each tooth in turn into position for filing.

Hold the file as shown, and use the full length of the file at each working stroke.

Filing the tooth face. File the tooth face first, merely cleaning this up, taking care not the change the hook angle. Keep the file parallel to the floor as this retains a square-across gullet.

Filing the tooth top. Take care to retain the original clearance and bevel angle. If the saw teeth were already OK for height, then the same number and weight of filing strokes on the teeth tops will keep them well within the ballpark.

Fig. 5.24
Typical plate saws from the Atkinson-Walker range, one a regular file-sharpened type and the two others with peg teeth of different pitches,.
Source Atkinson-Walter

If the sawblade has been stoned, then the number and weight of filing strokes will vary from tooth to tooth. Note how the stone mark shrinks at each filing stroke, then make shorter or lighter strokes to leave a stone mark of consistent width for the final, full-length pass of the file. Possibly then make a final filing pass to ensure a sharp point - but in this case the weight and length of the final pass must be consistent.

Filing technique

In filing the tooth tops slew slightly towards the heel by moving the file marginally sideways. This ensures that the heel of the tooth finishes lower than the point, thus guaranteeing an adequate clearance angle on the tooth top. Also the stoning mark can be seen without lifting the file clear - and no filing *'burr'* will form on the front cutting edge of the tooth.

Filing peg teeth

Peg teeth saws are filed using a triangular flat file. The main difference between filing peg and regular circular saw teeth is that the back of one peg tooth and the face of the following peg tooth are both filed at the same time. This cleans up the full tooth profile completely - so gullet grinding is never required - and only one filing stroke is needed. The usual practice with these saws is to file alternate teeth, then reverse the sawblade in the clamp to file the remainder.

To give the necessary front and back bevel angles to the teeth, the file is held at a slew angle to the saw body - when looking directly down onto the sawblade.

The file is held is horizontally when filing a cross-cut sawblade which has the front and back face bevels equal in angle. When the sawblade has a square front such as for a ripsawing, though, hold the file at an angle to horizontal.

Machine filing

Fig. 5.25
The Foley-Belsaw SF1000 is shown here filing a handsaw. The handsaw carrier can be removed and a cup and cone fitted for mounting peg-tooth circular sawblades of up to 600mm (24in) diameter.
Source - Foley-Belsaw

Machine filing of alloy circular saws is not a regular practice for gullet-type saws, these are normally machine-ground on an automatic grinder to form a continuous tooth profile - as detailed in the following Chapter. The exception is with peg tooth circular saws up to 600mm (24in) diameter which can be filed on such as the Foley Belsaw SF1000.

Chapter 6
Machine Grinding Alloy Saws

Ever since circular saws were first made, gulleting in one form or another has always been necessary as the teeth in a sawblade wear down through hand filing. In the early days, circular saws had a series of holes punched in a diagonal line just below the bottom of each gullet, allowing the filer to break through to the next hole when the tooth height was sufficiently reduced. It was also claimed at the time that these holes, about four in each series, helped to cool the rim.

The Saw Gulleter

Eventually the hand gulleter became readily available, and this allowed shallow gullets to be deepened by abrasive grinding. These machines are now virtually obsolete in most factories because alloy saws - where they haven't been replaced by tungsten carbide tipped saws - are mainly sharpened using automatic saw grinders.

Some parts of the world still use hand gulleters, though, so for the record I include some details.

Fig. 6.1
A typical manually-operated gulleting machine.
Source - Robinson/Wadkin

The machine has an abrasive wheel which can be tilted on a horizontal arbor to give a downward movement at the required hook angle, and an adjustable depth stop to ensure consistent gullet depths. Some machines are also able to grind top and front bevel angles. A vertically adjustable cone or bush provides a mounting for the sawblade.

The tooth shape conventionally used when a sawblade is intended for hand filing is the straight-back type - a tooth profile also suitable for hand gulleting. The continuous profile, or round-back, type can only be profiled satisfactorily on an automatic saw grinder.

Machine Setting
Mounting the sawblade

Sawblades are mounted either on a bush to fit the saw bore, or on a self-centering cone *(See Figs. 6.32 & 6.33)*. These assemblies rotate either on a plain or ball-bearing arbor. The actual method of mounting is not important provided the sawblade rotates easily and runs true to its centre bore.

To avoid vibration the vertical setting of the sawblade should be such that the bottom of the new gullet will be approximately 6mm (1/4in) above the supporting piece behind the saw *(see Fig. 6.2)*

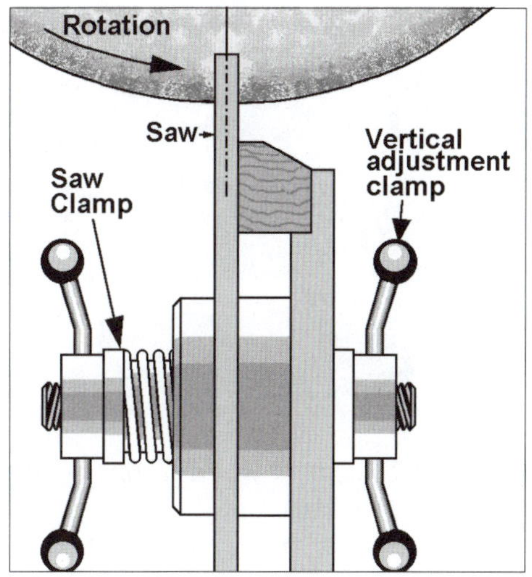

Fig. 6.2
Sawblades are often mounted on precision-ground bushes, as shown here and Fig. 6.3. Alternative bushes can be fitted to suit different saw bores.

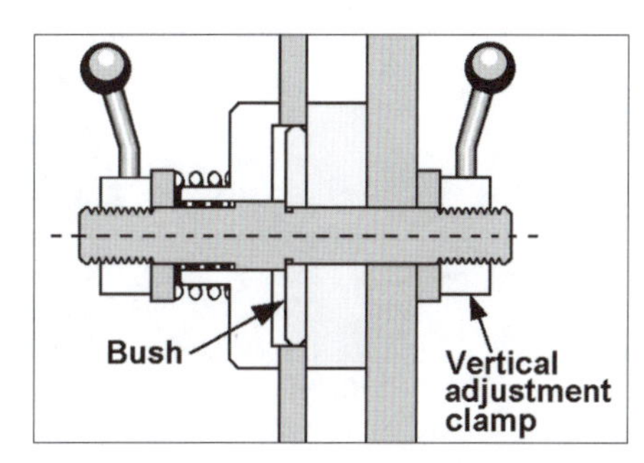

Fig. 6.3
This is a section through the bush-mounting shown in Fig. 6.2. Some types incorporate ball bearings for easier sawblade rotation.

As different thicknesses of grinding wheel can be fitted to most gulleters, make sure that the grinding wheel thickness and shape are suitable for the profile of the sawblade to be gulleted. For most alloy saws the gullet radius, determined by the grinding wheel thickness and its shape, should be 0·15 of the pitch. Fig. 6.5 shows how different wheel thicknesses affect the shape of the saw gullet.

It will be noted on Fig. 6.5 that radial lines, drawn from the tooth point towards the centre of the saw, pass through the gullet radius at different points. Ideally, the gullet radius should have its deepest point directly below the tooth point on softwood ripsaws - in line with a radial line from the tooth point. Of course there are slight variations with different hook angles, and this does not apply to sawblades with a negative hook angle.

Gullet shape

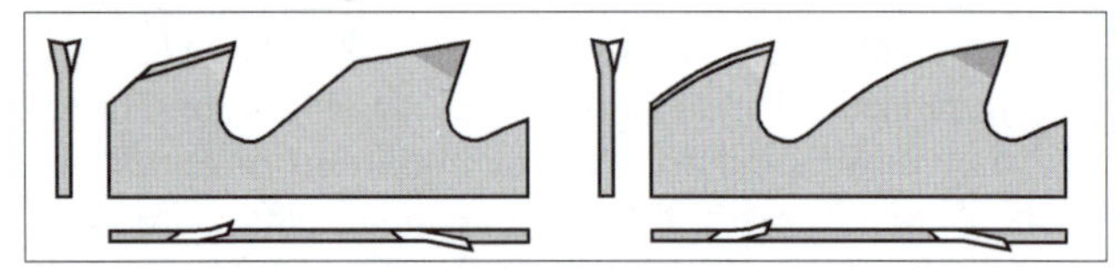

Fig. 6.4
Left, typical profile of a straight-back ripsaw tooth. Right, Typical profile of a round-back ripsaw tooth.

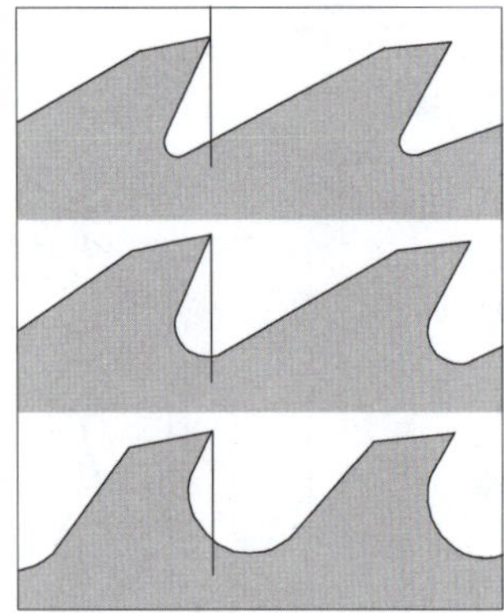

Fig. 6.5
Top: *This shows a tooth profile with a narrow a gullet formed by using too thin a grinding wheel, or one profiled with too-small a radius. Resinous sawdust will likely lodge in the gullet.*
Middle: *This profile is ideal, with a nicely-rounded and correctly proportioned gullet.*
Lower: *This profile has a large a gullet and a narrow and weaker tooth formed by using too-thick a grinding wheel.*

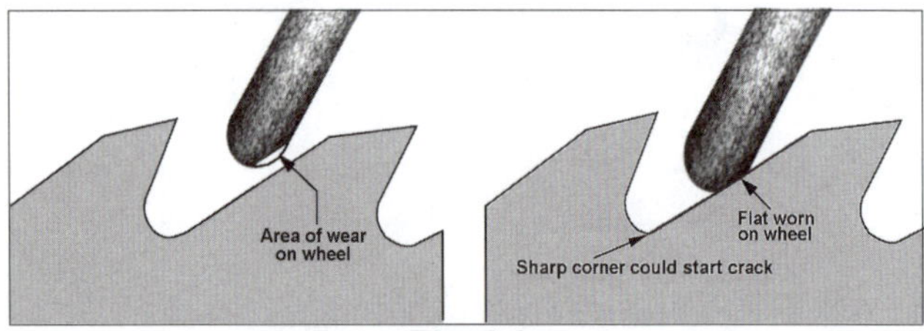

Fig. 6.6
Left, this shows where most wear occurs on a grinding wheel initially shaped to a full half-round profile. Right, if a half-round wheel is continuously run without dressing, the flat formed on it distorts the gullet profile to form a crack-inducing sharp corner.

With straight-back teeth excessive wear tends to form a flat where the grinding wheel meets the tooth back. If the grinding wheel is not dressed regularly this flat will eventually slightly narrow the gullet - though this in itself is not a problem. However, this flat also forms an internal corner in the gullet where stress concentrates and where a crack could start.

Grinding Wheel Dressing

When using a grinding wheel, either grinding, dressing or simply observing the operation, it is essential to wear goggles or a full face mask to protect the eyes from stray sparks.

To avoid wearing a flat on the grinding wheel, use one which is actually thicker than really needed. Form the half-round profile only at the left-hand side and blend this into a flat at the right-hand side dressed to the same angle as the tooth back. A flat at this point wears to a lesser degree than a radius, and also helps to maintain a straight tooth back when gulletting.

Frequent dressing and accurate shaping of the grinding wheel is essential to grind sawblades with a consistent gullet profile. Avoid sharp corners on the wheel as these produce crack-inducing sharp corners in the gullet.

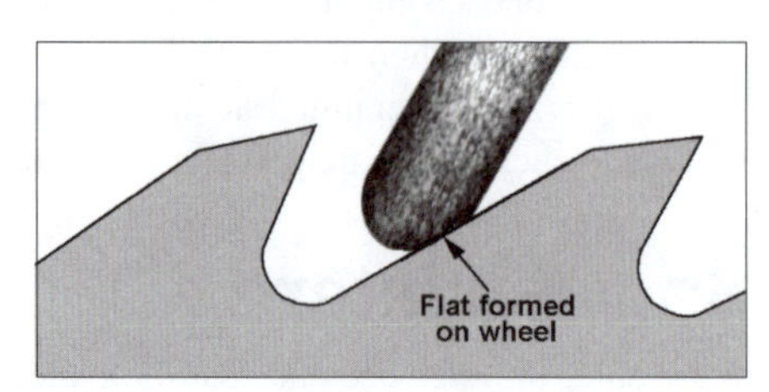

Fig. 6.7
To avoid wheel wear and subsequent gullet distortion, use a wider wheel with a flat to match the tooth back.

Many operators use a hand-held dresser which, with experience, can produce an acceptable wheel profile, but learners need practice to get this right. Preferably use a mechanically controlled dresser which will guarantee a consistent grinding wheel shape.

In addition to maintaining the correct shape, dressing is also needed if grinding becomes difficult because the grinding is glazed or loaded. Grinding techniques are fully dealt with in Chapter 7 - Hard Tips and Abrasives.

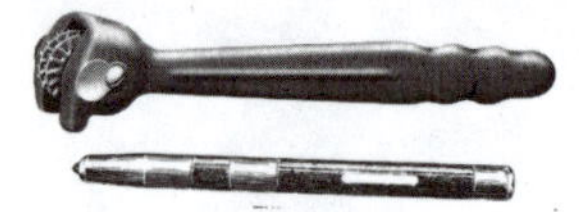

Fig. 6.8
These are typical hand-held dressers, upper, Huntingdon type with rotating, toothed wheels and, lower, a single point diamond dresser.

Centering the grinding wheel

Some gulleting machines allow the grinding head (or saw carrier) to be moved horizontally at right angles to the sawblade.

For regular gulleting the grinding wheel arbor should be directly above the sawblade centreline. Slight adjustment should theoretically be made to compensate for different thicknesses of saw body, but a small misalignment is not that important for this process.

Indicator marks sometimes show the correct horizontal settings for different saw body thicknesses. Where neither marks nor scales are provided for the horizontal movement the correct position can be easily found for a sawblade of average thickness:-

Draw a thin pencil line on the back of the wheel across its diameter (this is best done with the grinding wheel removed). With the grinding wheel fitted, bring the head down so that the wheel just clears the bottom of the gullet. Rotate the wheel so that the pencil line is vertical (use a plumb line or spirit level for accuracy, or place a straightedge against the sawblade.) Check that the pencil line is central to an average sawblade. If not, adjust the grinding head horizontally to suit, then mark the cross-adjustment movement at this setting.

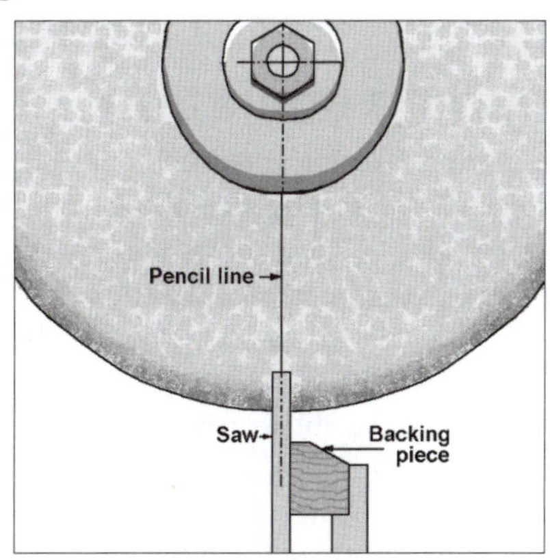

Fig. 6.9

Aligning the grinding wheel horizontally with the centre of the sawblade.

Setting the hook angle

A scale is usually provided as a guide when setting the hook angle, but on some older gulleters this may be unreliable.

To check the tilt angle scale, set the abrasive wheel vertical using this scale and place a straightedge against the left-hand wheel face. If this aligns with, or is closely parallel too a vertical line through the saw arbor centre, then the scale is correct. If not, then use the scale only as a general guide.

On some older machines the grinding wheel does not tilt, so the hook angle has to be set by shifting the saw mounting to one side. The amount of off-set for a given hook angle varies with the sawblade diameter.

The machine may have a plate listing typical measurements, but if none is supplied it is useful to record the actual amount of off-set needed for sawblades regularly gulleted. Do this by setting the machine using a protractor or a guide mark on the sawblade (see Figs. 6.11 & 6.12).

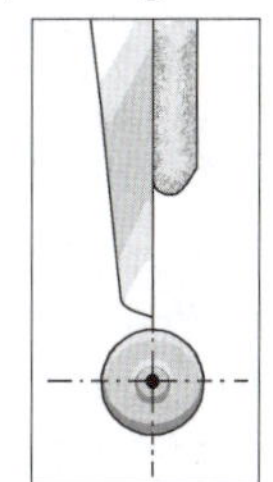

Fig. 6.10

Lining-up the grinding wheel with the saw arbor. Most gulleters use a flat grinding wheel, as the sketch shows.

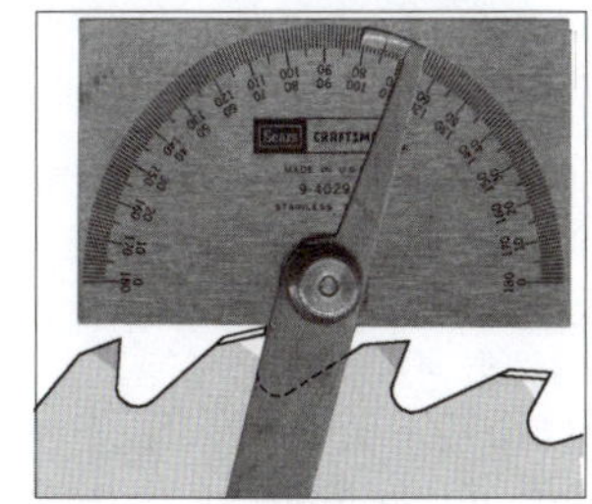

Fig. 6.11

This Sears 9.4029 Craftsman protractor, being much less clumsy than regular engineering protractors, is useful for measuring or marking saw angles. To measure the hook angle, for example, rest the protractor on the tooth being checked (with an equal gap between the underside of the protractor and the tooth points on either side) and set the arm along the tooth front.

If the tooth pitch is large, place the sawblade flat on a table with a narrow steel rule on the table and against the teeth, leaving an equal gap between this and the tooth points on either side. Place the protractor against the opposite edge of the rule and measure as before.

Marking the hook angle on a sawblade

See Fig. 6.12. To mark the hook angle on a sawblade, first draw a straight line from the tooth point **'A'** to the saw centre **'B'***. Count the number of teeth in the circular saw and divide by three. Starting at **'A'**, count this number of teeth in a counter or anti-clockwise direction to **'C'**. Draw a straight line from **'C'** to **'A'** to give an exact 30 degree hook angle **BAC.**

From point **'A'** measure along both arms equal distances of 116mm, then join the two as a baseline and divide this into six equal parts each of 10mm spacing. Join each dividing point to **'A'** to give 5 degree increments in hook angle, and from these choose the angle required..

With a measurement along each arm of 116mm the baseline should be exactly 60mm long, which is easy to divide into six equal parts. For greater accuracy on larger saws double all dimensions, i.e., set-out 232mm. along the arms to give a baseline of 120mm, then divide this up into six 20mm steps. (Metric measurements only are used because imperial measurements are not as convenient.)

*The exact centre of the sawblade is most easily found by fitting a saw grinding bush in the saw mounting hole. This will have a small central hole in it so, by measuring to this rather than the saw mounting hole itself, the chance of error is much less.

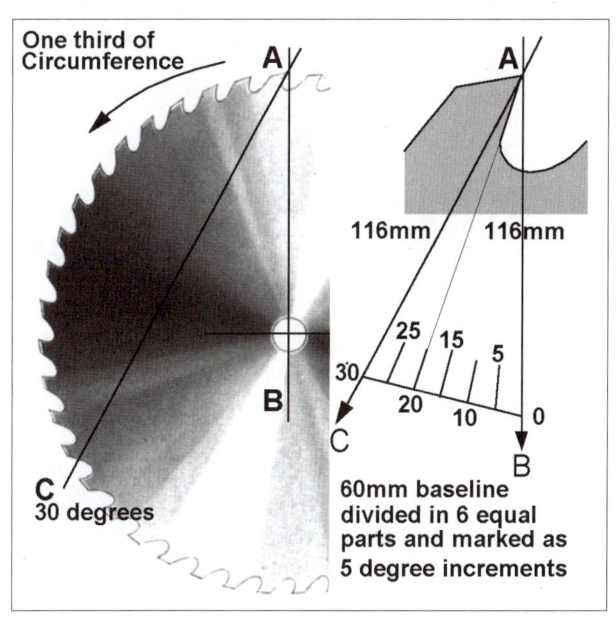

Fig. 6.12

One method of setting-out a hook angle is first to draw a 30 degree angle, then divide this up into 5 degree increments. The sawblade shown has a 20 degree hook angle and 44 teeth, one third of which is 14²/₃ teeth, as indicated at 'C'.

Mark the hook angle required by scribing a line from **'A'** through the appropriate point on the baseline scale. To check the hook angle set the grinding wheel just above the tooth point and place a straightedge against the left-hand side of the wheel.

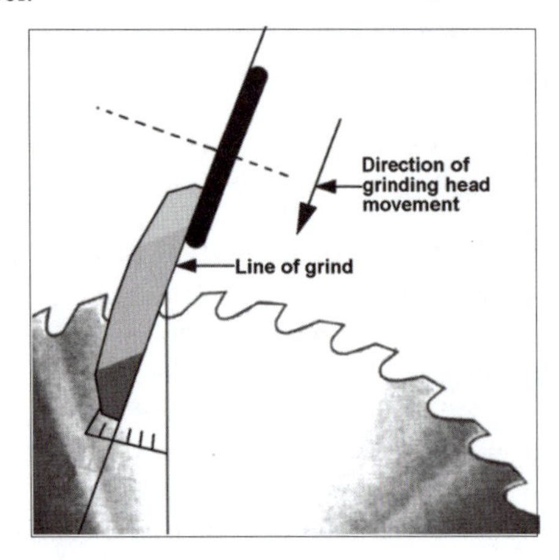

Fig. 6.13

Having marked the required hook angle on a sawblade, the gulleter is set by tilting the head until the grinding wheel aligns as shown, using a straightedge against the flat grinding wheel.

For a negative hook angle use the same general method, but count teeth in a clockwise direction to find the initial 30 degree angle.

An alternative way of finding the required off-set relative to the hook angle and diameter of the sawblade being measured is to use the Nomogram Fig. 16.74, Chapter 16.

Grinding a square tooth front

All gulleters will grind the tooth front square for ripsawing. In this case all teeth are gulleted at the same machine setting.

On those machines which allow a front bevel to be formed by swivelling the saw support in a vertical plane, make sure that the grinding wheel is square to the saw plate in plan view when grinding ripsaw teeth.

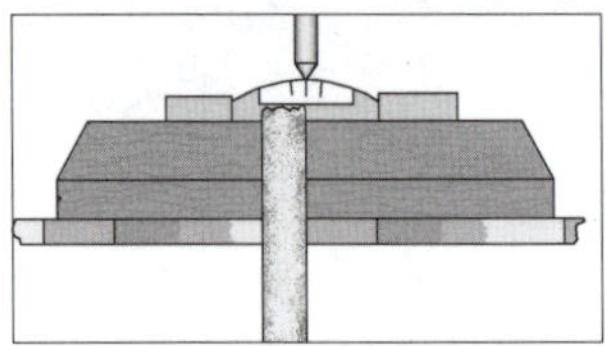

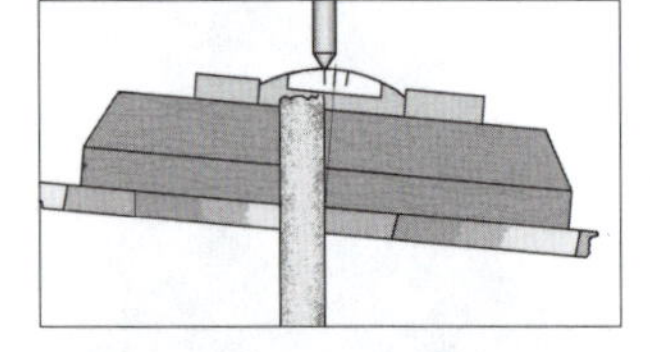

Fig. 6.14

Many gulleters swivel on a vertical axis to form a front bevel angle. Make sure that the machine is set square for grinding ripsaw teeth, as shown on the left.

For cross-cut saws the saw mounting is first swivelled by 5 degrees to one side and alternate teeth gulleted (right). It is then swivelled to the same setting, but to the opposite side, and the remaining teeth gulleted.

This shows a saw mounting which swivels on the sawblade centreline, other types may not.

Grinding alternate front bevel saws

When grinding sawblades with an alternate front bevel, such a cross-cut saws, every alternate tooth is first ground to one front bevel angle setting, then the saw support is swivelled in the opposite direction and the remaining teeth ground.

On most gulleters the support swivels on the saw centreline - if it does not, then a correction has to be made to the hook angle when setting for the alternate bevel - otherwise the hook angle on following teeth will be marginally different.

The grinding operation.

For all grinding and dressing operations it is ESSENTIAL to wear eye protectors, preferably the full-face type.

Before starting to grind ,dress the abrasive wheel to clean it and shape it to the correct profile - see previous notes and Chapter 7 for details of abrasive wheel truing and dressing.

When grinding a sawblade which already has a reasonable tooth profile - the result of regular gulleting - first bring the wheel carefully and lightly onto the heel of the tooth, then grind down the tooth back and into the gullet. *(By grinding down from heel to gullet the abrasive wheel tends to move the saw clockwise - so that the operators left hand merely has to steady the saw movement at this stage.)*

Fig. 6.15

To use gulleter, the right hand controls the grinding head and the left hand shifts and controls the saw.

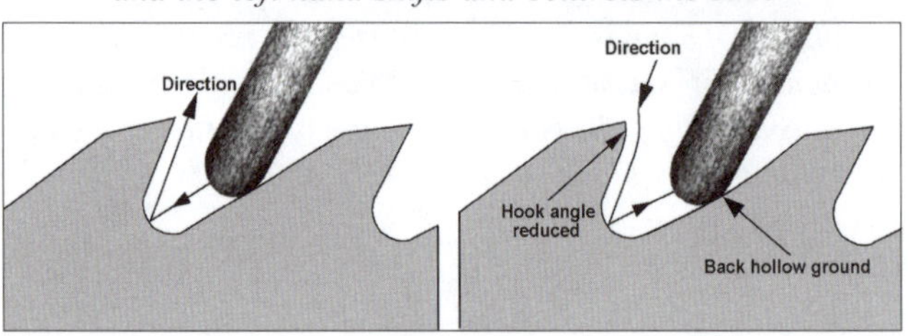

Fig. 6.16

Left, the arrows show the direction of grind recommended. To give the best gullet shape, grind down the tooth heel and up the tooth front clear of the tooth point.

Right, this shows the possible effect of grinding in the reverse direction to that recommended. The arrows show the direction of the grind, starting at the tooth point and finishing at the heel. By grinding in this way the tooth front tends to become rounded and the back hollow.

Once in the gullet, shift the sawblade with the left hand so that the front of the tooth lightly contacts the left face of the wheel. Holding the sawblade steady, allow the grinding wheel to rise to face-grind the tooth.

Make the movement down the back and up the front a continuous one - do not hesitate in the gullet.

Experienced operators quickly develop a rhythm in repeating the sequence described, and are able to form acceptable tooth profiles very quickly.

The method of gulletting practiced in some shops is an attempt to imitate the action of the automatic grinder. On these the grinding wheel is brought down the tooth front and up the tooth back - but this action is awkward when grinding manually.

Often the tooth front becomes rounded-over because a heavier grind commonly takes place where it starts - at the tooth point - and the back can become hollow. It also makes it difficult to form a consistent tooth profile - it demands perfect coordination between the left hand shifting the sawblade and the right hand controlling the grinding wheel.

Grinding Techniques

The abrasive wheel must never be allowed to dwell at any point during the grinding operation - otherwise the steel will overheat and change colour*. When dealing with a badly-shaped sawblade, re-form the gullets at two or more attempts. If excessive grinding is needed, do this as a series of small steps rather than one heavy-handed attempt - this will take longer but is worth the extra time spent.

**This is known as 'blueing' - although the actual colour varies from light straw to dark purple. The discolouring can easily be removed with acid or abrasive paper, but the local heating and rapid air-cooling this creates actually case-hardens the steel at this point.*

Blueing tooth tips makes filing virtually impossible, so a further light grind or heat treatment is necessary to remove the hard casing.

Bluing in the gullet creates the right conditions for starting cracks so, in all cases, grind lightly and take care to use a suitable grinding wheel. Incidentally, if blueing occurs too readily, even with a light grind, then you may be using the wrong grit or grade of grinding wheel - see Chapter 7.

Heavy gulleting and the excessive heating this causes can also stretch the rim of the sawblade. This reduces the tension slightly and may possibly make the sawblade "fast" under the tooth and cause it to "run" in the cut. In this case the saw has to be re-tensioned - creating extra work and incurring unnecessary expense. Light grinding avoids all these pitfalls.

Saws with badly shaped gullets

Fig. 6.17 shows a sawblade which has been badly file-sharpened and rarely gulleted. The profile can be corrected on a hand-operated gulleter - if care is taken - and provided the tooth pitch is still correct.

Sawblades which are badly out-of-pitch should be re-toothed using a tooth punch and dividing disc.

It is better to re-shape irregular gullets in two stages. First clean-up the tooth front by plunge-grinding down the face and into the gullet until the correct depth is reached. The tendency to blue the steel and glaze the wheel are both reduced if the grind takes places as several, slightly staggered grinding steps. Once at the correct depth set the depth stop. With the grinding wheel in the gullet, carefully shift the sawblade to make light contact between the tooth face and the grinding wheel then raise the wheel in a straight line clear of the tooth point to finally clean-up the front.

During the process of gulleting around the full saw, the grinding wheel will need dressing several times and will also reduce in diameter.

As a result the last gullet will be shallower than the first, so a final grind of all gullets is essential to bring them all to the same depth. (If the gullet depths are left uneven the sawblade will run out-of-balance.)

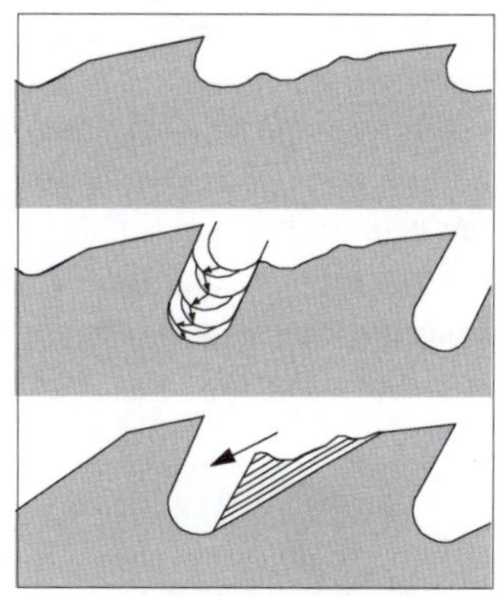

Fig. 6.17
Top. *This shows typical badly formed teeth.*
Centre. *The first stage in treating badly-formed teeth is to fully deepen the gullet at the tooth front. Plunge-grind this step by step, alternately to one side and then the other.*
Bottom. *Finally fully form the tooth back by a series of light traverse-grinds towards the gullet.*

The final operation is forming the backs of the teeth. Starting at the tooth heel, gradually remove the excess steel by several passes down the tooth back and into the gullet. Following this a full profile-grinding of the tooth back, gullet and front should be carried out.

For badly-shaped cross-cut saws follow the same general procedure, initially grinding the tooth fronts square, then form the front bevels as a final pass.

Top grinding

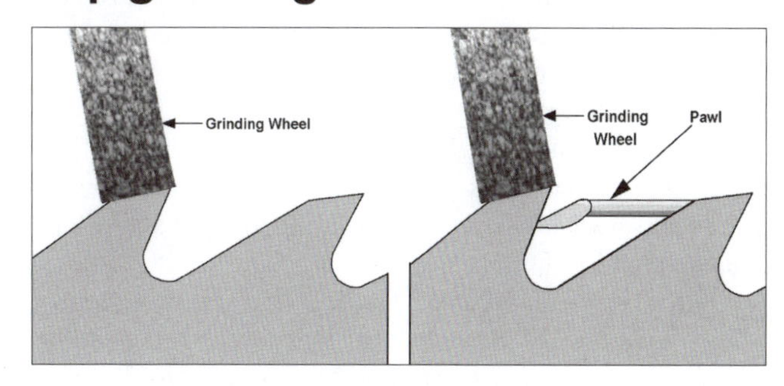

Fig. 6.18
Left, topping is possible by using a wide, square-edged grinding wheel. Although faster than filing, the process is prone to vibration, so only teeth with narrow tops should be attempted.
Right, using a pawl to position the tooth when topping makes the operation faster and much more accurate.

It is possible to actually grind the tooth tops using a wide stone with a flat face - but **only if one can be fitted and run safely, of course**. Grinding a wide area such as the tooth top can vibrate the saw plate excessively and possibly *'bounce'* the grinding wheel - chancing grinding wheel disintegration - in which case abandon the idea. If this method is acceptable and safe, however, the best way is to lower the sawblade slightly from the regular gulleting position and use a front clamp to reduce saw body vibration to a minimum. Set for the required clearance angle by tilting the head.

The hook angle graduations in this case actually show the clearance angle. The ground angle can be checked (after removing the sawblade) by using a simple protractor. Place the under edge of the protractor on the top of the tooth and point the left-hand edge of the arm towards the saw centre. (Mark a radial line to the ground tooth before mounting the saw on the grinder.)

Otherwise use a steel rule against the left-hand edge of the arm, positioning it radially by setting the contact edge a distance equal to half the bore diameter from its facing edge.

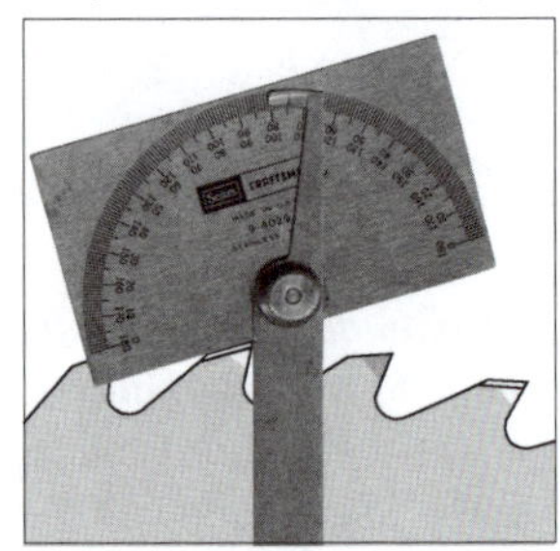

Fig. 6.19
Measuring the clearance angle. The left-hand edge of the arm should be radial - in line with the saw centre.

Using a pawl

To consistently position teeth, and so ensure tooth-height accuracy when using a regular depth stop, fit and use an adjustable pawl. Shift and hold the tooth front in contact with the pawl when topping - this is helped by the grinding action.

The pawl should be spring or weight-loaded, and set to contact the tooth front just below the point.

In positioning the sawblade for the next tooth, turn it so the pawl rides up the back of the preceding tooth to drop off the point and into the gullet space. It only then needs a small movement in the opposite direction to bring the next tooth front against the pawl ready for tooth topping.

As with gulleting, it is better to make a series of light grinds on the tooth tops rather than attempt a single heavy grind - and always give a final light grind to bring all the tops to a common height.

It is essential to dress the grinding wheel regularly when topping as this tends to wear a groove in the grinding wheel which, if left untreated, can round-over the tooth point.

If it is impractical to use a pawl, then the saw could be stoned and top-ground to the stone marks using a similar technique to that described for hand filing. In this case the depth stop setting is ignored and the witness marks are instead used as a guide.

Grinding alternate top bevels

To grind alternately top bevelled teeth first off-set the grinding head horizontally to the front in order to grind the first tooth top-bevel *'uphill'*.

The amount of horizontal offset needed to grind a specific top bevel angle varies with the grinding wheel diameter. As a general guide, off-setting a grinding wheel of 230mm (9in). diameter by 20mm (4/5 in) gives a top bevel angle of 10 degrees.

Grind all the teeth of one hand first, then re-set the machine for the opposite hand. It is not recommended to off-set the grinding head to the rear for the remaining teeth as the grinding wheel would grind *'downhill'* and create excessive vibration - a dangerous practice. To still grind *'uphill'*, reverse the saw, tilt the head in the opposite direction and fit the pawl to the opposite side of the grinding head.

Automatic machines grind quite safely 'downhill', but in their case the grinding wheel is very narrow, and the movement is across the tooth top instead of a plunge grind into it, so there is not the same tendency to vibrate.

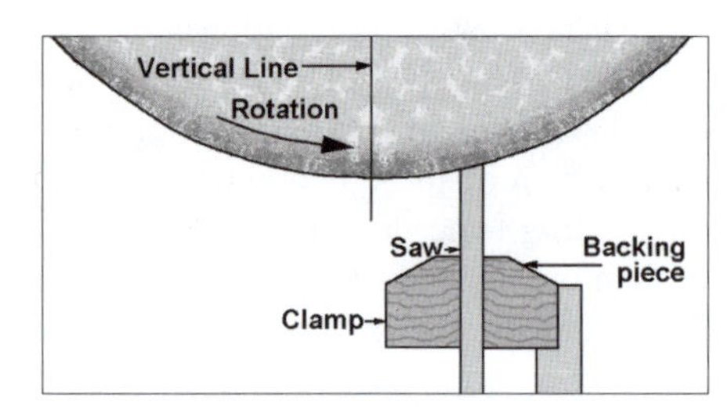

Fig. 6.20
Off-setting the grinding wheel horizontally to the front allows a top bevel to be formed.

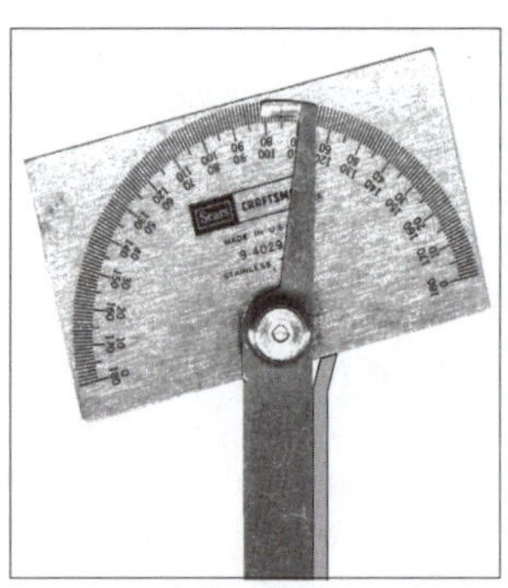

Fig. 6.21
Measuring the top bevel angle using a Sears protractor.

The two sets of teeth could be ground to a different height when dealing with alternate top bevel angles - to cause the saw to 'run' when cutting. To avoid this, stone the saw square across before removal, then slowly increase the grinding depth (by progressively adjusting the depth stop) so as to barely remove the stone marks on the first tooth of each hand.

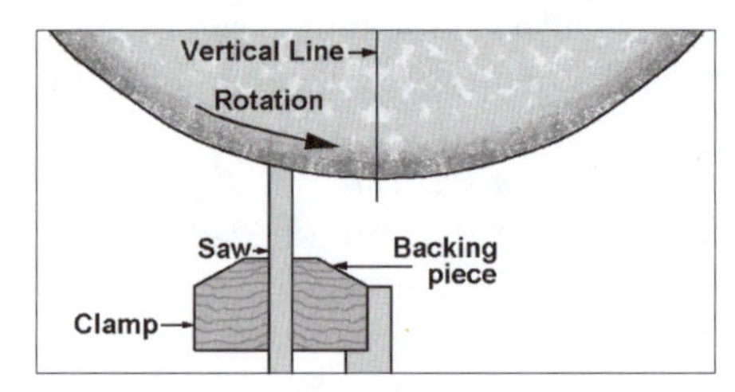

Fig. 6.22
Grinding 'downhill' for the opposite-hand teeth with the grinding wheel off-set to the rear causes excessive vibration and could be dangerous. Preferably reverse the saw in the clamp, tilt the head in the opposite direction, and fit the pawl on the opposite side.

As an alternative to stoning before the sawblade it is removed, mount the saw on the gulleter in the usual way and set the grinding wheel absolutely central to the saw body. Lower the grinding wheel against a depth stop set so the grinding wheel barely touches the tooth points then, holding it in this position, rotate the saw so the grinding wheel grinds several tooth tips. As this is equivalent to stoning the saw, setting the depth stop is then done relative to these 'stone' marks.

Automatic Grinding Machines

Automatic grinding machines for circular saws are made in many different sizes and styles. Alloy saws are much less popular than at any previous time, so few manufacturers make dedicated automatic grinding machines for them. However, automatic saw grinders intended for grinding wide bandsaws are still in wide demand and, as circular and bandsaw teeth are similar in many ways, some manufacturers adapt their wide bandsaw grinders to also grind circular saws.

Fig. 6.23
This Iseli AS circular saw grinder grinds both straight and alternate-bevelled teeth on sawblades of up to 1500mm (60in) diameter. Unlike some early machines of this general type, the AS grinds the tooth front straight and the top bevelled.
Source - Iseli

The conventional tooth for machine grinding has a continuous, round-back, profile, see Fig. 6.4. This wears the grinding wheel fairly evenly - so little wheel dressing is normally needed.

Some machines also allow teeth to be ground with a file-type straight-back profile. Although this gives a marginally stronger backing to the tooth tip, a straight-back is not essential - and it results in irregular grinding wheel wear.

It is not the intention to cover machines individually as they are well documented by their own instruction books - and, in any case, specific guides are limited to specific machines. However, the following general notes apply to almost any type of automatic saw grinder.

Sequence of operation

Automatic grinders mostly grind in one continuous movement down the tooth front to the gullet, up the tooth back and then finally across the tooth top. The grinding wheel then lifts to allow the pawl to move the following tooth into position.

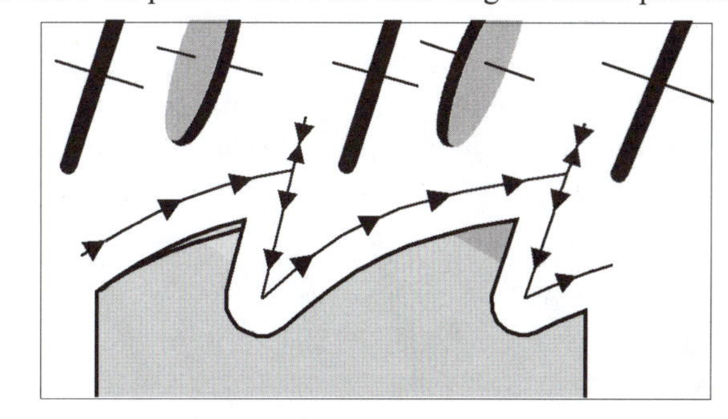

Fig. 6.24
This shows a typical sequence followed by an automatic grinder in grinding a round-back tooth profile.
The lines and arrows show the movement of the grinding wheel and its relative direction.
The sketch appears to show the grinding wheel moving across the sawblade - which it does not - the pawl moves the sawblade. This and other similar sketches are merely a convenient means of showing the mode of operation.
The grinding wheel swivels to form alternate top bevels, but remains square-on for the tooth fronts.

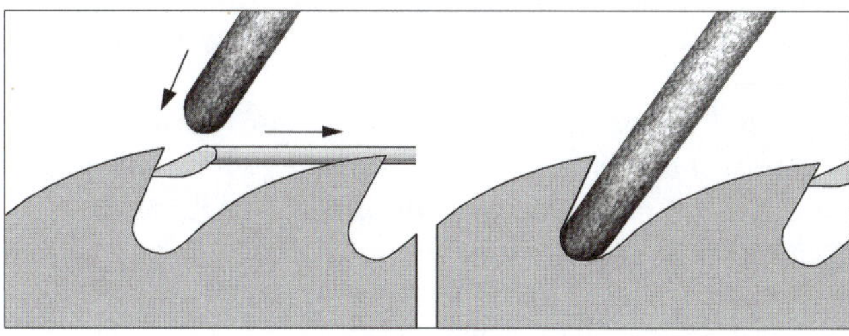

Fig. 6.25
Left, the starting sequence - the pawl has moved the next tooth into position and starts to withdraw, then the grinding wheel moves down to dress the front of the tooth. Right, the grinding wheel passes fully down the tooth front to form the gullet profile.

The grinding wheel has (usually) a straight line angular movement into and out of the gullet, together with a twisting movement when grinding alternate-bevel tooth tops. The rise and fall of the grinding head is coordinated with the traverse movement of the pawl to create the tooth back and top profile.

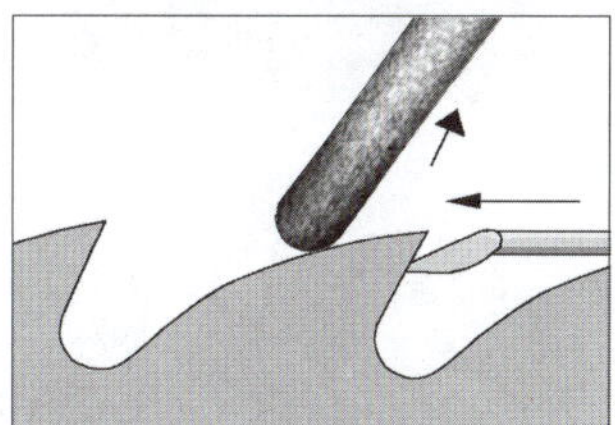

Fig. 6.26
Finally the tooth back and top are profiled under the dual action of the grinding wheel being raised, and the pawl moving the tooth to the left - both under cam or similar mechanical control.

The gullet shape itself is mainly determined by the grinding wheel profile - so it is up to the operator to keep the grinding wheel properly shaped as well as free-cutting.

The tooth profile is controlled by the grinding head and pawl movements. The two movements are linked mechanically to ensure consistency of tooth shape, and may be operated by one or more cams working together.

To change the profile the cams are either switched, adjusted or moved into or out of contact - how this is done depends upon the make of grinder. Fine adjustments are also provided to customize the tooth profile to individual sawblades and operator preferences.

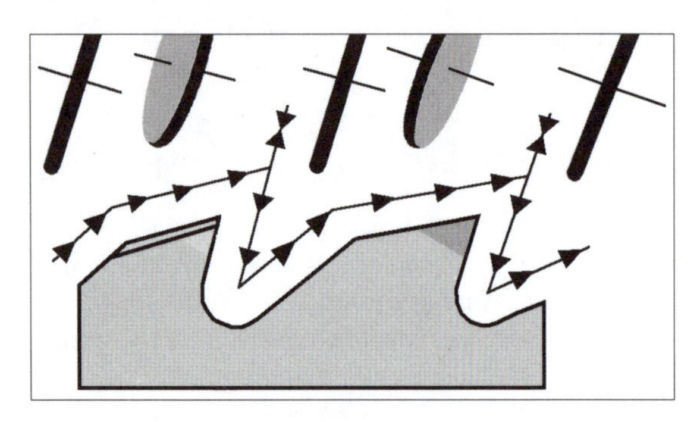

Fig. 6.27
Grinding a traditional straight-back tooth profile. Lines, arrows and grinding wheels are as in Fig. 6.24.

Some machines are simple to change-over, perhaps using levers to switch profiles, others not so easy, maybe needing cams to be physically switched - and not always all that conveniently!

Before purchasing an automatic saw grinder find out how easy it is to switch between the tooth profiles you mostly use, and how accurately these profiles can be repeated. Getting this basic requirement wrong can waste a lot of time, effort and sawblade life.

A further point to watch is how tooth pitch and depth is adjusted. On some machines one adjustment affects the other and makes getting the right tooth profile at the required depth and pitch very difficult indeed.

Fig. 6.28
This close-up of an Iseli AS shows the ready access to levers and controls which determine the tooth profile.
Source- Iseli

Grinding wheel types

Flat grinding wheels, as used on early machines, often wore unevenly at the contact point between the wheel and the tooth face. Wear tends to form a slight unevenness on the side of the grinding wheel to possible reduce the hook angle of the tooth over time, and eventually this may even begin to slightly radius tooth tips. To avoid this, regularly true-up the left-hand side of flat grinding wheels to slightly beyond the tooth depth - even though this also reduces their thickness.

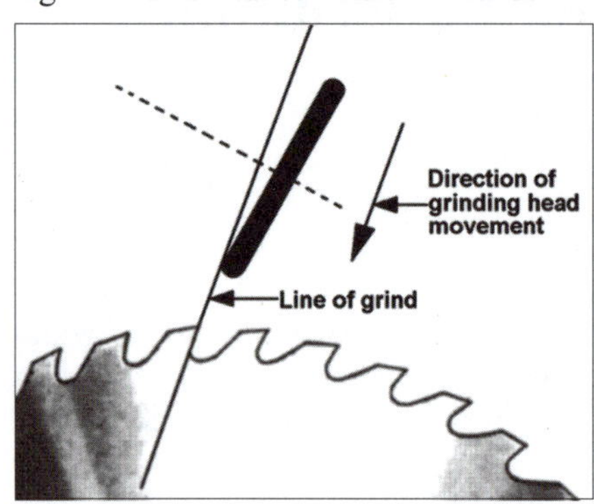

Fig. 6.29
Tilting the grinding wheel arbor gives increasing clearance between the tooth front and the body of a flat grinding wheel beyond the radius. This ensures that the grinding wheel never looses thickness through wear or regular dressing.

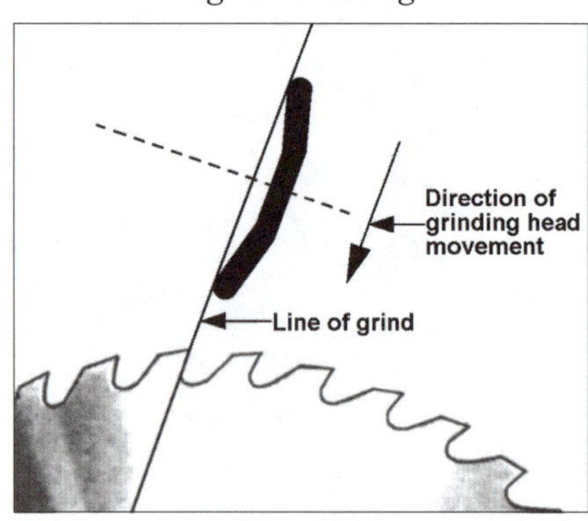

Fig. 6.30
Using a saucer grinding wheel also gives working clearance between wheel body and saw tooth front.

Machine makers realized the problem many years ago and have overcome it in one of two ways:- One is to set the grinding wheel arbor a few degrees out of square to the slideways on which the grinding head moves. This allows a flat grinding wheel to be used which effectively has a growing clearance between the grinding wheel and the tooth front beyond the wheel rim.

The other method is to set the grinding wheel arbor perfectly square to the slideways, but fit a saucer grinding wheel. This gives the same effect as an inclined-arbor machine using a flat wheel, but requires special grinding wheels. *(Flat or plain grinding wheels are obtainable from a wider range of suppliers than saucer wheels.)*

When grinding the tooth front, both types of grinding wheel make contact only with the rim - never with the inside face of the wheel. This has several advantages - it avoids irregular wear at the side, and the wheel retains an acceptable shape throughout its life - without losing thickness. Also, because of the growing clearance, it is in actual contact with steel for a shorter time - so the chance of blueing the sawblade is very much reduced.

The grinding wheel used should always be matched to the gullet profile it is intended to form. See Fig. 6.6, 6.37 & 6.38, also *'Grinding Wheel wear'*.

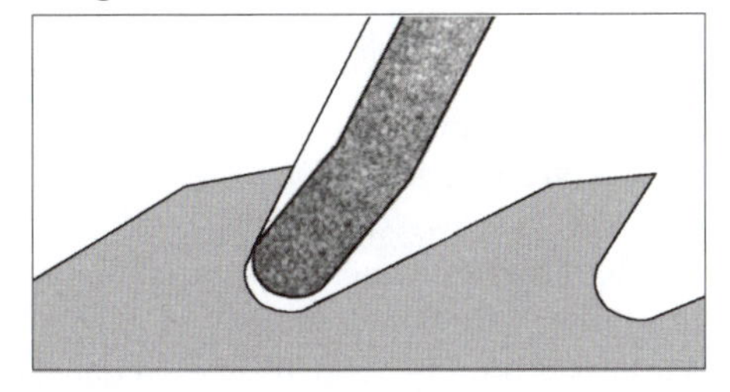

Fig. 6.31
Tilting the grinding wheel arbor or using a saucer grinding wheel gives the clearance needed to avoid wear on the side of the wheel

Mounting the saw

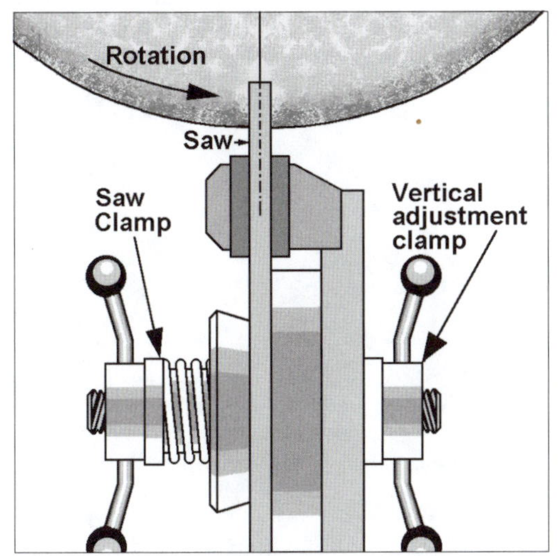

Fig. 6.32
Side view of a typical cup and cone. This type will accurate centre sawblades within a given range.

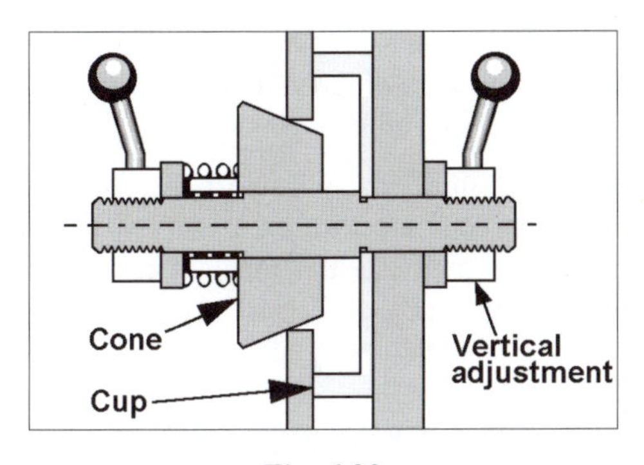

Fig. 6.33
Section through the cup and cone shown in Fig. 6.32.

An automatic grinder forms teeth absolutely concentric with the sawblade bore. For this reasons it is essential to ensure the saw bore is properly centered on the mounting device - which may be a hardened bush or a cup and cone. On some machines the mounting device rotates with the sawblade, either on a plain or ball bearing arbor. On others the sawblade bore rotates on a fixed arbor, see Figs. 6.2 & 6.3.

NOTE: Automatic saw grinding assumes that the saw mounting arbor on the sawbench is free of deviation or wear. If either is present the sawblade will not run true - and the only action to cure this is to make the machine mounting good. If nothing is done the sawblade will ***never*** *perform as efficiently as it should.*

With automatic grinders 'stoning' *is unnecessary as the grinding process guarantees that all teeth share the work equally. However, if machine setting is unreliable when grinding alternate-bevelled teeth, light stoning can be used to ensure they are ground to equal heights - see later details under* 'Grinding top bevelled teeth'

Centering the grinding wheel

The centre of the grinding wheel arbor *(when set square)* must be directly above the centreline of the saw body. If the setting is wrong then square teeth will be ground instead to a top bevel angle, and top bevelled teeth will be ground to unequal heights. In both cases the saw will 'run' in the cut - lead to one side when sawing.

Correct alignment is made by adjusting either the grinding head or the saw mounting - which depends upon the individual machine. Automatic machines require this setting to be precise, so correction is needed when fitting a sawblade with a different body thickness. Some form of scale or graduation is provided for this setting on most machines, but it is wise to periodically check this for accuracy, as follows:-.

Mount any convenient saw in position, and set the saw-gauge adjustment scale to its exact body thickness. Draw a thin pencil line on the outer face of the wheel across its diameter (this is best done with the grinding wheel removed). With the grinding wheel replaced, lower the head so that the wheel just clears the tooth top*.

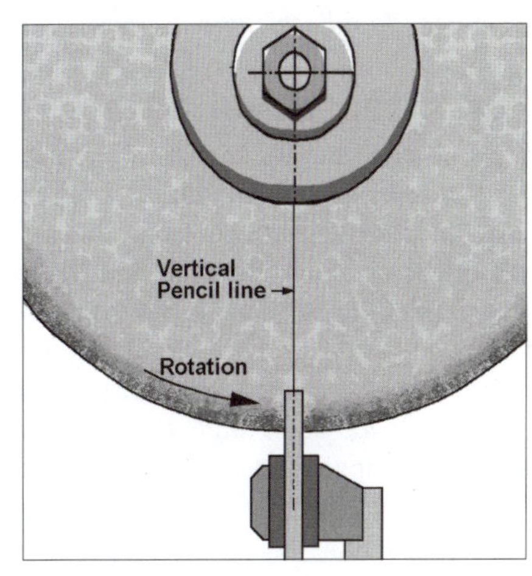

Fig. 6.34
Checking the alignment of the grinding wheel to the body of the saw

Rotate the wheel so that the pencil line is absolutely vertical *(use a plumb line, a spirit level, or a straightedge against the saw body)* then check if the pencil line is central to the saw body, Fig. 6.34 If the setting is correct then so is the setting scale, if not adjust the scale, if practical, or note the error for future setting.

**Most grinding heads rise and fall in a straight slide, so it doesn't really matter where you check alignment. On machines having an arcuate rise and fall movement from a pivot point, though, check only at the tooth top - this is where* *it really counts.*

The true test, though, is in the grind itself. After setting the machine as accurately as possible fully grind a 15^0 alternate top bevel saw, then make a test cut on a table saw using a piece of hardwood. (See Figs. 5.8 & 5.9) If both sets of teeth align squarely across, this shows that the machine is accurate on all counts - centering of the grinding wheel and equal top bevel angles from the twisting movement.

If the top bevel angles are seen to have unequal height, then check both the centering of the grinding wheel and the degree of twist of the grinding head. Most machines have a single adjustment to control the twisting movement in both directions, but others have individual controls.

Another way is to first stone the saw and note how the witness marks grind out. Start with a very light grind to leave the stone marks visible, then check the amount remaining on both left and right-hand teeth. If they are unequal make an adjustment to the machine settings and check again until they do match. When correct, reset the machine to remove the stone marks completely and grind as usual.

Setting the hook & clearance angles

On automatic machines the scale for hook angle setting is usually accurate enough to be relied on*. If in doubt, though, grind any suitable saw at any set hook angle setting, and then check the ground angle using either a protractor, Fig. 6.11, or the method shown on Fig. 6.12. The clearance angle setting may be a separate control or determined by the cams used and/or their fine setting adjustments. Finally check the clearance angle as shown on Fig. 6.19.

**In some cases the saw centre has to be off-set for the pawl to operate correctly. In this case the hook angle setting has to be corrected to take this into account.*

Setting the top bevel angle.

The top bevel angle is formed by a twisting action of the grinding head on an axis roughly in line with the hook angle. On newer machines the twisting action only takes place as the grinding wheel passes across the tooth top, returning it to square-on when grinding the tooth front - the correct combination for ripsawing. *(Some early machines did not have the square-on feature, and so formed an unwanted front bevel on all teeth ground to a top bevel angle.)*

The top bevel angle formed depends upon both the degree of twisting movement of the grinding head and the diameter of the grinding wheel. For this reason many adjustment scales give only a rough guide as to the actual angle formed - this actually increases as the grinding wheel wears. The accuracy of top bevel angle is not all that critical, though, but it is easily measured using a protractor, see Fig. 6.21.

Setting the pawl

Some operators set the pawl to index the tooth front next to be ground - this accurately aligns the tooth face with the grinding wheel even if the pitch varies tooth to tooth. By doing this heavy grinding of irregular-pitched sawblades is avoided - but any pitch variation remains indefinitely by doing this.

Variation in tooth pitch is common in alloy saws because of the way they are manufactured. Although a small pitch variation is actually no detriment to the operation of the sawblade - preferably, though, make the pitch regular.

Indexing the tooth next to be ground is practical with many tooth forms, but there is always the danger that the pawl will be caught and damaged by the grinding wheel - there is so little clearance between them.

To avoid this chance and to correct an irregularly-pitched alloy sawblade, set the pawl for a light face-grindwhile indexing one tooth back from that next to be ground. By grinding the sawblade through several revolutions in this way any initial irregularity will progressively be ground out. Thereafter use the same setting to maintain a regular pitch as the sawblade wears down and the pitch slowly reduces.

The front of each tooth must **always index to exactly the same position each time** - and there must be no backlash *(a slight backward movement of the sawblade when the feed pawl moves clear)*. If movement is noticed the sawblade clamp needs some attention - either to the clamp pressure, or by making good any springiness or looseness in it.

Fig. 6.35

The tooth profile is formed by a combination of grinding head and pawl movements. The pawl can be clearly seen on this Iseli AS. A magnetic clamp is used, so the indexing movement is guaranteed backlash free.

Source - Iseli

In place of a mechanical clamp, some makers provide a magnetic clamp. This does not allow backlash, and makes removing and replacing sawblades that much easier and simpler. It also leaves the front of the saw clear enough to allow a dividing disc to be used without problem.

Checking the pawl movement

Normally the pawl should set to contact the straight part of the tooth front just a little way down from the point (when set to index the tooth next to be ground). This position should be such to avoid the grinding wheel fouling the pawl as it grinds the tooth top.

With the head locked up in the rest position, start up the machine and note the action of the pawl in indexing the teeth. Check that, as the pawl returns in readiness for the next tooth, it remains well clear the grinding line when the grinding wheel passes over the back of the preceding tooth. If it appears too close, adjust the machine as necessary - otherwise a rogue tooth just might trap the pawl - especially when indexing the tooth next to be ground rather than one tooth back. Repeat this check when the machine begins to operate.

The contact point between the pawl and the tooth front should be consistent throughout the full indexing movement.

The pawl could index saw teeth badly if it moves up and down the face - or it could even slip off a tooth to cause a crash. A small movement is acceptable, but better if this can be eliminated altogether as the friction this causes will wear the pawl unevenly.

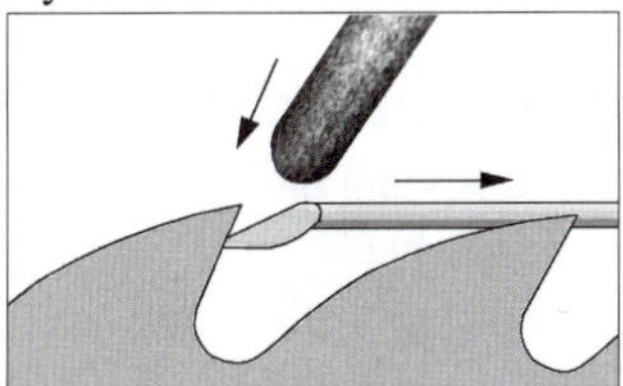

Fig. 6.36
This shows an acceptable point of contact between the pawl (when set to index the tooth next to be ground) and tooth. Check that the pawl moves as little as possible up and down the tooth whilst indexing.

The path of the pawl when indexing is sometimes controlled by a cam fitted alongside it. If so, there may be more than one cam available - each one to suit a range of saw diameters - so make sure the right one is fitted.

If the path of the pawl cannot be set to eliminate the up and down movement entirely, then at least make sure that the pawl contacts the same point on the tooth front at both the beginning and end of the indexing movement.

Off-centre sawblade mounting

Some machines have a horizontal* straight-line pawl movement only. With these the saw centre has to be shifted sideways** so that the contact points with the teeth at the start and finish of the pawl movement are in the same horizontal line.

The reason for this is to reduce, to an acceptable minimum, creep of the pawl up and down the tooth during the indexing movement.

**The pawl movement could, in fact, be at a slight angle - in which case the start and finish contact points on the sawblade teeth should likewise be at the same angle.*

***To do this off-set the mounting cone or arbor from its regular position, directly below the grinding wheel, instead towards the pawl. Record the amount of off-set needed for each type and diameter of sawblade used, also the amount by which this then alters the effective hook angle. This will give the necessary information to quickly re-set the grinder.*

Grinding techniques

Preferably grind all badly-profiled sawblades at two or more attempts - this is easier both on the machine and on the sawblade itself - and results in less grinding wheel wear.

Heavy grinding loosens grinding grit prematurely and leads to rapid loss of grinding wheel diameter. This starts a downward spiral - the grinding wheel needs re-shaping more often, grit loss accelerates, and repeated grinding is then needed to make good the uneven tooth heights that such excessive wheel wear produces.

However much is done to prevent it, grinding wheel wear always takes place progressively as any sawblade is ground - so the final tooth is always fractionally higher than the first. This is easily rectified by a second light grind to bring the teeth truly concentric with the saw bore - fail to do this and sawing efficiency with suffer.

Automatic downfeed

On some modern machines grinding wheel wear can be largely offset by an automatic downfeed system of the grinding head so the final tooth is always the same height as the first - at least in theory. The timing and amount of downfeed can be pre-set by the operator who, through experience, should be able to guesstimate what is needed to off-set wheel wear relative to the individual saw being ground, the amount being ground off and the grinding wheel used.

Automatic downfeed allows the grinder to operate at the maximum allowable depth of grind - without the concern that this could result in uneven tooth heights.

It also allows the operator, once the machine is working to his satisfaction, to see to some other urgent work leaving the grinder to get on with its task without further attention. .

The experienced operator will always be aware of the way the machine is performing. By listening to the repeating rhythm of the grind he will be instantly alerted to any change in pitch which might indicate that something is wrong.

After grinding check that all the teeth are properly ground and, if not, re-grind the saw fully - after making enough adjustment to finish-grind any section previously skipped.

Grinding wheel wear

If heavy grinding takes place, or if the machine is run without dressing the wheel, there is a chance of the grinding wheel wearing unevenly, perhaps to form a stress-inducing internal corner in the gullet. To avoid this dress the grinding wheel regularly to a shape to suit the gullet profile.

Preferably use a thicker grinding wheel than actually required for the gullet profile, this keeps its shape better and is much more economical in the long run. Shape the grinding wheel as described below:-.

Grinding round-back teeth

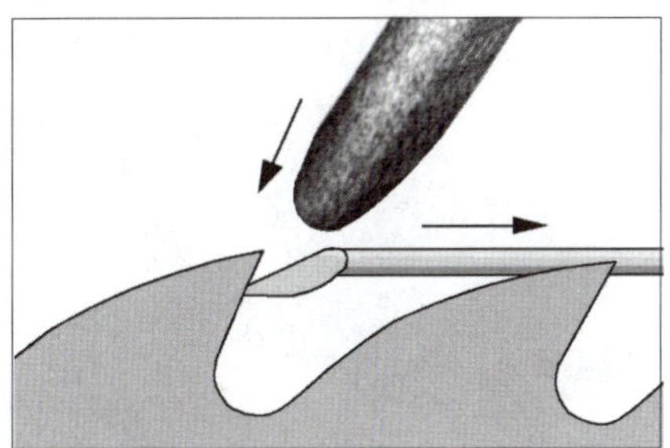

Fig. 6.37
A suitable grinding wheel shape for round-back teeth should be roughly 1/3 thicker that needed for the actual gullet profile, and with the radius blending into flats at both left and right-hand side..

Automatic saw grinders perform best when grinding round-back teeth because the grinding wheel retains its original shape for a longer period and with less dressing than needed for straight-back teeth. This is because the point of contact of the grinding wheel changes as it follows the curved back - so that at no point is there a straight section to wear the wheel excessively.

A grinding wheel for round-back teeth should be thicker than actually needed to suit the gullet profile, with a short flat blending into the radius at the left-hand side parallel with the hook, and a short flat or slightly curved section blending into the radius at the right-hand side. This type of shape will show least wear and need less dressing than a thinner wheel with a simple radius.

A grinding wheel with an 'ideal' profile will virtually become self-dressing* - needing only occasional attention from the operator. This is what should be aimed for - even if this then needs the gullet profile modifying slightly to suit.

**If in doubt as to what form a self-dressing grinding wheel takes, allow the machine to continue grinding an old saw with round-back teeth - without dressing the wheel.*

By doing this a grinding wheel profile will eventually form through natural wear which will thereafter remain more or less stable. When this occurs make a steel template of this profile for use as a guide for future dressing.

Grinding straight-back teeth

With straight-back teeth uneven wheel wear will mainly occur where the grinding wheel contacts the flat back. To prevent this again use a wider grinding wheel, and form a small flat parallel with the tooth back to blend in with the radius. This profile has to be dressed more than with a round-back tooth form as it is rarely self-dressing.

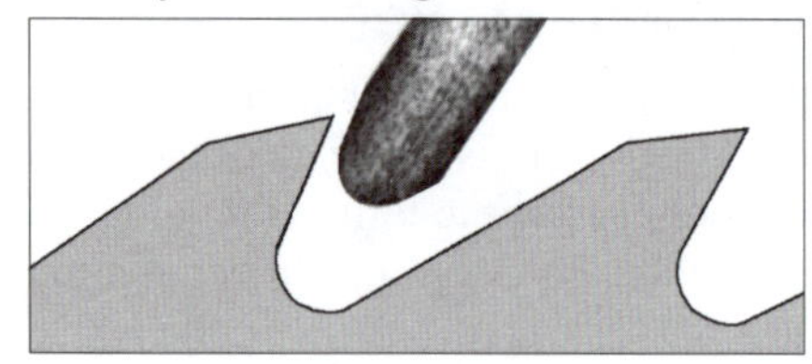

Fig. 6.38
A suitable grinding wheel shape for straight-back teeth has a central radius blending into two flats, the left-hand flat parallel with the tooth front and the right-hand flat parallel with the tooth back

Dressing the grinding wheel

Fig. 6.39
The Armstrong ShapeUp grinding wheel dresser has adjustments to control both the grinding wheel radius and its position on the grinding wheel. It gives precise control and an absolutely repeatable gullet shape.
Source - Armstrong

Dressing a grinding wheel to the recommended shapes needs skill on the part of the operator. Most machine makers recommend using a hand dresser for this - hardly the proper tool.

To be practical, though, dressing the grinding wheel should be mainly to remove glazing and filling - rather than to actual shape it (when grinding round-back teeth).

Square tooth grinding

For square teeth the grinding head must be set directly above the centerline of the sawblade.

Before starting the machine ensure that the grinding wheel is clear of any part of the sawblade tooth by taking it manually step by step through the operating cycle, making adjustments in the process, then lock the grinding wheel in the top rest position.

Start the machine *(or manually crank or jog it)* so the pawl indexes for the first tooth and then draws clear. Carefully lower the grinding wheel into the gullet to check its setting - it should barely brush the tooth front*. If it does not, then adjust the feed pawl accordingly and index and grind another tooth face until only light contact is made.

**When the pitch is already regularized - otherwise set slightly clear and progressively increase this as initial grinding takes place.*

Set the lowest position of the grinding wheel to barely clear the gullet base. Start the machine on automatic feed, carefully lowering the grinding wheel until it reaches the correct operating position.

During this time make sure that the grinding wheel does not foul the pawl, and check on how accurately the grinding wheel follows the tooth profile*. Make adjustments in small, careful steps until the grinding wheel makes light contact along the full tooth profile.

Once initial settings are correct, tweak them until full contact is made. Mark the following tooth to indicate when the sawblade has been fully sharpened.

**If the grinding wheel cannot be adjusted to follow the tooth profile precisely then another cam may be needed. On machines where cam switching is inconvenient, though, the easy way out is to grind teeth to suit the cams already fitted to the machine.*

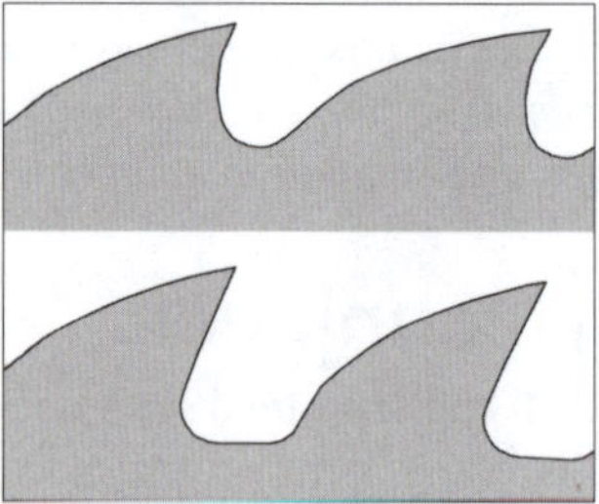

Fig. 6.40
These show two common grinding faults.
Top: *In this case the abrasive wheel has not reached the set depth down the front of the tooth before the pawl starts the indexing movement. This eventually forms a weak* back.

Lower: *Making the pawl start too late, or grinding too deep a gullet, creates this gullet shape - and a weak back.*

Alternate bevel tooth grinding.

This is basically the same process as that described for grinding square teeth but, in setting, make sure that the appropriate teeth are indexed relative to the top bevel angle grind for which the machine is set.

Check this first with the grinding wheel raised and the alternating swivel action engaged. If wrong, move the sawblade forward one tooth manually, then check again before starting-up properly. Machine start-up and operation is otherwise similar to grinding square teeth.

Grinding wheel centralization and the degree of twisting movement to form the top bevels both need to be absolutely correct when grinding alternate-bevel teeth. If either is wrong, then alternate-bevel teeth will have unequal heights to make the saw 'run' when cutting. See previous notes on 'Centering the grinding wheel'.

Maintaining tooth profile

Grinding should be such that the tooth profiles can be maintained without distortion - by correctly balancing the amount ground off the tooth face and back.

One way of doing this is to assume a line of grind that the tooth point must follow as grinding takes place over time. A suitable line might be roughly 15 degrees greater than the hook angle - roughly 50 degrees to horizontal for softwood ripsaws. Keeping to this line will maintain a true tooth profile. Fig. 6.41 shows acceptable lines of grind.

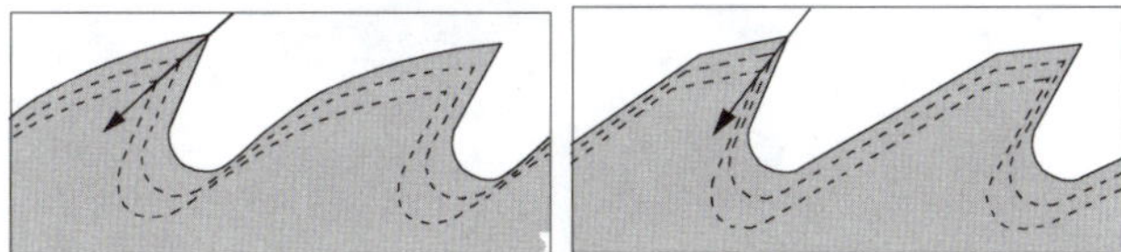

Fig. 6.41

When a suitable line of grind is taken on a round-back tooth, left, the correct profile will be maintained throughout the life of the sawblade. The correct line of grind for a straight-back tooth is shown on the right.

If the balance is wrong, then the tooth will eventually lose shape - even though the grinding wheel will still follow the correct profile. Figs. 6.42 & 6.43 shows what can happen.

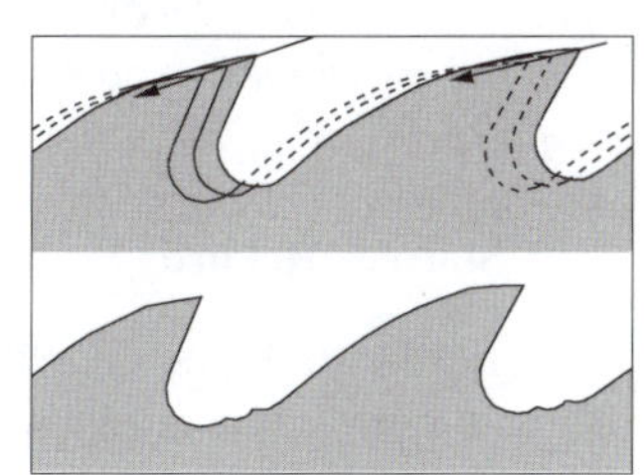

Fig. 6.42

If the line of grind is too low the tooth back will be missed to eventually become distorted as the lower sketch shows.

If the line of grind is too low. The shaded outline shows the original tooth shape for which the machine has been set. As the line of grind is too shallow, subsequent grinding follows the paths of the broken lines. These are still true to the original profile, but the grinding wheel contacts the tooth back only partially - to leave some sections unground. This gives both a slightly weakened tooth and a distorted gullet that is less efficient in ejecting sawdust.

Many operators alter the feed pawl setting when they see a black, unground section forming on the back of the teeth. What this does, though, is to stop grinding on the tooth front and begin to grind on the back only - to widen the gullet and so marginally weaken the tooth.

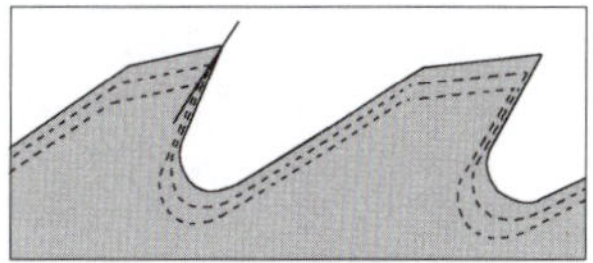

Fig. 6.43

Too steep a line of grind wastes sawblade life.

If the line of grind is too steep then a heavy top grind is needed to maintain the profile - and this wastes both sawblade and grinding wheel life.

It takes only a small amount of incorrect adjustment to lead to these defects. They do not show up with a single grind.

Only by consistently grinding too heavily on either the tooth top or the front do these problems actually show up.

Where doubt exists it is an advantage to scribe an acceptable line of grind on a single tooth, then check every two or three grinds that this is actually being followed.

Economic tooth profiles

Forming an economic tooth profile is much more important with alloy saws than with tipped types. An economic tooth profile is one which allows most grinding on the tooth face, with only a light skimming of the tooth back needed to maintain the profile. This reduces the sawblade diameter through grinding the top by the least amount, and so gives the longest practical life to alloy sawblades. Examples of extreme tooth shapes are shown in Fig. 6.44.

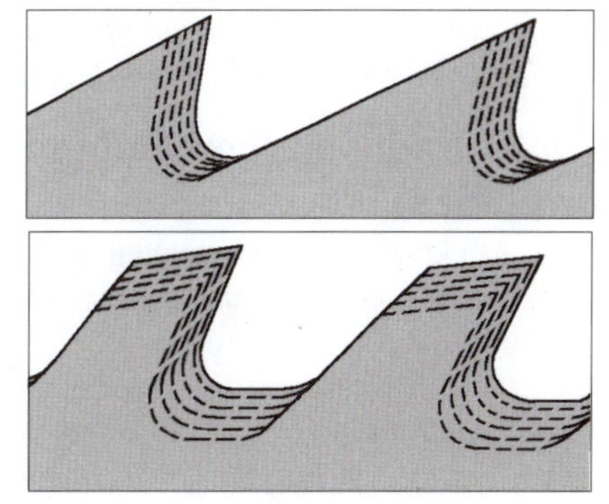

Fig. 6.44

Top, a straight tooth back running all the way to the point is the most economic profile for alloy sawblades. Most grinding is on the tooth front, so the sawblade reduces in diameter by the least amount.

Bottom, the most uneconomic tooth profile is when the tooth back is steep. This needs excessive grinding on the tooth top to maintain the tooth profile - and quickly reduces sawblade diameter.

Problem Saws

Grinding teeth with a varying hook angle

Normally the pawl should be set to engage the tooth front just below the point - the exception is when the sawblade to be ground has varying hook angles - for whatever reason. In this case set the pawl roughly halfway down the tooth front. In this position any difference in hook angle is averaged out between the tip and the root of the tooth.

Dealing with broken teeth

Broken teeth are not unusual in a rough mill. When teeth are indexed by a single pawl, however, a broken tooth may not be indexed at all by the pawl or, worse still, it may partially index and cause a crash.

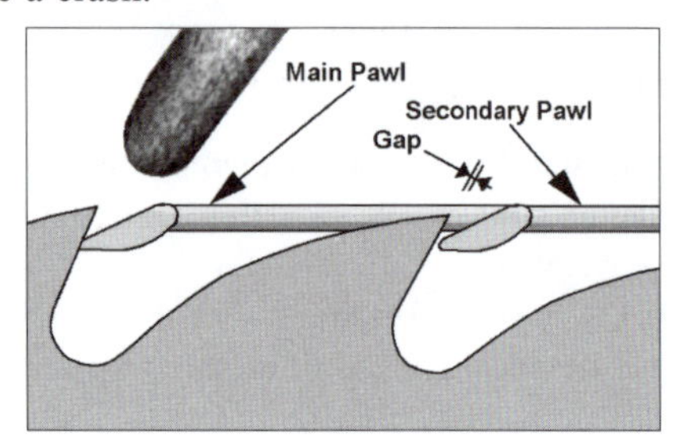

Fig. 6.44

When using a secondary pawl, make sure that a gap exists between this and the tooth it would otherwise index - checking this against the biggest tooth pitch.

Before grinding any sawblade check that all the teeth are sound and complete - this only takes a few seconds but could avoid costly damage. If the sawblade has one or more broken teeth then fit a secondary pawl to index one tooth back from the tooth indexed by the main pawl. This will continue the feed movement past the broken tooth without problem.

Make sure that the secondary pawl always has a slight gap between it and the tooth front it indexes when the main pawl is already in contact. If the sawblade has teeth which vary in pitch slightly, then make sure that the distance between the main and secondary pawl points is marginally greater than the longest tooth pitch in the sawblade.

Dealing with badly-pitched and group teeth sawblades.

Group teeth saws, and saws with an irregular tooth pitch, can be ground satisfactorily if the machine allows the use of a pitch plate.

Pitch plates are toothed like sawblades, and are fastened to the sawblade to be ground, allowing the pawl to index the pitch-plate teeth instead of the sawblade teeth thenselves. A pitch plate is regularly used for grinding metal-cutting sawblades, but few grinding machines for woodcutting sawblades have this feature.

The pitch plate must have the same number of teeth as in the sawblade, in which case the pawl is set to index each pitch-plate tooth. It could also have double or treble the number of teeth in the sawblade, in which case the pawl is set to index only every second or third pitch-plate tooth.

The pitch plate should be fixed to the sawblade to leave a slight gap between the two - so that the pawl can work correctly - and must also be smaller than the sawblade to avoid interference by the grinding wheel. The technique is similar to that when grinding a sawblade in the regular way - but using the pawl to index the pitch plate instead of the saw teeth.

Because the pitch plate may not correspond exactly with the existing saw tooth pitch when used for the first time, take care to avoid too heavy a grind on short-pitch teeth.

Initially set the machine to barely clear the tooth fronts and with extra clearance at the tooth back. Then, with the machine running, make careful and small adjustments to the pawl until all the tooth fronts begin to grind. Only then adjust the machine to begin to grind the tooth backs.

Caution is only needed when first grinding a badly-pitched sawblade using a pitch plate. Once the pitch has been regularized future run-in time is much less.

Some use a pitch plate periodically to regularize the pitch of sawblades which are pitched by indexing the tooth to be ground. Using a pitch plate should not be necessary when indexing one tooth back from that to be ground - indexing in this way will normally regularize the pitch.

Fig. 6.45

Using a pitch plate on a Rekord 400SR.
This machine will grind metal-cutting saws and bandsaws, also triangular-type teeth on woodcutting circular saws up to 35mm (1¹/4 in) pitch.
Source - Rekord

NOTE: Also see grinding Stellite saws, Chapter 8. This gives additional information on saw grinders and grinding.

Chapter 7

Hard Tips and Abrasives

Hard Facings

Many attempts have been made to hard-surface alloy saw teeth, but most have been unsuccessful. There are several surface hardening processes such as cyaniding, nitriding and tuftriding used in engineering. These change the thin outer skin of basic steel into carbon steel which is then hardened by heat treatment. These processes are not practical for sawblades, but they are used for surface hardening of machine parts that would otherwise wear too quickly.

Some sawblades in the past have been hot-sprayed using a variety of hard materials, but the friction of timber in sawing wears away such thin films fairly quickly and the tooth then reverts to the basic metal. Perhaps the only successful surface treatment is hard chrome-plating, which is used both on alloy and tungsten-carbide saws. It makes alloy teeth a little harder, but mainly it is used to reduce friction between sawblade and timber.

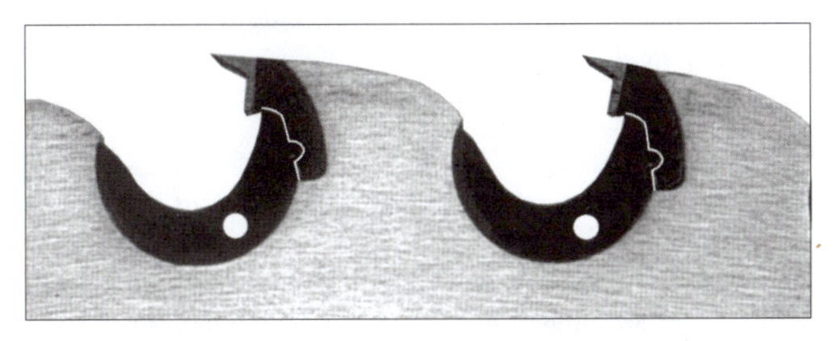

Fig. 7.1
The teeth of insert-teeth saws are of a different material to that of the saw body, being much harder and tougher. Different grades of tip are available.
Source - Simonds

Tip Hardening by Heat Treatment

Heat-treatment can be used to tip-harden alloy saws. The basic saw plate is hardened and tempered as part of the manufacturing process. *(Tempering actually reduces the hardness of the saw as a whole - without this it would be too brittle.)* It is possible, however, to heat-treat the tips only to make the teeth harder.

Any suitable heating method can be used, such as an oxy-acetylene blow torch, or resistance heating. The most controllable method is by using RF *(radio frequency)* heating via a coil around the saw tips *(very similar to the coils and method used to braze tungsten carbide tips - see later details)*. The process is highly successful with so-named hard-tip handsaws and flexi-back bandsaws.

Only the very tooth tips are hardened - heating too far down the teeth make them liable to snap off. When first hard-tipped, alloy saws actually have quite a long life, but it is in maintaining this run-time advantage that gives problems. Hard-tip saws have to be ground - filing is not practical. When ground, though, the hard surface is ground-off the fronts and tops of the teeth leaving only the sides hardened, so run-time is reduced. The main problem, though, is that hard tips cannot be re-set - they tend to break off - so set is gradually lost and the saw eventually starts to bind in the cut.

In order to make the process a repeatable one the tips first require softening by heating them above a specific temperature and then cooling slowly. The saw can then be sharpened and the teeth set before finally re-hardening. The full process, therefore, needs special heat treatment equipment - hardly justified when other types of hard tip give greater advantages and guaranteed results.

Hard Tips

Where the saw tip and body are separate metals held together mechanically, as in insert-tooth saws, the tips are always made of a harder material.

However, hard tips which are actually bonded to the saw body are now widely used rather than mechanically-held types *(see these details in Chapter 8)*. Several types of brazed-on hard tips are used, each with a different characteristic and each intended for a specific range of applications.

Which hard tip is the best to use ? This question concerns not only which type of tip, but in some cases which grade of tip to use within the chosen type. But the same answer applies - it depends upon the application.

Tip hardness and toughness

The wear resistance of any hard tip is in direct relationship to its hardness - no-one disputes this. The harder the tip and the longer the run-time *(the time period a saw can operate before re-sharpening is needed)*. A longer run-time reduces the production time lost when sawblades are changed - and more is produced before re-tipping becomes necessary.

However, hardness itself is of no use unless the tip is also tough. Sawing is a rapid series of impacts by the teeth into hard material - and this gives repeated shocks to the cutting edges. Without toughness a hard cutting edge is brittle and can rapidly break down whilst sawing and become incapable of further work. This effect is considerably worsened if the material being sawn contains hard or highly abrasive inclusions. All this adds up to the need of toughness in saw tips as well as hardness - more so as new, tougher materials are brought onto the market.

Saw tips vary widely in their hardness and toughness according to the materials used in their manufacture and in their production processes. Years of dedicated research has perfected the right combination of high hardness needed for long run-times, together with the essential toughness necessary for the arduous work of sawing modern materials. Not surprisingly modern hard tips demand complex and sophisticated production processes - and this is reflected in their higher cost.

It might be argued that the hardest and toughest tips alone should be used routinely - regardless of their high initial and ongoing costs. The reasoning is that, as these have the longest possible run-time, they must also return the greatest profit.

But this is not necessarily so, the gain in run-time is not consistent for all materials. The most sophisticated tips give the best performance when sawing the hardest and most difficult materials - where their increased cost can be fully justified. This may not be the case when sawing softer and more easily cut materials.

As a general guide, a harder and more costly tip is needed for sawing hard and tough materials, whilst for softer materials a less costly tip can be used - but there are other factors which affect the choice. These include the corrosive effect of liquids naturally present in some timbers which can attack some tips more than others to cause premature edge breakdown, and the convenience or otherwise of facilities for repair and re-sharpening of the chosen saw type. *See also 'Costing different saw types', following pcd saws.*

Tip Hardness Grades

Most modern saw tips are offered in different grades, ranging from a tougher tip with a lower hardness, to a harder tip with a lower toughness.

Hardness and toughness in the same basic tips always work in this way, as hardness is increased toughness is reduced, and vice versa. As different grades within a specific types of tip vary little in cost this factor need not affect choice - which should be made purely on correctly matching the grade of the saw tip to the job.

A harder grade tip normally gives a longer run-time - but only if it is also tough enough for the intended work. If using a hard tip which lacks toughness, the cutting edges would be brittle - and could breakdown so quickly that an even a shorter run-time would result than with a corresponding 'softer' tip of the same material. There is no universally suitable grade of tip for all materials, a grade suitable for one material could be inefficient for another, so there is no cut and dried answer to all sawing requirements - these simply vary too widely.

Choosing the right type and grade of tip has to be a careful balancing act between hardness and toughness. It has to be right for the material being sawn - too hard a tip and the cutting edges may break down prematurely, too soft a tip and the cutting edges will wear quickly. In either case the run-time will be less than expected - but the real loss will be to operating profit. To make the right choice for a specific application take advice from the sawblade manufacturers and talk to other users - but the following information should get you in the ball park.

In rising order of increased performance and higher cost, tips are currently made from high speed steel (hss), Tantung & Stellite, tungsten carbide and, the hardest and toughest of them all, polycrystalline diamond (pcd) and monocrystalline diamond (MCD).

HSS (High Speed Steel)

Steel is a highly purified alloy of iron mixed with bonded carbon, plus some alloying elements. Although it has been produced for centuries, manufacture was originally labour-intensive and expensive. From 1856 Bessemer converters began to produce cheap steel in quantity, but now even more efficient conversion methods are used.

After conversion, steel is remelted and further alloys added to make special steel for specific applications. Sheffield Crucible Cast Steel was the first carbon steel suitable for tools, but this was eventually replaced by other much harder and tougher steels.

HSS is a basic steel with added alloys such as tungsten, chrome, molybdenum, vanadium and cobalt. These all form hard and complex carbides within the steel as this is heat-treated by hardening, quenching and single or multiple tempering.

Fig. 7.2
This Woodsaw splitting saw can have hss, Tantung, solid Stellite, tungsten carbide or possibly pcd teeth. It is mainly used on moulding machines.
Source - Wadkin

There is not just one, but a family of high speed steels used regularly in industry. The most common ones for woodworking are T1, tungsten-based HSS, and M2, molybdenum-based HSS. There are other steel-based tips manufactured, for example HCHC, High Carbon High Chrome, but these types are rarely used. HSS is not for the general run of sawblades, it is mainly used for splitting saws on moulders, grooving saws and similar. Hardness, at around 56-62Re, and toughness, vary with the specific HSS used and its subsequent heat-treatment.

Tantung

In 1939 Tantung was developed by V R Wesson. It is a cast alloy, with 41% cobalt, 30% chromium, 16% tungsten, 3% carbon, 5% columbium plus some iron and nickel. Tantung is melted in an electric arc furnace and cast in chill moulds to give a fine grain structure and a hardness of 60 to 65 Re. It is brazed onto grooving saw blanks to form tough, sharp and abrasion resisting cutting edges.

Being a cast material Tantung is used a direct replacement for HSS on specialist saws, but not for the general run of saws.

Stellite

In 1912 Stellite, a hard wearing, weld-deposit or solid tool tip, was devised by Elwood Haynes. It was initially a by-product of a smelting process pioneered by Deloro Mining and Smelting, Ontario, Canada who are the largest processors of cobalt made from arsenical ore. Stellite was first used in cast form for hard-tipping metal cutting tools, and was only later used for tipping wide bandsaws and circular saws.

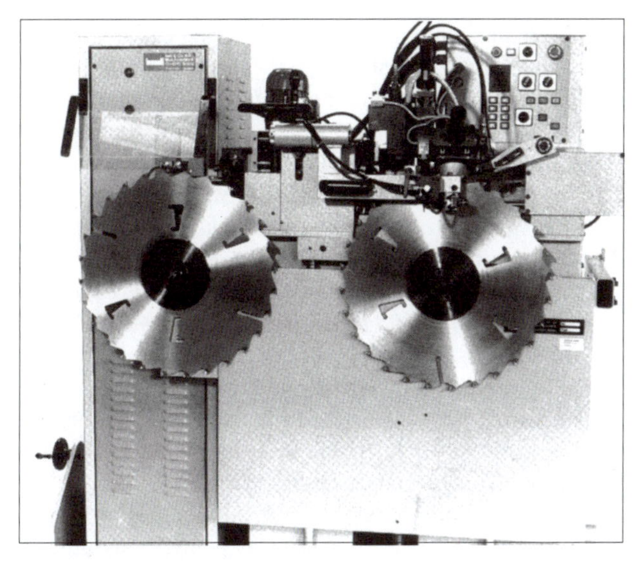

Fig. 7.3
This shows a Stellite tipping machine for circular saws.
Source - Iseli

Stellite 1 and 12 are both used for weld depositing. Stellite 1 has 52% cobalt, 33% chrome, 13% tungsten and 2% carbon and a hardness of 51 to 58 Re, Stellite 12 has 60% cobalt, 29% chrome, 9% tungsten and 1.8% carbon and a hardness of 47-51 Re.

Stellite 100 has slightly more chrome and tungsten, less carbon, a hardness of around 61-66 Re and is supplied as a cast material for brazing onto grooving saws, for example. Stellite 100 is a direct replacement for HSS tips, but is much harder and tougher and lasts considerably longer.

Stellite cast blanks are by their nature pore-free and harder than weld-deposited Stellite - and give a better performance for this reason.

However, recent developments in Stellite tipping equipment has vastly improved the quality of weld-deposited tips so that they now are close in performance to solid Stellite tips.

Treatment of Stellite and Tantung

Neither Stellite nor Tantung needs heat treatment to harden them - in this respect they are better than HSS which can vary in performance according to how carefully the heat-treatment process is carried out. Also they do not soften if overheated when grinding - in rare cases heating can affect the hardness of HSS.

In addition to lasting much longer than HSS, they have an excellent corrosion resistance, and are especially suitable for timbers where tip corrosion is a problem due to their naturally-occurring acids and other liquids.

Like HSS both alloys are homogeneous *(the same material throughout - not in the bonded particle form of tungsten carbide)* so the teeth can be ground to a very sharp edge - to give a longer run-time.

Tantung and Stellite-tipped saws easily out-perform both alloy and hss-tipped saws and, most importantly, they can use the same grinding wheels.

They do not require special grinding skills - grinding techniques already familiar to existing users of alloy saws give satisfactory results, but CBN actually gives better results.

In my opinion Stellite tip saws are greatly undervalued - this is especially so where carbide has proved unsuitable.

Tungsten Carbide

Although tungsten carbide originated in 1914 it was not used on woodworking saws until the 1950's. Since then the use of tungsten carbide has increased dramatically, virtually replacing alloy steel for both breakdown and finishing operations.

Unlike steel, tungsten carbide is not a continuously-grained material, but consists of fine, hard grains bonded into a matrix (bonding agent) usually of cobalt (some makers add titanium make the bond more impervious to chemical attack).The manufacturing process produces this composite material by sintering. The proper term is cemented tungsten carbide, but most users refer to it as tungsten carbide, carbide, hardmetal or simply wc.

NOTE: Other carbides based on titanium, tantalum and niobium are also manufactured, but are not normally used in woodworking. The sintering process, also called powder metallurgy, is commonly used for other products such as phosphor-bronze shell bearings.

Tungsten Carbide Manufacture

Tungsten carbide proper is a chemical bond of carbon and tungsten, impossible to produce as a cast metal because of the high melting temperature needed - 2500°C.

It is manufactured from thoroughly mixed powders of tungsten and carbon (as lampblack), and heated in a hydrogen atmosphere to chemically bond them. The grains formed by this process are then milled and screened.

Fig. 7.4
Manufacturing and quality control of cemented tungsten carbide components in a modern factory.
Source - Dymet Alloys

The carbide powder is mixed with powdered cobalt, together with temporary bonding and lubricating agents, and compressed as a paste into the required basic shapes. A pre-heat treatment at 700-800 degrees centigrade initially processes the relatively soft and chalk-like green tips which, following this stage, are sized and shaped as needed.

Finally tips are sintered at 1400 to 1600 degrees centigrade for about an hour in a non-oxidizing atmosphere. Between 10% and 50% of the mix becomes molten during this sintering process as the cobalt liquidizes to dissolve much of the carbide.

As cooling takes place the carbides re-form out of the cobalt as tungsten carbide crystals and the cobalt solidifies to bond them into a hard and dense structure.

Sintering shrinks the tips because the grains initially

space the powder by leaving tiny gaps. As the cobalt melts and compacts on sintering the material becomes extremely solid as no spaces remain in the finished tips.

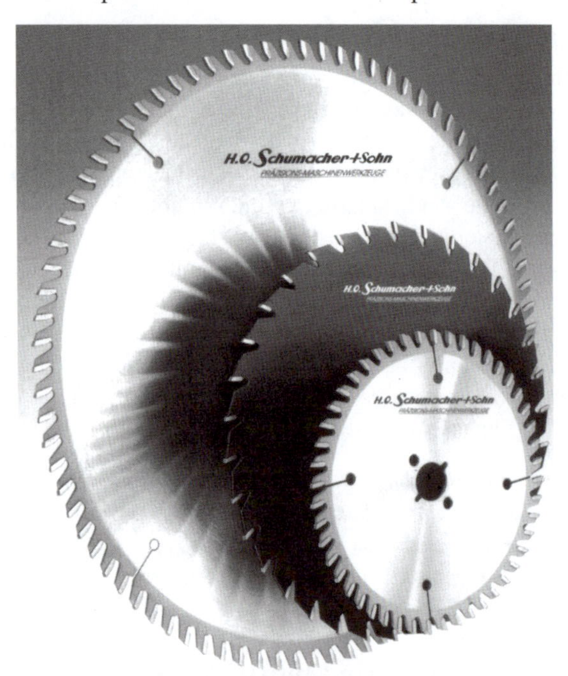

Fig. 7.5
Typical array of tungsten carbide tipped saws.
Source - Schumacher

Sintered tips are face, end and edge ground, perhaps treated by coating or pre-soldered, and finally brazed onto the sawblade. After this the tips are side, top and face ground. Finally the outer rim of the sawblade is sandblasted to remove surplus materials and to generally clean-up the teeth.

Initially carbide saws were used for sawing naturally abrasive timbers such as teak, and newer manufactured boards bonded by highly abrasive glues. Alloy steel saws were never entirely suitable for natural timber or boards with abrasive inclusions - but carbide saws simply took these in their stride.

The relatively large grains of the first carbides formed an irregular cutting edge which was never as keen as steel. This initially gave poor results on sawing softwood as carbide simply did not cut cleanly. For this reason carbide was initially thought unsuitable for general use, and was reserved for the abrasive materials that gave steel saws problems.

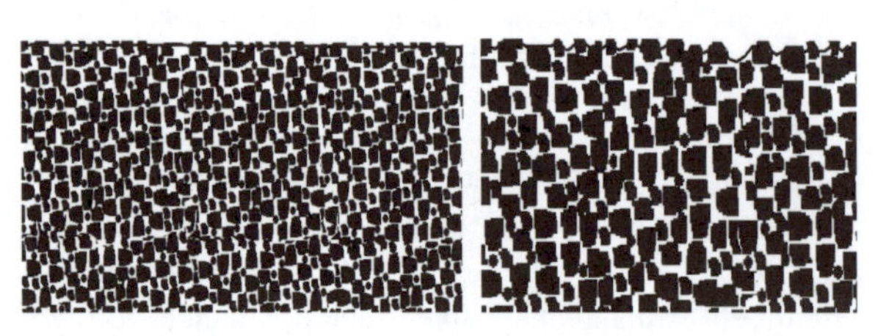

Fig. 7.6
Coarse tungsten carbide crystals, right, leave an uneven edge as the cobalt wears down - even though the crystals themselves remain sharp-edged. Finer crystals, left, give a much smoother and longer-lasting edge. Because of the structure of tungsten carbide the crystals at the cutting edge can be dislodged through running too-long after the cutting edge has become dull, or by poor grinding practices.

However, production processes steadily improved - and eventually finer grains of approximately half a micron were successfully manufactured to produce tungsten carbide with an exceptionally fine grain structure. *(In fact, half micron grains actually grow to about 3/4 of a micron on sintering.)*

Through being brought up in woodworking environment, microns never really meant anything to me - in fact I had never used the word. The smallest measure was a gnats's whisker, and a micron was obviously something only boffins and nerds dealt with. However, as a guide to size, a micron is 1/1000th of a millimetre, or 0.00004in. A regular piece of writing paper is perhaps 100 times thicker than a micron.

Modern carbide tips, when sharpened on specialized grinding machines using greatly improved grinding wheels and techniques, now have a consistent and much keener cutting edge - virtually comparable with steel - which is suitable for almost any wood and wood-based material.

Tungsten carbide is extremely hard and highly abrasion-resistant. It outlasts alloy saws by many, many times in all applications - from softwoods through hardwoods, to abrasive timbers and all manufactured materials such as regular and plastic faced plywood, particleboard, medium density fibreboard and similar materials.

Tungsten carbide is, in fact, the first choice for virtually all finishing conversion, and for breakdown of all materials save those containing liquids that attack the cobalt bond.

Treatment of Tungsten Carbide

Fig. 7.7
Grinding tungsten carbide teeth needs great care - no problem now with modern, highly efficient grinders such as this Vollmer CHP.
Source - Vollmer

Tungsten carbide has a structure unlike that of steel or the other cutting alloys. It is similar to that of a stone wall, with crystals of tungsten carbide surrounded and bonded by the cement of cobalt.

Because of its structure carbide has certain special needs. Like a stone wall the edge can be destroyed if crystals of carbide are ripped-out through the action of cutting. To avoid this adequate backing is needed for the cutting edge, ie, a ground or included angle of 50-55 degrees - much more than steel or other cutting alloys require. A 50-55 degree included angle is fine for most sawblades cutting wood and wood-based products, so this is no disadvantage.

The crystals at the cutting edge can also be dislodged through the pounding action **after** the edge has lost its initial sharpness. The *'wall'* then begins to crumble and rapid cutting edge failure follows. For this reason carbide saws should be re-ground **before** this begins to happen - failure to do so results in badly damaged edges which then need heavy, costly grinding and unnecessary loss of saw tip life.

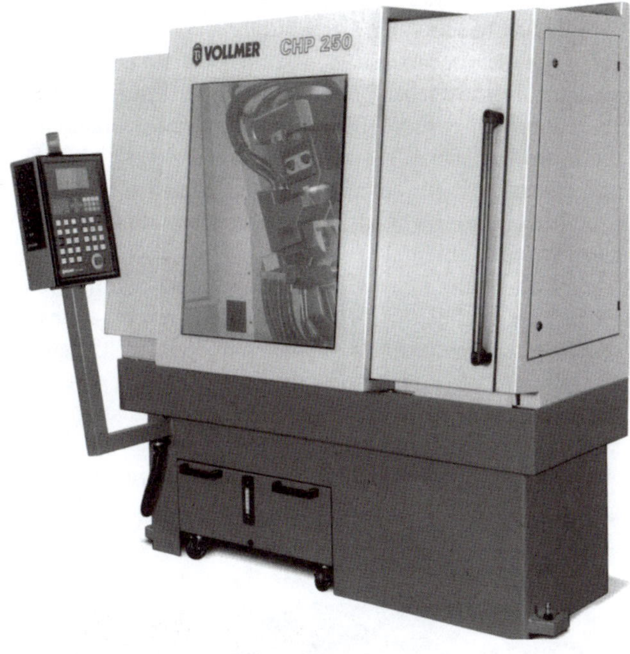

Fig. 7.8
Tungsten carbide-tipped saws need a precision grinding machine to maintain them, such as this Vollmer CHP 20H.
Source - Vollmer

Tungsten Carbide Grades

The hardness and toughness of the tip depends upon the percentage of cobalt to carbide, and the size of the carbide particles.

Cemented carbide grades used for woodworking are classified by the letter 'K'. There is no fixed ratio in the tungsten carbide to cobalt for 'K' type products, the classification used is only an application guide. These range from mixes with only 3% cobalt - the hardest and most brittle grade, to mixes with over 30% cobalt - the softest and toughest grade.

Hardness is constantly checked in manufacture against agreed standards, but there can still be differences between different makes - and some makers use a grading system different to that given, which makes comparisons difficult.

The right grade of carbide will be provided by reputable saw makers when the material to be cut and cutting conditions are stated. In general terms, the ideal tip for converting the stated material in regular cutting conditions is that made from the hardest grade practical to give long run-times - but which is also tough enough to resist premature edge breakdown.

Grinding Tungsten Carbide

Great care has to be taken in grinding carbide because of its unusual structure - and only diamond grit is entirely suitable for this.

Green grit (silicon carbide) wheels are still used by some for rough shaping, but they tend to overheat and crack the carbide. Also they quickly lose abrasive grain sharpness, and the extra pressure and heat this causes can dislodge or loosen carbide grains at the cutting edge - again leading to early cutting edge failure. This can also happen through force-grinding with a glazed diamond abrasive wheel. Overheating and quenching in between grinds can micro-fracture the carbide, so a continuous flood coolant is essential during grinding. Grinding must be at a slow rate, and with the correct grinding pressure. See further details later in this chapter.

Diamond tips

Polycrystalline Diamond (PCD)

This is a relatively new and highly-specialized cutting material manufactured as Compax by The General Electric Company (USA), as Syndite by the De Beers Industrial Diamond Division, and under various other names by a few other makers.

Fig. 7.9
Diamond saws are capable of working long and hard in the toughest operating conditions.
They outstrip the performance of tungsten carbide saws by many, many times.
Source - Supreme Saws

Fig. 7.10
This shows an actual diamond saw tip. The diamond face, uppermost, is supported by a tungsten carbide substrate.
Source - De Beers Industrial Diamond

PCD comprises of a polycrystalline aggregate of manufactured diamonds approximately 0.6 - 0.8mm thick bonded to a substrate of cemented tungsten carbide 1 - 2.7mm thick. It is produced by sintering together micron diamond particles at ultra high pressure and temperature in the presence of a solvent/catalyst metal, usually cobalt.

The composite tip combines the hardness and high thermal conductivity of diamond with the brazeability of tungsten carbide. The micron-sized manufactured diamonds are randomly-oriented to give consistent wear-resisting properties in all directions. In this way polycrystalline diamond tips ideally combine the abrasion resistance of diamonds with the toughness and impact resistance of carbide.

The Development of PCD

PCD was introduced by The General Electric Company, Worthington, Ohio USA, in 1973 as a replacement for the single-crystal natural diamond and other tools used for machining highly alloyed metals at high speeds. PCD tools give exceptional improvements in engineering tool life and productivity, provide better control over the quality of finish, and maintain close product tolerances.

The application of PCD tools to woodworking was always a possibility, but it was only in 1978, when fine-grain PCD blanks were introduced, that a convincing demonstration on resin bonded particle board and solid plastic panels proved the point beyond any doubt. This demonstration gave an increase in saw life of 125 times more than carbide. In this case the tool life increase related directly to the abrasion resistance factor of PCD which, in the grade manufactured then, was 125 times that of cemented carbide.

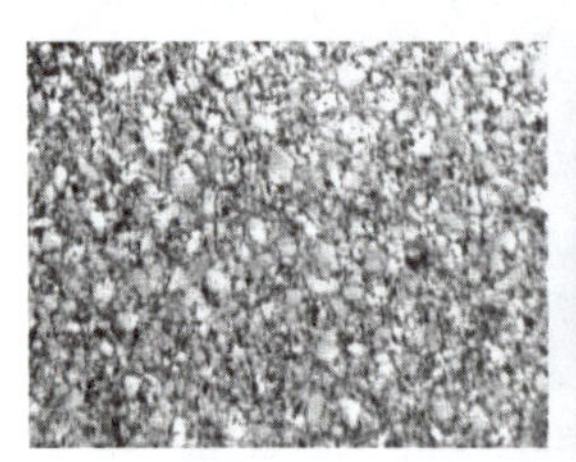
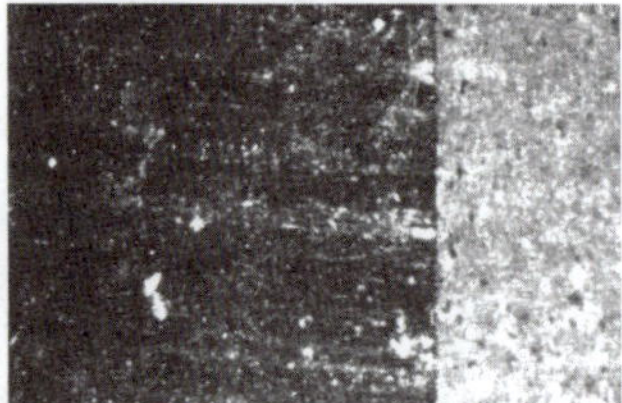

Fig. 7.11
This shows an enlarged view of the face (left) and a section (right) of a PCD tip. The section shows the polycrystalline diamond mounted on and securely bonded to the tungsten carbide substrate
Source - General Electric

Subsequent experience showed that tool-life gain varies widely according to the machine in use, the material being worked and the cutting edge geometry - from a low gain of 70 to a high one of 1 200 times. The average gain is between 150 and 250 times.

These figures, of course, all refer to materials which are difficult to saw, abrasive and exotic hardwoods, blockboard, plywood, fibreboard, laminated or plain chipboard, MDF and compressed wood and plywood.

An initial doubt about using diamond tools in woodworking was that they might prove too fragile, but the diamond crystals are very securely bonded to one another and to the substrate of tungsten carbide.

NOTE: A diamond cutting edge is no more susceptible to accidental damage than regular carbide, and in the right application, and with the correct geometry, diamond saws are highly successful. Because of their high cost, great care is essential in handling them and when setting the machine - but properly used they run for years rather than the hours, days or weeks of regular saws - and give excellent service.

Treatment of PCD

Type	High Speed Steel	Stellite	Tungsten Carbide	Polycrystalline Diamond
Included Angle	40-45°	40-45°	50-55°	70-75°

Fig .7.12
The different tip materials require different included angles - this shows the minimum allowable angle in each case.

PCD inserts are available in all the shapes and sizes needed for sawblade tips, and are brazed into position in a manner similar to that of carbide tips, with the diamond face at the tooth front - or more recently on the clearance face. As with carbide, the tips are then sized and finished.

For the same reasons that carbide needs a larger ground angle, PCD also needs a ground *(or included)* angle of around 75 degrees, much more than carbides and considerably more than alloys and steel. This does not restrict the use of PCD in any way as the hard and tough materials which need diamond saws also require a small hook angle - which allows for a large included angle.

Re-forming PCD edges is not currently within the capability of smaller woodworking grinding rooms - nor their regular grinding machines. However, in-house PCD sharpening will almost certainly become an established practice in the years to come - something that has already started with some forward-looking manufacturers.

With complex woodworking tools it is not practical to grind diamond, the re-sharpening process is one of electro-discharge spark-eroding - see Chapter 13 for more details.

Fig. 7.13
This Vollmer electro-discharge machine QR20P, shown here in close-up, is used for original manufacture and re-sharpening of diamond tipped saws. The machine is quick, accurate and has guaranteed repeatability.
Source - Supreme Saws

All modern electro-discharge machines are computer-controlled. Some are general purpose machines with up to 7 axes capable of completing almost any type of PCD tool tip at a single handling - but these types are mainly for sharpening a whole range of PCD tools, not merely sawblades. Newer types have similar features, but are manufactured at much lower cost as dedicated machines purely for sawblade grinding alone, with perhaps only three axes.

Polycrystalline Grades & Tip Types

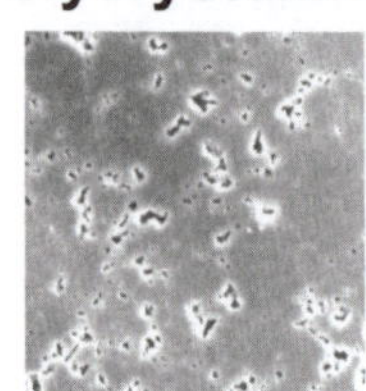
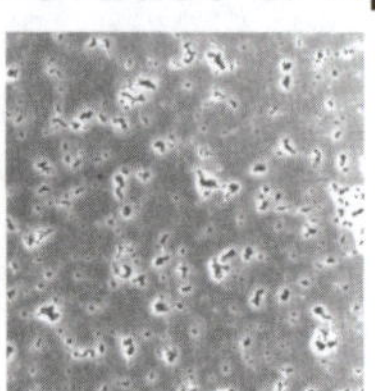
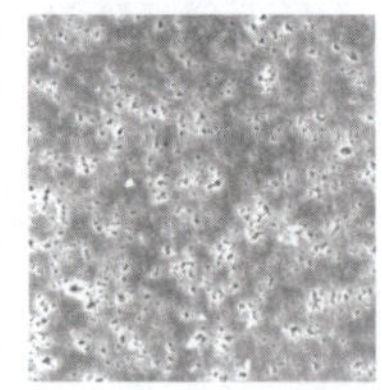

Fig. 7.14
These are micro-photographs taken of etched surfaces of PCD which show the tiny diamond crystals interlocking to form an almost complete surface of diamond. The different grades of the Syndite PCD shown here are CTB025, left,

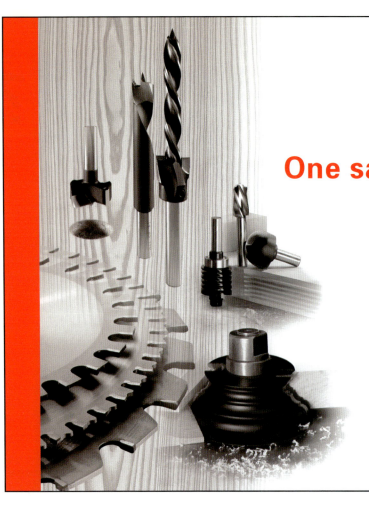

CTB010, centre, and CTC002, right. Of the tips shown CTC002 has the finest grains - the type used almost exclusively in woodworking applications.
Source - De Beers Industrial Diamond

As with other hard tips, PCD tips are available in different grades - which relate to diamond grain size. In general the coarser *(larger)* diamonds have the greatest wear resistance, but finer *(smaller)* diamonds give a better surface finish. For the majority of woodworking applications the finer grains are the most suitable but, for abrasive wood products such as HPL, medium and large grain types give a better performance - and are now being used much more widely.

PCD is produced initially as discs of different diameters and thickness according to the intended application. These are then divided up by wire erosion into the smaller pieces required for tips - which can be for sawblades or for the many of the other tools now using diamond as the cutting agent. Because of the high cost of producing PCD discs they are carefully divided to give the maximum number of usable tips.

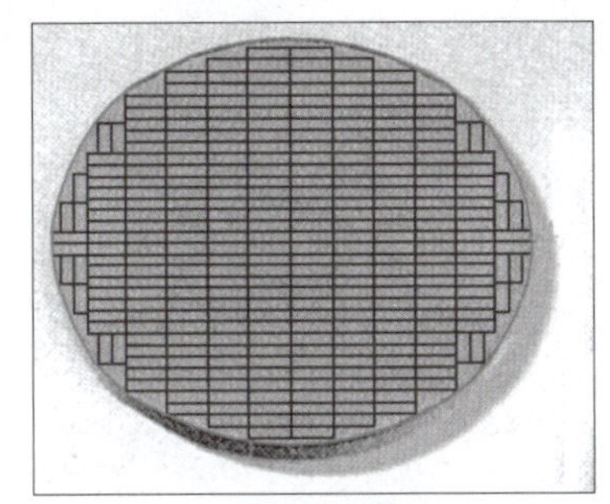

Fig. 7.15
This shows the method of dividing up thin discs into sawblade tips by wire eroding. The orientation of the tips is of no consequence because the diamond particles are themselves randomly orientated. This show tips of the same size being produced, but often a mixture of sizes and shapes are produced from a single disc by nesting them - usually with a computer determining the optimum cutting pattern.

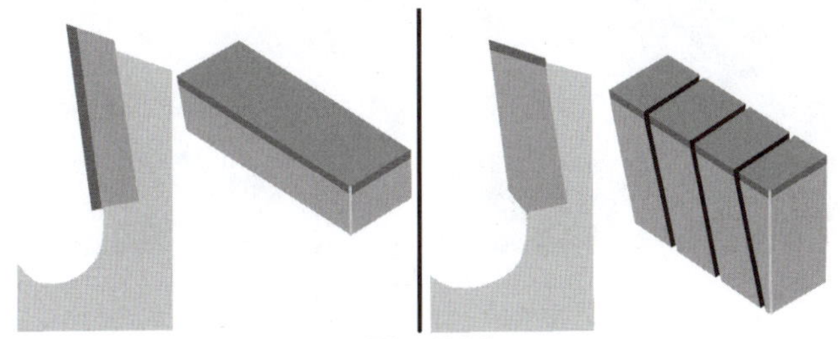

Fig. 7.16
The original and regular method of producing PCD tips for circular saws has the diamond facing on the tooth front (left). Sharpening takes place on the tooth top only. The 'new concept' in producing sawblade tips is aimed at lowering the cost of tips (right) by forming them from a thicker disc cut in such a way that the diamond facing is on the tip top.

There is continuing research into improving PCD tips and, in particular, in bringing down the cost. The traditional way is shown in Fig. 7.16, left, with tips that can be re-ground between 8 and 10 times* before replacement is necessary *(depending upon the degree of wear and the refurbishment-method being used)*. Initially only small discs could be manufactured, but larger discs are now being made that lower the cost-per-tip.

**Because the body of a sawblade suffers from the repeated and intense use that diamond tips make possible, sawblades made with PCD tips should be considered worn-out after completing the maximum number of re-sharpenings the tips are capable of. In practice the cost of a new sawblade body is a fraction of the cost of tips and re-tipping - so it is false economy to re-tip well-used and suspect bodies.*

Fig. 7.16 shows an alternative method of producing 'new concept' tips made in such a way that the PCD lies on the top instead of on the face of the tip - in this way the amount of diamond used is about 1/3 of that needed for conventional tips - so the tip cost is much lower.

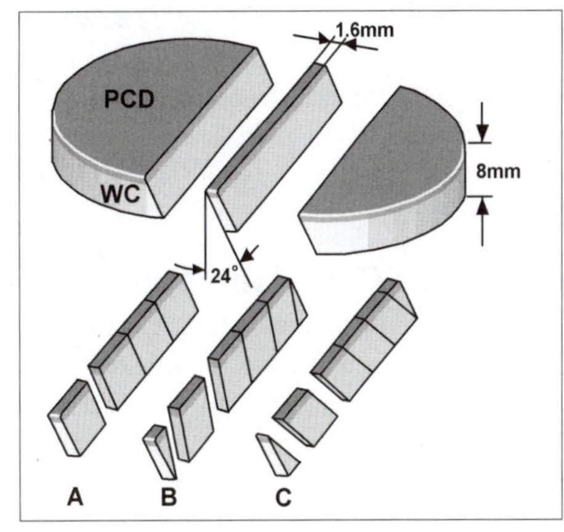

Fig. 7.17
Discs for 'new concept' tips are sliced (wire-eroded) from thicker than normal discs - and at an angle to give the required hook and clearance angles. The tips are then sliced a second time to width. At present flat-topped tips only are produced as at A, but it is possible to produce alternate-bevel tips by slicing at a second, reverse angle on alternate strips, as B & C.

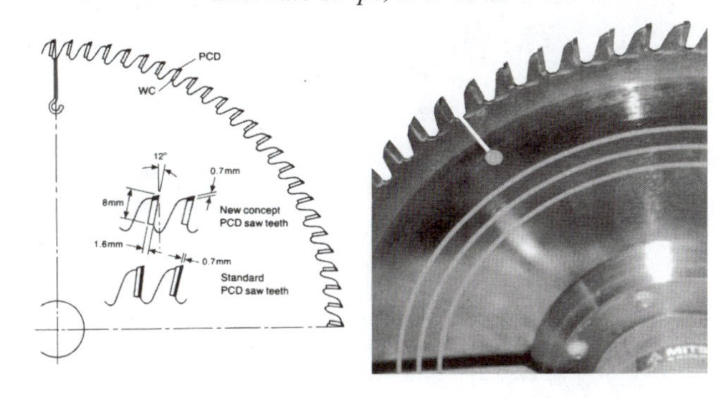

Fig. 7.18
Details of the 'new concept' tips are shown above (left) with a 'new concept' sawblade in use on a Giben panel saw cutting stacks of three 19mm thick laminated chipboard.
Source - De Beers Industrial Diamond/Giben

The different positioning of these tips makes no difference to the tip life between grinds - the cutting edge is positioned and cuts as before - but the number of grinds possible is only about half that of conventional tips. Re-grinding is by spark erosion *(as for a conventional PCD saw)* but on the face of the tips only. No change in saw geometry is necessary.

Initial test on 'new concept' PCD tips on a 400mm diameter, 72-tooth blade mounted on a Giben panel saw (used to size 19mm thick laminated chipboard in stacks of three) cut 150 000 linear metres without need of re-grinding, compared to 3 000 linear metres with a tungsten carbide tipped saw.

Fig 7.17 shows how the tips are produced from manufactured blanks to provide the necessary hook and top clearance angles. So far flat-topped tips only have been used, but it is possible to produce alternate-bevelled tips.

These 'new concept' tips provide a lower-cost means of producing sawblades for less intense applications, where there is a high risk of damage, or where a throw-away sawblade becomes a viable proposition - DIY perhaps?.

The new concept tips promise an increasingly wider application of diamond tooling for woodworking applications.

It is, however, essential that these tips are brazed with great precision radially on the sawblades - correction to the height after manufacture - easily achievable with regular PCD tips - is impossible with tips of this type.

With the rapid advances in manufacturing machinery made recently, though, there is little doubt that such accuracy is possible on a production basis for anyone who wants to seize the opportunity.

Production and Cost Advantages of PCD

Whilst PCD saws are much more expensive than other hard-tip types, many other factors need to be taken into account. The high purchase price is merely one factor, when the overall costs are considered there can be a considerable saving.

An important factor is the higher efficiency of PCD saws. Because the cutting edges are so keen and tough a PCD saw can run faster, cut deeper and give cleaner results than carbide. By combining high-quality sawing with long run-times and fewer downtimes, these sawblades give production figures which well exceed the best results from conventional tc sawblades by many, many times.

Fig. 7.19
*Twin PCD sawblades on this Schwabedissen S1001panel saw give an impressive performance in cutting 200mm depths in high volume sawing of chipboard. Carbide-tipped sawblades had a maximum life of 70 000 linear metres of cut and a re-grind life of 5-6 times - compared with the Supreme Syndite CTC002 sawblades which cut 865 000 linear metres **between re-grinds** and have a re-grind life expectancy of 20 times.*
Source - De Beers Industrial Diamond/Supreme Saws

The wear rate of PCD cutting edges is not only much less than carbide - but also much more even - PCD wears by erosion rather than edge breakdown, see Fig. 7.20. As a result diamond cutting edges remain sharp for a much greater proportion of their useful run-time - so the quality of the sawn surface is consistently higher and less subject to deterioration.

So, to list the factors that need to be taken into account for PCD, first the advantages - the higher production rates possible, the drastically-reduced down time which keeps machines working longer, and the need of fewer sawblades.

And the disadvantages - the higher initial cost and the cost and inconvenience of re-sharpening out-of-house when compared to conventional tools.

PCD is not the answer to every problem, no single tip material is, but it does have very definite advantages in converting certain hard and difficult manufactured materials - although not entirely suitable for low-quality chipboard.

As a guide, a PCD saw costs perhaps 15 times that of a tungsten carbide saw, lasts between 150 and 250 times longer, and has a re-grind cost of between 15% and 20% of the original purchase price.

The makers of diamond saws now have considerable data on cost savings when using PCD, and will give sound advice on their viability over a wide range of sawing applications.

Call them in - you may be pleasantly surprised at the cost savings you could make by changing to PCD.

Fig. 7.20
This close-up shows two Syndite CTC002 teeth after cutting 310 000 linear metres of MDF with a 95 mm stack height. The sawblade, made by Leuco, is 400mm diameter, has 36 teeth and cuts at 23M/min. It shows uniform tip wear on the cutting edges with no chipping.
Source - De Beers Industrial Diamond/Leuco

Fig. 7.21
This Supreme 450mm diameter sawblade (left) is tipped with Syndite teeth for cutting 7-deep 32mm melamine-faced chipboard at 45M/min.
The machine (right) is a dedicated Selco WNA angle plant for processing boards automatically and with full optimization. The sawblade has 60 teeth of 4.5mm wide on a 3.5 saw body. With fewer teeth and larger gullets than the type of sawblade originally used on earlier slower-running beam saws the WNA replaced, it has ample gullet capacity to remove the greater amount of waste produced at the high operating speeds of the Selco plant.
Source - De Beers Industrial Diamond/ Supreme Saws/Selco

Costing different sawblade types

All factors must be considered in order to get the true picture of which type of sawblade has most advantage for each specific application. These must include the initial and re-sharpening cost of each type of sawblade, the volume and rate of production between sharpenings, the number of possible sharpenings before re-tipping or replacement is needed, the percentage of downtime when changing sawblades, and the minimum number of sawblades needed to maintain production whilst repair and re-sharpening is carried out.

These factors, when properly evaluated, will clearly show the relative cost per metre or per foot of board materials converted for each of the different sawblade types checked - and the answer in these terms can often be surprising.

Don't continue using regular sawblades simply because they have been traditionally used, diamond sawblades may have some surprising advantages.

For example, in comparing the relative cost of carbide and PCD in one instance quoted by Supreme Saws, the results were quite staggering. Carbide saws ran 2-3 days and cut 14,000 linear metres between sharpenings when converting wood panels, whilst PCD saws ran for 6 months and cut 865,000 linear metres between sharpenings.

The cost of Re-grinding PCD saws twice a year was approximately half the ***monthly*** *cost of Re-grinding carbide saws. This 96% reduction in annual re-grinding costs more than offset the higher initial purchase price of PCD saws. Added to this, of course, was the higher productivity - which even further emphasized the advantage of changing from carbide to PCD.*

Other diamond tips

Monocrystalline Diamond (MCD)

MCD is a diamond product similar to PCD, and produced in a similar manner, but consisting of a single diamond crystal. Unlike PCD, MCD is non-conductive and has to be pressure-ground using a diamond grinding wheel and specially-developed dielectric fluid,

The cutting edge with these tools is not interrupted by the diamond grain boundaries, so extremely smooth and sharp cutting edges are possible. The material has a high hardness coupled with brittleness, so the included or sharpness angle has to be large, and application is restricted to those with a low chipflow. Some examples include machining plexiglass and aluminium to give a highly polished edges, machining the highly abrasive top layer of laminate flooring, and machining of the matching edges of cement-fibre panels for dry house construction.

Laminated flooring is growing in popularity, but proved too difficult for regular PCD saws and tools. MCD proved to be the only tooling capable of efficient operation. It is composed of a number of layers - a composite material - an overlay added to a central layer bearing the decorative pattern and secured to both sides of MDF or HDF (high density fibreboard).

To achieve high wear-resistance the composite includes melamine resin with particles of aluminium oxide or silicon carbide - with the proportion of abrasives as much as 30% in the highest-grade panels.

It has been established through research that the harder the material the cleaner the cut with MCD - added to which the extremely good cutting edge quality achievable with MCD results in long tool life between re-grinds - but for which specially-developed machinery is needed. See Chapter 10 for further details.

Fig. 7.22
The above shows a Prewi MCD edge-milling cutter (above), a close-up of the patented safety insert pocket for the saw teeth (lower left) and Monodite cutting tool blanks used for this and other MCD tools (lower right) - for which a special brazing technique was developed.
A regular PCD tool machined 24 000 linear metres of floor panels between re-grinds, compared with 2 million linear metres for the MCD tool in the same period - plus a re-grind life of 15 times.
Source - De Beers Industrial Diamonds/Prewi-Schneidwerkezeuge

Chemical Vapour Deposited (CVD) Diamond

CVD diamond is produced in a different way to both PCD and MCD, through a heated chemical vapour which deposits a thin, multi-layer of diamond crystals to the surface of a carbide substrate. As yet this type is confined to the metal-working industries, and even here to the more exotic materials being produced for aerospace and weaponry research. As such materials filter down to more mundane applications it is likely that CVD diamond will be needed to machine them. It is possible to make CVD diamond electrically conductive, consequently such blanks can be wire eroded.

Grinding Abrasives

Grinding wheels consist of fine abrasive grains bonded together by a matrix. The grains have naturally sharp edges which cut fine slivers of material away from the material being ground in a process called abrading. The matrix, or bond, holds the grains firmly in position whilst they remain sharp, but releases them when they loose sharpness to expose fresh, sharp grains underneath.

This process, continual restoration of the effectiveness of the grinding wheel, is called grain loss or 'shed'. This ongoing process depends for its effectiveness upon the extra

pressure and heat created when grains no longer work effectively.

Although grinding wheels appear to be made of many different materials, in grinding steel the main type is aluminium oxide.

Aluminium Oxide

Initially natural emery and corundum, both forms of naturally crystallized aluminium oxide, were used in making grinding wheels.

Natural abrasives vary widely in performance, though, and have largely been replaced by manufactured abrasives which can be closely controlled as regards toughness, friability, grain size and shape. Most are now made by refining and fusing either bauxite ore *(clay)*, or by processing the aluminium oxide residue from aluminium manufacture *(this is created in electric arc furnaces during the smelting process)*. In both cases the processed grains are crushed, sieved and graded.

The least pure form is made from clay and varies in colour from light grey to brown or blue, according to the origin of the clay and the processing temperature. To increase toughness when intended for general-purpose grinding wheels for high-strength steels, this type may have a small amount of titanium oxide added.

The purest form is made from the natural residue of aluminium smelting. It is almost white in colour and it cuts fast, economically and with least heat generation. It also has high friability - the ability to fracture when the initial edges become dull - to form fresh cutting edges rather be dislodged.

Some makers colour grinding wheels. However, apart from identifying a particular grit or grade, such colouring serves no real purpose. The exception is so-called pink wheels. These have chromium oxide added to white bauxite to make the friable grains slightly tougher and less likely to crumble - but this is mainly used for grinding highly alloyed HSS.

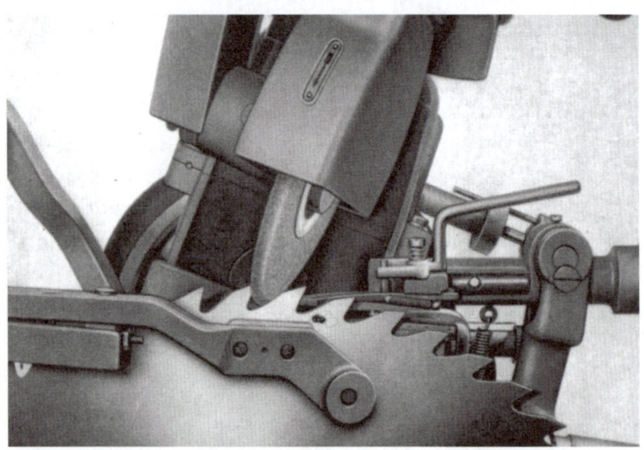

Fig. 7.23
A shaped regular-grit grinding wheel in use on a saw grinder profiling alloy steel saw teeth.
Source - Wadkin

Silicon Carbide

The other regular abrasive is silicon carbide. This is a manufactured material which does not occur naturally. It was discovered by Dr. E G Acheson in 1891 and was initially used as a substitute for diamond dust in polishing gemstones. It is harder and more brittle than aluminium oxide and varies in colour from black to light and dark green.

All silicon carbide abrasives have similar physical properties and, of these, light green is the purest form. This type is known as green grit, and is sometimes used for rough-grinding tungsten carbide.

Silicon carbide grains are sometimes mixed with aluminium oxide grains to give more 'bite' than aluminium oxide grains alone.

Grinding Wheel Specifications

Although grinding wheels of different specifications behave differently to one another, the hardness of the actual abrasives is more or less consistent. The differences in performance are mainly because of the grain size, the proportion of bonding material to abrasive, the structure of the grinding wheel, variation in the mix of abrasives used, and the quality of the grains - the latter achieved by controlling their toughness and friability in processing.

The specification of each grinding wheel is printed on the wheel itself or on the blotter (a compressive paper washer fitted to both sides of most grinding wheels). Not all makers use the same code, but most give some identifying details and these usually include the following, often in the order listed:-

Abrasive type.

Usually letters are used, e.g., A for aluminium oxide and C for silicon carbide. Numbers usually indicate a mixture of two abrasives or the addition of fillers.

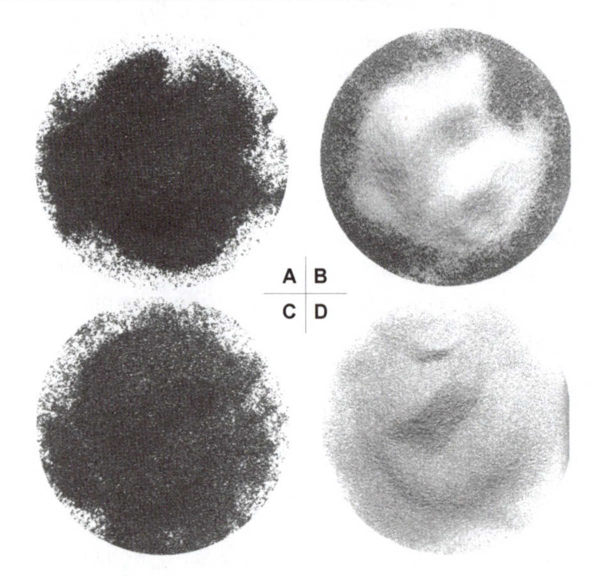

Fig. 7.24
These are the four main types of abrasive grains used for regular grit grinding wheels:- A - Brown or blue aluminium oxide sometimes called bauxite, B - White aluminium oxide, C - Silicon carbide and D - Pink aluminium oxide.
Source - Universal

Abrasive size.

Abrasive sizes are usually between 8, large and 600, small. The numbers used approximate to the number of openings per linear inch in the sieving mesh used. As an example, grit size 60 has an average grain size of 0.01in (0.25mm).

Grinding is an abrading process with each particle cutting a narrow track across the material being ground, so grain size makes a difference to the way a grinding wheel performs.

Large grains give a fast grind but a poor finish. Small grains give a good finish but a grind only slowly.

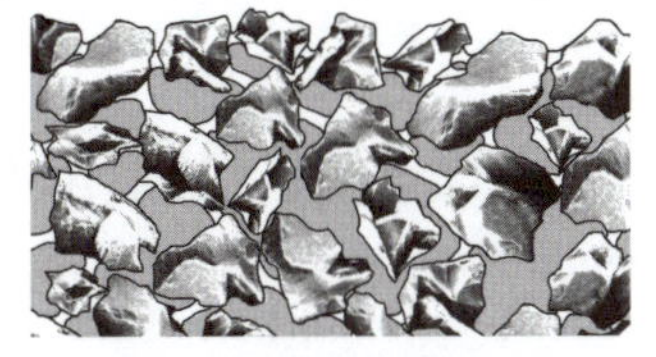

Fig. 7.25
The two enlarged sketches compare a large grit size, above, used in making a fast-cutting wheel which gives a poor finish, with a small grit size, below, used in making a slow-cutting wheel which gives a good finish.

Fig. 7.26
A further enlarged view of the grains used in aluminium oxide grinding wheels shows their sharp edges and relative uniform size. Each of the SG grains shown here contains submicron particles which break off in grinding to form additional sharp edges.
Source - Norton

As a guide for manual grinding, 8-36 grit is coarse, 46-120 medium, 150-320 fine, and 400-600 very fine. Usually grains of different size - within a specific range - are mixed in most wheels to give better compaction and are shown as, for example, 46-60 grit.

Grain size is not an absolute guide - it depends on how grinding is carried out. Coarse wheels give a better finish than manually-controlled fine wheels with machine-controlled grinding, for example.

Abrasive Wheel Grade.

The proportion of bond to abrasive determines the grinding wheel grade. A wheel which has a higher proportion of bond is described as hard, and one which has a smaller proportion of bond is described as soft.

Hard wheels give a better and more consistent finish and wear down more slowly because they have a better grip on the grit. Wheels which are too hard retain their grit long after they have lost their effectiveness to then become glazed, a condition that requires dressing to restore their former bite.

Soft wheels have a weaker grip on their grit and are more open, and so readily *'shed'* to expose fresh grains beneath. In this way they retain their excellent grinding ability without need of frequent dressing. If too soft, though, they wear too quickly and subsequently cost more in abrasive terms.

The bond strength of grinding wheels varies between E which is very soft and Z which is very hard. The correct wheel grade is always the one that releases grains only when they become dull - and not before. Wheels harder than this give grinding problems, softer wheels escalate costs.

In general, harder tips need softer and finer wheels, and softer tips need harder and coarser wheels. To further complicate matters, though, different tip materials also may need different wheel grades to grind them satisfactorily.

Whilst grinding wheel hardness primarily depends upon the grain to bond proportion, the rim speed at which the grinding wheel runs also affects its apparent hardness. Wheels with a faster rim speed are *'harder'* than slower running wheels of the same grade.

This is an important point with plain wheels which grind on their rim - when profiling saw teeth, for example. As these lose diameter the wheel appears to be softening - wearing more quickly than when new - simply because the rim is moving more slowly. To counter this the arbor speed should be increased to maintain the same speed at the rim.

Abrasive Wheel structure.

This refers to the closeness of the grains. All wheels contain some gaps to allow the grains to work effectively and to contain material as it is abraded (and coolant when used).

The closest wheels, a low number on the scale 1-15, have over 60% by volume of closely-packed abrasive. These give a steadier and more controlled free-hand grind when dealing with a small contact area.

An open-grade wheel, with a higher number on the scale indicating a smaller proportion of abrasive, has additional gaps left by fillers which burn away in processing. This type is freeing-cutting and suitable for grinding a large contact area.

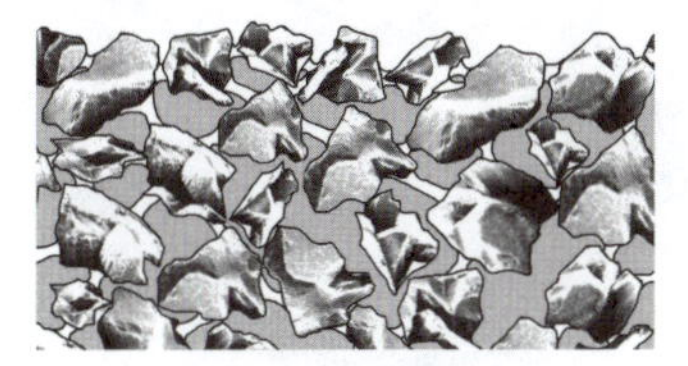

Fig. 7.27
With vitrified grinding wheels the grains are held together by bridges of glass. All contain gaps (the porosity) which varies in size according to the structure it is given - small gaps make a close structure, large ones make an open structure.

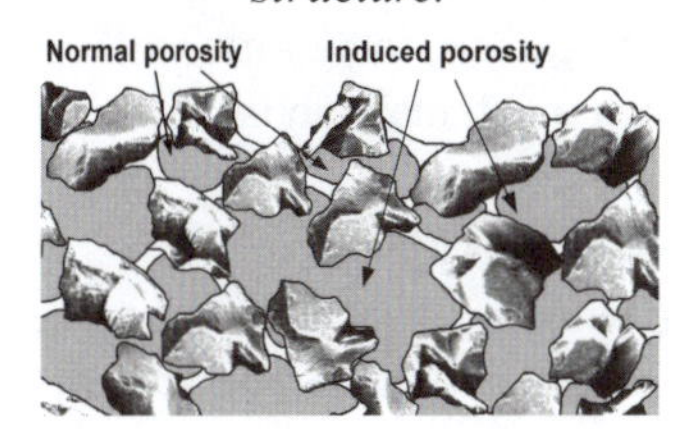

Fig. 7.28
This shows a very open wheel where extra spaces are induced in vitrified wheels by including volatile materials that burn away when the wheels are fired.

Abrasive Wheel Bond

The bond is what holds wheels together. Four types are used, vitrified, resinoid, shellac and rubber.

Vitrified wheels are the most common. They have a porous structure with grains held by bridges of glass formed by firing in kilns at temperatures in excess of 1,000°C.

These wheels are rigid and, whilst they can shatter under the wrong pressure, they are ideal for steel and Stellite grinding. With sulphur added they are suitable for wet grinding, with wax added they should be used dry.

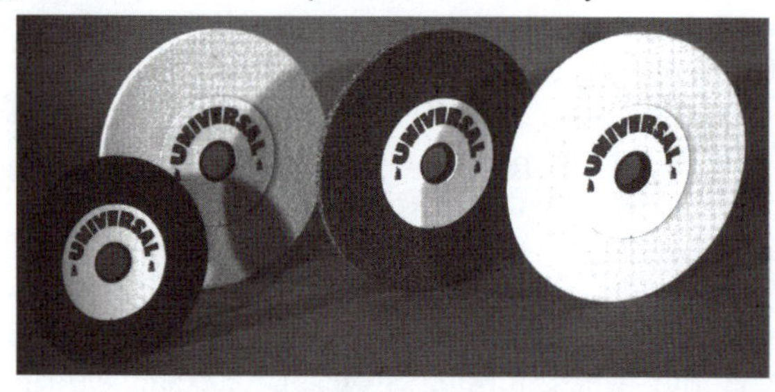

Fig. 7.29
Plain grinding wheels are produced in many diameters and thicknesses. Those for saw grinding usually have a vitrified bond.
Source - Universal

Resinoid wheels, using phenol formaldehyde as the bond, form a more compact but flexible grinding wheel which gives a better finish on surface grinding than vitrified wheels. Resinoid wheels are mostly used for dry grinding. Wet grinding resinoid wheels are made - but choose the coolant oil carefully as some attack the bond.

Fig. 7.30
Typical cut-off wheels. These may have a resinoid, rubber or shellac bond.
Source - Universal

Resinoid, shellac and rubber-bonding is mainly used for cut-off and parting wheels, sometimes with a fabric reinforcement. These stand much more abuse than vitrified wheels.

Abrasive Wheel Types

Edge-grinding wheels are flat and are intended to grind only on their outer rim or periphery, and are arbor-mounted between flanges. Never grind on the side of plain wheels, particularly thin vitrified wheels, as these can shatter if pressured.

Cup or dish wheels are shaped, as their names suggest, with parallel sides in cup form and with angled, possibly tapering, sides in dish form.

Cup wheels grind with their radial face edge.

Dish or saucer wheels can grind either with their outer rim as an edge-grinding wheel, or with their face edge as with a cup wheel. Both types are mounted to overhang the arbor and, in the case of cup and deep saucer wheels, normally with the nut and outer flange recessed below the abrasive level.

Safe use of Abrasive Wheels

Many safety regulations apply to grinding wheels, but these vary from one country to another. Make sure that the local regulations are understood and observed when fitting and using grinding wheels - they can be dangerous.

Always handle grinding wheels with care. Never use a grinding wheel that has been dropped, and never be tempted to roll them along the floor.

Grinding wheels should only be handled and fitted by a person competent to do so, and a visor or goggles must be worn by the person using the grinding wheel, and by anyone else in the area.

Storage

Many grinding wheels are brittle and easily damaged. Carefully store them in medium temperatures, away from strong sunlight, moisture and high humidity. Leave them undisturbed, storing thick plain wheels vertically on a two-point cradle, and large wheels in individual compartments. Thin plain wheels can be low-height stacked with corrugated paper between. Low-height stack cup and dish wheels (of the same dimensions) face-to-face and back-to-back with corrugated paper between.

All grinding wheels deteriorate with age, some faster than others, but in all cases use grinding wheels within two years of purchase. Keep small stocks only, use older stock first and re-order regularly from your stockist who, with having a big turnover, can supply fresher wheels.

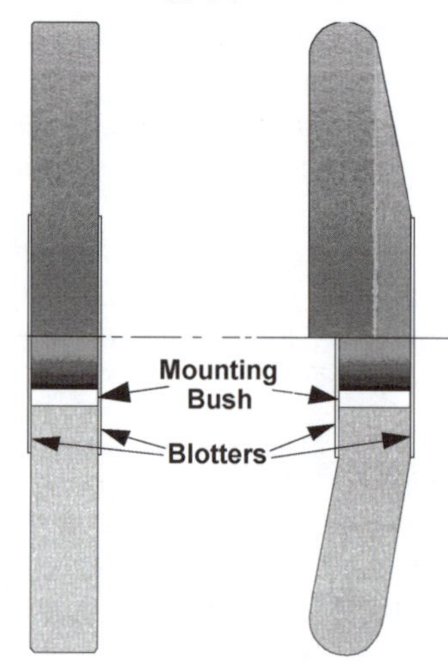

Fig. 7.31
Sketches of a plain, edge-grinding wheel, left, and a shallow dish wheel, right.
The blotter should remain on the wheels when mounted. A bush in the bore prevents the grinding wheel from scoring the arbor and provides a more stable mounting.

Fitting

Before fitting a grinding wheel carefully examine it for cracks or damage, and discard it if either is present. To test a vitrified wheel for cracks support it vertically and loosely through the bore. Tap it lightly with a small piece of wood *(never metal - this could cause the condition you are attempting to detect!).* A crack-free grinding wheel will *'ring'* quite clearly - a cracked wheel will sound cracked. The ringing tone varies with the type and dimensions of the wheel. *(Resinoid, shellac and rubber wheels do not 'ring', only visual inspection is possible.)*

Before mounting the wheel check that the wheel and mounting are clean and free of grease, oil and loose dirt, and that the wheel flanges run true and are free of warp or damage.

Flanges for plain wheels should be equal in diameter, with equal bearing surfaces which grip the wheel remote from the centre hole (they are recessed for this purpose) - and measure at least one third of the wheel diameter.

Periodically check that over-tightening has not warped the flanges. Use a straightedge across the inner face of the flanges to ensure that the contact faces near the rim are perfectly flat and that the recessed face remains well clear.

The grinding wheel should be a push-fit on the arbor - do not force-fit a wheel which is tight on the arbor, and discard wheels that are a sloppy fit.

Always fit blotters (never plain paper washers) between the wheel and the flanges - on both sides - where each should cover the entire contact area between wheel and flange. These cushion the clamping action, give a better grip and reduce the chance of cracking the wheel on tightening.

Tighten the securing nut evenly and progressively - but not excessively - merely enough to ensure a positive drive to the wheel.

Many wheel nuts are self-tightening, so apply light pressure only and never use an extension bar. With a multi-screw flange fitting tighten the screws progressively, crisscrossing to give a uniform pressure.

Avoid unnecessary grinding wheel changes - damage is often caused when mounting or removing a wheel. Some grinding machines have an arrangement where each grinding wheel is permanently mounted on a separate *(usually a taper-lock)* arbor which is switched, together with the grinding wheel, as a complete unit.

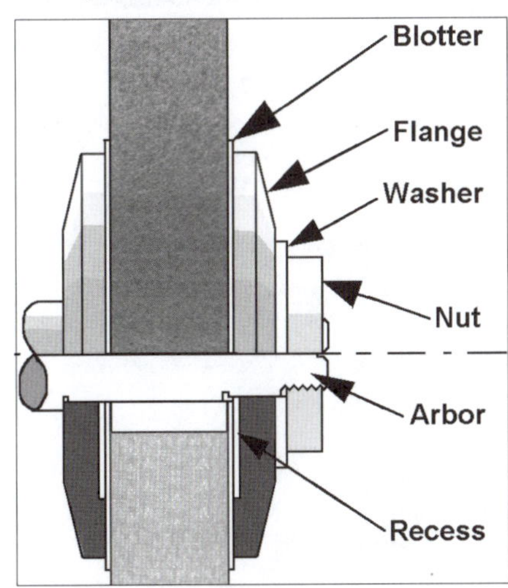

Fig. 7.32
Always place blotters between the wheel and the flanges (except on types of grinding wheel that specify otherwise - mostly these are not used in saw grinding). Flanges for plain wheels should be recessed as shown. The left-hand, fixed, flange should be keyed, shrunk or pressed onto the shaft and preferably trued-up on its own bearings after assembly.

Before switching-on the grinder replace all guards. These prevent accidental contact with the grinding wheel (as far as practical), deflect swarf and grinding grit, and should contain the pieces of an accidently broken wheel - without themselves breaking. Check that all parts are secure and that the grinding wheel turns freely. Make sure that anyone around is wearing goggles or a visor, and that no-one is standing in line with the grinding wheel. Allow a newly-fitted grinding wheel to run freely for about a minute before attempting to use it.

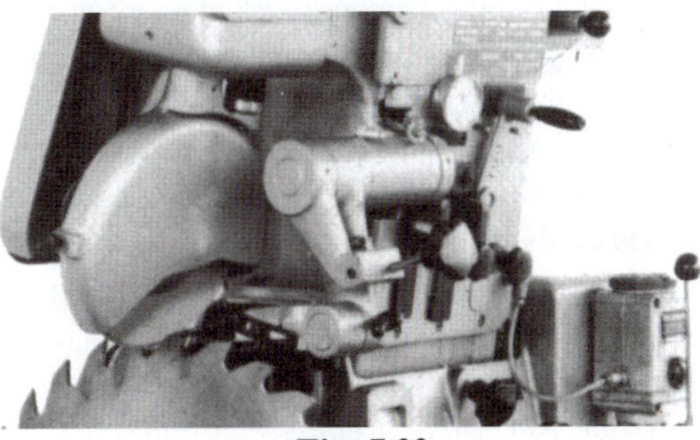

Fig. 7.33
Before starting, make sure that grinding wheels guards are properly fitted and secured, and that the grinding wheel is free to rotate, as this AS grinder.
Source - Iseli

On some older grinders using heavy wheels the wheel flanges are prepared to add or remove screw-in weights. These are used to balance the grinding wheel assembly. If the grinding wheel, flanges and arbor can be removed as a unit, then the assembly can be balanced between knife-edge rollers or bars, otherwise remove the drive belt, if any, and check by turning the arbor gently by hand.

All the balance screws are first removed and the assembly turned slowly. If it finally rests always at the same point this shows an out-of-balance condition - to compensate add screw-in weights to the topmost portion of the flange when stationary. Repeat until the assembly comes to rest in a different position each time. Check then that the balance screws are fully secured.

Operating Speed

All grinding wheels have a maximum surface or rim speed (often marked on the blotter) which should never be exceeded - check that this is the case. Recommended surface speeds are 4,500 - 5,400 ft/min (25 to 30 M/sec) for vitrified wheels and 6,000 ft/min (33M/sec) for resinoid-bonded wheels - but check the blotter or follow the makers recommendations.

With plain wheels which grind on the outer edge, or periphery, the rotary speed should be increased as the grinding wheel looses diameter to maintain the correct surface speed at the rim - but make sure that any speed increase gives a surface speed at or slightly lower than the maximum recommended. If speed is not increased as the grinding wheel wears, it will apparently become *'softer'* to wear away at an increasing rate.

As a guide, a vitrified grinding wheel will grind as though it is one grade softer if the rim speed drops by 750 ft/min (14 M/sec).

NOTE: Grinding wheels can, in fact, be made to operate as though of a higher or lower grade simply by varying the running speed - but remain within the speed recommendations.

Abrasive wheel defects

Glazed wheels. If a grinding wheel is too hard, the grains are retained long after they loose their sharp cutting edges to round-over and take on a glazed appearance. The grinding wheel then no longer works effectively, so grinding slows down, manual grinding becomes harder, the ground surface becomes polished, and enough heat may be created to burn the metal or thermally damage the saw tip.

Loaded wheels. Loading happens when material abraded from the tool lodges in and fills the pores of the grinding wheel instead of being thrown clear. This physically stops the grinding wheel working properly - an effect similar to that of a glazed wheel, but with a different cause. Loading is less likely if using flood coolant, although too high a mix of oil to water, or the wrong type of oil could encourage loading - check with your supplier.

The temporary cure for both glazing and loading is simply to dress the wheel but, if either continues, a permanent solution is needed. Use a softer wheel to cure glazing and a more open wheel to cure loading - but note that both cures also result in faster wheel wear.

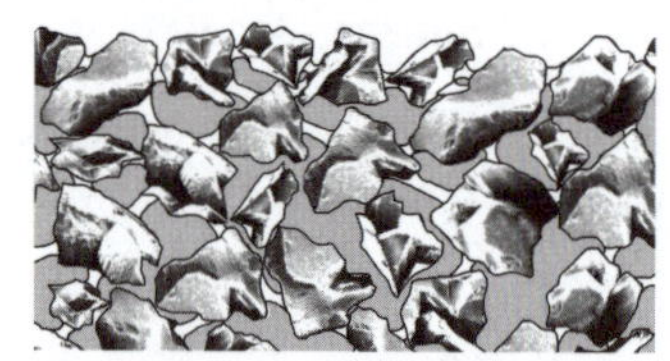

Fig. 7.34
The surface of a free-cutting wheel should have sharp grains and be free of grinding swarf.

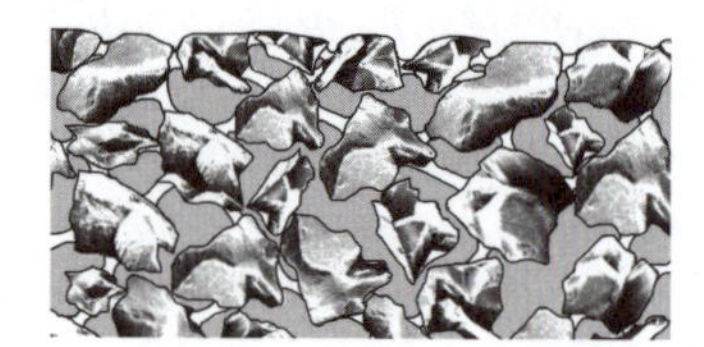

Fig. 7.35
A grinding wheel that is glazed has too hard a bond so that, instead of breaking away when they lose their cutting edges, the grains are still retained to become worn, rounded-over and take on a glazed appearance.

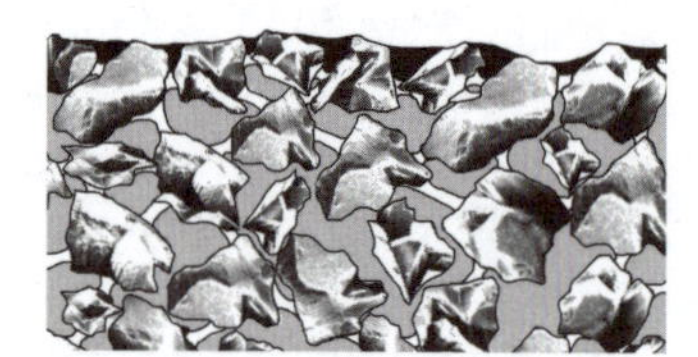

Fig. 7.36
A grinding wheel that is loaded has the surface pores clogged with grinding swarf so that the still-sharp grains are unable to grind properly.

Abrasive Wheel Truing and Dressing

When truing or dressing always wear eye protectors.
Truing is the process of making a grinding wheel run true and restoring its profile by removing unwanted abrasive.

Dressing or conditioning is the process of treating the grinding wheel to restore its cutting ability by removing metal deposits and eroding away bonding material - so exposing sharp grains beneath - without actually dislodging abrasive grains in the process.

With regular grit wheels both processes are carried out at the same time using an abrasive dressing stick, a rotary dresser or a single-point or cluster diamond.

Two types of rotary dresser are available. One type uses an abrasive wheel and the other type uses metal discs. Both rotate, usually on a ball bearing arbor, in basically what is a grain-crushing operation.

Abrasive wheel dressers have the abrasive wheel mounted either on the end of a handle or between two ball grips

Metal disc dressers have metal discs on a ball bearing arbor at the end of a handle. The regular metal type, with star-shaped and plain discs, leaves the vitrified grinding wheel open and fast-cutting. Wavy-disc dressers leave a slower-cutting surface but one which gives a better finish.

Both dressers are used dry and with the wheel running at normal speed. In use the dresser is traversed across a straight wheel or around a shaped grinding wheel - whilst at the same time being supported firmly. Although many filers originally used these types, diamond abrasive sticks and dressers have now largely replaced them.

Fig. 7.37
A rotary abrasive wheel dresser mounted between two ball grips.
Source - Wadkin

Fig. 7.38
A disc type wheel dresser. The one shown has alternate star and plain metal wheels.
Source - Wadkin

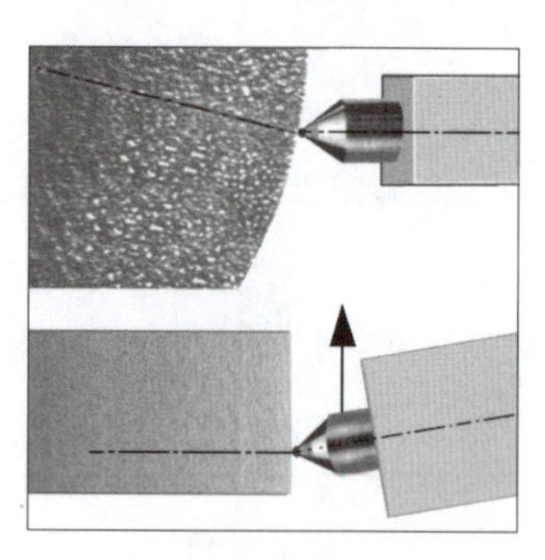

Fig. 7.39
When using a single point diamond dresser this should preferably be held at the drag angles shown.

Regular Abrasive stick dressers are supplied in various shapes, grades and sizes and are often used freehand without any support - although this is not recommended by any maker.

Diamond dressers produce a close surface on the wheel which gives an excellent ground finish on the edge of the sawblade. Both cluster and single-point diamond dressers are used.

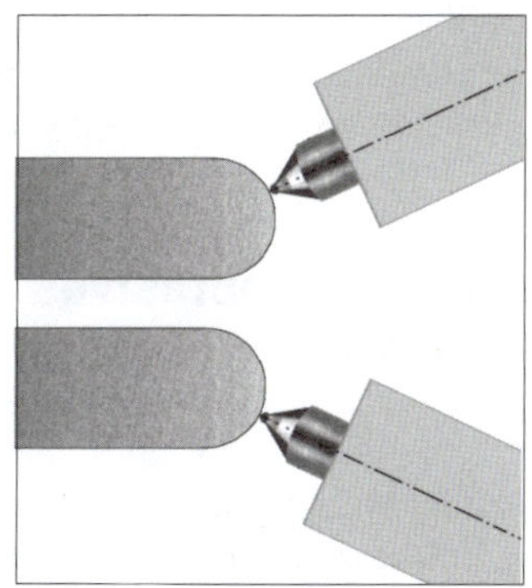

Fig. 7.40
When shaping abrasive wheels it is better to hold a single point diamond dresser at the recommended drag angle, but with mechanically-controlled dressing, as used in woodworking, the diamond often remains square-on in order to shape the wheel whilst moving in either direction.

A single-point dresser should be mounted at a drag angle of between 5 and 15 degrees, and should be used with plenty of coolant - if practical. Free-hand dressing is practical only when using cluster diamond dressers - and then only provided this is given rigid support and a carefully controlled movement. Far better results are possible, however, when both single-point and cluster diamond dressers are mechanically-controlled.

With both types traverse slowly across a flat wheel or around a shaped wheel to give the best finish, and remove only a small amount of abrasive at each pass. Systematically rotate a cluster diamond regularly to avoid forming flats.

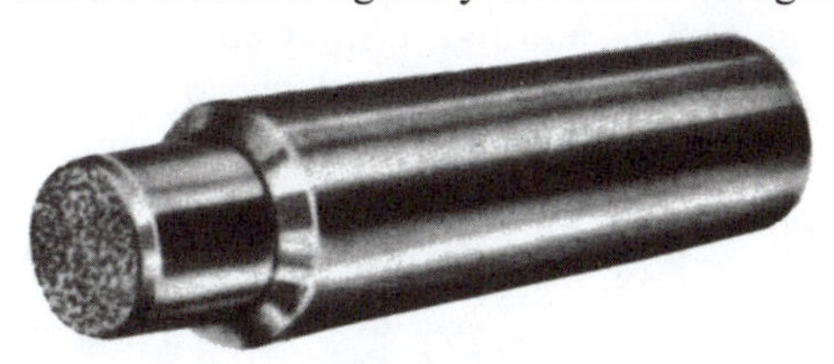

Fig. 7.41
A cluster type diamond dresser.
Source - De Beers Industrial Diamond, Winter

Diamond abrasive sticks are available in a wide range of shapes and sizes. They are a direct, but much more efficient, replacement for regular-grit abrasive sticks. The Formset diamonds manufactured by General Electric, for example, are polycrystalline in structure and do not have the weak cleavage planes common to natural diamonds.

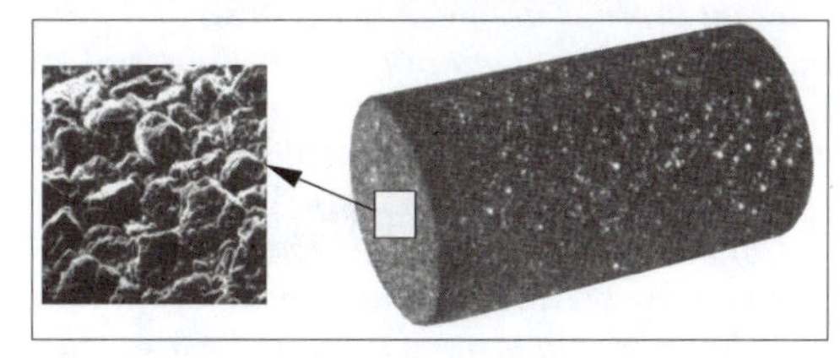

Fig. 7.42
A Diamond Abrasive stick. This type, the Formset, is polycrystalline in structure.
Source - General Electric.

Because saw grinding is an old-established practice, dressing evolved into a manual, hit and miss, operation. The user was given minimal guide by machine makers as to how dressing should be carried out, and no-one then supplied a mechanical dresser. As a result most used a dressing stick or rotary dresser free-hand and without proper support - and wheel profiles changed a lot.

This need not be so. There are quite simple mechanical dressers available for some saw grinders (see Fig. 6.39) and similar dressers are fitted on other woodworking grinders.

Superabrasives

In addition to regular abrasives, two extremely hard superabrasives are now widely used. One is natural or manufactured diamonds for grinding tungsten carbide. The other is cubic boron nitride *(CBN)* sold by **General Electric Company** as **Borazon** and by **De Beers Industrial Diamond** as **ABN** *(Amber Boron Nitride)* for grinding HSS, Stellite and Tantung.

Natural industrial diamonds, the remains of mined diamonds from which the best has been selected for jewelry, are crushed and sieved as part of the manufacturing process. Although size is sorted by the sieving process, control of shape is impossible.

Industrial quality natural diamonds - used as-mined - do not necessarily have the most suitable shapes for abrasives. They may have unsuitable cleavage faces *(the way a diamond splits apart under pressure)* or they may also contain impurities which reduce their hardness and grinding efficiency.

Manufacture of Superabrasives

As long ago as 1797 it was shown that diamonds are merely a high-density form of carbon *(other forms are oil, coal and graphite)*. But it was 1954 before the General Electric Company of Worthingon, Ohio, USA, perfected a commercial method of converting graphite into a synthetic diamond using extreme heat and pressure.

In an extension of the technology in 1956, General Electric were able to form a solid phase of boron and nitrogen *(that does not occur naturally)* as cubic boron nitride.

The greatest impact of these remarkable manufacturing achievements is in the ability to key special mechanical properties into the abrasives during synthesis *(synthesis is the process of producing synthetic materials)*.

These properties include a modified impact strength, absence of the random cleavage planes of weakness common to natural abrasives, specific planes of cleavage suited to the application, and isotropic form *(equal stress-resistance in all directions)*.

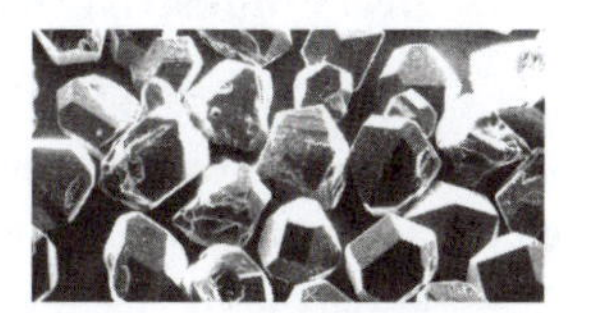

Fig. 7.43
An enlarged view of manufactured diamonds. This shows G-11 , described as a blocky, tough diamond that gives high material removal rates and good finish.
Source - General Electric

Blockiness is another in-built feature. This describes a

superabrasive grain *(usually called a particle)* which contains numerous edges so that, no matter how it is positioned when bonded, it always presents at least one sharp cutting edge.

These abrasives, in fact, can be tailor-made to match the application rather than simply accepting what nature provides. It is for this reason they are called superabrasives.

Fig. 7.44
An enlarged view of CBN abrasives. This shows amber-coloured CBN which, without coating in this form, is termed Amber Boron Nitride ABN300.
Source - De Beers Industrial Diamond.

In many instances superabrasive particles are coated during manufacture with different materials, usually metals. These, in providing a better grip in resin-bonded wheels, securely hold friable grit particles to give a more precisely-controlled breakdown, and this effectively increases the 'G' ratio by as much as 100%.

Metal coating also acts as a heat-sink to prevent the bond being weakened by the excessive heat produced in grinding. A ferrous bond allows long, thin particles to be set in a specific direction when bonded in a magnetic field.

One long-standing problem with the first superabrasives was that each particle had only a single effective cutting edge and, once this had gone, rising heat and pressure then loosened the particle to waste it and expose another beneath.

The effective life of some diamond and CBN particles can, however, be substantially increased to give a more effective and cost-efficient grinding process. By using a closely-controlled manufacturing system, these particles sub-fracture along designed planes of cleavage to expose fresh cutting edges when the initial edges no longer abrade efficiently.

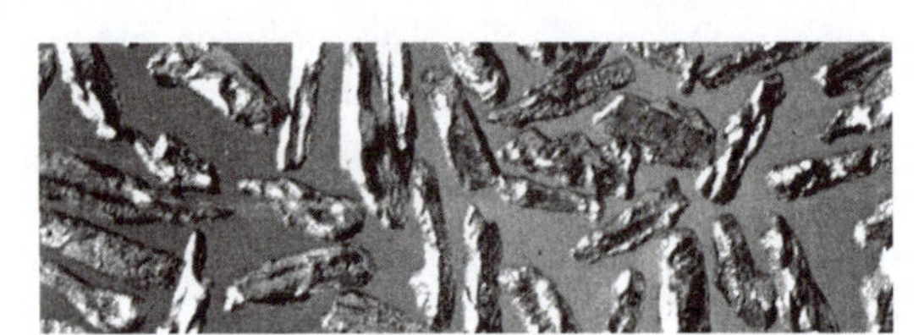

Fig. 7.45
Diamonds can be grown into long needle-like shapes. In peripheral grinding wheels metal-coated particles are oriented by a strong magnetic field to bond radially to the grinding wheel centre.
Source - De Beers Industrial Diamond

Another method is by growing them into a long shape which fractures across the thickness to expose new and exceptionally sharp cutting edges. These are not bonded together in a random way as by doing this their special properties would be wasted. Instead the particles are coated with a ferrous *(an iron-based mix)* and bonded to the matrix whilst in a strong magnetic field.

The magnetic field makes the particles lie along the magnetic lines - which are arranged to be square to the grinding face - and so they are bonded with the long particles permanently in this position.

As the wheel wears each particle first exposes a sharp tip then, as further wear takes place, the tip dulls and breaks off to expose another sharp edge, and so on, until the last sharp edge of the particle dulls and it is finally dislodged. In this way these particles have an extended and very effective life.

Advantages of Superabrasives

Superabrasives wear very slowly indeed, for example CBN outlasts aluminium oxide by a factor of 100 - 200 times. Although CBN wheels initially cost more, they are often less expensive over time when taking into account down-time, wheel replacement, dressing etc. In some cases, of course, CBN is essential for efficient production grinding of, say, Stellite tipped sawblades.

Although CBN wheels are less aggressive than aluminium oxide wheels at regular wheel speeds, they perform equally well or better than them at the higher surface speeds they are intended to work at. They grind cool and fast, maintain their shape well and need minimal dressing.

When replacing regular grit wheels with CBN wheels a finer grit can be used, for example a 40-60 aluminium oxide wheel would be replaced by a CBN wheel of between 80-120 grit. Specify the same wheel size - a smaller wheel would run and grind too slowly and would have a much reduced 'G' factor.

A hidden advantage is the more consistent performance of CBN. With plain aluminium oxide grinding wheels diameter loss gives a progressive change in performance. In contrast cbn wheels do not effectively reduce in diameter, and so never vary in performance for this reason.

They should, however, always be run at the recommended surface speed to give the best possible and most consistent performance.

Fig. 7.46
Typical superabrasive wheel bodies. The six on the left are diamond abrasive wheels for grinding tungsten carbide saw tips. Different profiles are needed because of differences in grinding machinery and in tooth profiles and sizes.
Third from the right is an edge-grinding diamond wheel used for top grinding tungsten carbide saw tips on some grinding machines.
The two on the extreme right are cbn wheels for grinding Stellite-tipped saws. Narrow wheels have a solid tip, wider ones a thin layer of cbn around a formed wheel. Half-round profiles only are shown, but alternative profiles can be manufactured.

As regards diamond grinding wheels, there is no other viable choice for grinding tungsten carbide. Some still insist on using green grit wheels *(silicon carbide)*, but this is false economy as these wheels quickly fill and distort - and can cause thermal cracking of the tips.

Superabrasive Wheel Bodies

Superabrasives are not made as a solid wheel as are regular abrasives, but as a relative thin layer of diamond or CBN particles, perhaps up to $^{1}/_{8}$in (3mm) thick, bonded to a metal or phenolic-resin wheel body.

Solid wheel bodies, usually of aluminium, are used for delicate wheel sections where strength is essential, for wheels produced only in small quantities, or where excessive heat is generated.

Diamond wheels are prone to breakdown if grossly overheated when grinding. Solid metal wheels are used when this might be a problem as they dissipate heat - spread it outwards and away from the grinding area - more efficiently than resin or resin-bonded types.

Whilst a solid wheel body helps to prevent overheating, an ample supply of coolant - directed where it is most effective - is also necessary.

Resin-bonded, aluminium powder wheel bodies are produced using individual dies. They cost less to manufacture in quantity and so are used for regular grinding wheel bodies. This type of body is easily worked, and is the preferred type when the wheel has to be wasted during its life (the backing dressed-away to prevent this fouling as the abrasive wears down).

Phenol-resin wheel bodies are less rigid than metal wheels of whatever type, and they can flex slightly in operation. Because of this they give a better finish for the same abrasive grit and grade than aluminium bodied wheels.

Electrostatically deposited wheels have only a single layer of particles metallically-bonded onto a pre-formed wheel body, and for this reason they are low in cost compared to resin-bonded types.

Often they are used instead of resin-bonded wheels for one-off applications. This is because turning a wheel body to an individual shape for electro-depositing is relatively simple and cheap - compared to the need of making the mould necessary to produce resin-bonded wheels.

Fig. 7.47
Enlarged view of electrostatically deposited Amber Boron Nitride 300 suitable for grinding steel.
Source - De Beers Industrial Diamond

This type of wheel is also extremely tough, grinds freely, needs less power and generates less heat than most other types - but it also has a limited life.

These wheels never effectively lose their profile, but when the thin abrasive layer eventually wears away at any point they have to be discarded. Re-shaping is not possible, but as loading can occur dressing or conditioning may be needed periodically.

With regular superabrasive wheels both diamond and CBN is formed in a much thicker layer using a metal, vitrified or resinoid bond - usually the latter for woodworking applications. The thickness of superabrasive material bonded to the wheels varies according to the wheel thickness and the application.

These wheels have a much longer life than electrostatically-deposit wheels, but cost more and they eventually wear in use. Whilst it is possible for the makers to re-shape them, most users simply continue with worn wheels but in a less critical application - see later notes.

Superabrasive Wheel Types

Superabrasive grinding wheels are made in similar styles to regular abrasive wheels, Plain, Cup and Saucer. Where required the wheels are profiled, for example, plain CBN wheels shaped to a half-round profile to grind Stellite-tipped saws.

Superabrasive Wheel Grades

As with regular abrasives, superabrasives can be made with various particle sizes, ranging from a coarse 60 to an extra fine 1,200, and with different bond strengths. The effect of using these alternative particle grades are similar to those described earlier for regular grit-type abrasive wheels.

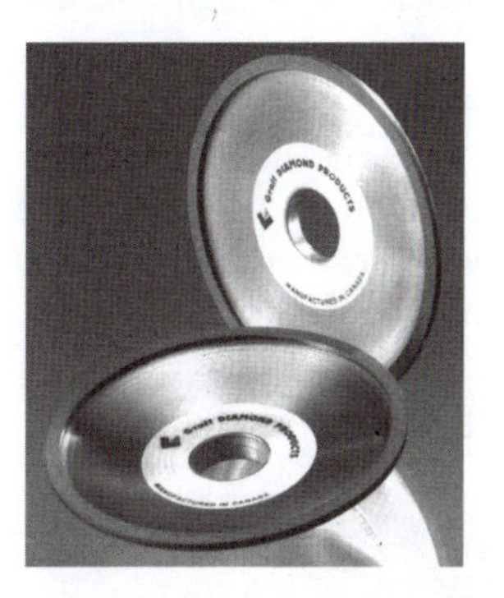

Fig. 7.48
Left, typical shallow saucer diamond grinding wheels.
Right, typical deep saucer diamond grinding wheels.
Source - Graff

An added factor is the concentration, the ratio of particles-to-bond. The average is 100 concentration, a ratio considered to give the best 'G' result for wet grinding. For dry grinding a concentration of 75 usually gives the best 'G' ratio.

The concentration can be lower at 75 or 50 to give a lower wheel cost - and faster wheel wear - but this concentration is suitable only when using very fine grit wheels or where there is a large wheel-to-tool contact area.

A higher concentration is essential for heavy-duty grinding. Some grinding wheel makers use a different grading system for grinding wheels, so rely on the machine makers advice for the best type of grinding wheels to use.

NOTE: Many machine makers recommend that only grinding wheels marked as suitable for their machines should be used. Usually such grinding wheels are supplied exclusively by the machine maker, and a suspicion is that they are merely regu-

lar grinding wheels that are being sold at a higher price than identical wheels available elsewhere at lower cost.

However, there has been a considerable amount of research into grinding wheel specification for the modern, highly efficient cnc grinders that have recently become available. In these cases it is essential to use the recommended grinding wheels as they have been developed, along with the grinding machines themselves, to raise the grinding efficiency to a far higher level than was previously possible. Obviously this puts extra strain on the grinding wheels - hence the strong recommendation to use only the grinding wheels sold or recommended by the machine makers.

Wet or Dry Grinding?

Resin-bonded Superabrasive wheels are normally manufactured for wet grinding as this gives by far the best results - a better finish and a longer abrasive life. Grinding wheels made for wet grinding should always be used wet, otherwise efficiency will suffer. Coolant can be either a soluble oil and water mix, or a neat, thin oil, As a general rule grinding wheel efficiency improves as the oil content (of an oil/water mix) is raised, although too much can cause wheel clogging.

Most modern cnc grinder makers now recommend the exclusive use of grinding oil rather than a water and oil mix. This gives a higher grinding efficiency, keeps the machines cleaner and well-lubricated and causes no rusting of machine parts or the sawblades themselves. This oil can, however, be a fire risk, so machines using it should preferably have an automatic, built-in fire extinguisher to immediately quench any ignition - especially if the machines are run overnight unattended - as is the modern practice.

For machines which do not have coolant facilities, dry grinding superabrasives are available. In all cases check the grinding wheel specification to make sure you use the wheels correctly - failure to do so will give poor results.

Working with Superabrasives

Fig. 7.49
A shallow diamond saucer wheel being used on an automatic carbide saw grinder CHP.
Source - Vollmer

Diamond grinding wheels should only be used on tungsten carbide. Never use them to grind the tooth backing down - there are other ways, see Chapter 10. Where grinding both tungsten carbide and the backing steel is unavoidable it is necessary to use a type of diamond specifically produced for this purposes, such as the De Beers PDA321N5 - provided the proportion of steel to carbide is 40% or less. Other makers use a mix of diamond and CBN to grind both tips and backing at the same pass.

Diamond wheels are usually self-dressing when grinding tc tooth tops , i.e., the grinding wheel wears evenly because of the grinding action.

When grinding the fronts of teeth, or in some extreme cases when grinding tooth tops, superabrasive wheels can wear unevenly. When this happens the user has the choice of returning the wheel to the makers for re-dressing, or continuing to use it perhaps for rough grinding only - for example initially dressing replaced tips prior to the final regrind using a new wheel.

CBN grinding wheels are most suitable for high alloyed steels, weld-deposit Stellite and solid brazed-on Stellite or Tantung tips. They can be used to grind both the steel body and the tip on Stellite-tipped saws - where the normal practice is to fully profile the teeth in the same way untipped saws are ground.

Superabrasive wheels are dressed to shape by the makers, often using a regular abrasive wheel rotating at a controlled speed - brake grinding - only rarely are they shaped by the actual user.

It is a common fallacy that CBN wheels keep their shape until worn out, in fact they do not, wear does take place - though much, much more slowly than with regular abrasive wheels.

Plain cbn grinding wheels edge-shaped for profiling Stellite-tipped saws eventually distort as wear takes place. Worn wheels can be reshaped by the maker, but this is inconvenient, increases cost and looses costly grit. Unless a worn wheel produces an unacceptable gullet profile there is no good reason not to continue using it. In some cases a worn wheel can be reversed to even out the wear to a certain extent - but only on certain machines. Usually a worn wheel is reserved for rough grinding badly-shaped sawblade teeth prior to finish-grinding using a new grinding wheel.

Grinding Machine Requirements

The design and condition of the grinding machine affects the quality and efficiency of grinding when using superabrasives. Regular abrasive wheels are much more tolerant and will produce acceptable results even if the machine is not up to scratch - although it may need some coaxing - older operators and filers are expert at doing this.

Superabrasives demand machines that are solid and vibration-free and with bearings, arbor and flanges that run true and in perfect balance. These requirements do not only apply to the grinding wheel itself, but also to the sawblade and its mounting. The coolant supply must be ample, dependable, with an efficient filtering system and able to direct coolant to where it is most effective.

Old and worn machines cannot be upgraded by switching to superabrasives - attempting this could be an expensive mistake. The machine should preferably be reasonably new - or of good quality manufacture and properly maintained.

Ideally use a machine specifically designed and built for superabrasives - these are heavy and well-engineered, with arbor speeds and wheel mountings to take full advantage of the special qualities of superabrasives.

Fig. 7.50
Typical of modern, dedicated tungsten carbide machines is this Woodtronic NC3 numerically-controlled saw grinder.
Source - Walter

Keeping the Grinder Clean

Grinding is a dirty operation compared to most others in woodworking, and cleanliness is essential to reduce wear and to keep machines working efficiently. To maintain a healthy atmosphere in the grinding room airborne grinding dust and coolant particles should be kept to the immediate area of the grinding machine itself.

On all manually-operated dry grinding machines make sure that guards and the essential exhaust units are properly fitted, set and working efficiently. When using coolant on manual grinding machines, and on automatic machines where no enclosures are provided, adjust the rate and direction of coolant to avoid forming a mist, and set any baffles and guards to contain the mist as much as possible within the immediate area of the machine.

Mounting Superabrasive wheels

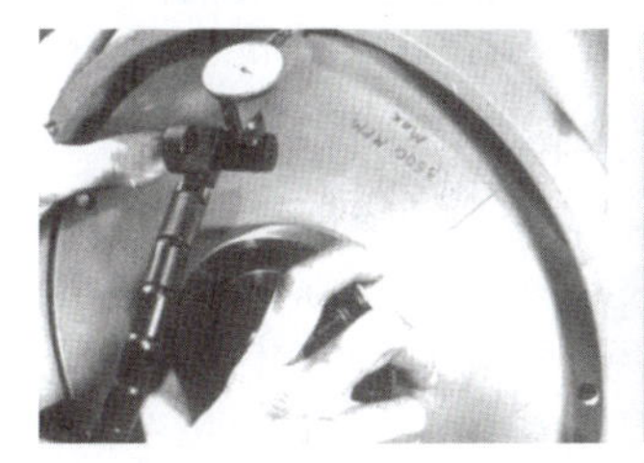

Fig. 7.51
Using a dial indicator to check the concentricity and side run-out of a superabrasive grinding wheel.
Source - General Electric

Great care is needed in handling and mounting superabrasive wheels as they can easily be damaged - keep them in their containers when not in use. Do not use blotters, and make sure that the mounting faces are clean and free of any burr, grinding grit or damage. Tighten the wheel generally as described earlier for regular abrasive wheels.

After mounting check the wheel for roundness and truth. It is not practical to check the abrasive itself, so check for roundness and truth against the body of the wheel. Some makers form a groove for this purpose and true the abrasive to this in manufacture.

Side run-out is possible if the flanges are distorted or damaged, or if grit is trapped between the wheel and the flanges. If everything is clean and apparently correct but side run-out still persists, then check the fixed flange and the arbor for truth, using a magnetic dial indicator. If either shows run-out this must be made good before using superabrasive wheels.

A superabrasive wheel that runs-out - for whatever reason - will sound and grind irregularly. It can bounce and vibrate the sawblade to give a poor surface finish - and possibly damage the wheel. Vibration is sometimes present with new wheels before they are run-in, see later notes but, if vibration remains after running-in, stop the process before damage takes place as there may be a basic fault.

Slight vibration actually helps regular grit wheels to shed - lose grit at the proper rate. With superabrasive wheels, though, vibration is a deadly enemy that can cause endless problems and inefficiency.

Run-out at the rim only might be improved by relocating the grinding wheel at a different rotary position.

Another way, regularly used by engineers, is to partially tighten the wheel and lightly tap the high spot with a rubber mallet before fully tightening it. But take care, superabrasive wheels are so easily damaged.

This practice needs to be repeated each time a wheel is fitted - a routine procedure in engineering - but not the case with woodworking applications. Woodworkers, saw doctors and grinders much prefer to simply fit a wheel and expect it to be correct - without fiddling about in this way - see notes on running-in.

Operating speeds

Diamond wheels are suitable for roughly the same surface speeds as regular abrasive wheels of between 3,600 - 5,400 ft/min (20-30M/sec), so they can directly replace regular grit wheels - provided the machine and application is suitable, of course.

CBN wheels can run at surface speeds greatly in excess of regular-grit speeds - up to 10,000 ft/min (60m/sec). At these higher speeds there is faster stock removal and an improved 'G' ratio. Machines built purely for CBN wheels take full advantage of the higher running speed their use makes possible. The high running speed of CBN wheels and the fine grit they allow gives a greatly improved finished to the ground teeth. This not only gives a better finish to the sawn material but, more importantly, a longer sawblade run-time.

Running-in Superabrasive wheels

Grinding wheels for grinding woodworking tools are supplied ready shaped and trued to their bore, and so should never require truing. New wheels, however, do need running-in - a form of truing - but this simply means grinding slowly at first, using a light grinding action. This naturally wears away any minor high spots during the running-in period.

It is best to leave superabrasive wheels in place after first fitting and running then in. This is not practical on grinders where wheels are routinely switched - unless each grinding wheel has its own removable arbor.

Otherwise file a shallow reference mark on the fixed flange. Fit each new wheel when this reference mark is at the top, and make a corresponding mark opposite this on the wheel itself. When refitting each wheel bring the flange mark uppermost and set the wheel mark in line with it before tightening. This should reduce any minor run-out there might be.

Dressing or Conditioning Superabrasives

All superabrasive particles are initially buried in the bond and, as particles above them wear-out and are dislodged, the bond around them has to be eroded away for the particles beneath to cut effectively.

This may be done wholly or partly by the grinding action itself - but only when the wheel is perfectly matched to the grinding operation. In most cases dressing or conditioning the wheel is still necessary from time to time. This is evident when the wheel no longer grinds cleanly, when it labours when grinding, and when the wheel appears to take on a glazed and solid appearance.

The condition is simply cured by using an abrasive dressing stick, usually an A220G for cbn wheels up 170/200 grit size *(use finer sticks for finer grit wheels)*, and a C100E for resinoid bond diamond and CBN wheels. By forcing the abrasive stick against the grinding face of the wheel, whilst running, it breaks-out the high bond and dislodges any metallic deposits within the pores of the wheel face. This it does without dislodging the actual abrasive, or at least with only minimal abrasive loss.

As conditioning takes place most wheels gradually lighten in colour. Conditioning is complete, and the grinding wheel restored to its original condition, when the dressing stick begins to wear down at an increasing rate. Always dress the wheel on the down-cutting side, and hold the stick firmly at a downward angle to avoid vibration and snatch.

Fig. 7.52
Using an aluminium oxide abrasive stick to dress a cbn grinding wheel. The stick should be forced into the wheel quite hard, usually by hand, but better if a holding device is used. The wheel rotates clockwise, and coolant is used to improve the dressing action.
Source - General Electric

Grinding Coolant

All grinding wheels intended for wet grinding must be used with a suitable grinding fluid, normally a water and soluble-oil mix, or a neat oil for some grinders using superabrasives. The term used is coolant.

Its purpose it to prevent overheating whilst grinding, to lubricate the cutting action, to prevent wheel loading, and to flush-away swarf.

In its simplest and earliest form, coolant was applied by drip feed to a pad in contact with the grinding wheel. A better alternative was an oil mist system where coolant was sprayed onto the grinding wheel to coat this - and just about everything else - with coolant. The main attraction of this system was its low-cost and general simplicity - but it requires an air supply, constant replacement of the oil - and it puts carcegenic oil particles into the air. It is not now recommended for this reason.

Flood coolant

Fig. 7.53
Grinding a Stellite-tipped bandsaw using flood coolant on a Vollmer CAS 44U.
Source - Vollmer

Where suitable, a flood coolant system is the best and most efficient type. This circulates coolant or oil from a reservoir, through a pump and pipes which direct it to the point of grind, after which it is collected in a receiving tray, filtered and returned to the reservoir. Coolant works better if properly filtered on an on-going basis. The filter can be a fine metal mesh or a replaceable filter paper. Alternatively the reservoir tank can be fitted with baffles to settle-out heavy particles, plus a top baffle to keep back floating scum.

A magnetic separator is sometimes fitted to continuously clean coolant. This can be permanent magnets placed in the tank and which need cleaning periodically, or a mechanically-driven magnetic cylinder rotating in the return flow to the tank. The cylinder picks up a mixture of magnetic particles and grinding waste to deposit these onto a collecting tray via a scraper.

With all these systems the coolant is continuously recirculated and only requires topping-up occasionally *(usually when the coolant turns frothy)* by adding water only to an oil/ water mix (neat oil doesn't evaporate). Coolant needs replacing and the tank and system flushing-out when the oil loses its effectiveness - perhaps four times a year, but check makers recommendations.

Using Flood Coolant

In most cases direct the coolant spray nozzle at the saw tooth rather than the grinding wheel itself. This is usually more effective as coolant directed at the wheel can often create an excessive and ineffective oil-mist spray. Most coolant pipes are now poppet-type fittings that can be shortened or lengthened as required and which, though very flexible, stay in position once set. There are also different types of single and multiple end nozzles that give the user lots of options.

Often the air flow created by a rotating grinding wheel can actually deflect a coolant spray directed at it. Some wheel guards incorporate a baffle to stop this airflow With these it is possible to direct coolant onto the wheel beyond the baffle. In some situations this helps to make coolant cling to the wheel and work more effectively. To maintain efficiency the guard must be regularly adjusted to suit wheel wear.

Fig. 7.54
Modern coolant pipes are of the poppet type which can easily be changed in length or nozzle fitting, and stay in place where positioned. A single pipe is shown on this Woodtronic NC3 carbide saw grinder.
Source - Walter

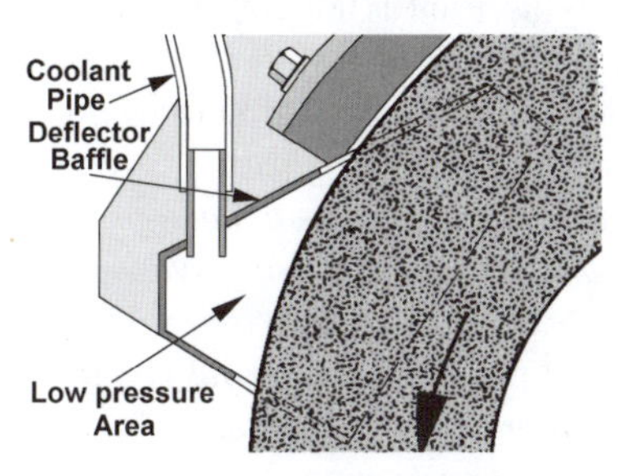

Fig. 7.55
To deflect the airstream created by the wheel rotation, an enclosure can be fitted to the wheel guard. This allows coolant to be fed into a low pressure area beyond and so make it cling to the wheel.

Flood coolant should always be ample and continuous. An irregular supply is worse than none at all as this can cause thermal cracking of tips during grinding.

Vitrified wheels are porous and can absorb a great deal of coolant. This can gravitate to the lower half of the wheel when stopped. Then, when the wheel is restarted, excessive vibration will occur because the wheel is unbalanced. Running quickly throws coolant from the wheel but, in exceptional circumstances, the vibration caused before coolant is thrown off can damage the wheel or its bearings.

To prevent this happening always start the wheel before starting the coolant, and switch-off the coolant before stopping the wheel.

Advantages of using Coolant

Coolant gives a better ground finish, improves grinding efficiency, allows a faster grind, greatly reduces the danger of thermal damage *(cracking or de-tempering)*, and gives a cleaner environment - if used properly. It can also significantly extend the working life of the grinding wheel when compared to a similar wheel made for dry grinding.

Types of coolant.

Originally an aqueous coolant was used with saw grinders consisting of a mixture of water and soluble grinding oil. When in the correct mix - take note of suppliers recommendations - this provides effective cooling at the point of grind with a high heat dissipation rate, reinforces the cutting accuracy of diamond grinding wheels, absorbs cobalt released in grinding and prevents corrosion.

It does, however, need constant monitoring of the water/oil concentration to keep the two in correct proportion. It requires regular cleaning of the machine, which should preferably be totally enclosed and with a coolant cleaning unit fitted. Choose coolant or oil carefully, use a grinding coolant or oil - not cutting coolant - and ensure that it has no harmful effects. *(Some coolants have a very short life, some damage human skin or cause infection, others actually attack the paintwork or metal they are supposed to protect - and virtually all oils are carcegenic).*

The current trend is to replace aqueous coolants with synthetic grinding oils. These give effective cooling at the point of grind although twice the volume of oil is needed because of the lower thermal capacity. Using a pure oil protects the working parts of the machine and is obviously highly effective in preventing corrosion. It efficiently and continuously cleans away carbide grit to prevent grit build-up, and so requires less machine cleaning.

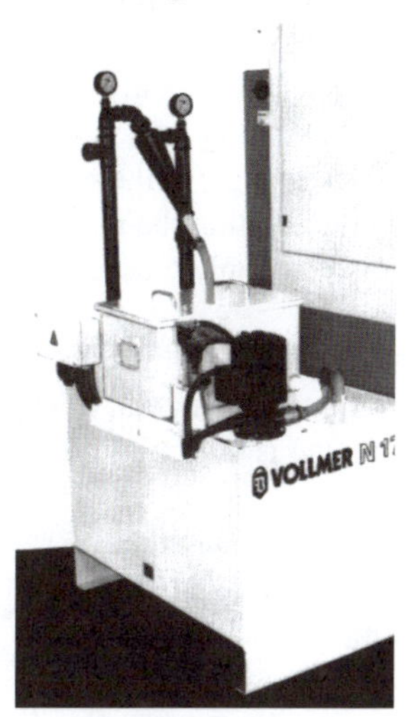

Fig. 7.56
Coolant cleaning system for flood coolant type grinding machines, N170.
Source - Vollmer

Unlike aqueous coolants, grinding oils do not evaporate and, due to the high ageing resistance, long exchange intervals are possible - but always use the oil recommended. However, grinding oil is combustible and requires by *(German)* law a full machine enclosure with special sealing elements and a drip tray, built-in fire extinguisher and an extraction device with downtream electrostatic filter and pressure equalizing device. Although the use of pure oil requires a high investment cost, those already using it believe the higher efficiency it achieves more than compensates for this.

Keeping the Grinder Clean & Wear-Free

Grinding is a dirty operation compared to most others in woodworking, and cleanliness is essential to reduce wear and to keep machines working efficiently.

To maintain a healthy atmosphere in the grinding room, airborne grinding dust and coolant particles should be kept to the immediate area of the grinding machine itself.

On all manually-operated dry grinding machines make sure that guards and the essential exhaust units are properly fitted, set and working efficiently. When using coolant on manual grinding machines, and on automatic machines where no enclosures are provided, adjust the rate and direction of coolant to avoid forming a mist, and set any baffles and guards to contain the mist as much as possible.

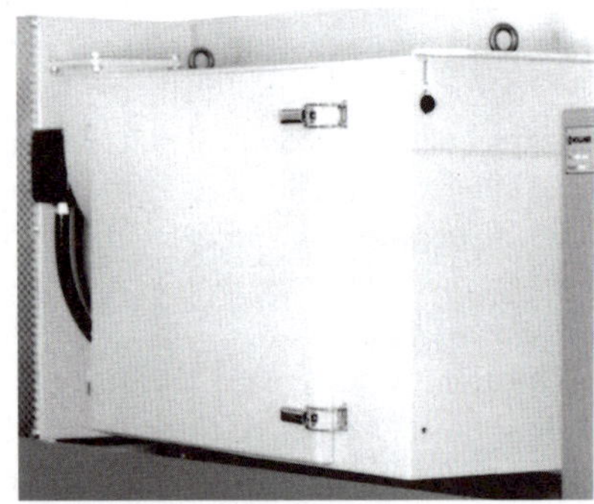

Fig. 7.57
Extraction system for removing coolant mist from the fully enclosed version of the CHC grinder. This protects operators from the health hazard of air-borne coolant particles.
Source - Vollmer

Automatic machines which are totally-enclosed give the best protection as they allow the operation to be seen but effectively prevent waste escaping. On these it is possible to keep even the air-borne waste within the machine enclosure - but make sure that doors are firmly closed when the machine is operating.

Clean the grinder thoroughly after use, and always at the end of the day. Abrasive material, if left overnight, hardens so that later removal is difficult - especially when using an aqueous coolant.

Swill away abrasive material and swarf using an extended hose from the coolant system, if fitted, otherwise wet-brush material clear.

Regularly check and clean-out the cleaning unit or coolant tank and filters. Replace these regularly so that abraded grit does not re-circulate to the point of grind. Regularly check that seals on totally-enclosed machines are well maintained and free from damage or wear.

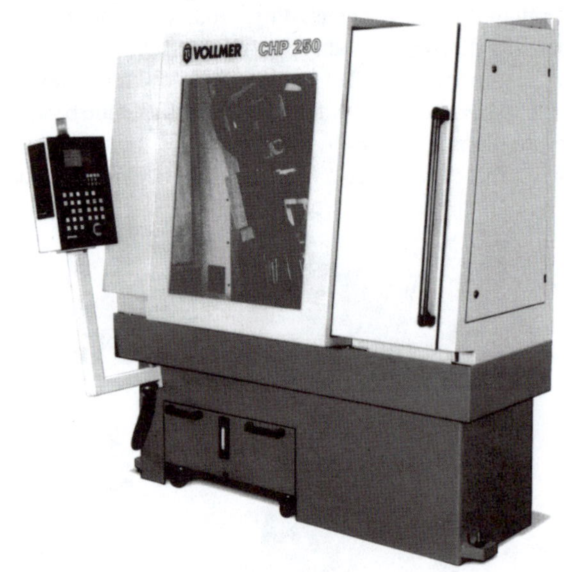

Fig. 7.58
A good example of a totally-enclosed tungsten carbide saw grinder, CHP 250.
Source - Vollmer

If the dry grinder has an extraction system make sure that this is properly connected and without leaks, and regularly empty the container.

Dismantle, clean and oil the parts that separate in normal use, the saw mounting for example. Keeping working parts lubricated where indicated by the machine maker, but wipe off surplus lubricant as this collects abrasive dust to make a damaging grinding paste.

On a wet grinder wipe-off surplus coolant after use, and in particular from working parts.

Coolant is intended to be a metal protective, but some aqueous coolant mixes are less effective than others, and impurities in the water can itself cause problems.

Periodically check that aqueous coolant is not causing rust - an indication that the oil content is too low - and that the paint and bright parts are not being damaged. If they are, then check with your supplier - another brand or type of oil could cure this.

NOTE: The latter part of this Chapter has dealt with superabrasives for regular tungsten carbide sawblades. For grinding PCD sawblades, see Chapter 10.

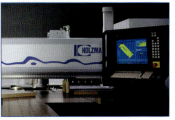

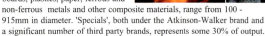

Chapter 8

Stellite-tipped & IT Saws

Stellite-tipped saws

Stellite-tipped bandsaws have been successfully used for many years, but the advantages of Stellite-tipping for circular saws have largely been overshadowed by carbide tipping. This is because Stellite-tipping originally was an in-house operation that relied heavily on the skill of the saw doctor - and could give highly variable results.

In contrast to this tungsten carbide had superb back-up by specialist companies from the very start - and in their hands tc saws always gave a consistent and highly impressive performance. In effect Stellite-tipping had a rather old-fashioned image, that of a low-tech method overtaken and eclipsed by high-tech carbide.

Modern Stellite tipped saws, however, are in quite another ball-park - as this chapter will clearly show. Before detailing the new equipment and methods currently used, though, it is useful to look at the origins of Stellite tipping.

First the advantages. Stellite is a hard-wearing and tough material that gives a run-time at least ten times that of regular CV (chrome-vanadium) saws. The tips can be ground to an excellent cutting edge and, as Stellite holds its form far better than either spring or swage set cv teeth, it is practical to use less set and a narrower kerf than would be used on conventionally cv saws of the same body thickness.

Because of their much longer run-times Stellite-tipped saws need grinding less often, saw diameter loss is considerably slower than with cv saws, down-time is less and fewer saws are needed to maintain production. Because of all this the grinding room is under less pressure to keep the mill in production. This allows more time to be spent on preparing sawblades - a well-proven advantage, as sharp and well-prepared sawblades fully pay back for their extra attention in longer running times and better quality of sawing.

It was, and is, possible to Stellite-tip saws with fairly basic welding equipment, also to grind Stellite-tipped saws using regular grinding room equipment. In this respect adoption of Stellite for in-house saw maintenance in place of regular cv saws was less costly than changing to tungsten carbide saws - as these need new skills and special grinding equipment.

And the disadvantages. In spite of the advantages detailed, when Stellite was first used for circular saws it was not widely adopted - instead tungsten carbide quickly became the accepted hard tip. The reasons are not hard to find. Traditional Stellite tipping is a laborious process which relies rather too much on operator skill. The Stellite deposit needed to form the tip in traditional gas-welding - unless applied with exceptionally care - can vary wildly in size and quality. Even with an experienced welder the tip has to be well over-size to ensure that it makes the required kerf width and tooth height - so heavy grinding is always necessary.

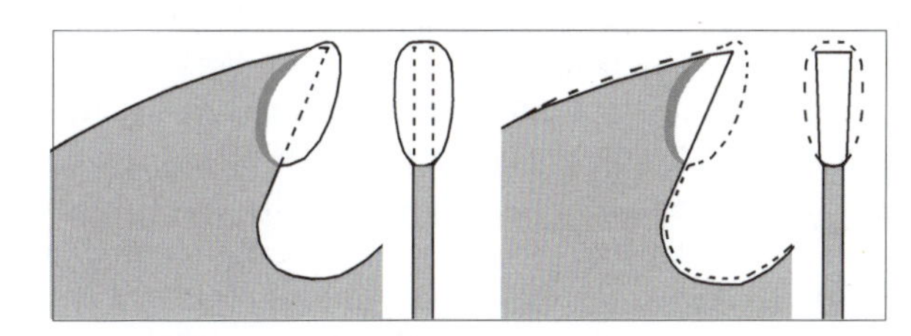

Fig. 8.1

Gas weld-depositing of Stellite, even with a skilled welder, results in a ball of Stellite in the swage cup which must be well oversize to ensure it makes the correct width and profile at the tip. The original tooth profile is shown dotted on the left on a Stellited tooth, with the original Stellited outline shown dotted on the right on a finished tooth.

To make matters worse, regular grinding wheels wear quickly when grinding Stellite - so that both tooth width and height increase from start to finish of the grind. Sawblades left in this condition cut badly and inefficiently, so both need to be regularized by side and profile grinding a second or third time in a process termed sparking-out. This takes excessive time and erodes profit.

When first used, Stellite-tipped circular saws needed almost as much time and trouble in preparation as did carbide saws - but the gain in run-time was nothing like as great. The potential was there but the methods and equipment available, at that time, could not compete with tungsten carbide - Stellite tipping simply was not viable.

The current position. With modern techniques and machines Stellite-tipping of circular saws has improved beyond recognition, making Stellite tipping a thoroughly practical and efficient alternative hard-tipping process. Details of this equipment is covered in this Chapter, but first an outline of the original methods of Stellite tipping for those whose budget does not extend to the sophisticated and highly efficient equipment used by current leaders in this field.

Stellite-tipped profiles

Stellite-tipped saw-tooth profiles for fast-feed conversion of natural timber are based on regular cv swage-set saw profiles. These have flat tops and half the number of teeth of alternate-bevelled saws intended for the same class of work. They also have considerable more gullet capacity - which allows them either to cut deeper and faster, or to be given the shallower-gullet/strong-back profile that fast-feed edging demands.

For fast feed sawing, it is better to form a tip flush with the original tooth top rather than one projecting above. In this way maximum back support is always given to the tooth. (See earlier notes on tooth profiles.)

The general remarks made in Chapters 2 & 3 regarding hook and clearance angles, numbers of teeth, etc., for cv saws, apply equally-well to Stellite-tipped saws. Although Stellite-tipped saws are traditionally flat-topped for fast feed sawing, they can be formed with an alternate top bevel to give a better finish. Modern Stellite-tipped saws may also have a modified tooth form, and for some applications could have expansion slots, strob slots, or cluster-teeth.

Expansion slots are narrow cuts made from the base of 3 or 4 teeth radially inwards by perhaps a tenth of the sawblade diameter, to end in a hole that is perhaps four times greater in diameter than the slot width. Expansion slots take up any expansion caused by the rim heating-up when tipping - without them the tension of the sawblade could be upset. *Modern methods of Stellite tipping keep heat to an absolute minimum so that expansion slots are not normally required, but in any event sawblade tension is normally checked and corrected after tipping.*

Fig. 8.2
The Stellite-tipped saw shown here has both expansion and strob slots.

Strob slots are wider rectangular slots generally cut in a radial direction from the base of two or more teeth. The length of these slots may be up to a quarter of the sawblade diameter. They usually end in a small radius, and often slope backwards at a slight negative angle. Closed strob slots are also used, extending from close to the saw collars outwards to perhaps the depth of the expansion slots. Both types serve the same purpose - to disturb and expel the sawdust and fibres that would otherwise build-up on the saw body. They can be plain slots, but are more effective if tipped with tungsten carbide.

Without strob slots sawblades may be prone to blistering and warping - also called burning-up because the condition shows up as a burn mark on the saw body. This is a particular problem with deep-cutting gang rip-saws. It is caused by the saw body overheating locally through friction between the sawn timber faces and the resin, sawdust and fibre build-up on the saw body. Burn-up plays havoc with production by virtually stopping sawblades in their tracks. Badly burned sawblades need removal for levelling and re-tensioning, and in severe cases may have to be scrapped.

Cluster saws have deep gullets which act in a similar way to strob slots, and are the type preferred on double slabbers.

Fig. 8.3
Cluster saws as used on double slabbers have groups of teeth as shown here.
Source - Bennett

Problems of Stellite Tipping

Gas weld-depositing is the original method of Stellite tipping circular saws. By heating the tooth and melting part of a Stellite rod a quantity of Stellite is deposited on and bonded too the tooth tip. Machine methods of Stellite tipping now use other means, but all involve heat - and heat can cause problems.

The heat created when depositing or welding Stellite causes steel atoms in the saw body *(atoms of iron, carbon, chrome, vanadium, etc.)* to vibrate more than they do at normal temperatures.

The temperature used is necessarily high, and the vibration this causes actually breaks the natural bonds of some steel molecules. The released atoms immediately re-combine, but into molecules which crystalline into a hard but brittle steel when cooled quickly. However, if cooled slowly in a process called annealing, these molecules break apart during the cooling process and then immediately re-combine in their preferred bond as a softer but much tougher steel.

At the surface of the steel the detached steel atoms may instead form a bond with any suitable atoms close by. This can be oxygen from the surrounding air to form iron oxide - an unwanted and thick surface scale.

If another metal such as Stellite is in close enough contact - as at the mating faces of a Stellite tip - then this process instead forms the steel/Stellite mixture necessary to give a secure bond on cooling.

If too much heat is applied - or if heat is continued over too long a period - then the two metals become much more inter-mixed in a process called diluting.

Diluting can be extensive enough to spread into the cutting edges, and so cool into an alloy that is considerably less efficient than pure Stellite alone. Diluting greatly reduces the performance of a Stellite-tipped sawblade. *For Stellite tipping to be successful these three problems must be controlled.*

Tooth hardening

Hardening of the base metal of the saw teeth is an unavoidable side effect of the heating process needed to ensure sound bonding of the Stellite tip - regardless of the method used. If left in this hard state saw teeth would be too brittle and could snap off when sawing - to possibly strike following teeth and cause even further damage.

Tooth hardening is simply cured by annealing the teeth at a second heating cycle. This essential process makes Stellited teeth tough enough to absorb the high impact stresses of sawing without damage. Reheating the teeth to anneal them causes no additional complications because the Stellite tip is unaffected by heat.

Annealing must be carefully carried out by precisely controlling the degree of heat and the timing cycle, but it is practical to use any one of several different methods of heating.

Oxidizing of the steel surface

Oxidizing forms an iron oxide surface scale which prevents the steel and Stellite bonding properly. When gas-welding this can be a serious problem as oxidizing can happen before the steel and Stellite have time to bond, particularly if the wrong type of flame is used - see Fig. 8.9. It can be avoided in machine-controlled welding by reducing the welding arc to an absolute minimum, or by surrounding the area with an inert gas that prevents oxygen reaching the steel.

Diluting

Diluting is more of a problem when weld-depositing Stellite manually using a gas torch, but should not be a problem with modern Stellite-tipping machinery - provided the equipment is correctly set and used.

Stellite Tipping Methods

Traditional Gas Welding

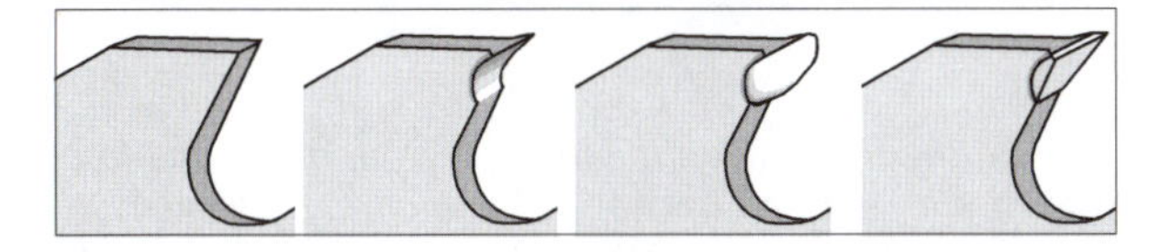

Fig. 8.4
This shows the Stelliting sequence for a typical tooth profile, left to right; ground square; swaged; Stellited, then finally profile and side ground.

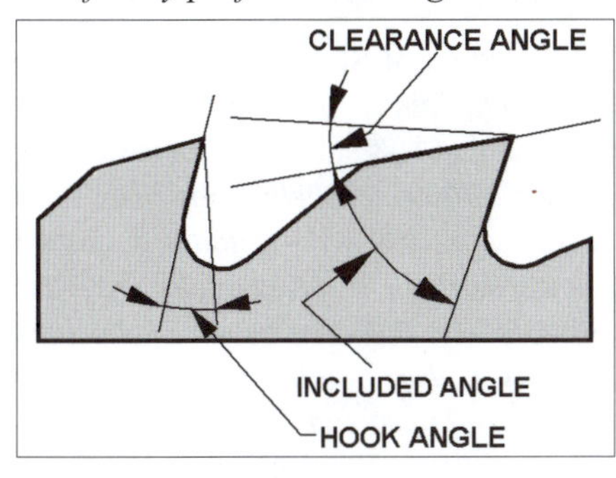

Fig. 8.5
This sketch shows the way of measuring the included or sharpness angle, which should be 44 - 45⁰.

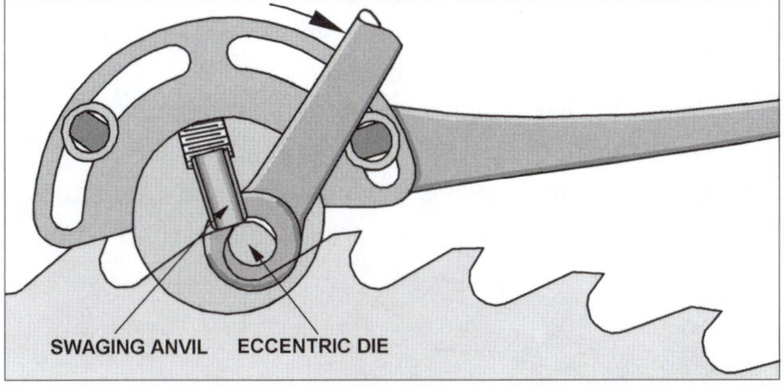

Fig. 8.6
This part section of a typical swager shows it correctly positioned with the anvil sat on top of the tooth. The hardened, eccentric die is rotated into the swage position shown by moving the lever in the direction of the arrow and against the front stop. The effect is to squeeze the tooth face inwards and sideways to produce the cup form required for Stellite depositing. Back and front stops adjust to control the amount of swage. During swaging the swager is securely clamped to the sawblade (not shown).

Prior to Stellite-tipping, saw teeth need to be ground square-across. It is impractical to tip teeth that have been punched but not ground as punching leaves an unsuitable surface for welding. To reduce the chance of the welding heat starting stress cracks, make sure that the gullet base is regular, smooth and free of any sharp corners. *This, in any event, is normal, good practice - but a smooth and regular gullet is even more important when heating is applied to the rim.*

The tooth should have a suitable profile for swaging, i.e., with a 44 - 45 degree included or sharpness angle. A typical softwood tooth has a hook angle of 25 degrees, a clearance angle of 20 degrees, and a sharpness angle of 45 degrees. If a smaller hook angle is needed, then the clearance angle must be increased by the same amount to leave the same included angle.

Fig. 8.7
Armstrong bench type swager for circular saws. To steady the swager it is tied via a bar to the workbench. The left lever clamps the sawblade and the right one forms the swage.
Source - Armstrong

After grinding, the teeth are swaged in the same way that regular swage-set teeth are treated, using, for example, the Armstrong bench type swage. Side dressing is not normally necessary, although some operators prefer to Stellite-tip fully swaged and side dressed teeth. For side dressing use a regular side dresser *(shaper)*.

Fig. 8.8
Armstrong 6900 bench type shaper.
Source - Armstrong

Swaging and shaping equipment is readily available for bandsaw teeth, but the practice has been long out of favour for circular saw teeth - and only specialized makers such as Armstrong still offer this equipment.

It is essential to take the normal, recommended precautions when welding manually. Use industrial gloves and protective clothing and wear a purpose-made welding face mask. For full precautions note the factory or local State regulations - together with those of the equipment makers.

Mount the sawblade on an arbor which allows it to turn, but with some means of securing it so that the tooth face to be

Stellited is near horizontal. Each tooth is brought into this same position in turn, perhaps using a spring-loaded pawl to index the next-but-one tooth.

Use a 100 litre welding torch with a nozzle orifice of about 0.75mm *(0.029in)*. The oxygen pressure should be very low and the acetylene adjusted so that the flame has an inner, white core 6mm *(1/4in)* long and an outer flame 2 - 3 times this length. Use a Stellite No 12 rod of 2.4mm *(3/32in)* diameter.

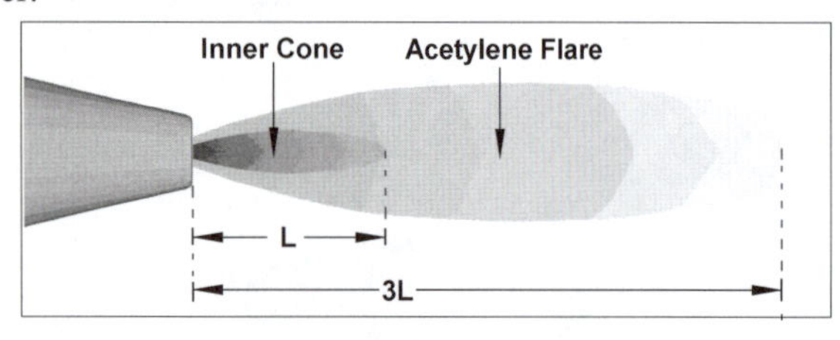

Fig. 8.9
The correct oxyacetylene flame suitable for weld-depositing Stellite.
Source - Deloro Stellite.

Bring the end of the Stellite rod into the flame until it becomes red hot, and then apply the flame to the face of the tooth until this becomes bright red. This will carburize the surface of the steel to form a thin coat *(water-clear in appearance and called 'sweating')* which is essential to give a good bond. Hold the torch almost horizontally, and with the Stellite rod placed to protect the tooth tip from overheating.

Keep the inner cone of the flame just touching the back of the tooth, but with the flame still concentrated on the tooth where Stellite is required. Maintain this until a sufficient drop of Stellite is melted from the rod and deposited onto the tooth face to spread well up to the tooth point. Immediately this happens withdraw the torch. The process should be carried out quickly or the steel will oxidize to prevent the Stellite from spreading properly. Take care not to melt the base steel or the tip will dilute, and do not withdraw the oxyacetylene too quickly as this causes shrinkage in the weld.

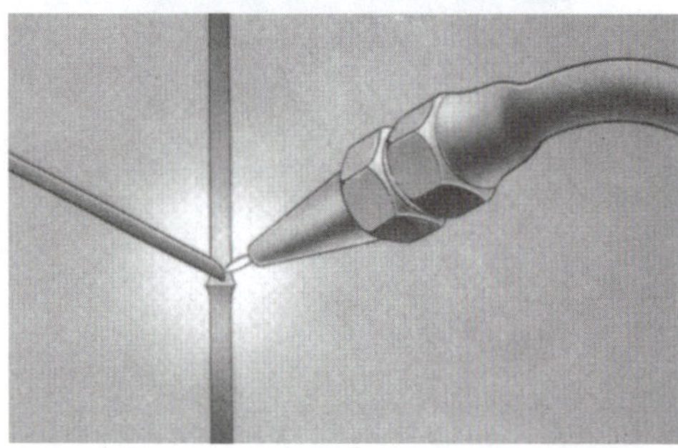

Fig. 8.10
Stellite tipping - the correct way to position the welding torch and welding rod.
Source - Deloro Stellite.

As already mentioned, weld-depositing Stellite hardens the steel body and makes the teeth brittle.

To anneal the steel each tooth should be heated a second time in a torch flame of 450 ^{0}C (840^{0}F), and held at this temperature for about 2 seconds before allowing it to cool.

Judging the correct temperature is the difficult part. Many users apply a temperature-sensitive liquid or crayon to each tooth before annealing. This changes colour when the correct temperature has been reached and tells the user when to hold for two seconds.

The process needs skill and experience to get the timing correct and to form a Stellite deposit of the right size - not so small that it forms too-narrow a kerf or short point, and not so large that excessive grinding is needed.

Jig Welding

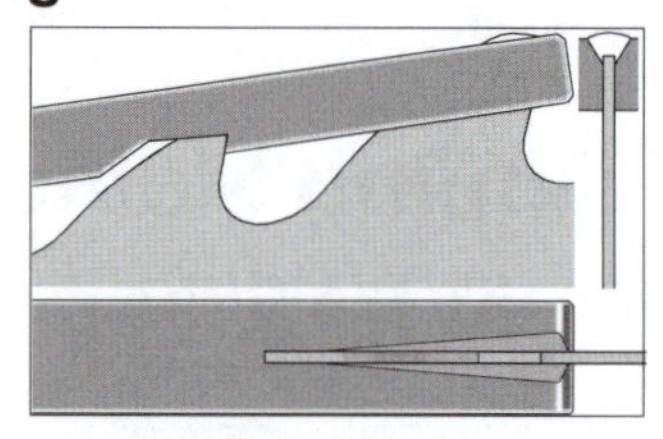

Fig. 8.11
The welding jig rests on the tooth in front of the one to be tipped but needs an anchor point and an adjustment to correctly angle the jig (neither is shown). The jig should be positioned to give the proper amount of tooth top projection above the vee - this is seen in the right-hand section. Below, in plan view, the welding vee and tooth top (without the Stellite deposit) are both clearly seen.

An alternative method of Stellite-tipping circular saw teeth uses a welding jig. This gives a more precise shape to the Stellite deposit so that less side grinding is needed.

The method described below has been used successfully by the saw doctor who originally devised it, though mainly on bandsaws, and it was eventually recommended by Deloro Stellite. Others have not been so successful, but it is reprinted here as a possible alternative where gas welding only is still the normal practice.

A steel jig, perhaps 25 x 6mm. (1 x 1/4in) in section and 200mm (8in) long, is slit through its thickness to fit the saw body accurately, and then filed to a 60° vee about 15mm x 3mm (1/2 x 1/8in) at the end where the Stellite deposit is needed.

In order to locate the jig, some sort of notch needs filing in it to engage with one tooth back, also a hinge or pivot has to be provided at the remote end.

With this particular method the teeth are not swaged, but instead the tooth tips are filed or ground down by roughly 3mm (1/8in) and to a reverse top angle of about 15 degrees.

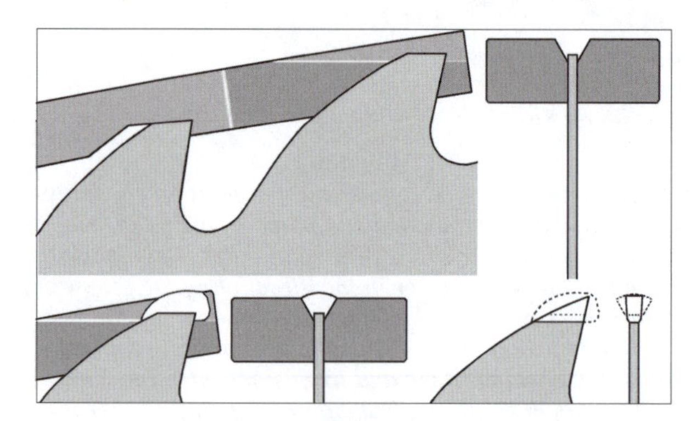

Fig. 8.12
This shows, above, a front section through the jig and an end section. Below are front and side views, left within the jig and, right after side and profile grinding.

Mount the sawblade on a freely-rotating arbor and set the jig pivot height so that, with the jig correctly positioned, the bottom of the vee is more or less horizontal.

The correct position for the jig is with the notch against the following tooth, with the hinge engaged, and with the tooth to be Stellited projecting slightly above the vee base.

Use the welding technique previously described. As soon as sweating conditions are reached on the saw body, drop molten Stellite into the vee and around the tooth, applying enough Stellite to make the correct tooth width and profile when ground.

To release the jig lift it off the hinge, moving the opposite end downwards. Relocate the notch on the next tooth-but-one *(in a counterclockwise direction)* and move the jig and sawblade until the hinge can be re-seated.

Whilst this method gives slightly better control of the Stellite deposit width, so that the least amount of side grinding is needed, it does not absolutely control the size of deposit - this is still up to the skill of the welder.

Modern Stellite-tipping

Stellite tipping machines

In contrast to manual gas-welding, modern Stellite tipping machines use highly-controllable electrical heating to take all the guesswork out of the process. They produce dense and consistent Stellite tips with superior edge-lasting properties.

Saw teeth can be newly Stellited without special tooth preparation such as swaging or reverse angle grinding being needed - in most cases. Stellite can also be machine-welded onto worn-out tips with the minimum amount of preparation. This avoids the need of completely grinding-out the remaining Stellite - which is an essential process when gas-welding.

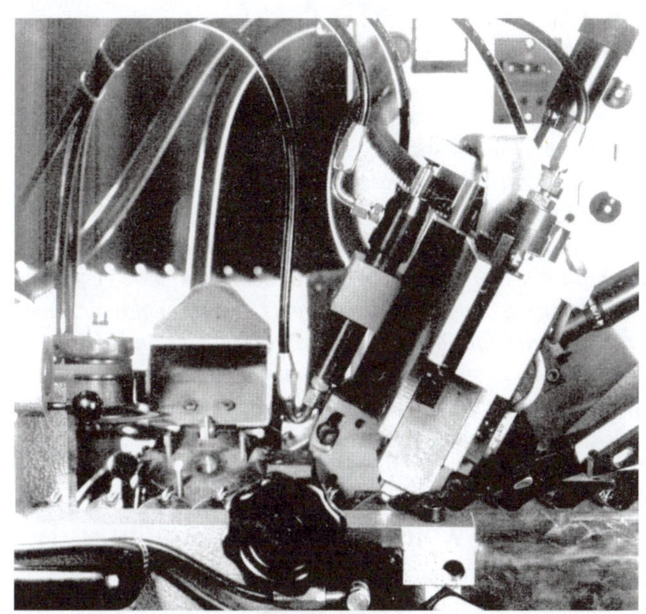

Fig. 8.13
Modern Stellite-tipping machines are highly sophisticated, fast, cost-effective and efficient. The close-up shown here is an Iseli SAA for resistance-welding sintered Stellite rod - in this case onto bandsaws.
Source - Iseli

In machine welding, tips are precisely and consistently sized - so the absolute minimum of grinding is required. When using regular grinding wheels this enables spark-out to be reached more quickly and at considerably less cost than with gas-deposited tips.

All tipping machines are capable of forming different sizes of tip, and are fully controllable in respect of the welding and/or annealing heat.

Some makers offer a range of machines of similar construction for manual, semi-automatic or fully automatic operation - at different cost levels to match different budget restrictions and operating requirements. All give the same high quality of hard tip - the differences between them are only in the degree of operator attention they need.

Sophisticated equipment of this type allows for individual experimentation - the saw doctor can match the tip precisely to what in his experience is needed for his particular circumstances.

There is no absolute guide as to the correct proportions of a hard Stellite tip - not surprising in an industry where opinions vary so widely - but some saw doctors and filers do have strong views about this.

Stellite tipping machines do not use gas heat for tipping - heating is instead by a carefully controlled electrical current using one of two basic methods - either plasma welding or induction welding *(also called resistance or flash-butt welding)*.

In neither case is it necessary to prepare the sawblade other than by grinding a regular tooth profile. Because neither swaging nor tooth top reverse angle grinding are required there are no restrictions on the hook and clearance angles, nor in the general shape of the tooth - machine welding in fact gives the operator much greater degree of control.

Plasma Welding

 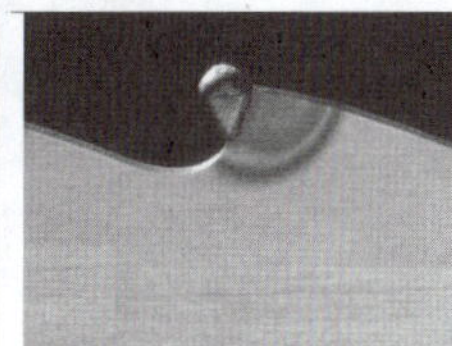 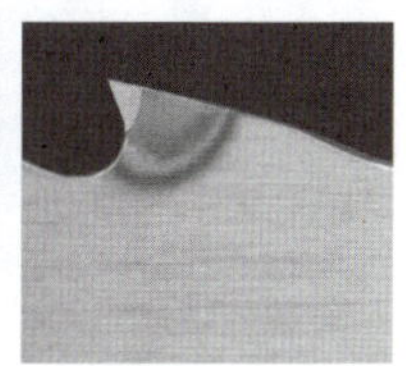

Fig. 8.14
Left, it is possible to plasma-weld Stellite tips onto an existing tooth profile manually and without either swaging the tooth or using shaping jaws provided for machine welding. However, the excessive ball size (centre) needed to guarantee the correct tooth width and profile then requires more grinding to produce the finished tip (right).

Plasma welding is the conventional and well-established way of depositing Stellite onto a variety of base materials. In the case of circular saws the plasma arc both pre-heats the base metal of the tooth and liquefies the Stellite rod.

Plasma welding equipment creates an electrically-charged plasma arc within a conductive gas. This produces an extremely high temperature to quickly heat the tooth tip and melt the Stellite rod. Liquid Stellite is then poured onto and around the saw tip between two water-cooled shaping jaws. Because the process is very fast only the tooth tips are heated, and this avoids dilution of the Stellite.

The Stellite *(in cast rod)* is fed into the welding area within the enclosing inert gas which, for thinner saws, is virtually a pure argon. This acts both as a pilot to promote the plasma flame, and as a shield to prevent oxidizing of the steel.

The argon mixture for thicker saws includes a small amount of hydrogen to give a higher heat and so ensure excellent filling of even the deepest shapes.

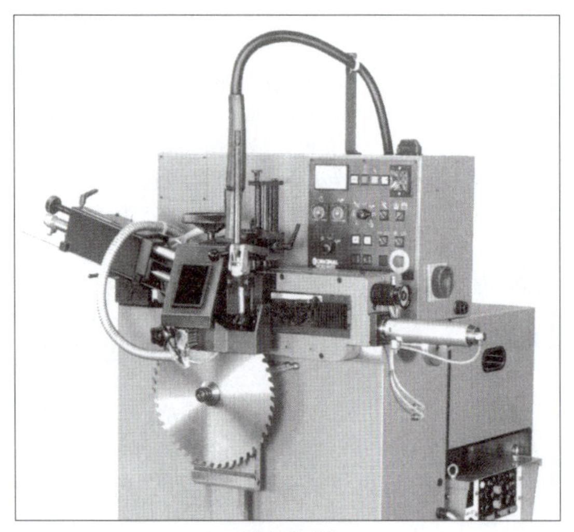

Fig. 8.15
The GPA machine shown here is a fully automatic plasma welding machine using extruded Stellite rod.
Source - Vollmer

Fig. 8.16
This clearly shows the plasma torch and the water-cooled shaping jaws of the GPA.
Source - Vollmer

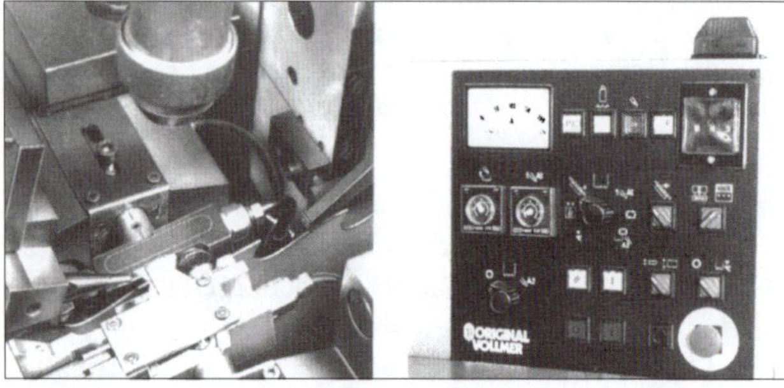

Fig. 8.17
Left, after the fluid Stellite is poured into the water-cooled jaws the tooth top is formed automatically.
Right, the simple and logical arrangement of the Stellite tipping controls on the control panel of the GPA.
Source - Vollmer

The shaping jaws determine the amount of side clearance given to the tip - with switchable jaws to suit different tooth profiles. The jaws are pressed against the sides of the sawblade during the welding cycle. They precisely control the widthwise profile of the Stellite tip - including suitable radial and back clearance angles. The profile of the tooth top is formed automatically immediately after weld-depositing - so all grinding tolerances are closely controlled and minimal.

Precise and adjustable control of the Stellite deposit makes it possible to vary the tip proportions to a certain extent.

For example, the amount the Stellite tip projects beyond the original tooth face or top can be increased so that subsequent grinding never exposes the base metal at the tip. (See grinding techniques later in this Chapter.)

A single size of extruded rod can be used to weld most tips. Although other sizes of rod can be used if preferred, the regular rod size is 3.2mm ($^{1}/_{8}$in) diameter - variations in tip sizes are catered for simply by using alternative jaws.

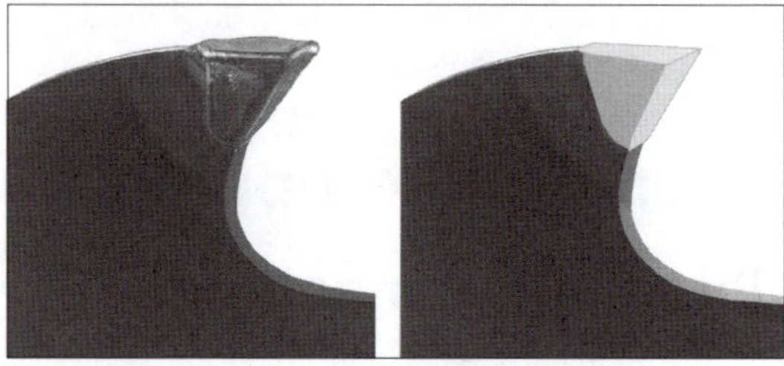

Fig. 8.18
These views (left) after Stelliting and (right) after grinding, show how closely the grinding tolerances can be controlled when plasma welding

Annealing can be a second operation using the plasma burner at a different setting.

Resistance welding.

For these machines the Stellite used is of a sintered and densified form. It is produced in different sizes and profiles to suit a variety of applications, either as a rod or as individual tips.

Only the minimum amount of side and profile grinding is required when using precisely-shaped rods or tips - and the tip proportions can be chosen to match exactly the saw doctor's requirements.

Sintered Stellite rods and tips are produced by powder metallurgy. A fine powder of Stellite, together with temporary bonding and lubricating agents, are formed under heat and pressure into a green product of the size and shape required. This is then sintered at a much higher temperature to produce the final sintered material. Sintered Stellite has superior density and edge-retaining qualities when compared to standard cast Stellite rods.

Resistance-welding sintered Stellite rod

With machines using Stellite in a shaped rod form the rod is butted against the tooth front and held there under pressure. By passing a low-voltage electric welding current between them, the Stellite rod is securely welded to the tooth front. The tiny initial gap between the Stellite and the saw body creates the necessary electrical resistance to form the ideal short welding arc.

Unlike gas and plasma welding in which Stellite is actually melted, resistance welding retains the Stellite in its original, solid form.

During this process the Stellite actually displaces the

steel at the tooth face - so that the tooth can retain virtually its original profile. This is possible without any preparation other than regular tooth profiling. To a limited extent the Stellite and base metal do melt into one another, but only at the adjoining faces where it is essential to make a secure bond - well away from the tip.

Fig. 8.19
Close-up detail of the GWM resistance-welding machine. The cut-off grinding wheel can be seen on the right.
Source - Vollmer

Fig. 8.20
This semi-automatic GWM Stellite tipping machine uses a sintered Stellite rod which is resistance-welded onto the teeth.
Source - Vollmer

Resistance welding is fast and the welding arc is short. This guarantees that virtually no dilution of the Stellite takes place - so only pure Stellite remains at the cutting edge even when the tip is worn to the point where replacement is necessary. The short arc also ensures that the chance of producing a poor weld through prior steel oxidizing is almost zero - in fact there is no need for a shielding gas with this process.

An abrasive cut-off wheel parts the welded tip from the remaining rod *(for further use)* once the Stellite has been welded in position - leaving the tip with a precise amount of side or top projection and the correct clearance angles. Following this the next tooth is moved into place for hard tipping in the same way.

When rod is fed from the side it is rotated through 180 degrees after cutting-off so that the cut-off angle on the rod is correct for the opposite side clearance angle of the tip next to be welded.

Annealing of the tooth can be carried out as a second operation using the resistance welding current at a lower setting, or at a separate station using an alternative heating method.

Fig. 8.21
Left, after weld-depositing, the remaining rod is separated from the welded tip using an abrasive cut-off wheel. Right, annealing is a second operation. In the case of wide bandsaws, as shown, gas burners can be used.
Source - Vollmer

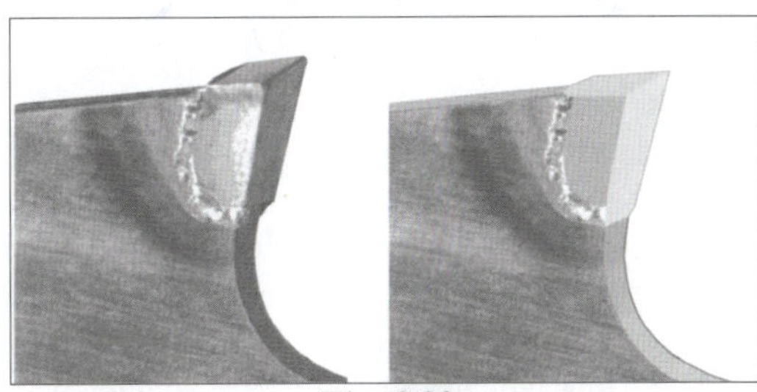

Fig. 8.22
The grinding tolerances of resistance-welded Stellite tips, like those of plasma-welded tips, is closely controlled so that only the absolute minimum of grinding is required.

Resistance welding sintered Stellite tips

All the advantages described above for resistance-welding preformed sintered rods, also applies to machines designed for resistance-welding individual sintered tips - no dilution of the Stellite, a strong, substantial bond and close grinding tolerances. Usually machines of this type are also fully automatic in operation, leaving the operator free for other work.

Because Stellite tips are so securely held, it is not necessary to use tips of the same proportions as tungsten carbide tips - which tend to be long relative to their width. *(Following early problems with brazing, long tungsten carbide tips were subsequently adopted as these have a correspondingly bigger brazing surface to make up for their weaker bond.)*

The small Stellite tips that good bonding makes viable means less Stellite is needed - and also lowers costs for profile and side grinding.

One machine in this class, the Armstrong AUTOTIP, is an hydraulically powered and heavy duty Stellite tipping machine suitable for tipping individual, preformed tips not only on circular saws, but also - and with a change-over time of just five minutes - on band and sash gang saws.

The machine both welds and anneals each tip before

passing to the next tooth - so avoiding the need of a second pass or a separate process for annealing. With microprocessor control machine operation is fully automatic and has a capacity to place up to 100,000 tips per year.

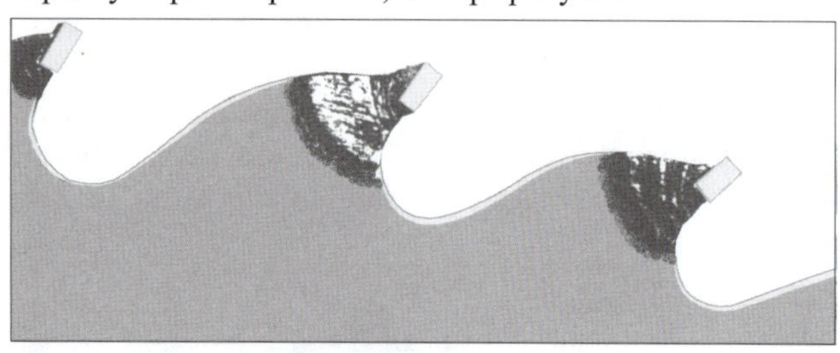

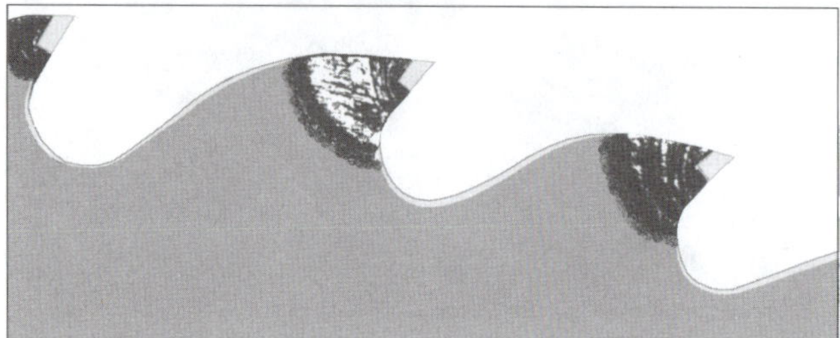

Fig. 8.23
Although individual, preformed tips are relatively small - to reduce material and grinding costs - they are so securely held that tip loss is rare. The top illustration is after tipping, and the lower one after grinding. Square-section tips are shown, but tips are also available with the clearance angle already formed.

Individual Stellite tips are available in a wide range of profiles and dimensions. For example, individual Stellite tips available from Armstrong Manufacturing vary from the regular sizes (in mm.) of 2.5 width X 1.9 thickness X 3.3 length, and 6.3 w. X 3.8 t. X 7.6 l., to special longer tips for alternate top bevel saws of from 3.2 w X 2.3 t X 6.1 l to 4.20 w X 3.2 t X 7.6 l. These tips, which place the precise amount of Stellite each time, are claimed to be half the cost of alternative Stelliting methods.

Individual Stellite tips are also available from Deloro Stellite Inc., Belleville, Ontario, Canada, in tungsten carbide dimensions and style for brazing, or dimensioned in the chunkier form preferred for resistance welding.

Brazing sintered Stellite tips

It is possible to braze sintered Stellite tips onto circular saws, but not in the chunky form supplied for resistance welding. The tips should be in the size and proportions corresponding to tungsten carbide tips used for similar sawblades. The brazing process is similar to that used for tungsten carbide tip brazing - which is covered in the next Chapter. As resistance-welded Stellite tips give a superior bond and have lower tip and grinding costs, however, brazing only applies where the requirement is for too-few Stellite tipped saws to justify dedicated Stellite welding machines, and where carbide tip brazing is already an established practice.

Annealing tipped teeth

Annealing can be carried out using a lower setting of the welding current, either immediately after the welding process or at a second pass on the same machine. Separate annealing stations are provided on some machines which may use alternative methods of heating, such as a radio frequency induction heating loop, or butane or propane gas flames.

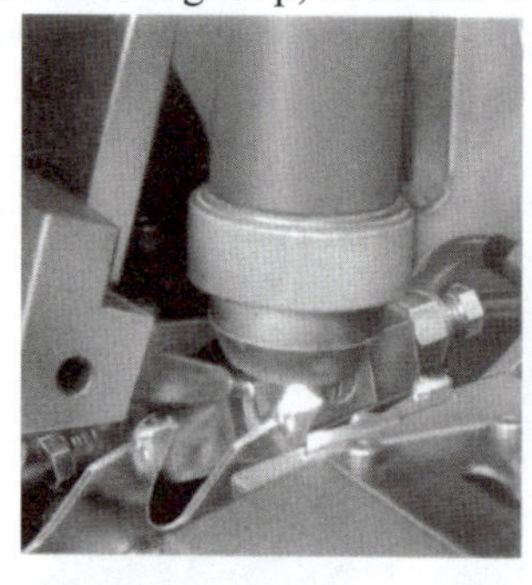

Fig. 8.24
Two alternative methods of annealing are shown here. Left, annealing using the plasma burner at a second pass and right, annealing using an induction loop - both illustrated on a wide bandsaw.
Source - Vollmer

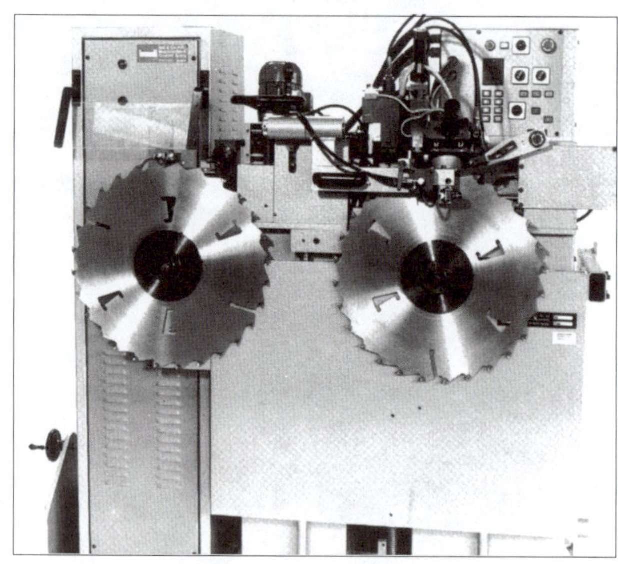

Fig. 8.25
This automatic Stellite tipping machine, type SAM, has two separate and independent stations. One station resistance-welds Stellite tips from a rod, and incorporates an abrasive cut-of wheel. The second station anneals tipped saw teeth.
Source - Iseli

The annealing temperature is critical. Because of variations *(in the tip size and shape and in the saw body thickness)* the annealing equipment needs to be very carefully set to condition the teeth correctly. Failure to do this can result in broken teeth and unwanted saw damage,

One method is to use a temperature sensitive liquid or crayon on one or more teeth to test whether or not the setting is correct for the particular saw being tipped. Once properly set the machine can then operate automatically. (It does not matter if, during set-up, several teeth are wrongly treated - provided these are reprocessed once setting is correct.)

NOTE: all annealing methods, regardless of thc mcans of heating, give the same result - the differences between them is merely in the speed, cost and efficiency of their operation.

It saves time, and is more convenient, if annealing is carried out where welding actually takes place - the cycle time per tooth in this case being the welding time plus the annealing time.

When annealing automatically at a second station on the same machine, the time cycle is little more than that taken for welding itself - although the faster welding cycle may actually have to be slowed enough for the annealing process to be fully effective.

Annealing immediately after welding at a second station can be difficult with circular saws because of their curvature (bandsaws have a longer, in-line working length to make this process a practical one).

When completely separate welding and annealing stations are provided on the same machine both operations are possible at the same time - each working independently of the another on different saws.

Cost comparison

Both plasma and resistance welding of Stellite produces excellent hard tips - but the two basically different processes vary in cost.

To get a true picture, any cost comparison must include the investment cost of the machine against its estimated life, the saving in down-time when servicing the sawblade, and the cost of running the machines - including that of all consumables.

Plasma welding. This uses various diameters of cast rod for tipping, but requires the use of an inert gas which must also be included in the list of consumables, together with the occasional replacement of the welding clamps.

Resistance welding. The sintered Stellite and densified rods and tips used in resistance welding are more expensive *(by volume)* than the cast Stellite rods used in plasma welding.

As the shape and size of the rod or tip must correspond to the tip size needed, this method also requires a range of Stellite rods or tips to be stocked - but these are the only consumables needed. Tips and rods are readily available, do not deteriorate over time - and change-over times between different tip types and sizes is rapid.

Because heating and consumable costs vary throughout the world, a factual comparison between the two methods cannot be given in a book of this sort. However, makers of these machines can give very specific cost comparisons for different countries - and these, together with the comments within in this text, should give a reasonable guide as to the most suitable equipment for any part of the world.

See further notes on Stellite-tipping under 'Stellite Tip Profiles' later in this Chapter.

Grinding Stellite teeth

Traditional Grinders

Stellite-tipped saws can be profile-ground using any conventional automatic saw grinder, and any side-grinding machine of the type used for regularizing swage-set teeth on cv saws.

When using all grinders wear eye protectors.

Regular aluminium oxide grinding wheels can be used on both these machines, for example WA46KBV for profile grinding and MA800V for side grinding.

Because of the hardness of Stellite and the high degree of wear likely with regular grinding wheels, accuracy of tooth width and height cannot be guaranteed without sparking-out.

Grinding the tooth profile

The first operation after Stellite-tipping is normally to profile-grind the teeth. If using a regular grinder designed and manufactured for grinding regular cv saws this will probably be of the dry grinding type using normal grit wheels.

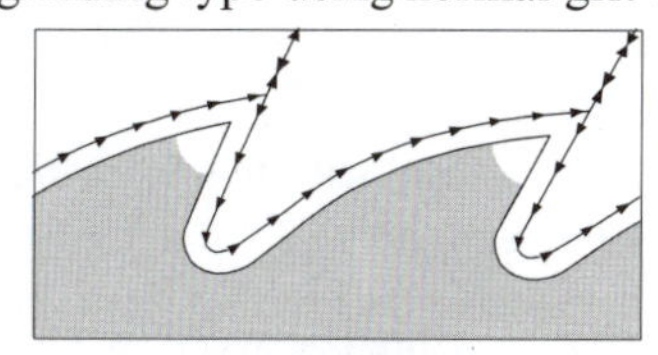

Fig. 8.26
Most older saw grinders are powered by cams designed to give a constant-speed traverse throughout the grinding movement. They grind down the face and into the gullet and then, with the saw being indexed by the pawl, up the tooth back and finally across the tooth top. Here the grinding wheel is raised clear of the tooth whilst the pawl shifts the next tooth into position.

Set the saw grinder for a slow movement and a light grind to first regularize the tooth profile. Grind completely around the sawblade several times, tweaking the setting each time the first tooth re-appears. Before finally sparking-out the grind carefully check that all the teeth have a fully formed point - and will also make the required kerf width when side-ground. If they will not, then grind the sawblade until these conditions are met.

Setting the feed pawl

It is preferable to index the tooth next to be face-ground - but only if there is no danger of trapping the pawl with the grinding wheel - in which case index one tooth back. Immediately the pawl is withdrawn the grinding wheel grinds down this face into the gullet and then, with the pawl indexing the next tooth, up the back and across the top. See previous notes on pawl setting, also additional notes under tc saws.

The regular practice is for the pawl to contact the original tooth front or face. The Stellite deposit itself may have an irregular face and, if indexed against, could give an uneven tooth pitch resulting in some teeth being wrongly ground.

Caution! There could be a problem with the pawl with weld-deposited Stellite processes which form a flash projection outwards from the seating. In some instances the pawl could hook this flash on the return movement and drag the tooth back after indexing it - to cause a crash. A pawl that returns without tripping over the tooth top is safer in use in these conditions.

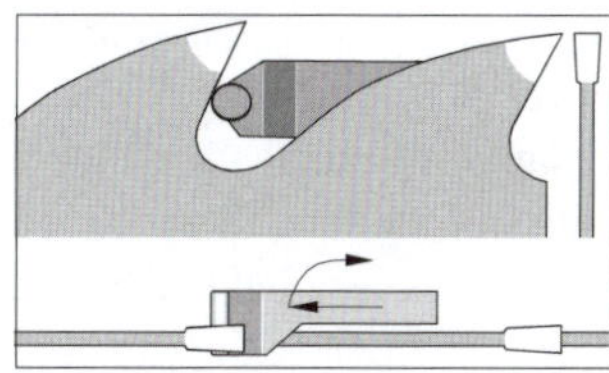

Fig. 8.27
The pawl must index the tooth next to be face-ground, or one tooth back. With Stellite tips more consistent tooth indexing results if the pawl contacts the original tooth face - not the hard tip. This sketch shows the pawl swinging sideways on the return stroke to avoid the chance of hooking the projecting flash - the Armstrong SideWinder is one example of this type.

Grinding Side Clearance

Following tooth profiling the sides of the tipped teeth can be ground to the required kerf width and suitable back and radial clearance angles. If heavy side grinding is needed, make two or more light passes until the kerf width is correct - before finally sparking-out.

Grinding side clearance is necessary once only after first Stellite-tipping a saw - any subsequent re-sharpening is by profile-grinding the teeth only.

Radial clearance is measured between two lines drawn from the tooth tip and in line with the hook angle, one line parallel with the saw body and the other along the tip itself. The tip should taper-in from the cutting edge towards the saw body at an angle of up to a 5° (but normally around 1.5°).

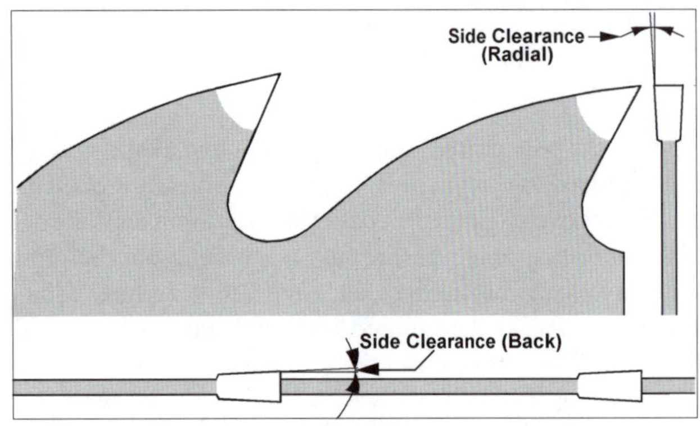

Fig. 8.28
When grinding side clearance two independent angles are formed as shown here, radial and back clearance, the latter sometimes called tangential clearance. The amount of clearance, and the clearance angles themselves, vary between hard and softwoods.

Back clearance is measured between two lines drawn from the tooth tip and in line with the clearance angle, one line parallel with the saw body and the other along the tip itself. The tip should taper in from the cutting edge backwards towards the following tooth at an angle of up to a 5 degrees.

Although many side grinders allow a maximum of 5 degrees on both clearance angles, most makers recommend the use of smaller angles. Large clearance angles give a clean cut - but the tips then quickly lose kerf width. Larger clearance angles are usually needed for wet and soft timbers and less for dry and hard timbers. In all cases, though, use the minimum clearance angles practical so that kerf width is only slowly lost through grinding.

If in doubt use small radial and back clearance angles to begin with, then gradually increase these if the sawblades take excessive power to drive them.

The same degree of radial clearance angle should be formed on both sides of the saw, likewise the back clearance angles should match - but the radial and back clearance angles need not be, and are usually not, the same.

Kerf width is the actual cutting width of the sawblade, the amount the teeth project sideways beyond the body, plus the saw body thickness.

The relationship between the thickness of the sawblade body and the tip clearance varies for different sawing conditions. Take, as a guide only for newly-tipped sawblades, the figures under H1, H2 & H3 in Chapter 12 under 'Side Grinding' of carbide-tipped saws. As the sawblade is repeatedly ground the tooth tip clearance at each side will reduce, and with it the kerf width. However, these figures take such normal wear into account.

As always, individual saw doctors and filers have their own ideas regarding both the amount of side clearance and the clearance angles themselves - so these figures are not set in concrete. Experience will quickly show what suits each circumstance. However one factor is beyond dispute - the amount of clearance of the tip beyond the saw body must be exactly the same at both sides - if they are different the sawblade may run to one side.

Fig. 8.29
It is essential that Stellite teeth have equal clearance at each side. To check this, accurately measure the side clearance using some form of precision gauge, such as this magnetic dial indicator, the Mag-Dial.
Source Armstrong

Setting the feed pawl

When side grinding Stellite-tipped teeth the pawl could be set to index the actual Stellited face of the tooth face to be ground - at this point profile grinding should have formed a regular and smooth face.

Alternatively, of course, some filers prefer to index below the tip and against the original tooth front. Whichever indexing point is chosen when side grinding, though, the same choice should also be made for following tooth profile grinding - otherwise slight variations in kerf width could result.

The Sparking-out Process

Regular grit grinding wheels cease to grind properly when the first-ground tooth is reached a second time - because by then the grinding wheel will have lost some grit and form. If the sawblade is left in this condition the teeth may have an uneven width or height - and will then could cut badly and inefficiently.

To correct this adjust the machine for a light grind only on the first-ground tooth, then grind the sawblade fully a second time around. Repeat the process until just a few sparks are thrown off when grinding all teeth - this guarantees that all the teeth are ground equally. Continue the grind, but without further machine adjustment, to full spark-out - the point when no more sparks are produced. By following this grinding sequence any uneven tooth width or height resulting from the initial grind will be made good.

Sparking-out is essential both with side grinding to ensure a consistent kerf width, and with profile grinding to ensure an even bite per tooth. The need for grinding more than once around depends upon the initial depth of grind, the grit size and hardness of the grinding wheel used, and on the physical size of the hard tip.

Such an idealistic method is not always practical - especially in a hard-pressed grinding room where too long spending grinding a single sawblade could hold up production.

In these cases the saw doctor or filer must make a reasoned judgement as to when the sawblade is in an acceptable condition, tweaking the grinder during this process to guesstimate the grinding wheel loss.

In fact, many experienced operators can judge the amount of wheel wear by the sound and sight of the grinder when it reaches the first-ground tooth a second time. This is the advantage of experienced operators, they save time and cost without sacrificing quality. They will tweak the machine only enough to even-out grinding wheel wear without the extensive sparking-out process described above.

When grinding side clearance the grinding wheels are self-dressing and need no attention - unless they become glazed or filled.

When grinding the tooth profile, though, the grinding wheel will probably have lost both diameter and form and needs to be re-shaped before starting a further grinding sequence.

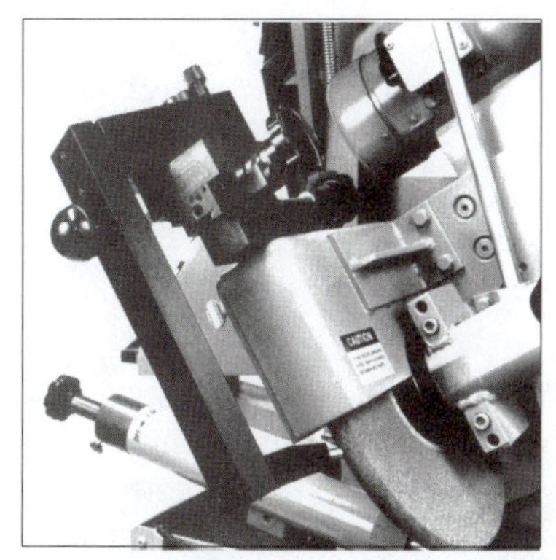

Fig. 8.30
Regular grinding wheels are likely to wear unevenly when profiling Stellite tipped teeth. A grinding wheel profiling attachment, such as the one shown here, the ShapeUp, will maintain a precise wheel shape indefinitely. Made by Armstrong it can be retrofitted on most Armstrong saw grinders.
Source - Armstrong

Dealing with worn tips

Saws tipped with Stellite can be profile-ground several times before the kerf width is so reduced that the saw needs re-tipping.

When this is needed with manual gas-deposited tips, any remaining Stellite must be completely ground-out before the process can be repeated. The saw can then be either swaged or simply tip-ground - and the tipping process repeated.

With machine-welded Stellite tips it is not necessary to grind-out the worn-down tips before re-tipping - new tips can often be welded onto existing tips with the minimum of preparation.

Modern Stellite Saw Grinders

Since Stellite tipping was first used grinding techniques have vastly improved - and grinding machines of advanced design are available from several makers. Some machines are dedicated Stellite grinders, others are capable of grinding regular, Stellite-tipped or tungsten carbide tipped saws with equal precision and speed.

The best results, measured in terms of efficiency, speed and quality of the grind, are only possible on the newer grinders which use the latest grinding technology of fast-running, wet-grinding, cbn wheels, and plc or cnc machine control.

These machines use CBN (Cubic Boron Nitride) grinding wheels. These have a grit size of around 80, run at the grinding wheel makers recommended rim speed of around 60m/sec (12,000 ft./min.) and operate with full flood coolant within an enclosed machine.

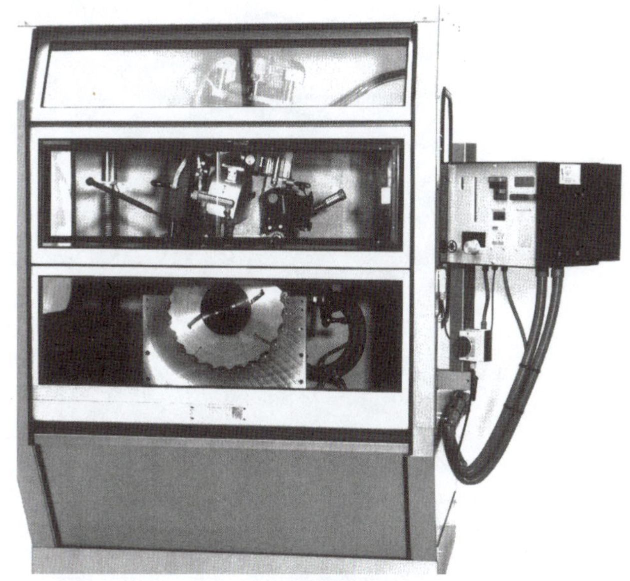

Fig. 8.31
The AS-2 circular saw profile grinder has all the attributes needed for wet grinding Stellite-tipped teeth, but is equally capable of grinding cv and tc saws by changing the grinding wheel type and operating mode, see later detail in Chapter 10.
Source - Iseli

The fast grind, combined with a fine grit and wet grinding action, forms an excellent finish on the teeth to produce cutting edges which are superior to those formed by regular, slower-running, dry-grinding wheels.

This gives two distinct advantages. The obvious one is the very sharp cutting edges which extend the run-time of the sawblades, produce a superior sawn surface and absorb less power. The second advantage is the smooth and clean surface formed within the gullet. This lessens the likelihood of cracks starting - and also ejects sawdust more efficiently to reduce the chance of sawblades burning-up.

The greatest advantage of cbn wheels - when operating under ideal conditions - is their slow rate of wear and excellent profile retention. In fact, the wear per circular saw of cbn grinding wheels is virtually zero. The costly and time-consuming process of sparking-out is hardly ever needed - although a second grind may be needed if grinding a damaged or completely re-tipped sawblade. For regular profile re-grinding a precise height of teeth can be virtually guaranteed from a single grinding sequence.

If a heavy grind is needed with regular wheels the normal practice is to grind fully around more than once - be-

cause otherwise the teeth can overheat and become brittle - then finally fully spark-out. The combination of CBN wheels, wet grinding and precise machine control avoids these problems - so a heavier grind is possible than regular wheels allow - and without their excessive wear. This drastically reduces re-grinding times to well below those previously needed with conventional grinding machines and wheels. As a result less time is spent grinding - and sawblades are back in production with the least delay.

Grinding the tooth profile

Modern Stellite grinders have precise grinding head and pawl movements through highly versatile plc or cnc controls.

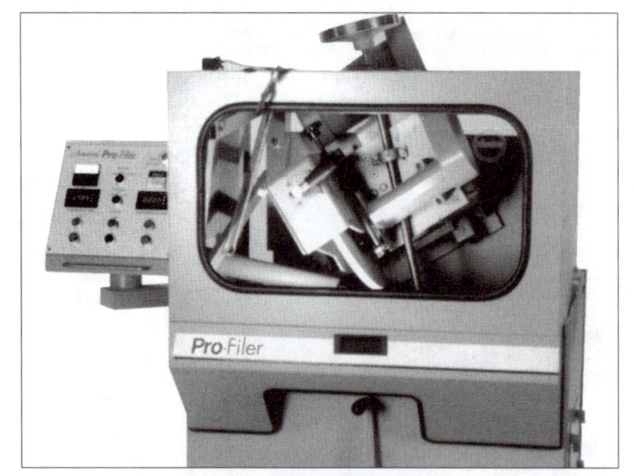

Fig. 8.32
Typical of the modern approach to saw profile grinder design for dedicated Stellite tipped saw grinding, the Pro-Filer has infinitely variable speed control for both grinding wheel and tooth feed, dual pressure clamp, digital inputs and readouts and full flood coolant.
Source - Armstrong

PLC controls usually work in conjunction with electric motors, cams and air or hydraulic cylinders which, together with manual adjustments, physically determine the shape and size of the tooth profile. Electronic control of tooth counting stops the machine at the final, pre-set tooth number.

CNC machines have computer-controlled movements of the grinding head and pawl - giving ultimate and precise control of the tooth profile through number processing and continuous-path grinding. The depth of grind, for example, can be increased by small and precise increments, and the speed and direction of traverse can on some machines be varied at different points along the tooth profile.

The cnc computer can hold a memory bank of standard tooth profiles - any of which can be quickly recalled and instantly applied. Alternatively, any new tooth profile or grinding requirement can be programmed into the computer in a simple and readily-understood way - or an existing tooth profile can be duplicated and the copy modified to suit a different requirement. The tooth shape and grinding data are usually shown on a built-in screen as diagrams and figures, and alterations to these are input via an adjacent keyboard.

Grinding two such different materials as the base steel and the Stellite at the same pass poses a problem on older grinders as each material ideally needs a different traverse speed. Traditional grinding machines could sometimes be slowed-down to grind at a more appropriate speed for the Stellite tip - but grinding the full tooth at the same slow speed not only slows-down the process, but can also case-harden the gullet to chance starting stress cracks.

To overcome this problem, some modern Stellite grinders can be programmed to automatically slow the traverse speed when grinding the hard Stellite tip itself, then speed up when grinding the softer body steel. This gives the best possible edge finish on the tip without slowing down the sequence or compromising the grinding conditions.

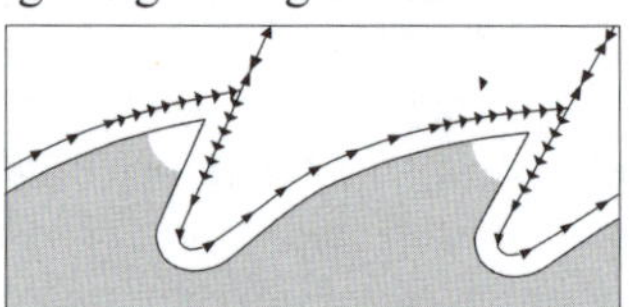

Fig. 8.33
On modern Stellite grinders the traverse speed can often be varied along the grinding path, slower for the tip and faster for the gullet and tooth back.

Dealing with a grinding burr

A long-standing irritation to machine operators is the grinding burr which forms during any grinding process - a thicker one when dry grinding heavily with a coarse grinding wheel - a thinner one when wet grinding lightly with a fine grinding wheel.

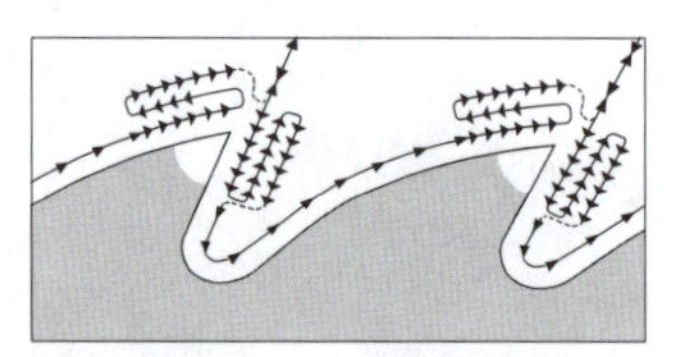

Fig. 8.34
Some grinders incorporate controls which allows back-tracking on the tip top and/or face to give an even better finish and virtually eliminate burr.

Fig. 8.35
The AS-2, shown here with the enclosing hood raised, has a dual pressure clamp and back-tracking facility to remove any grinding burr.
Source - Iseli

The grinding burr normally forms along the edges of the tooth profile, both on the Stellite tip and on the saw body itself. This is because the pressure and heat of grinding forces the outer skin at the exit face into a right-angle bend *(a burr)* instead of abrading the metal fully away. This burr is most prominent at the exit side of the grinding wheel but can also form on both entry and exit faces in some conditions.

Most burrs are weak and brittle, and quickly break off when the sawblade is first used. Unfortunately, when they do break off, the burrs can also take with them part of what would have been the cutting edge, leaving this in a damaged condition - to give a poor cut and a reduced run-time. A tougher burr can actually remain in place and mark the sawn surface unevenly. Some operators wire-brush saw teeth to remove burrs, but this cure is not fully reliable and could also damage teeth.

With modern Stellite grinders the combination of a fine grinding wheel and a fast, wet, grinding action produces only a small grinding burr. But even this can be completely eliminated on those machines that are programmed to back-track the grinding wheel across the top and/or face of the tooth. The effect of this repeated grind - if necessary with a tiny extra depth of cut - is to so reduce the thickness of the burr where it joins the tooth so that it breaks off cleanly to leave a perfectly sharp edge.

Modern Stellite grinding machines usually have infinitely variable feed-speed control to match this precisely to the sawblade profile and thickness being ground - whilst still retaining the slower and faster traverse speeds needed between the tip, the gullet and the tooth back. Saw clamping can be oil or pneumatic, with a higher pressure to ensure accurate face grinding, and a lighter one when grinding the tooth back- when movement of the sawblade is needed to form the tooth-back profile.

Some machines include a grinding wheel wear compensation facility. This trips-in after a predetermined number of teeth have been ground to lower the grinding wheel by a pre-set amount. Based on his experience by noting the number of teeth, tooth width and weight of grind, the filer can usually guesstimate the amount of down-feed needed - and its timing - to precisely balance-out grinding wheel wear.

This technique is mainly used with regular grinding wheels, both when grinding heavily on cv saws and when grinding Stellite-tipped saws. Automatic down-feed is not usually needed when grinding Stellite-tipped saws using cbn wheels.

Setting the Pawl

Regardless of the shape and size of Stellite tip, the operator usually has the choice of where the pawl actually contacts the tooth - it could be against the tip itself or against the original tooth face. The latter is probably the best point of contact as here there is no interference from the tip itself and the saw pitch will then remain as originally produced. (See also previous notes on setting the pawl under 'Traditional Grinders'.)

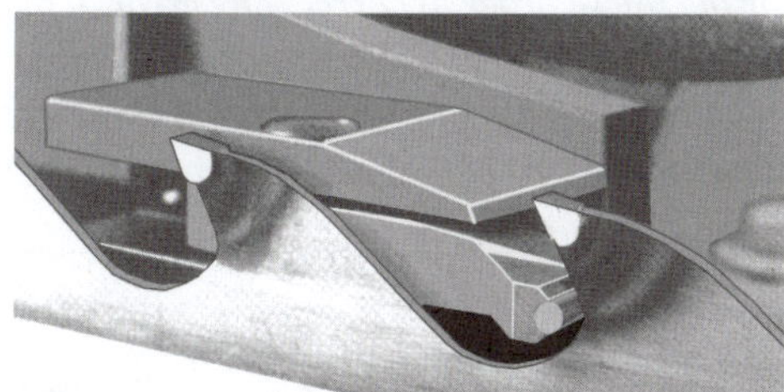

Fig. 8.36

The tungsten carbide-faced contact rod shown here on the SideWinder pawl is fitted as standard on their Pro-Filer, or as an add-on for other Armstrong saw profile grinders. It contacts below the Stellite tip and side-winds around the tooth when resetting - so the feed finger never catches the burr that forms through some Stellite tipping procedures.

Source - Armstrong

Note: More data is given under tungsten carbide saws regarding the use of pawls and in particular the arc traversed when indexing the tooth.

Stellite Tooth Profiles

Modern Stellite-tipping machines allow wide control of the amount and size of tip. There is no hard and fast rule regarding the actual proportions and shape of the Stellite deposit or tip - nor in the precise machine settings for the grinding process.

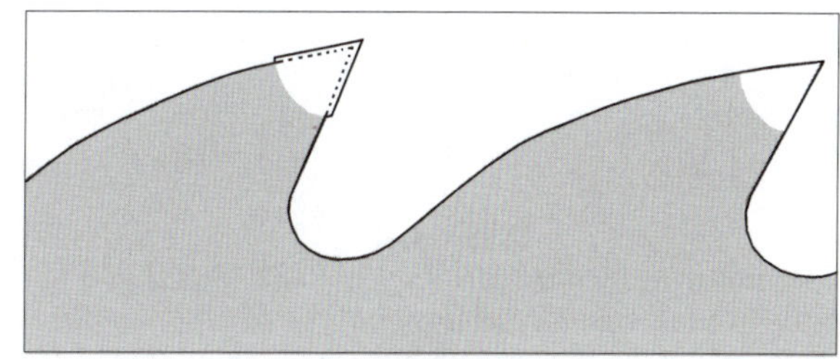

Fig. 37

A plasma-deposited tip, which forms Stellite around an existing, regular tooth profile, can be fully profile-ground as shown - but there is chance that the Stellite at the tip may be diluted - even before the first grind.
On this, and other following drawings, the untreated tip is on the left and the finish-profiled tip on the right.

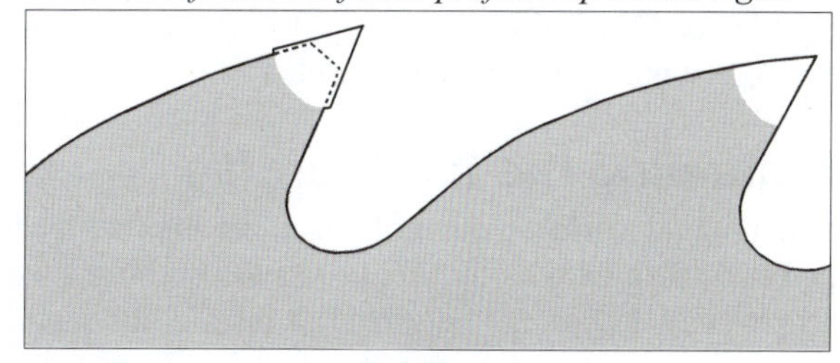

Fig. 8.38

One method of ensuring that a fully-profiled plasma-deposited tip is pure at the point both initially and throughout grinding is to grind a bevel on the tip prior to Stelliting.

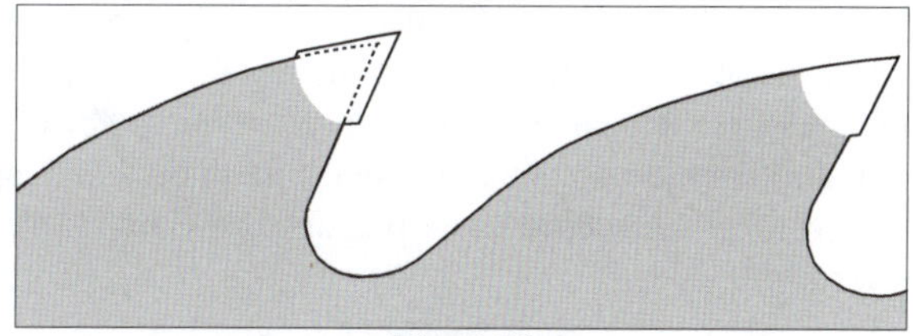

Fig. 8.39

An alternative method (used to avoid the need of bevel-grinding with plasma-deposited tips) is to form the Stellite with a front face forward of the original tooth-front. This gives a tooth profile not unlike that of a tc tipped saw - but with a tip of undiluted Stellite. The sketch shows Stellite proud on the face alone, but it is possible to form Stellite proud of both face and top.

When plasma depositing, the tip can initially project above the original tooth top and/or forward of the original tooth face - very much like a tungsten carbide tip in fact (this is done to avoid dilution at the tip.)

In this case the grinding wheel follows a regular tooth profile but may initially only grind the tooth face and top - missing the tooth back and gullet altogether. Eventually the whole tooth will be ground - though by this time the tip will probably be ready for replacing.

If grinding only the tooth top or tooth face, it is possible to use a grinder having straight-line grinding movements only, the type originally made for saw gulletting or gumming, or the simple, hand-operated machines originally made for grinding tungsten carbide teeth). Full profile grinding will still be necessary at some time, though, and this will always have to be carried out on an automatic profiling saw grinder.

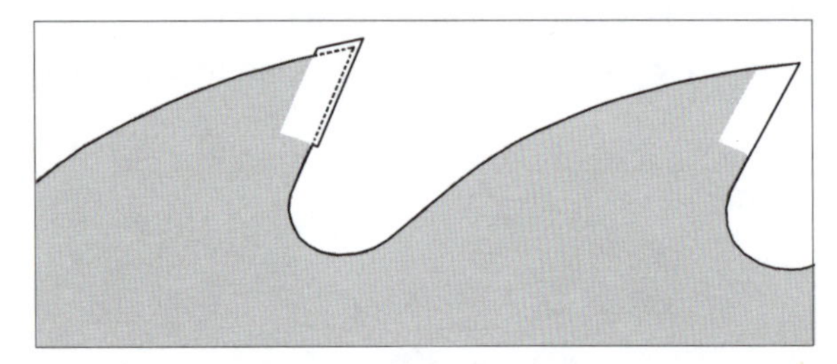

Fig. 8.40
Resistance-welded tips displace the base metal and can then be ground virtually as a regular cv saw. As no dilution takes place the tips remain pure throughout their life.

Resistance welding of solid tips enables the solid tip to displace the base metal of the tooth front, so virtually maintaining the original tooth profile. In this case the whole tooth is profiled with the machine set as a regular cv saw grinder. As with plasma depositing it is also possible to weld a solid tip to project beyond the tooth face or top if preferred, in which case only top and face grinding is needed.

Maintaining the tooth profile

Maintaining the correct tooth profile is more important with tipped saws than with un-tipped alloy saws. Because a hard tip is so small and cannot be spread or reset in the way cv saws are treated after sharpening, the saw kerf can quickly be lost through poor grinding practice. Following the wrong line of grind can quickly wear-out the hard tip, as this Chapter describes. *Also read the earlier notes under this heading in Chapter 6.*

It is important that the sawblade is firmly gripped when grinding the Stellite tooth face to ensure a precise grind. The clamp pressured is then normally reduced as the sawblade is moved through the clamp in grinding the tooth back and top. Most modern grinders have a dual-pressure saw clamp to meet this need.

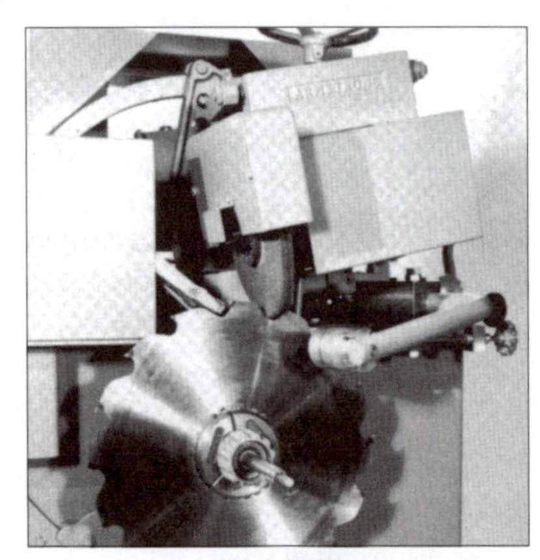

Fig. 8.41
The dual-pressure saw clamp shown here is fitted on a modern grinder - but can also be retrofitted on earlier Armstrong grinders.
Source - Armstrong

The normal practice in profile-grinding the teeth is to set the machine as accurately as possible before starting up. Once the machine is up and running, an operator will usually tweak the settings to get the right balance between the weight of grind on both tooth face and top. Conventionally the pawl is first adjusted to get the proper weight of grind on the face, then the depth setting is adjusted to get the required weight of grind on the tooth back and top. Once these are correctly set the tooth count can be restarted and the grind allowed to continue for the full sawblade.

If a second grind is needed, perhaps because all the teeth are not fully formed or wear has not been fully ground out, a second grind is undertaken once the necessary adjustments to the machine settings have been made.

Once all the teeth have been initially ground, continuing the grind without further machine adjustment can have different effects according to how the pawl is set. If the pawl is set to contact the hard tip face, then grinding will continue on the front face but will cease on the tooth top. If the pawl contacts the tooth below the tip, then the full tooth profile will still be ground - but only lightly. In resetting for a further grind the operator should, therefore, take the pawl setting into account.

Setting the line of grind

The weight of grind relative to the face and tooth top is a matter of opinion rather than fact - but getting this completely wrong will grind tips away prematurely.

Using a fixed line of grind gives dependable results. This can be shown by scribing a line on at least one tooth, then checking periodically that the tooth point remains on this line.

The correct line of grind for plasma-welded tips splits the hook and clearance angles. Follow this line by grinding more or less evenly on both top and front of each tooth.

The correct line of grind for resistance-welded tips (either using rod or with individual tips) is that drawn diagonally between the tooth tip and the opposite corner of the tip. Follow this line by grinding the tooth top slightly more heavily than the tooth front or face.

The above describes the ideal conditions, but most operators merely note how a tip wears down over a grind or two, then make whatever adjustments are needed to the current grinder setting. Some operators avoid grinding either the face or the top in order to correct any error in the line of grind. This is not good, for the best results grind both face and top each time - correcting any creeping error by slightly adjusting the relative weight of grind between face and top.

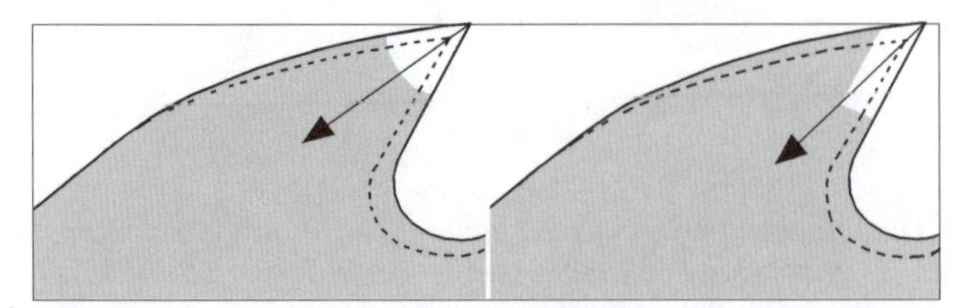

Fig. 8.42
These shown typical lines of grind for plasma-deposited tips (left) and resistance-welded tips, (right). The dotted lines show the grinding paths when tips are partially worn.

Failure to keep a good line of grind - in either case - will more quickly loose kerf width and the sawblade will need re-tipping earlier than would otherwise be the case.

Modern Stellite Side Grinders

Newly Stellited and profile-ground tips have to be side-ground to the correct kerf width for the body of the sawblade, with appropriate radial and back clearance angles and equal tooth projection either side. See previous notes under ' Traditional Grinders' earlier in this Chapter.

Modern machines use cbn grinding wheels and a wet grinding action to give all the advantages of dedicated Stellite grinding, such as one-pass grinding, precise control of kerf width and clearance angles, and a fine ground finish.

Methods of Grinding Side Clearance

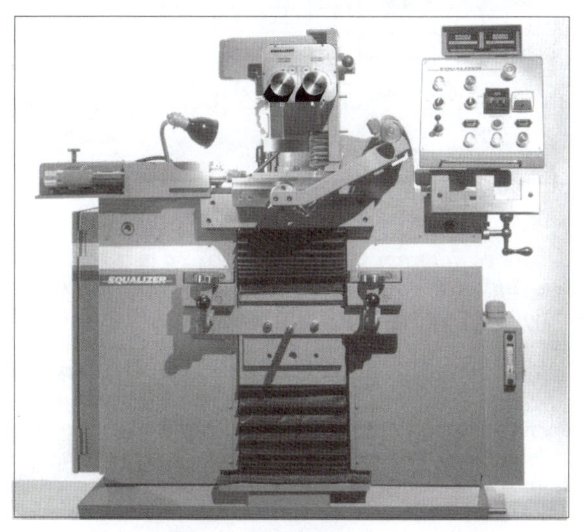

Fig. 8.43
This No. 90 Equalizer is a dedicated Stellite side grinder with built-in setting gauges, optional cbn wheel speeds, digital readouts for wheel positioning and full flood coolant. Capable of grinding 20 teeth per minute, the No 90 can also grind as few as 2 teeth per minute for 'single pass' grinding of tips with excessive side projection.
Source - Armstrong

Fig. 8.44
This type EBW side grinder uses cbn grinding wheels and powerful, enclosed water coolant system. Also available is an optional variable speed feed. In addition to circular saws the EBW can grind log and gang saw blades.
Source - Iseli

There are two basic methods of grinding side clearance on both Stellite and carbide tipped saws. The most common method is to use a pair of opposing cup-type grinding wheels which are angled to create the required back clearance, and move closer towards the sawblade centre to match the radial clearance angle needed. Another type uses opposing cbn edge-grinding wheels which can either move parallel to the saw body or taper in slightly when grinding.

In both cases the amount of side clearance and the clearance angles themselves are controlled by the line of grind and the setting of the grinding wheels relative to the tooth point. The pros and cons of both methods are discussed in greater detail in Chapter 12 under 'Side Grinding'.

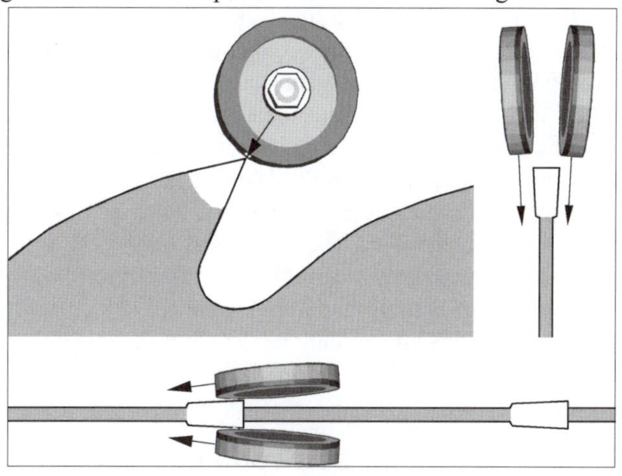

Fig. 8.45
The most common way of side grinding is by using cup grinding wheels. This sketch and that following merely give a general impression, in practice the grinding wheels are much larger than actually shown.

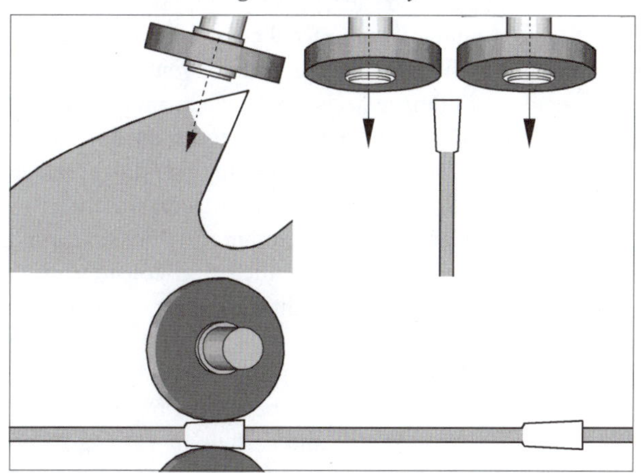

Fig. 8.46
Another way is by using edge-grinding wheels.

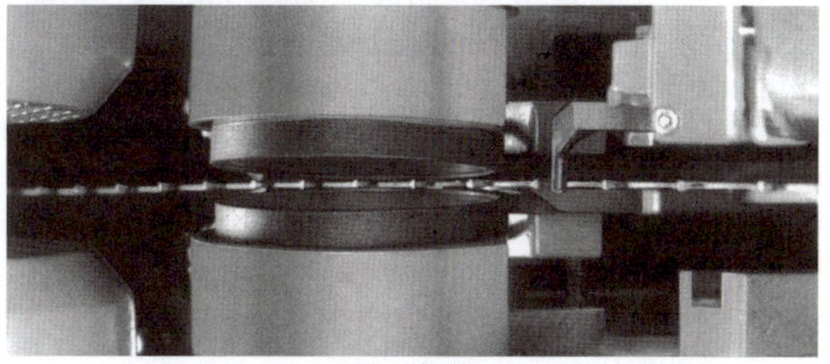

Fig. 8.47
The two opposed cup-type side grinding wheels can be clearly seen on this EMS side grinder (shown grinding wide bandsaw teeth).
Source - Vollmer

Summary

Machine welded Stellite tips are tough and well-secured, and so are ideally suited to the rough and tumble of fast-feed timber conversion. They damage less than carbide tips when striking hard knots and metallic inclusions, and they stand up well to the high impact stresses of excessive bites. When converting natural timber this quality makes possible higher feed speeds with bigger bites than was previously thought practical for carbide-tipped saws.

Modern Stellite grinders can grind saws faster and with deeper stock removal than was previously possible - this is because of the higher rim speed of cbn grinding wheels, wet grinding and the advanced design features of these machines. At a single pass the saws are ground to much tighter tolerances, with sharp and longer-lasting cutting edges, and smooth gullets which give more efficient ejection of waste.

In all, modern Stellite-tipped saws achieve far better results than the original gas-welded types. These include more efficient tipping and grinding operations, better consistency of the saw tips, increased feed speeds and run-times of perhaps 10%, reduced power consumption of perhaps 15%, reduced chance of burning, and improved quality of sawn finish. Taking all this into account modern Stellite-tipped saws are clearly in quite another ball-park.

Stellite tipping for softwood conversion

Stellite-tipped circular saws were originally intended for converting hard and abrasive timbers - very much the area also targeted for tungsten carbide saws. Carbide saws were eventually found also to be suitable for softwood conversion, and gained and retained wide acceptance in this field.

Stellite tipping was also tried out for softwood conversion, initially with gas-weld tips and conventional grinders using regular grinding wheels. Only when modern Stellite tipping and grinding machines began to be used, however, were the improved benefits of Stellite-tipping finally realized.

One factor that favoured Stellite was that carbide saws were found to under-perform on certain softwoods, returning a shorter run-times from those anticipated considering the hardness of their tips and the traditionally easy-sawing characteristics of these timbers.

That carbide has a longer run-time than Stellite on harder materials is beyond dispute, but the results from some specific tests comparing Stellite and carbide on sawing natural softwoods actually reverse this order in a very convincing way, as Fig. 8.48 shows.

The factor mentioned earlier of the poorer performance of hard tips on softer and more stringy softwoods has some bearing on this, but other factors also contributed.

A second factor is the more secure bonding of resistance-welded Stellite when compared to brazed-on tungsten carbide tips. Corrosion and downgrading of the bond can occur between a tungsten tip and the base metal of the tooth due to electrolytic action - something that does not happen with welded Stellite tips.

It is interesting to note that an earlier problem with green European redwood (Scots Pine) of rapidly blunting steel saws was also attributed to an electro-potential set-up across the sawblade during the cutting process - in this case markedly increasing when the moisture content was high.

In addition to the electrolytic action, some timbers naturally contain corrosive liquids which have no detrimental effect on Stellite, but attack the cobalt bonding of tungsten carbide tips to cause early cutting edge failure.

This type of damage is limited to certain softwoods - primarily unseasoned Wester Red Cedar (Thuja plicata) - and is assumed to be the result of plicatic acid attack, the major extractive found in the heartwood. Some experiments showed that the corrosive effect of cedar extractives, which are highly reactive compounds capable of forming organometallic complexes, varied considerably both with temperature and the rainfall level in which the trees were grown.

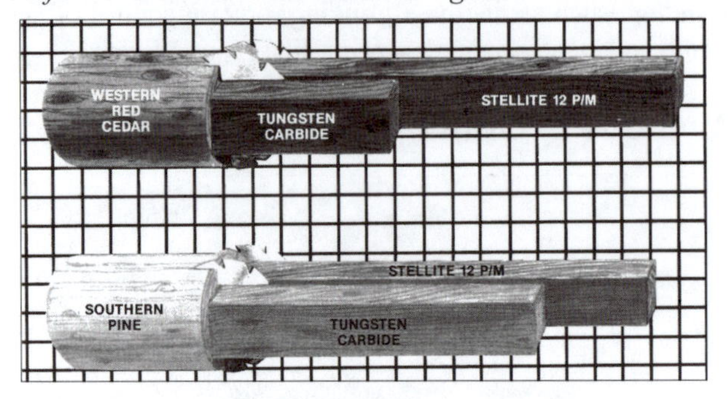

Fig. 8.48
The above diagram shows the improved performance of saws with Stellite 12 p/m tips when compared to saws with K6 tungsten carbide tips. The results are from a series of tests runs by ***Deloro Stellite*** *which showed that the wear on Stellite teeth was independent of the timber species sawn, whereas the wear of tungsten carbide teeth was highly dependent upon the species sawn. Of the species tested it was found that the greatest advantage in running time was with, in rising order, Southern Pine, White Spruce, Douglas Fir and Western Red Cedar.*
The Stellite 12 p/m used in these tests is Stellite 12 produced by powder metallurgy as a solid tip and brazed-on as a direct replacement for a tungsten carbide tip. Currently Stellite 12 p/m is also produced either as rods or individual tips for resistance-welding. This gives a more secure bond to allow the use of tips which are shorter and less in initial and maintenance costs than the original brazed-on types. Brazing is described under tungsten carbide tipped teeth

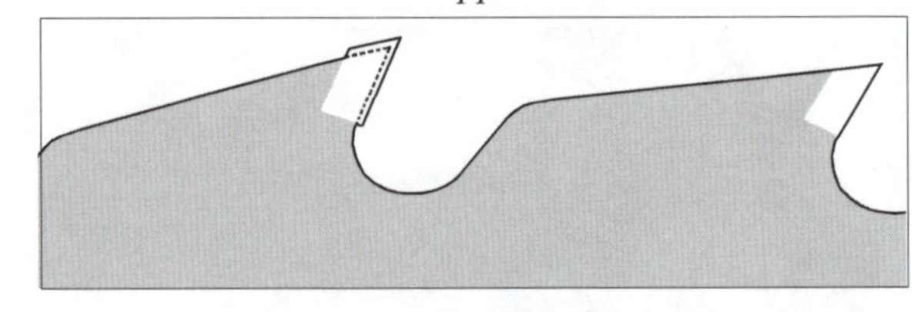

Fig. 8.49
A typical Stellite-tipped saw for fast-edging of softwood has a relatively small gullet, a strong tooth back and a tip ground flush with the tooth back. The tip on the left is before grinding.

Primarily the cobalt bonding in new carbide tips is attacked to cause premature crumbling of the cutting edges, but when tips have been in use some time it was realized that the likelihood of lattice damage to the carbide particles through the stress of cutting actually accelerated the rate of material loss - then extending partially to the tungsten itself.

The possible use of nickel alloys as the bonding agent for tungsten carbide instead of cobalt was a suggested solution for machining Western Red Cedar, but as Stellite appears satisfactory this seems the better solution.

So, if softwood conversion is your business, it might benefit you to look at the advanced equipment that has given Stellite-tipping a brand-new lease of life.

Insert-Tooth Saws

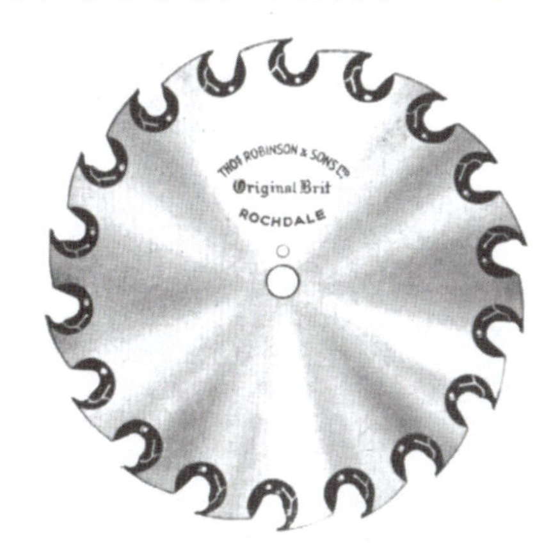

Fig. 8.50
Typical medium diameter insert tooth saw.
Source - Robinson/Wadkin

Insert tooth saws were once the main sawblades used for converting small logs and for edging boards. They are the John Wayne of sawblades, rough, tough and ready for anything - but they have an unacceptably wide saw kerf.

At one time an excessive saw kerf was not considered important - speed was the measure of efficiency - but now it is viewed as profit loss and unnecessary wastage of a valuable resource. For this reason insert tooth saws have largely been replaced in their traditional role by thin-kerf bandsaws or by thin-kerf tungsten carbide tipped saws.

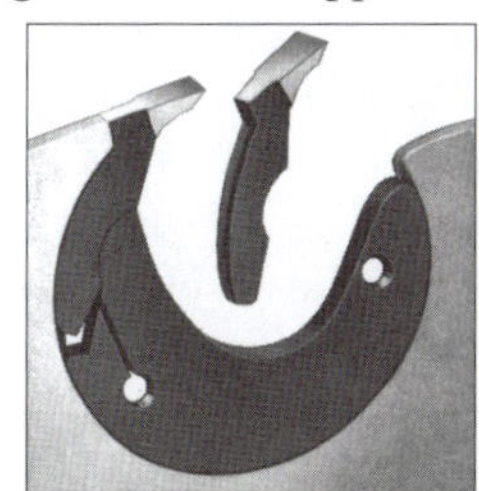

Fig. 8.51
A single-circle insert tooth, pattern 3, together with a separate bit.
Source - Spear & Jackson

They are ideal in some applications, such as forest sawblades and in agriculture - hence their inclusion in this book. They are also suitable for converting reclaimed timber which may contain metal parts that would badly damage other sawblades - with IT saws the tips can be easily replaced.

They remain the same diameter throughout their life, their rim speed does not vary, and they always retain the original gullet capacities. As a result the performance of an IT saw is consistent and dependable. In spite of all this, though, even these sawblades still need the care an attention that others demand.

Insert tooth saws have fewer teeth and are capable of a much bigger bite than other types. The actual number of teeth depends upon the sawblade diameter and on the tooth holder pattern - alternative types have different characteristics. The maximum allowable number of teeth is given in Chapter 3, although it is quite common for them to have fewer teeth.

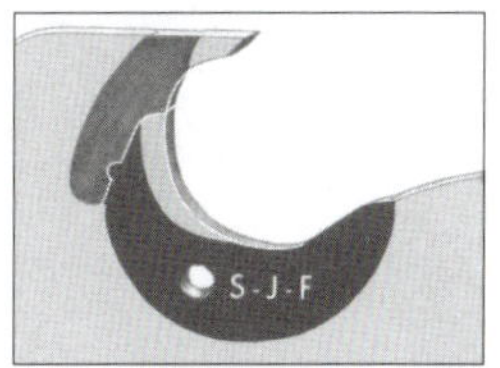

Fig. 8.52
Two-circle insert tooth, pattern F.
Source - Spear & Jackson

All insert teeth consists of two parts:- the bit, and the shank that holds the bit in place. Different widths of bit can be used with the same saw body thickness. Each shank is actually a spring, and it is the tension within this spring that holds the bit firmly in place. Both bits and shanks have a vee groove that locates on the inverted vee of the sockets they seat in.

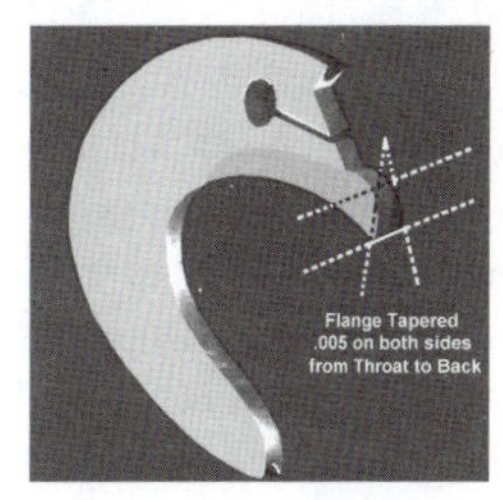

Fig. 8.53
This shows the taper grinding on a shank.
Source - Pacific Hoe

Two main types of insert tooth fittings are made, the original single-circle type designated by numbers where the bit seats against a shoulder near the rim, and a later, and considered superior, dual-circle type designated by letters where the bit seats against a second curved seating. Insert teeth are strong and tough, partially due to their low clearance angle and the excellent backing by the saw body. They also have a large hook angle to give a clean and easy cut.

The cutting action of an insert tooth saw can actually be changed by mixing and matching the various types of bits and shanks available. Damaged bits can be easily and quickly switched, often without need of removing the sawblade - so hold-ups through damage are rare.

Fig. 8.54
This is a one-circle shank and bit separated to show the arrangement more clearly.
Source - Pacific Hoe

Too-narrow a kerf for the sawblade body increases the loading on the saw motor to a point where the feed speed may need reducing to an unprofitable level, and the tooth kerf reduces as the tooth is filed or ground down. For this reason new saws should be ordered with a wide enough kerf to give an economic life before replacement is needed - or the teeth must be swaged when worn (not all bits allow this).

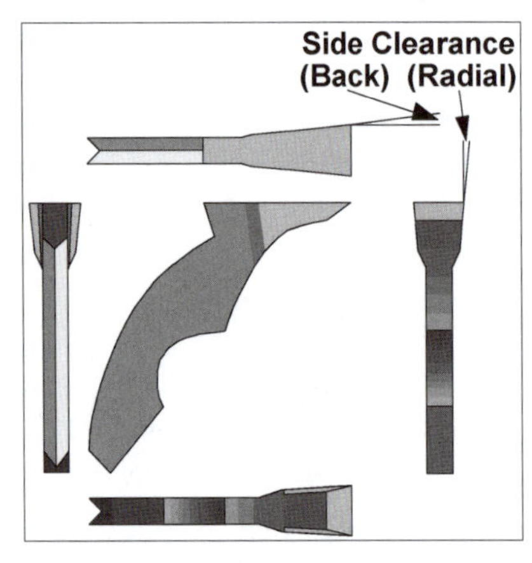

Fig. 8.55
These front, top and side sketches show the clearance angles ground into all bits.

Regular bits, the original-type, are made from steel that can be filed and swaged. It is also possible to buy tips made from solid and hardened high speed steel, or with an inlay of hardened steel, Stellite or tungsten carbide. These required grinding to re-sharpen and cannot be swaged.

All bits have built-in top, radial and back clearance angles formed during manufacture, and have the correct amount of side-projection. As a result no further preparation is needed after fitting new bits - other than correcting any height difference or sideways mis-alignment of replaced teeth.

Shanks are produced in different gauges to match the gauge of sawblade. Some are manufactured to a swaged section close to the bit (wider than the saw body but less than the kerf width) which more efficiently retains sawdust within the gullet. Another regular form is a frost shank for use with regular teeth to give better running when sawing frozen timber - *with regular shanks frozen timber waste bunches up and clogs the gullet.*

Fig. 8.56
Typical small diameter insert tooth saw for edging.
Source - Spear & Jackson

Fitting teeth.

New saws are provided with fitted shanks and bits to the users requirement, but at some stage they have to be replaced. An insert tooth saw is tensioned in the same way as other circular saws and the shanks and bits form an integral part of this tension. For this reason it is important always to both tension and run a sawblade with all the shanks and bits in place - never run or attempt to tension a sawblade with one or more teeth missing. Also, there is a wide variety of both shanks and bits, so take care that replacements are identical to the originals.

Shanks and bits are removed and replaced using an insert tooth wrench - which locates by a pin through a hole in the shank. To remove a shank and bit, secure the sawblade and, with the wrench engaged, move the lever end in the same direction as the saw's rotation.

Fig. 8.57
A wrench as used for removing and inserting shanks and bits.
Source - Pacific Hoe

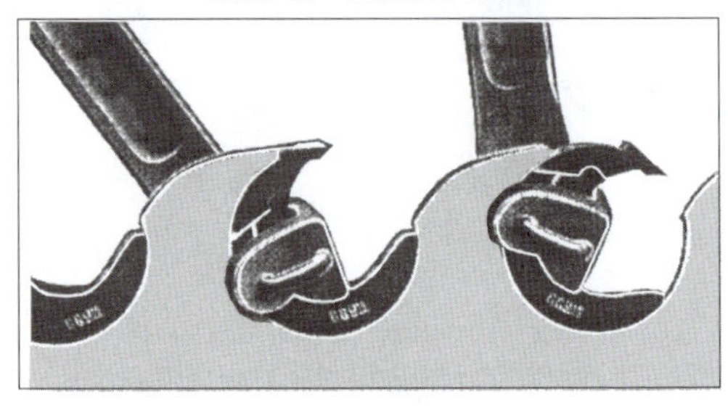

Fig. 8.58
Using a wrench to insert a shank and bit. The sawblade must be firmly clamped when doing this.
Source - Spear & Jackson

Before replacing the bit and shank clean and rub oil onto the seatings, this will make fitting easier and will also reduce wear of the socket.

Move the wrench in the opposite direction to refit the bit and shank and, as the fit should be tight, use one or more firm backwards and forward movements to ensure the bit fully seats.

If replacing all the bits in a saw some filers only partially seat each tooth initially, then fully seat them by going around a second time.

Note - remove and replace each bit and shank in turn, do not remove several shanks and bits at the same time as this could distort the sawblade.

After fitting new shanks and bits check that single-circle teeth are in line with the saw body, if not tap into place (they are the ones most liable to mis-align). Tooth sockets themselves can become out of alignment over time with either type of bit fitting - and the sawblade would then cut badly if left in this state.

Possibly use a click gauge to check that sideways projection of all the bits is the same, but for greater accuracy use a dial indicator mounted independent of the sawblade. To correct socket misalignment place a saw set on the saw body immediately behind the mis-aligned tooth and spring it back into place.

After replacing individual teeth - make sure these are the same - file or grind the front down to the same dimensions as existing teeth.

This should ensure that the new bit is identical in both cutting width and height to existing bits. If this is not done the new teeth will take a bigger or wider cut - to the detriment of the cutting action, the quality of cut and the life of the new bit.

Sharpening teeth.

Insert tooth saws do not need stoning down in the machine to keep them round - accuracy both in the milling of the seatings and bit manufacture ensure that the cutting edge of new and identical bits are always concentric to the bore.

Regular steel teeth can be hand-filed in the machine without need of removing the sawblade. Use a flat, round-edged file against the face only of the tooth only - never file the tooth top. Make absolutely sure that the original hook angle is retained. The tendency over time is to reduce the hook angle - and this results in the bits having a less aggressive bite which takes more power and cuts less cleanly.

The hook angle of insert teeth saw, at about 25 degrees, is suitable for most softwoods, but in some cases it may be an advantage to alter the hook angle slightly if a harder timber is being sawn. Unlike other saw types, the hook angle on insert teeth is easily altered.

The regular hook angle is too big for dry hardwoods so, if regularly sawing this type, it is better to intentionally reduce the hook angle slightly - otherwise the saw will cut roughly and tend to vibrate and chatter. For both hardwood and softwood frozen timber, though, the original hook angle gives the best results.

When filing make absolutely sure that the tooth face is filed perfectly straight across. If the front is filed to an angle the sawblade will run to the leading point side, tend to snake and generally create feeding problems.

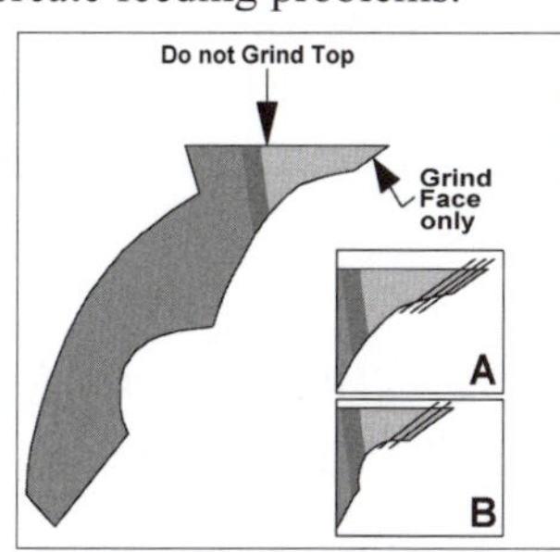

Fig. 8.59

*For softwoods make sure that the original hook angle is maintained, as shown by the sharpening lines at **A**. File or grind only the face, never the tooth top. Worn bits can have extra life beyond that shown in **A** if the inside face is relieved using a half-round file, as shown at **B**.*

For hardened solid and hard-faced teeth a grinder of some type is needed. Traditionally a jockey grinder was used to grind sawblades in situ, and this still remains the best option. This type actually sits on the sawblade and is located by a finger similar to the pawl used on a machine grinder.

Early models had a flexible shaft attached to the grinding arbor - driven by a bench-mounted electric motor or even a hand crank. Modern ones have electric or air drive and are much quicker to use and much more accurate. Regular grit and diamond grinding wheels can be used on them all.

The sawblade could be removed for face-grinding on a regular saw grinder or gulleter - but this is slower and holds up production.

Bits have radius formed below the cutting face to allow the front to be filed or ground clean across using a flat file. Eventually, though, the front will be filed back such that, in order to grind the bit even further, a small extra gullet needs to be formed in the bit face itself, *see B in Fig. 8.59.*

By this stage of wear, though, the kerf may be so narrow that the saw would bind in the cut - and so would probably need complete re-toothing.

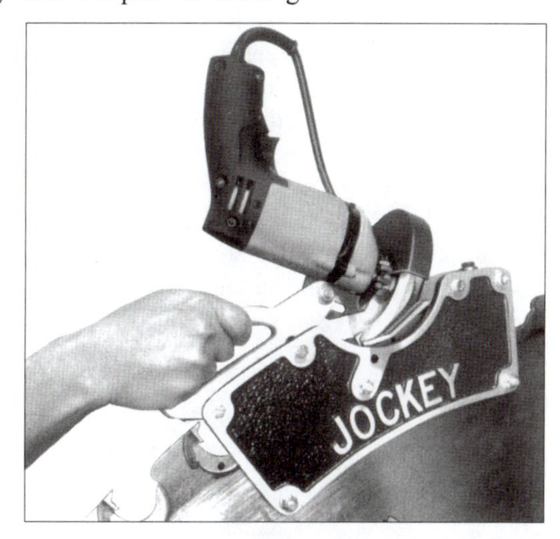

Fig. 8.60

A jockey grinder sits on the sawblade, so allowing re-grinding without the need of removing it from the sawbench. The unit is pushed along from tooth to tooth and, as it does not need clamping to the sawblade, a complete re-grind is possible within a mere five minutes. The unit shown is electrically-driven, but is also available air-driven and in a different form for edger saws.

Source - Hanchett

Maintaining the shanks.

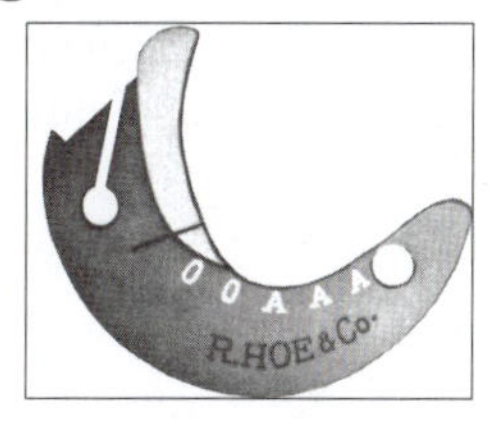

Fig. 8.61

To stretch a shank hammer as marked at AAA equally on both sides, hammer in addition at OO if still not satisfactory. Do not hammer beyond the straight black line.

Source - Pacific Hoe

Sockets can eventually wear or lose some tension to make the shanks a loose fit. Loose shanks hold the bit less securely and can adversely effect the tension of the sawblade.

To compensate for badly worn or re-milled sockets, oversize shanks can be fitted. However, the shank can be stretched slightly to make good a loose-fitting socket that has become only slightly worn. To do this remove the shank and place it flat on an anvil and, using a round-faced or ball pein hammer, make three to four hammer blows close to the centre inside edge of the shank. Turn the shank over and repeat the same amount of hammering from the opposite side.

Shanks help to retain sawdust in the gullet - to then discharge the contents when the tooth clears the timber. The action is effective only so long as the shank gullet remains perfectly square across.

Because shanks are used through many changes of bits, the constant stream of abrasive sawdust into the gullet can wear the edges rounded. This then can direct sawdust between the saw body and the sawn timber - with predictable results.

To avoid this check the state of the shanks from time to time and either file the gullet area of the shank using a half-round file, or lightly grind it on a hand gulleter - taking great care not to overheat it in the process.

Do not to file or grind the shank excessively as this will weaken the essential spring action, and make absolutely sure that it is filed or ground perfectly square across - filing at an angle will direct sawdust to one side of the saw/timber gap.

As with other sawblades the saw body must be checked periodically for flatness and tension using the methods shown later. With these sawblades this must be carried out with all the shanks and bits in position and fully seated.

Fig. 8.62
Tough new carbide bits for fast-feed sawing.
Source - Simonds

Modern IT Bits and Shanks

Modern tips are manufactured in various patterns and in different styles, including the following from the Simonds range.

Blue tip - Hardened and tempered tips which can be re-sharpened with either a file or a grinder.

Stand-all Bits - All season bits which break up sawdust and expel it quickly to ensure the sawblade runs freer and cooler. Especially suitable for frozen timber.

Si-Chrome bits - Chrome-plated on the cutting edge for increased hardness, edge holding capacity and reduced friction.

Dominator Carbide tips - Manufactured on a special forged body and with a tough carbide tip.

Simo-Brite Bits - Plated in an extra smooth chrome surface over the entire bit to greatly reduce heat-generating friction - and is extremely wear-resistant. Suitable for frozen timber.

Tungsweld Bits- Based on regular blue tips but with an inlay of tungsten cobalt alloy.

High Speed Steel Bits - Made from a hard and wear-resistance hss for abrasive hardwoods - which hold their edge up to ten times longer than regular bits.

Most bits are made in several types allowing saws to be customized for particular cutting conditions.

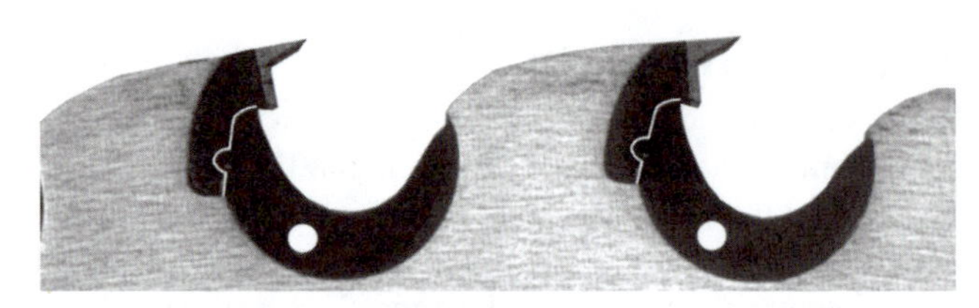

Fig. 8.63
A modern, insert-tooth saw with two-circle shanks and stqand-all bits.
Source - Simonds

Shanks are used interchangeably with various bits and are in different patterns:- Regular Shanks are full-swaged (spread) to keep sawdust in the gullet - with flat side grinding to improve sawblade stability. Super Shanks have a full end to disperse chips, prevent clogged gullets and bit creepage. Wintar Shanks include a flat chipbreaker for breaking frozen chips. Simo-Jet Shanks can be of any style but with a hard chrome finish.

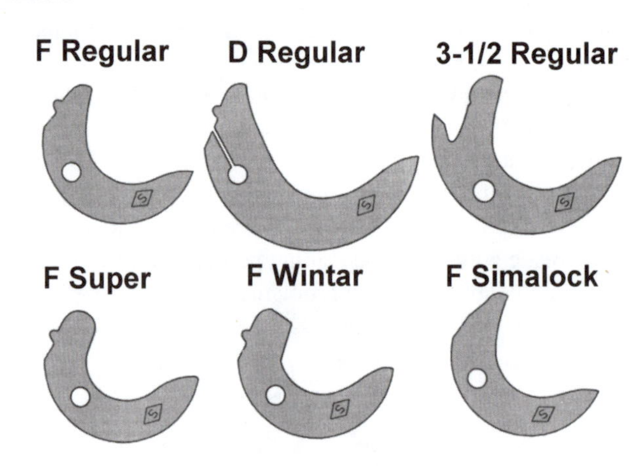

Fig. 8.64
A selection of modern shanks.
Source - Simonds

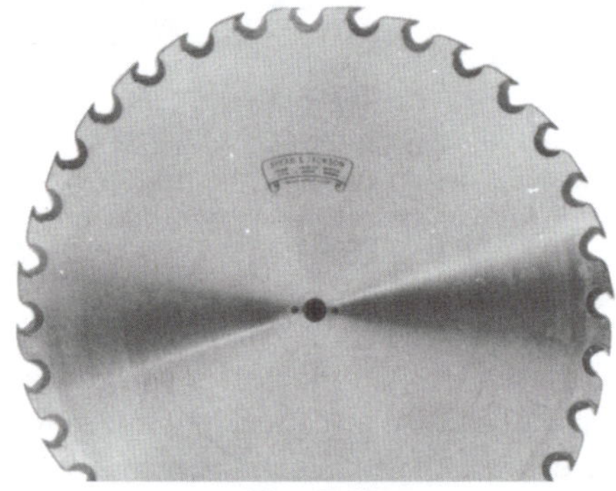

Fig. 8.65
Typical large-diameter insert-tooth saw.
Source - Spear & Jackson.

Fig. 8.66
This shows two modern bits a Si-Chrome, Stand-All type (left) and a carbide-tipped bit with a recessed face for easier sharpening (right).
Source - Simonds

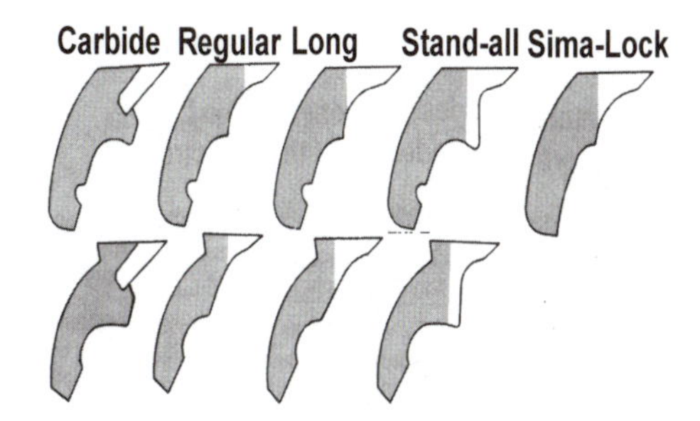

Fig. 8.67
This shows typical modern bits The upper range are B, D & F styles, the lower ones 2 1/2 - 4 1/2 styles
Source - Simonds

Insert-tooth cut-off saws

Traditionally IT cut-off saws are fitted with straight teeth. These are set to left and right alternately, and have alternate front and top bevels similar to regular CV sawblade teeth.

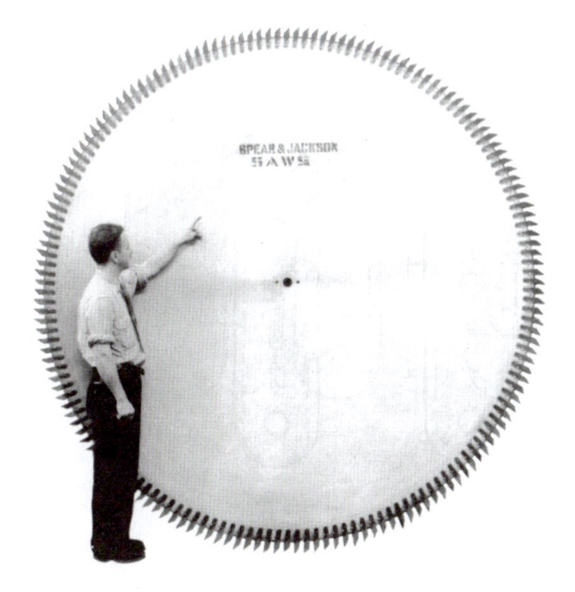

Fig. 8.68
Cut-off saws such as this were regularly used in sawmills, shingle and pulp mills in the days when large diameter logs were common harvested in the Northwest of America and Western Canada - now regrettably no longer the case. This is probably one of the largest produced around the 1960's
Source - Spear & Jackson

Like the single and double-circle insert-tooth ripsaw teeth, the teeth of cut-off saws can likewise be replaced when worn out, though in their case they teeth fit in parallel vee ways and each is secured by a tempered rivet.

The regular tooth pattern for cutting hemlock, spruce and fir has a parallel body which is bent outwards to the high point side by hammer-setting - in a similar way to plate-type cross-cut saw teeth.

A special S & J pattern once produced for cutting cedar is forged to be wider towards the point to give a stronger tip but still retain working clearance.

The milled seating of the saw body fixes the hook angle at slightly negative.

This is maintained consistently as most bits have a leading edge parallel with the seating. Some, however, are available with a modified face angle intended either to increase or decrease the hook angle of the sawblade according to the timber to be cross-cut.

Regular teeth are made from similar material to regular insert-tooth ripsaw teeth and, like them, are also available inlaid with hard alloy or furnished with a tungsten carbide tip - both are a direct replacement for regular teeth.

Fig. 8.69
Typical, regular cut-off type insert teeth
Source - Simonds

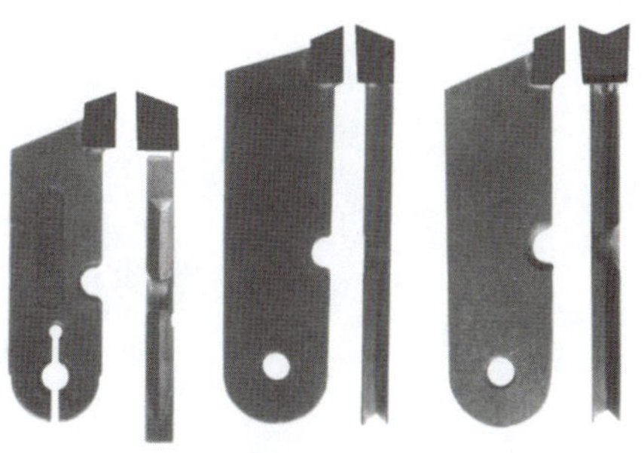

Fig. 8.70
Carbide-tipped cut-off teeth, regular styles made with left and right top bevels (left & centre) and with Future 2000 tips (right). Centre-tracking of the latter tips reduces both crack-causing vibration and power consumption.
Source - Simonds

The most recent type of tungsten carbide tipped cut-off teeth, Future 2000, have a slightly hollow face and a vee-shaped top to give a double-pointed tip.

These shear wood fibres instead of tearing them and, by cutting both sides simultaneously, only half the number of teeth are required when compared to regular alternate-set styles.

The double cutting action cuts freer and tracks better than regular teeth and reduces crack-causing vibration. In addition, only one type of tooth is needed - and the brazed-on carbide tips can actually be removed and replaced without having to remove the teeth from the sawblade.

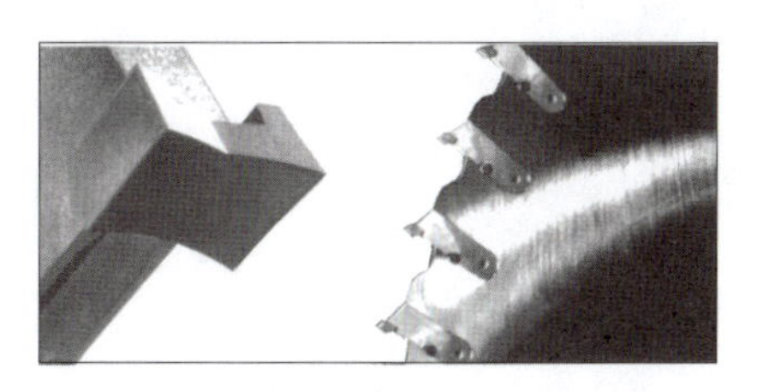

Fig. 8.71
Future 2000 carbide tips in close-up (left) and fitted to a cut-off sawblade (right).
Source - Simonds

Fig. 8.72
Carbide tipped cut-off teeth can also be supplied for single-circle type saws.
Regular style with left and right hand top bevels (left) and Future 200 style (right).
Source - Simonds

Double-pointed carbide teeth provide clean and fast cross-cutting with low power requirements. They also eliminate the need of separate left and right-hand bits.

Swageability of insert-tooth tips for ripsaws. Some of the softer grades of insert-tooth bits for regular applications are suitable for swaging. This may be necessary when the kerf width becomes so reduced through grinding that body heating and high power consumption becomes a problem. This process spreads the kerf to restore the original kerf width.

Simonds offer two swagers, one for 10 gauge sawblades and heavier and a second for 11 gauge sawblades and lighter.

An upset swage differs from a regular swage as used on bandsaws and regular cv circular saws in that there are no moving parts. The upset swage has a hardened internal vee which the tooth should be made to fit exactly (or perhaps with a fractionally smaller angle).

Swaging is carried out by placing the upset swage firmly onto the tooth point and striking its end with a light blow from a hammer, repeating this if necessary to spread the point as required.

Ideally teeth should be removed and firmly clamped in a vice when swaging. Using an upset swage with the teeth still in the sawblade can damage both - a common practice with old hands - but care should be taken and hammer blows should be light.

Chapter 9

Tungsten Carbide Tipped Saws

Tungsten carbide tipped saws are the real workhorses in most wood and wood-based industries, almost to the exclusion of other types in many applications. Other considerations apart, tungsten carbide tipped saws have one enormous advantages over steel saws - length of life between sharpening - varying from 10 to 200 times, depending upon the material being sawn and the cutting conditions.

The Industry was slow at first in accepting tungsten carbide saws for general woodworking applications - originally they were used only for abrasive hardwoods and abrasive-bonded boards. This was because initially, due to the granular structure of tungsten carbide, they were not as sharp as steel saws and so did not give a clean cut - especially in softwoods. Also, tungsten carbide saw tips were so-easily damaged when compared to steel saws and so had to be handled with a great deal more care than steel saws in an industry that had a rough and ready image.

Both these concerns are now water under the bridge. Modern carbide saws are offered with exceptionally keen cutting edges and in a wide variety of dependable saw types suitable for converting anything from natural softwood through particle and densified boards to aluminium.

However, care in use is still of paramount importance. To get good value and service from TCT saws it is essential to understand and cater for their somewhat peculiar characteristics - also see earlier remarks in Chapter 7 regarding the manufacture and upkeep of tungsten carbide saws.

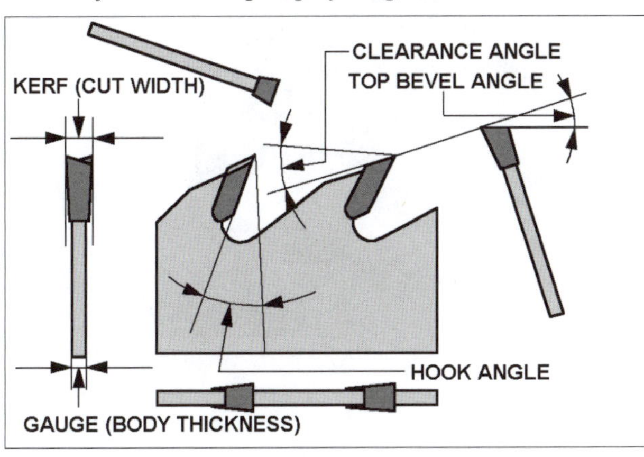

Fig. 9.1
This shows the various angles on a typical alternate-bevel tungsten carbide-tipped softwood ripsaw.

Tungsten carbide is not a metal, it is a sintered material produced by a powder metallurgy process pioneered in Germany in the 1910's - a composition of the elements tungsten and carbon produced as a fine powder and bonded together by cobalt. These tips are brazed onto the faces of saw blanks and then ground to the correct hook, top and side clearance angles.

Hardness and toughness of carbide tips varies according to the application grade, stated as a K number. As a general rule a softer and tougher grade of carbide is used for hard materials, and a harder and less tough grade is used for softer materials. The choice is not so simple, of course, as the operating conditions are also an important factor. Basically the working compromise is to use a tip tough enough to withstand the shock loads of an intermittent cutting action, yet hard enough to give a long working life.

Selecting the right tip for each sawing application is normally the saw makers problem, so the best guarantee of suitability for the purpose is to purchase sawblades from a reputable tool company who specialize in woodworking. For the same reason a sawblade bought for a particular use should be used for this application alone - using a sawblade for some other applications is a surefire way of getting poor results.

Tungsten carbide saw applications

In order to get the best results from carbide saws, the machines using them must be in good condition, heavily built, vibration free, generously-powered and with bearings in good condition and all running parts carefully balanced. Balance is particularly important as rim speeds are now higher than previously was the case and vibration from whatever source spoils the sawn finish and shortens the life of sawblades.

Note: the following recommendations do not apply to tungsten carbide tipped saws alone - they apply in general to machine-ground saws of all types. The reason that this detail is included under tungsten carbide saws is that they are a much more precise tool - and a higher standard of fitting is essential to get the best results. The spindle bearings must be free from play and the spindle run true. The saw should be a good fit on the spindle end, and the collars should run true, be in good condition and free of burrs or other damage. Engineers work within specified tolerances - two dimensions between which the part being measured for size is accepted as being correct, or as a maximum measurement when checking run-out.

With tct and other machine-sharpened sawblades, consistency of tooth height is reliant on the sawblade being ground accurately to the centre bore, and on running accurately on the machine. If any play exists between the sawblade and the arbor, or if the sawblade is a good fit but the arbor runs out-of-truth then, however accurately it is ground to the bore, it will still run eccentrically to give poor and irregular sawing.

If sawblades run with a sideways wobble only a few teeth would take most of the load, and contact with the sawn faces would only be with the teeth protruding at that side - giving a very poor finish. Sawblades are precision ground for side clearance relative to the saw body, so saw collars must run absolutely true. They should give maximum possible support to the sawblade at between one half and one third of the sawblade diameter, depending on the application.

As part of the routine maintenance of machinery - which includes ensuring that the shaft and bearing surfaces are free of dirt, nicks, gouges and wear - Gary Metzger of North American Products Corporation recommends that the following checks are also made, see Figs. 9.2,3 & 4 :-

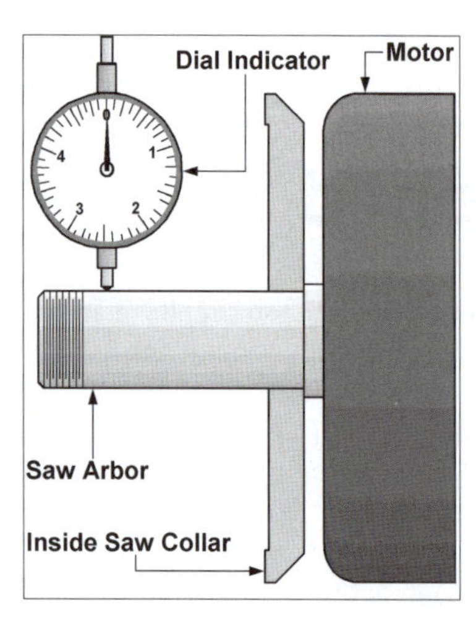

Fig. 9.2

Testing the saw arbor for run-out by slowly rotating it whilst a dial indicator is in contact. Replace the bearings if run-out exceeds 0.025mm (0.001 in.). Ensure that the arbor in good condition, and with its diameter between 0.0125 and 0.025mm (0.0005 and 0.001in.) of its true size. If not replace the arbor, or have it chromed and re-ground.

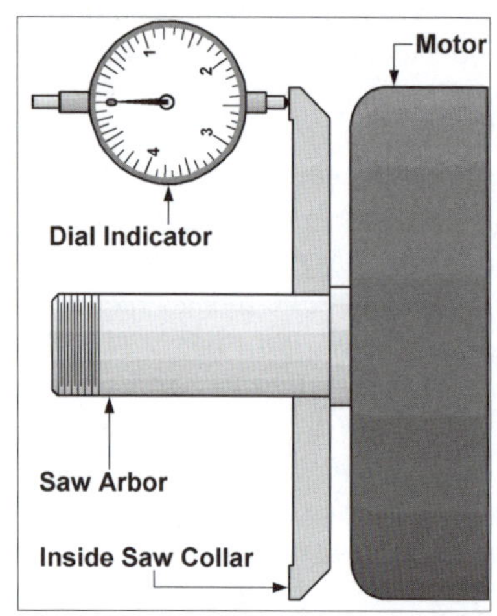

Fig. 9.3

Testing the saw back collar for run-out. With a dial indicator against the contact face of the saw collar, slowly rotate the arbor. The maximum run-out should be less than 0.025mm (0.001in.).

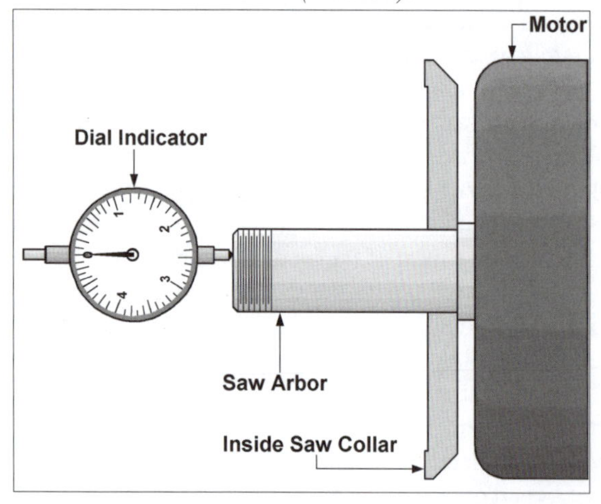

Fig. 9.4

Checking the saw arbor for end play (saw wobble could also be caused by end-play in the arbor bearings). With a dial indicator against the end of the arbor, push and pull the arbor manually to check for end float. If more than 0.025mm (0.001 in.) is found, replace or adjust the bearings and check the arbor mounting.

Support around the sawblade must be kept in good condition. Always ensure that both the fixed and sliding tables - and all inserts - are absolutely level with one another to give solid support to the work. *An uneven or inadequate support gives poor quality sawing however well-prepared the sawblade might be.*

With traversing-carriage panel saws check the condition of the table regularly and make adjustments wherever and whenever needed. Sliding-table panel and dimension saws often have replaceable lip on the fixed table - which needs replacing when worn. The sliding table itself usually has no such parts, but there may be some adjustment possible to the sliding table support or rollers, or to the fixed table - which can correct any out-of-level on older machines.

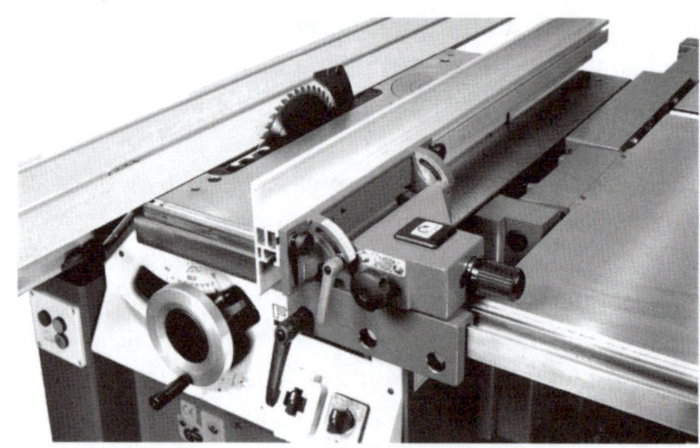

Fig. 9.5

The table insert, as on the fixed table of KFS-37, keeps the sawblade gap small, replace it when worn.

Source - Felder

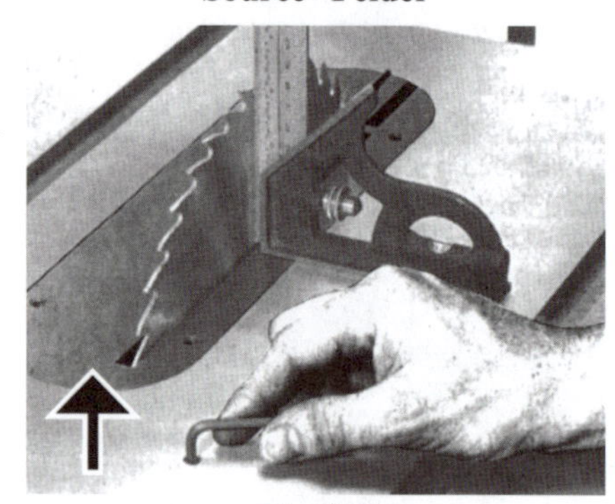

Fig. 9.6

Make sure that the table insert (arrowed) is correctly fitted to lie perfectly level with the table and so give good support when sawing.

When sawing material likely to chip or splinter, fit a hardwood, MDF or synthetic insert, then cut through this by raising the sawblade under power.

The operation shown is setting the vertical position on a 10in tilting arbor sawblade.

Source - Delta

Manual-feed general-purpose saws with a tilting arbor and fixed tables have a removable pocket *(insert)* for access when changing or adjusting the sawblade. Make sure that the correct insert is fitted to give the narrowest practical gap around the sawblade *(most machines have two or more inserts for use with tools other than saws).*

When converting double-faced boards on a single-sawblade machine they are likely to chip or splinter on the underside. To minimize this effect replace the existing table insert with a blank made of hardwood, MDF or a synthetic composite material. Slot the insert where the riving knife will project through.

With the sawblade upright, or tilted to the required bevel angle, wind it down clear of the underside of the insert and fix the new insert securely in place - making sure that the upper surface is flush with the table top. Start the machine and wind the sawblade up to the absolute minimum height needed for the material thickness to be sawn - no higher. The sawblade will cut through the insert and, by doing this, gives maximum support to the underside of material being sawn and the smallest possible table gap.

When sawing, pressure the board into close contact with the table insert at the point of cut - perhaps by using a roller pressure - otherwise chipping and break-out may still occur. If the insert wears in use, wind the sawblade slightly higher to restore support where it is needed.

On non-tilting saws a narrow hardwood fill-in piece is sometimes fitted at the infeed to keep the gap around the down-cutting part of the sawblade to a minimum. In the same way as the regular table insert is replaced, this fill-in piece should also be replaced at regular intervals to give the best support when converting faced boards

Fig. 9.7
The cross-cut and mitre fence on some machines, such as this KS-30 sliding table saw, have a wooden inset (arrowed) to give edge back-up to material being sawn. The mitre fence can be adjusted lengthwise to re-align a blank inset with the sawblade, either after swivelling it or when the inset wears.
Source - Casadei

When cross or angle-cutting natural timber or boards using a fence that has no provision for a wooden insert, it is useful to fit a hardwood backing to its stock-side face. This serves two purposes. It provides back-up for the material being cut to prevent break-out at the trailing edge. Also, when the method of working is to cross-cut workpieces which are pencil-marked individually for custom work, it gives an precise guide - the initial saw-cut - for accurately positioning them. Although the wooden back-up can be of any suitable depth, it should be notched out at the sawblade end to a depth only marginally more than the workpiece thickness - and at a distance from the sawline sufficient to clear the saw top guard.

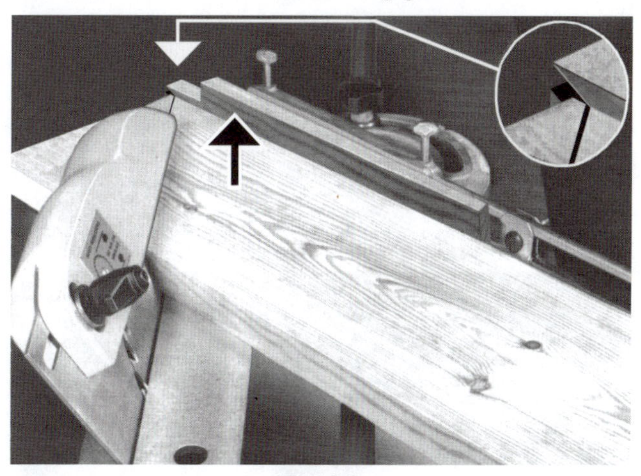

Fig. 9.8
On fixed table saws the mitre gauge slides in a machined slot, as shown on this AGS tilting arbor sawbench. It is useful to fit a hardwood fence (arrowed) to give back-up to the cut and a more stable support to the workpiece. The notch at the sawblade end is to avoid fouling the saw top guard. The first sawcut also gives a precise point against which marked-up workpieces can be positioned - as shown in the enlarged inset, top right.
Source - Wadkin

Sawblade specifications

Tungsten carbide saws are suitable for cutting a wide range of materials and, with an excellent choice of specifications, a tct saw can be precisely matched to virtually any material or type of operation.

The general rule is that easy-to-cut and abrasive materials saw best with low rim speeds, the greatest tooth pitches, large hook angles and big bites, and materials which readily chip or splinter normally require fast rim speeds, together with smaller tooth pitches, hook angles and bites. But it is never quite as simple as that.

Fig. 9.9
Typical tct saws are shown here, left, a fast feed ripsaw for natural timber and, right, a fine-tooth saw for cutting board materials.
Source - Leuco

Sawblade diameter

The working position of a sawblade in part determines what diameter is most suitable. Regular alloy saws for ripsawing natural timber are conventionally set so their teeth cut almost at right angles to the timber - as this takes less power and gives a longer saw life - and the same applies when using tungsten carbide saws. So for ripsawing natural timber on double slabbers and circular re-saws large diameter saws are the norm.

Straight-line-edgers and single-arbor gang ripsaws normally operate with the sawblade projecting just clear of the timber.

Double-arbor gang ripsaws barely overlap in the centre of the timber to divide the cut between them. In both cases use smaller diameter sawblades for thinner timber - provided they can still be run at the correct rim speed.

Fig. 9.10
Panel-cutting sawblades should barely break through the top surface. On this FW computer-controlled panel saw the sawblade height is automatically and continuously adjusted to the correct setting.
Source - Schelling

For most other applications, and for all board materials, the sawblade is normally set to just break through the top surface - this gives the best possible cut and least amount of break-out on the underside. Normally a sawblade projects through by no more than 10 per cent of its diameter, but when cutting very thin sheets the teeth should barely project through. For these applications a suitable sawblade diameter depends upon the thickness to be cut, the machine mounting arrangements and the choice of rim speeds.

Most manual fixed and sliding-table sawbenches have enough height adjustment to set even the largest sawblade to the recommended height regardless of the application. This is not always the case with some older machines - a sawblade small enough to be set to the correct height may run at too-low a rim speed. In this case use a larger diameter sawblade set to run at the correct speed - but at the lowest possible setting - and with a shorter tooth pitch than recommended for the material being sawn (doing this partially compensates for the squarer cutting action).

Unlike steel sawblades which lose considerable diameter through wear, tungsten carbide saws do not - they reduce by only 5-6mm *(1/4in)* during their life. For this reason it is viable to use tct saws of the smallest practical diameter for the work in hand. They cost less to buy and maintain, are more stable in the cut - and usually have a narrower and less wasteful saw kerf. It is false economy to use a large diameter, general purpose tct saw if much of the work is cutting thin material.

Rim Speed

There is a wide range of rim speeds used with carbide saws. The correct speed depends upon the material to be cut, two examples being 20 - 30 M/sec. for mineral board and 60 - 100 M/sec. for natural timber.

Fig. 9.11
Typical of modern sliding table saws, this saw unit houses both the main saw, the scoring saw and the drive motors, etc. It has a cast-iron frame to guarantee both vibration-free cutting and consistent alignment of the main and scoring sawblades - essential for sustained quality sawing. The main sawblade (right) has a three-speed drive, and the complete unit tilts for angle cuts.
Source - Casolin

The rim speed is governed by the sawblade diameter and the saw arbor speed. In the case of sliding table and panel saws - which are intended for converting many different materials - there may be a choice of two or three saw arbor speeds or, on high-specification machines, an infinitely variable speed drive to the main sawblade.

Note the saw arbor speeds provided on the machine. Check which of these is suitable for the sawblade diameter chosen and material to be cut, using the nomogram in Chapter 16. If more than one speed is within the range indicated, then use the higher speed as this will give a cleaner cut or allow a faster feed speed.

To use higher speeds the machine must be rugged and well-built - or vibration may set in to give poor cut even with a new sawblade - a common problem with older and lightly-built machines. Check also that the speed selected does not exceed the maximum running speed of the chosen sawblade - now often marked on the saw body itself.

Number of Teeth & Tooth Pitch

The number of teeth in carbide saws relates both to the sawblade diameter and the material being cut - *see previous notes in Chapter 3*. The tooth pitch is determined by the material to be cut - so sawblades of different diameter intended for the same class of work have the same pitch but different numbers of teeth.

There are some exceptions. For cutting single boards, for example, a sawblade would have a fine pitch and be fed at a relatively high speed. For cutting these as a stack a coarser pitch saw would be used - but possibly with a slower feed speed to give roughly the same bite (see pages 275 on.)

A similar logic applies when cutting natural timber.

Sawblades for deep-sawing natural timber have a large pitch and few teeth. Tooth profiles are similar to corresponding alloy saws, with a big gullet to remove the large amounts of sawdust generated. In contrast, sawblades for fast-feed edging have long backs to give the strong tooth needed for big bites, but a narrow gullets with enough capacity only to remove the smaller amounts of waste generated.

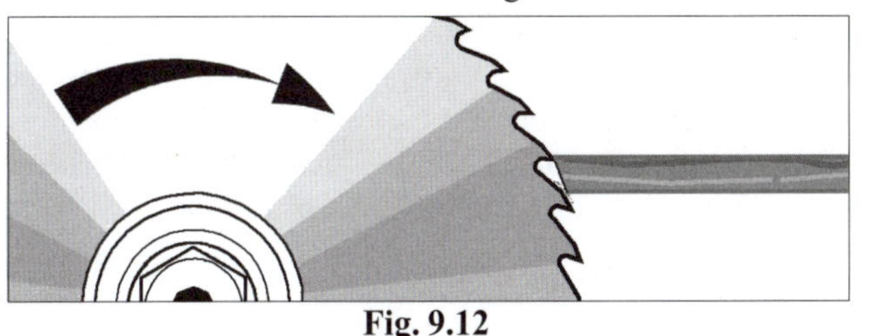

Fig. 9.12

When cutting thin material make sure that there are at least two teeth in the cut - not the case here with a sawblade set at its maximum height.

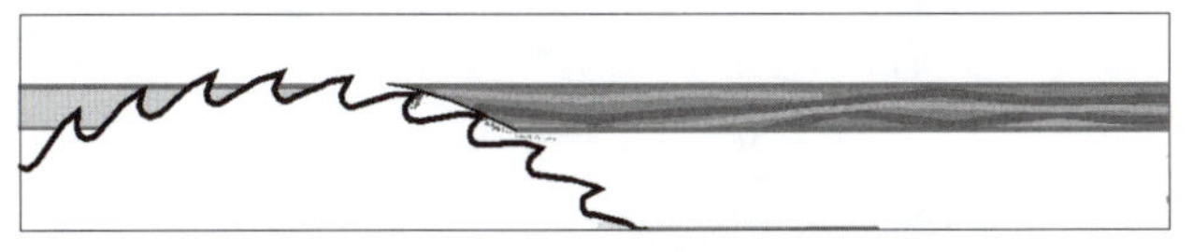

Fig. 9.13

Lowering the sawblade - to the recommended setting for board materials - brings more teeth into the cut.

One other factor has to be taken into account when feeding manually - the thickness of material being sawn - at least two teeth should always be in the cut. So small pitches are essential for thin materials, whilst larger pitches are allowable for thicker material.

The pitch of tooth needed for converting panel material must be matched to their ease or difficulty of sawing. Easy-cutting panels can be cut with medium pitches, brittle and difficult-to-cut materials such as acrylic plastic need the smallest pitches.

With the most difficult-to-cut materials a small feed-per-tooth is also essential. This requires a slow feed and a small tooth pitch. A small pitch tooth has a proportionately smaller gullet capacity, but as the amount of waste the gullet needs to hold is also correspondingly less this is not a problem.

Feed-per-tooth

The feed-per-tooth - the distance moved by the material between the passage of one tooth *(or tooth group)* and the one following - is one of the prime factors in deciding the suitability of any sawblade for the work intended.

A large feed-per-tooth is fine for natural timber, but a small feed-per-tooth is essential for quality conversion of difficult-to-cut boards. The nomogram in Chapter 16 gives details.

The feed-per-tooth is governed by three factors - the rim speed of the sawblade, the pitch and type of teeth and the feed speed. Additional nomograms show how all these factors relate to one another.

NOTE: The maximum and minimum values for the feed-per-tooth refers to groups of teeth, not individual teeth. One, two or more teeth may form a group to complete a cut. For example, an alternate-bevelled saw has two teeth per group.

In theory, any tct saw could be used for cutting (almost) any type of material - provided the saw arbor and feed speed combine to give the recommended feed-per-tooth. But this is not always so. For example, if a large pitch saw is used to cut a brittle board, the feed speed needed would be too slow to be practical. Conversely, a small pitch saw could not deep-cut natural timber because, even if the feed speed could be made fast enough to give the right feed-per-tooth, the sawdust generated would almost immediately pack and clog the small gullets.

The feed-per-tooth is only one factor, others which affect the efficiency of a tct saw are the hook angle and the tooth type.

Hook, Clearance & Bevel Angles

Hook angles vary from a positive 25 degrees to a negative 5 degrees (more for trimming and some mitering machines). Large hook angles give easy cutting of softwoods. Small hook angles are needed for hardwoods and quality conversion of panels - the smallest for the most brittle and difficult-to-cut materials. The regular clearance angle is around 15 degrees, but can vary slightly. Front bevel angles are zero for ripsawing and 10 degrees or more for cross-cutting. Top bevel angles are up to 15 degrees, but more for trimming on edge-banders, etc.

Kerf width

Kerf width is determined by the sawblade body thickness and the amount of tip clearance at either side of the saw body.

Although body thickness and kerf width must be kept small for reasons of economy, it is essential that the saw body is thick enough for the sawblade to remain stable in the worst conditions it is likely to meet - and thin sawblades are less stable and more easily wrecked than thick-bodied sawblades of the same diameter. Thin sawblades must be treated with great care to give good service over their expected life period.

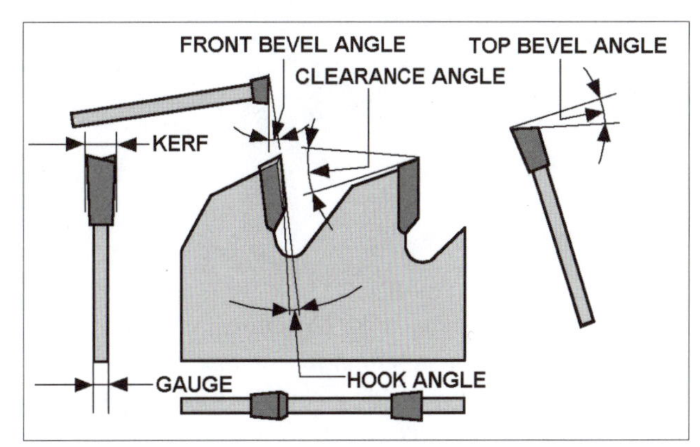

Fig. 9.14

Negative hook angle tct saws are used on cross-cut saws, for scoring sawblades and for cutting difficult materials. The fronts can be bevelled or ground square across.

There is no ideal body thickness and kerf for tct saws. Large diameter sawblades need thicker bodies and more tip clearance, as do those for tough work such as fast-feed edging, or when used as the outer sawblades in hogging sets. Sawblades can have a thinner body and less side clearance if they are small in diameter, intended for light work or intended for thin materials only.

Leading manufacturers offer their standard sawblade

types in more than one body and kerf-width combination. Usually there is a regular, recommended kerf width suitable for the bulk of work for which the type of sawblade is intended - with alternatives of thinner kerfs for less arduous work and thicker ones for tougher work.

NOTE: further information on body thickness to kerf width is given in Chapter 12 under 'Side grinding'.

Specifying sawblades

When ordering new sawblades state the requirements clearly:- diameter, arbor speed, material to be cut, depth of cut, and the type of cut (long-grain or cross-grain for natural timber) and, of course, the saw arbor details. Long-established and reputable makers will then provide a sawblade matched exactly to these needs and which will work long and efficiently. Some general purpose tct sawblades are suitable for converting a range of similar materials, but most are made to suit quite specific applications.

Feeding conditions

The feed should be smooth and continuous. Avoid hesitating in the cut - either when changing hands when manually feeding, or through some problem with a mechanical feed. Hesitation tends to dull teeth prematurely - particularly when cutting abrasive materials. Always use a push-stick or push-block on manually-fed machines so the cut can be completed safely and without interrupting the feed.

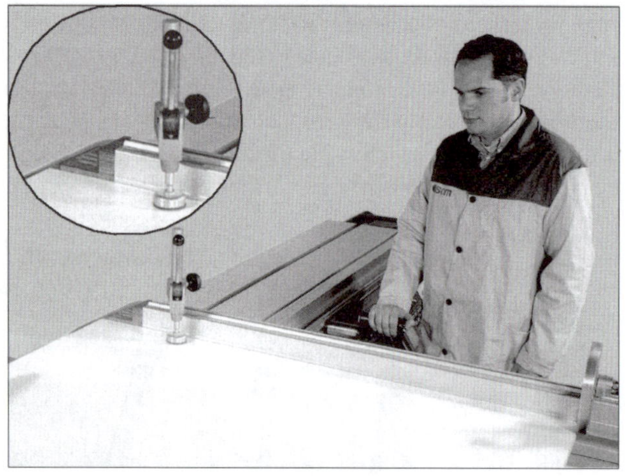

Fig. 9.15
The eccentric clamp on the SI 320 sliding table saw keeps material firmly in place and in full contact with the table when cross-cutting. Using a clamp is a safer way of working. It gives more control of the cutting action as the operator can use the rear handle alone in moving the table. Note the back-up piece shown in the inset.
Source - SCM

The rip fence should be absolutely parallel with the saw - *'lead'* to keep the timber up to the fence should be avoided. For the same reason check that the mechanical feed gear or sliding table feeds the timber absolutely parallel with the sawblade.

The material being fed should be firmly held-down to avoid vibration setting in. Thin materials should be carefully controlled on hand-fed machines by using a roller or other hold-down. *(In some cases a hold-down roller is mounted on the top guard and, though intended primarily to raise this automatically when material is fed in, may also steady the cut.)* Particular care is needed to keep brittle material in contact with the table insert - or with a scoring saw when fitted. When cross and angle cutting on a sliding table saw clamp the material to the table.

Contact with the up-cutting part of the sawblade when ripsawing natural timber can be avoided by fitting a riving knife 0·2mm. (0008in) thicker than the saw kerf. For cross-cutting, converting panels and multiple ripping on moulders, however, the riving knife should be greater than the thickness of the sawblade body - but must be less than the saw kerf.

Tooth styles

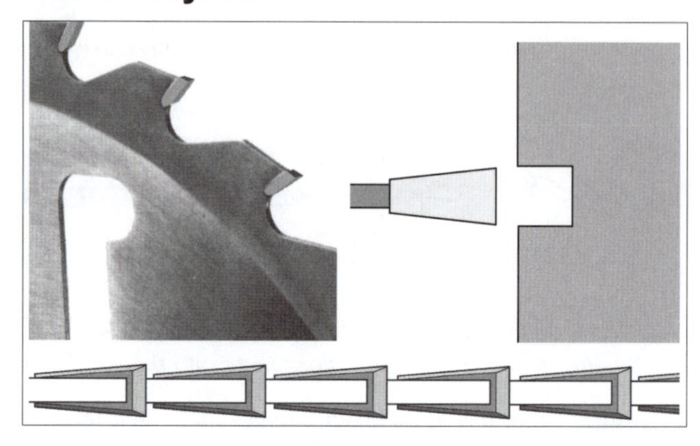

Fig. 9.16
Typical straight-topped tct teeth. In this and following diagrams the left illustration shows a typical tooth form, the sketch below the tooth sequence, and the right-hand sketch the cutting profiles.
Source for illustration - AKE

Straight teeth: These teeth are formed without top bevel *(flat-topped)* and are primarily used for ripsawing natural timber. Although the quality of the cut is poor compared to most other tooth styles, flat tops are better for fast-feed conversion, but are also used for general-purpose work on many other materials.

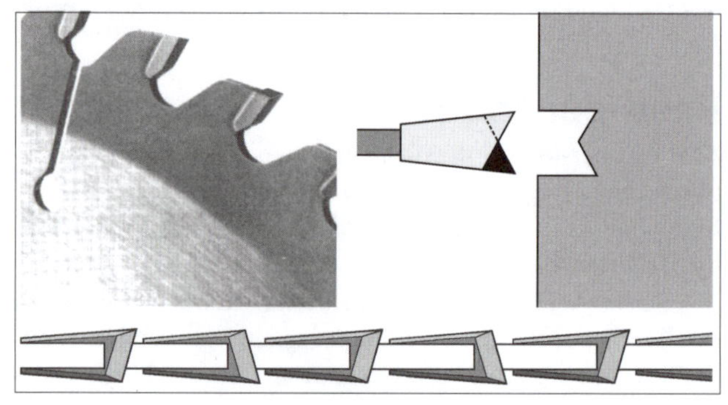

Fig. 9.17
Alternate top-bevelled tct teeth
Source for illustration- AKE

Alternate bevelled teeth: Sawblades with alternate top-bevelled teeth are regularly used for secondary conversion, finishing sizing of natural timber in joinery and cabinet shops, and for converting some panel materials.

The teeth are alternately top-bevelled at an angle, usually between 5 and 15 degrees.

The two leading tooth points sever the fibres marginally ahead of the remaining cutting edges - to give a better finish.

An alternate-top-bevel saw needs double the number of teeth to feed at the same speed as a flat-topped saw. This greatly reduces the total gullet capacity - and with it the depth of cut that can be sawn.

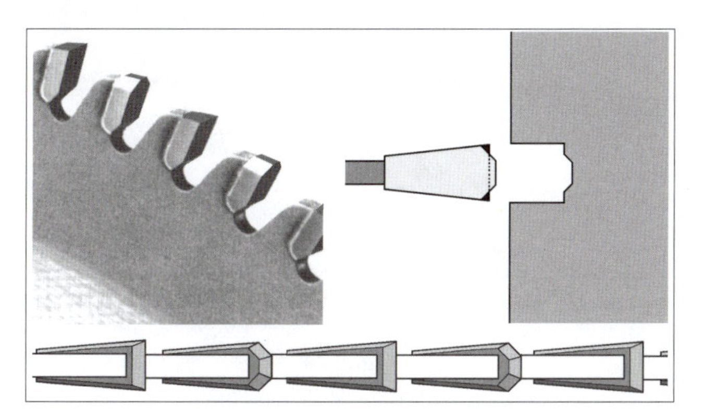

Fig. 9 18
Triple chip tct teeth alternating with flat-topped teeth.
Source for illustration - AKE

Triple-chip teeth: Sawblades with this tooth combination have an exceptionally clean cutting action, and are the preferred type for converting most panel materials. A pair of teeth complete the cutting action - first a tooth with a flat top and 45 degree side bevels, followed by a flat-top tooth ground lower than the flat on the first tooth.

The flat-topped teeth merely cut with their extreme corners, and it is these that form the clean edges. The triple chip teeth remove most of the waste, and in so doing protect the flat topped teeth from premature dulling. *If the same saw used flat-topped teeth alone the tooth points would dull at a much faster rate.*

There are other versions of this general style, for example a triple chip form but with side bevels of 25 or 30 degrees, sawblades with a pyramid first tooth, i.e., a triple chip style but without the flat top, and sawblades with double triple-chip teeth.

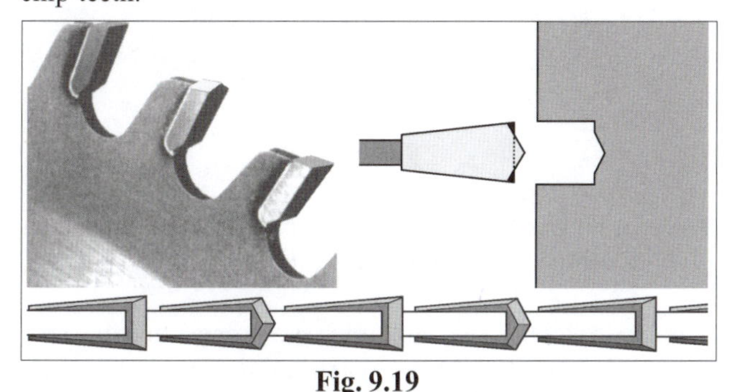

Fig. 9.19
Pyramid (or inverted vee) style teeth alternating with straight teeth.
Source for illustration - AKE

Hollow-front teeth: This is another type suitable for converting panel materials. The teeth have a flat top but a hollowed-out face formed by a small diameter grinding wheel grinding down the tooth front. This in effect forms two points on the tooth top which act as alternate bevel points - more effectively than flat-topped teeth on triple-chip saws - but formed on a single tooth rather than a pair of following teeth.

In some instances all the teeth are hollow-ground, but most saws of this type have flat-topped hollow-front teeth alternating with triple-chip or pyramid teeth. As before, the latter protect the extreme points of the hollow-front teeth to keep them sharp longer.

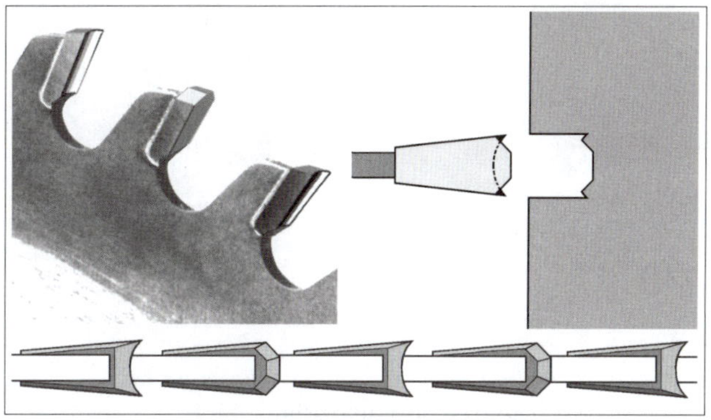

Fig. 9.20
Hollow-front straight-topped teeth alternating with triple-chip teeth.
Source for illustration - AKE

Cluster or Group-teeth: These are the tct equivalent of the alloy cluster saws, having five ripsaw teeth followed by a double-tooth width gap. They usually have flat tops and are used mainly for double slabbers, see Chapter 16.

Single-side top-bevelled teeth: Sawblades with these teeth, plus a thick body and wide kerf, are used as outside sawblades of hogging sets. Hogging sets reduce off-cuts to small enough waste for the regular extraction system to handle, and consist either of a gang of saws, or a hogging unit with toothed or saw-like segments.

They are normally used on double-end tenoners and certain types of panel edging and sizing machines.

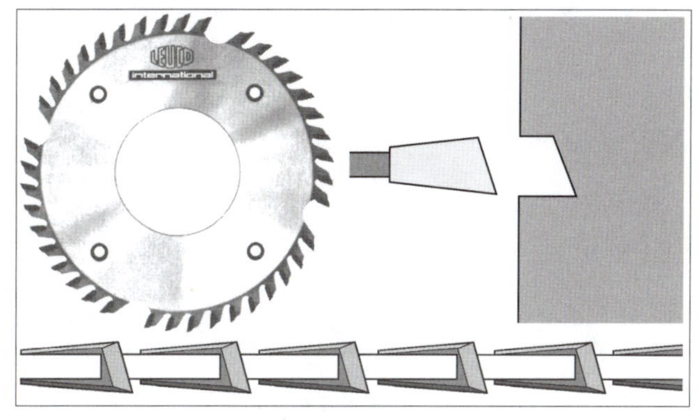

Fig. 9.21
Single-sided top-bevelled teeth - for use with hogging sets.
Source for illustration - Leuco

Negative hook angle teeth: Negative hook angle teeth are used on cross-cut and mitre saws, on scoring saws, and on saws for converting the most difficult and brittle materials.

On manual-operated radial and pull-out cross-cutting machines a negative hook angle is essential - as otherwise sawblades tend to run out of control.

For cross-cutting and mitering, the teeth are alternately top-bevelled - so that the leading point scores across the grain to give a cleaner cut than would otherwise be the case. Cross-cut teeth can have straight front faces, or preferably alternately front-bevelled faces to enhance the scoring action.

Negative hook angle teeth used on scoring saws can be flat, alternately-bevelled or of the triple-chip type, but tooth fronts are formed straight to ensure they keep to a straight and true sawing line. *In fact all saws except those for cross-cutting and mitering have straight faces.*

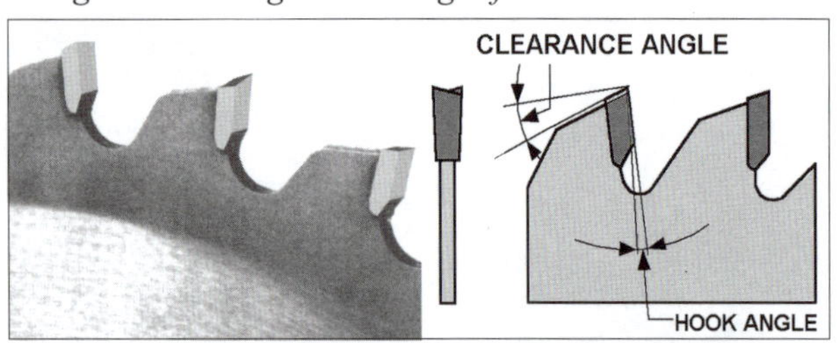

Fig. 9.22
Negative hook angle tct teeth with alternate-bevel tops and straight faces as used on cross-cut saws.
Source for illustration - AKE

Alternative tooth styles

There are several variations of the tool styles used above and which have certain advantages. They include:-

Double triple-chip teeth. Regular triple-chip teeth, but with 45 degree bevels on both following teeth. In this case alternate teeth are of different heights, the lower one also with a reduced width of bevel to give longer-lasting finishing corners than with regular flat-top teeth of a triple-chip/flat-top combination.

Hollow-front/pyramid teeth. This is an alternative version of the hollow-front/triple-chip tooth style but which gives an even cleaner cut.

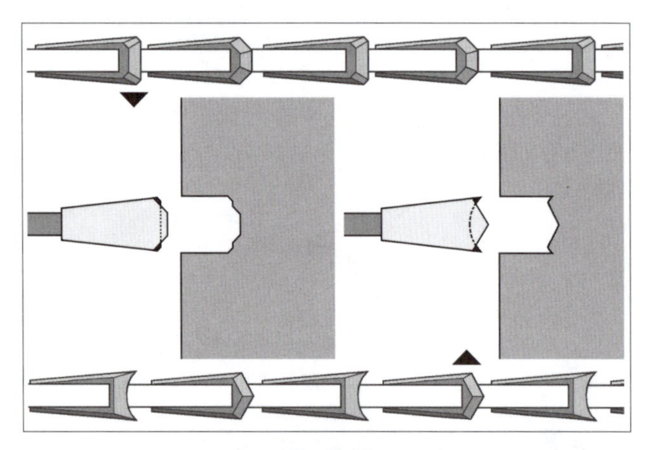

Fig. 9.23
Top and left: double triple-chip teeth. This is an alternative version of the triple-chip/flat-top style, but in this case the 'flat top' teeth have small 45 degree bevels to prolong tip life when cutting abrasive materials.
Right and bottom: Hollow-tooth/Pyramid style. This is an alternative version of the hollow-front, flat-top/triple-chip style which gives a better finish than regular flat-top/ pyramid teeth.

Novelty teeth: This is the tct equivalent of the original novelty-type steel saw, a five-toothed pattern consisting of straight-topped cleaner teeth following a wide and deep gullet, and followed in turn by four alternate top-bevelled teeth *(to scribe ahead of the cleaner teeth)*. This style was originally used for cutting plywood - but then normally made as a hollow-ground steel saw. It is now virtually an obsolete form - tct alternate bevel and triple-chip saws are entirely suitable for all the original applications of the novelty tooth pattern.

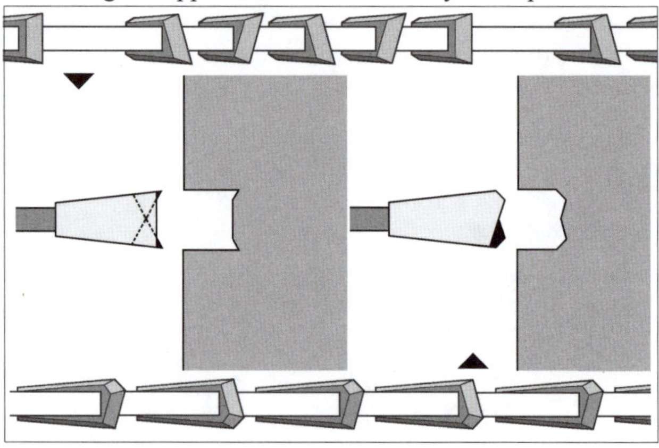

Fig. 9.24
Top and left: Novelty teeth.
Right and bottom: Broken-tip alternative top-bevel teeth. This is a variation of alternative bevelled teeth, but with a small 45 degree bevel on the tips. The bevels give a stronger tip which is more able resist breakdown when cutting abrasive materials.

Broken-tip alternate-top bevel teeth. This is a variation of the regular alternate top bevel teeth, but with a 45 degree bevel on the points to make them stronger and retain sharpness in extreme cutting conditions.

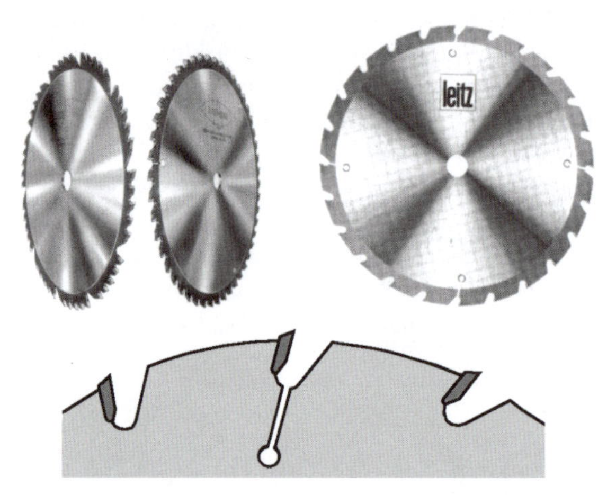

Fig. 9.25
Left: Novelty Tooth sawblade, Centre: Typical regular sawblade with expansion slots. **Source - North American**
Right & below: A close-circular design saw. **Source - Leitz**

Expansion slots: Tungsten carbide saws conventionally have expansion slots in the rim to counteract rim heating when brazing teeth. Without these the saw may loose tension - according to the first tct saw makers, that is.

Modern methods of tipping create less widespread heating, so expansion slots in their traditional form are probably now obsolete - although they have been, and still are, an firmly-established feature of most tct saws.

Strob slots: These are open or closed slots milled into the sawblade body and faced with a strip of tungsten carbide only fractionally less in width than the saw kerf.

These slots are essential for both single and double-arbor multi-rip sawblades when converting natural timber. Their purpose is to continuously wipe the cut clear of sawdust, chips and splinters etc., and so reduce the chance of resinous build-up on the sawblade body.

They usually are formed to a negative angle to give a fanning effect which further improves sawdust ejection - and also cools the sawblade.

Fig. 9.26
Typical strob slots (chip clearance or wiper slots).
Source for illustration - Leitz

Cooling slots: These take the form either of round holes, or as closed slots cut into the body of the saw at a negative angle to encourage dust ejection - similar to strob slots but without carbide facings. They are mainly used on saws for deep-cutting stacked panels.

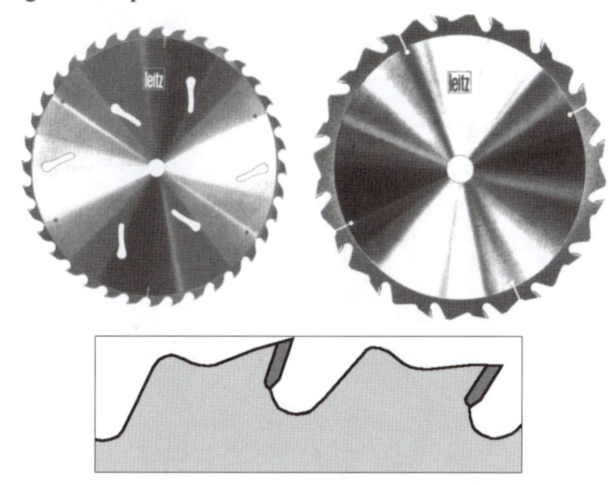

Fig. 9.27
Cooling slots in a saw for trimming and sizing cuts in stacked panels (left). Anti-kick-back shoulders in a sawblade for manual sawing timber (right and lower).
Source - Leitz

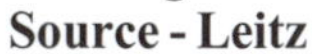

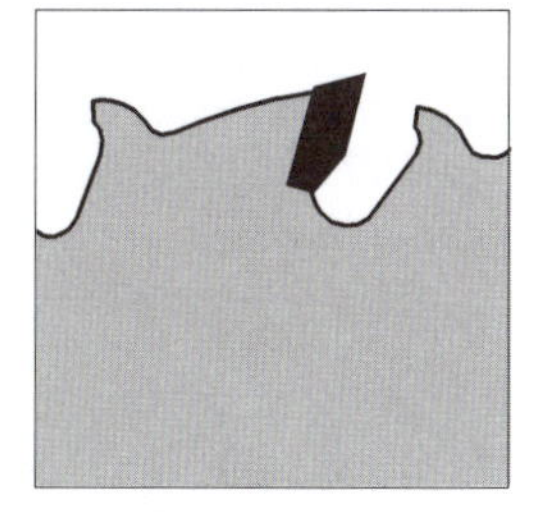

Fig. 9.28
This anti-kick back sawblade is for use on site, either on a table saw or radial arm saws.
Source - Freud

Anti-kick-back shoulders: These are precision-ground shoulders formed behind each tooth to limit the forward movement of the timber into the sawblade during the cut to avoid *"snatch"* and possible kick-back. They are intended mainly for manually-fed saws, but are also used on power-fed saws to give a steadier cut and prevent tear-out of such as dead knots, etc.

Fig. 9.29
Typical regular tungsten carbide tipped saws, outer - a long-pitch alternate-bevel type, inner - a short-pitch flat-top/pyramid type (Left).
Carbide saws are also being used increasingly for light duty and DIY applications - where low initial cost, a major factor, is achieved by using thin carbide tips. The above are examples of this type of general-purpose sawblade. They are so-priced that replacement of sawblades with worn-out tips is a better option than re-tipping (Right).
Source - Atkinson-Walker

Scoring Saws

Scoring saws are used on panel saws in front of the main saw, and on double-side or end machines in front of a hogging unit. They normally have straight or alternate-bevel teeth, and with a pitch roughly corresponding to that of the main saw. Because they climb-cut *(rotate with the feed)* they cut a clean, spelch and chip-free groove.

Fig. 9.30
An under-scoring saw combined with a hogger unit on a double-end tenoner for trimming panels to width.
Source - Leuco

For Double-end tenoners. Scoring saws under or over-score boards or natural wood parts with a cut depth of perhaps 1-2mm and operate ahead of the hogging unit. The hogging cut must be set fractionally to the outside of the inside scoring saw line by perhaps 0.1mm. - so that the hogger teeth do not make contact at this point. The alignment between the hogging unit and the scoring saw has to be exact - and the feed must be precise and without side movement.

For panel saws. Scoring saws cut 1-2mm. into the underside of faced panels in front of, and in line with, the main saw. No chipping occurs on top of the panel because the main saw cuts in a downward movement. No chipping occurs on the underside of the panel, either, because the main saw cuts within the under-scoring sawline - ***provided the two sawblades are correctly matched.***

To be effective, the scoring saw must have a saw kerf fractionally wider than that of the main saw - by some 0.2mm. or so. Although this gives a saw cut which is not perfectly square, the small step it forms is more acceptable than the alternative of chipping the under-face.

To keep the step as small as practical, the kerfs of both sawblades have to be very accurately matched to one another, and alignment between them must likewise be absolutely accurate.

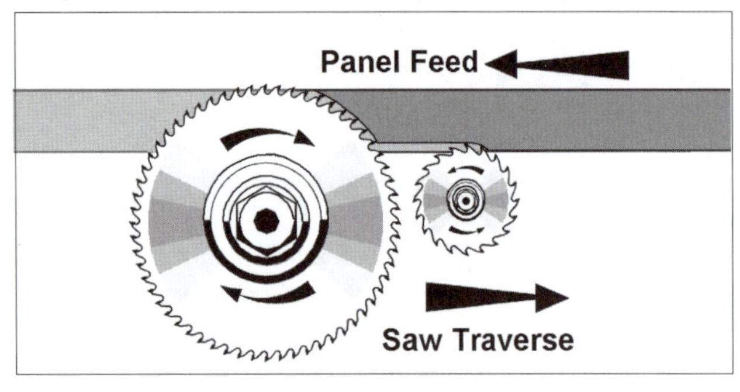

Fig. 9.31
On panel saws the scoring saw precedes the main saw which then cuts within the scoring saw's kerf.

Main and solid (one-piece) scoring saws are available as a matching pair ground to appropriate kerf widths.

Keeping two sawblade kerf widths in step, however, can prove difficult in practice - the rate of dulling and degree of wear is never equal between the two.

In an attempt to keep their relative kerf widths in step, both sawblades need to be face and top-ground when only one has lost its edge - and by precisely the same amount. This necessary practice shortens the long-term life such sawblade combinations and, even with the greatest care in grinding, they can still get out of step.

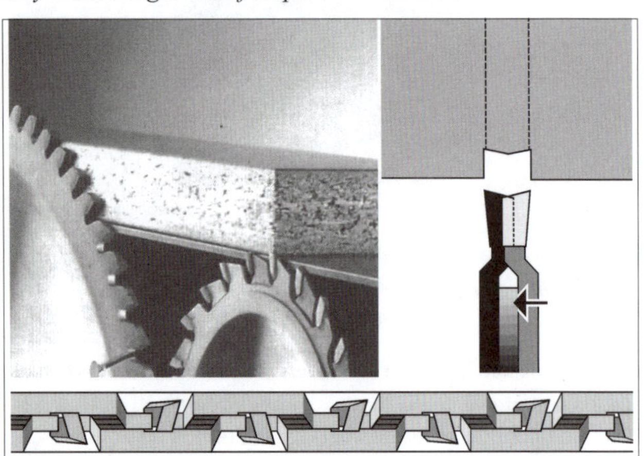

Fig. 9.32
Split scorers are adjustable to match the main saw by substituting spacers of different thickness (arrowed) between the two halves.
Source for illustration - Bennett

The regular practice is instead to use a scoring saw which is either split or tapered. Both types allow their effective kerf width to be varied to correspond with the kerf of the main saw. In this way both main and scoring saws can be re-ground without the worry of trying to keep their kerfs in step.

Two-part, or split, scoring saws consist of two thin-bodied saws with off-set tips and every second tooth removed. When clamped together the off-set teeth alternate and overlap one another to effectively give a combination sawblade of the required tooth pitch.

The kerf width is varied by switching the spacers between the two halves. Spacers can be thin precision-ground rings, or thin peal-off metallic rings.

Determining the correct spacer thickness *(after grinding)* is possible through trial and error by fitting them and then cutting partially through a scrap piece. By noting the relative kerf widths a suitable spacer can be selected and substituted - but this is a time-consuming process which holds-up production.

An alternate way is to first accurately measure and note the kerf width of the main sawblade *(at the square tooth with triple-chip teeth)*. Then mount the two-part scoring saw on a suitable arbor in the tool room, separated by the spacer previously used, and carefully position a flat metal plate to just touch the nearest-side teeth. *(The plate **must** be parallel to the sawblade both radially and tangentially.)*

Rotate the tooth clear. Place a dial indicator *(with a disc end)* to face the plate and, with the two in contact, zero the dial measurement. Withdraw the indicator clear of the plate, move an opposite side tooth in line with the dial indicator disc - and note the measurement when the two are in contact.

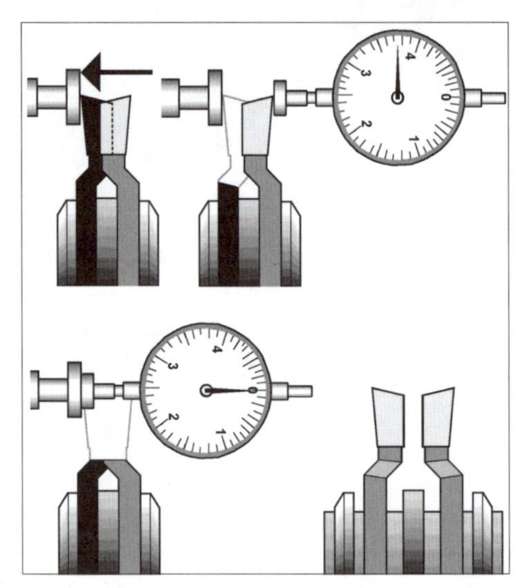

Fig. 9.33
One method of accurately measuring the cutting width of split scoring saws. A metal plate (arrowed) is first set against the nearside tooth, top left. Zero the dial indicator against this plate, bottom left. Measure the kerf width against the opposite tooth, top right. The sketch, bottom right, shows the unit disassembled.

The scoring saw kerf should be equivalent to the main saw kerf - plus 0.2mm. If it is not, then substitute another spacer, or add or subtract peel-off spacers.

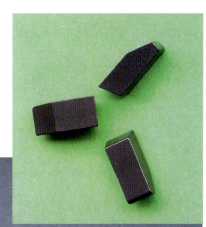

In practice the 0.2mm. stated is a guide, as the difference between the main and scoring saw kerfs should be the absolute minimum consistent with giving near-square-cut, com-

bined with clean and chip-free conversion of boards. Experience will show if a greater or less kerf width difference is acceptable.

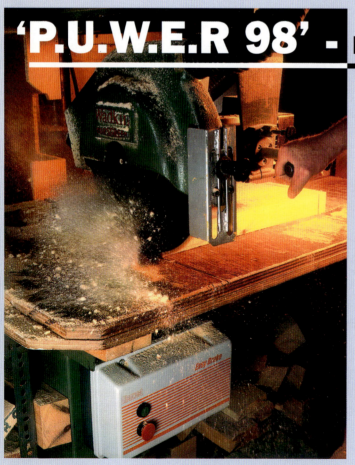

Once the main and scoring saws are properly matched, any mis-matching takes place only gradually. The point when the spacers next need switching can be based on experience - after noting the squareness and quality of the cut immediately following each sawblade sharpening.

A more convenient method, of course, is where the scoring saw can be infinitely adjusted whilst still on the saw arbor, see Fig. 9.33.

Tapered scoring saws have reverse-taper tct tips. The tips are ground to a regular back clearance angle, but with a reverse radial clearance of about 5 degrees - so the tip increases in width towards the saw centre. The kerf at the tip of the teeth is less than the kerf of the main saw, but at the base of the tip it is wider.

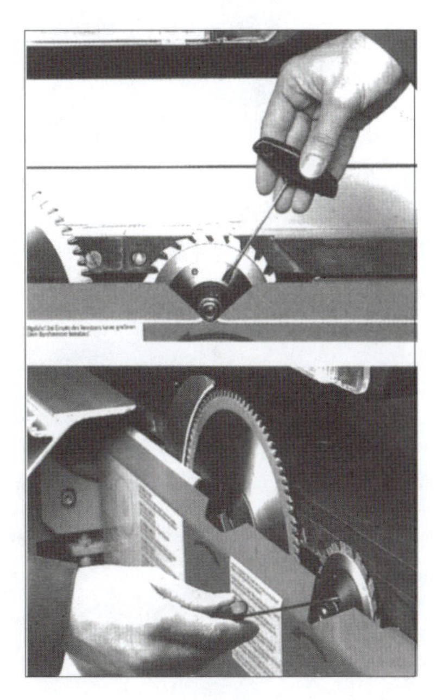

Fig. 9.34
On their C-Class machines split scoring saws do not require individual spacers, instead the two halves can be adjusted infinitely in an axial direction without removing them. This is both a greater convenience and a substantial time and cost saver.
Source - Altendorf

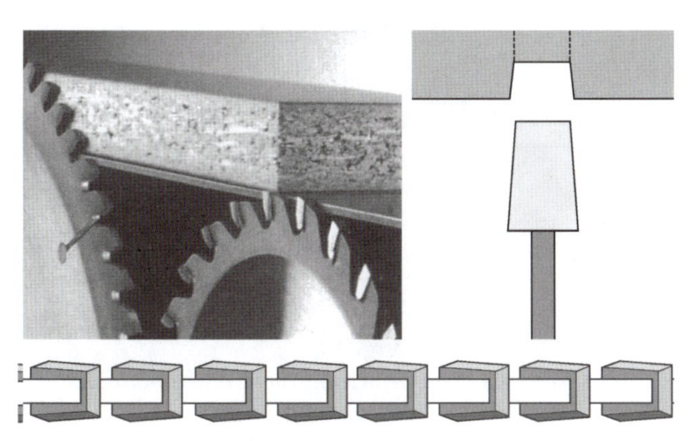

Fig. 9.35
Tapered scoring saws are one-piece, and are adjusted vertically to match the kerf of the main saw
Source for illustration - Bennett

It is essential to carefully set the height of a tapered scoring saw so that the kerf width it forms on the underside of the board is barely wider than that of the main saw. At this setting the main saw will make no contact at the underside of the board - and will not chip or break-out. As with the split type scoring saws the cut is not perfectly square, but the main square cut will blend into the scoring saw taper almost invisibly.

Because of the taper it is less obvious how deep the scoring saw actual cuts - so check squareness with an engineers square, or note the depth of the tapered cut from the slight change which usually shows up in the appearance of the sawn faces.

Although a tapered scoring saw has to be adjusted in height to maintain the quality of the finish, penetration should be kept to within 1.5 and 2.5mm. If chip-free sawing is not possible within this range, or if the scoring saw forms a noticeable step, switch to another tapered scoring saw with a more suitable kerf width.

Scoring saw alignment

With all scoring saws it is absolutely essential that the carriage supporting the main and scoring saws is sturdy and vibration-free, and that the two sawblades are in perfect alignment.

The saw carriage must travel in a precise and straight line or irregular chipping of the underside may take place. With tapered scoring saws it is also essential that the height of the saw carriage is absolutely consistent relative to the supporting table throughout the full movement.

If the machine gives unsatisfactory results with tapered scoring saws, then change to split scoring saws as these are not height-sensitive to the same degree.

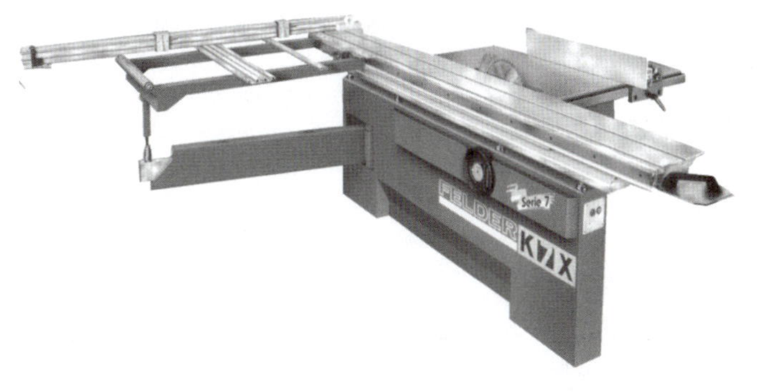

Fig. 9.36
Sliding table saws such as this K7X can handle large panels - for which a scoring saw is essential. This machine is one of a series made by Felder for precision cutting of panels of any type.
Source - Felder

When using any main/scoring saw combination on manually-fed panel saws it is essential to maintain full contact with the machine table at the point where the scoring saw operates - especially when using tapered scoring saws. In fact, two-part scoring saws are probably more suitable for manual-feed panel saws for this reason.

Edge-banding, soft and postforming

Edge-banding is the process of trimming the end or edges of faced panels and then gluing a narrow strip to these, perhaps plastic edging of the same colour and pattern as the main facing.

The overhanging projections are then trimmed flush with the facings and usually eased to remove the otherwise sharp corner. Solid, thin wooden edging can also be applied in a similar way - to be trimmed flush with the main facings and the corners then possibly rounded-over.

Kitchen, bathroom and office furniture tends now to be fully round-edged or edge-profiled both for appearance and practical reasons - square-edged doors and tops, particularly the corners, are more susceptible to accidental damage than a rounded or shaped edge.

There are two basic manufacturing processes used, softforming and postforming.

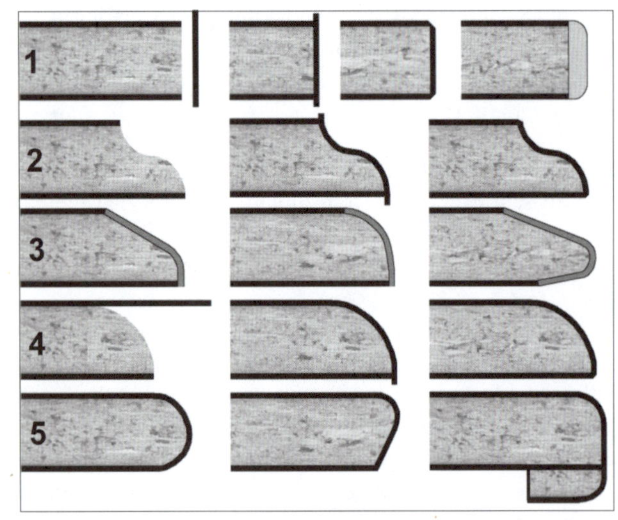

Fig. 9.37
Line 1 shows regular edge banding, first with the face board trimmed, then the edge added and finally trimmed. The extreme right sketch shows a solid edging. Line 2 shows the process with softformed edges, edge moulding, plastic application and trimming. Line 3 shows different postformed profiles, with the postformed section lighter and, in the centre, edge-butting the main facing. Line 4 shows postformed edgings, preparation, wrap and trim. Line 5 shows different types of postformed edges.

Softforming In softforming the edges of veneered or plastic-coated panels are trimmed and profiled to accepted a separate edging material to be glued in place. This is then edge-trimmed level with the original facing.

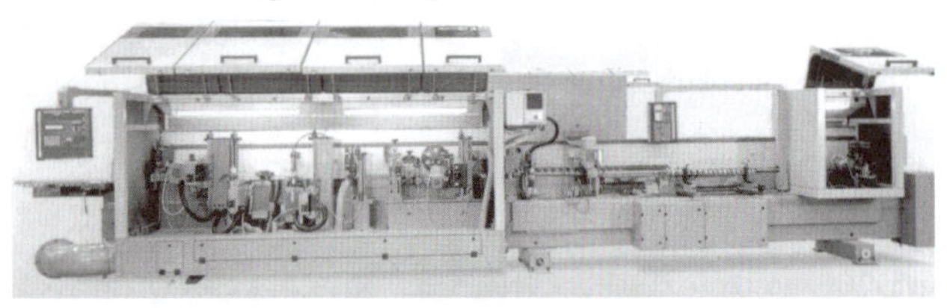

Fig. 9.38
One of two basic models for postforming, the VF/P operates a postforming technique in which directly-coated panels are used.
Source - Homag

Postforming In postforming panels are faced with deformable plastic. After trimming to length and width the core material is removed down to the plastic level along the edges to be post-formed, an d often rounded in the process.. The overhanging plastic or melamine is then glued and wrapped around the edge either as a L-profile or a U-profile to provide a completely seamless coating.

All the above can be applied to one or more edges of sized panels, sometimes passing through all the stages on specialized machines in a continuous production process.

Parts such as kitchen tops are often produced in long lengths, often with D-edges and upstands, to be cut to length or mitred at a later stage.

Fig. 9.39
Typical postformed items of the type used for kitchen worktops.
Source - Leitz

Fig. 9.40
This shows a postformed top being cross-cut square, but with the saw carriage tilted to 45 degrees. The main saw is to the left. The scoring saw, on the right, has been raised automatically from its regular low position to cleanly score the trailing rounded edge.
Source - Casolin

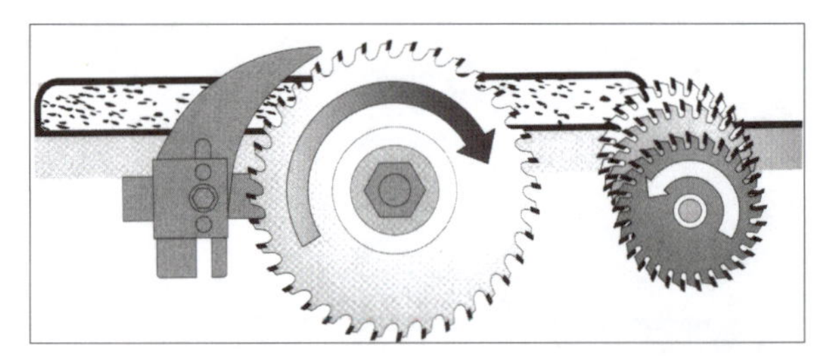

Fig. 9.41
This shows the general arrangement of the main and scoring saws in Fig. 9.39.
Source - Casolin

A regular scoring/main saw combination tends to chip where the postformed edge trails in cross and mitre-cutting such tops.

To avoid this, some sliding table saw manufacturers fit a large diameter scoring saw with an automatic powered rising movement. This initially scores the panel underside to the regular scoring depth, then rises as the trailing end passes to score through the section where the main saw would otherwise chip.

Ultra-narrow-kerf sawblades

TCT sawblades are available with ultra-thin kerfs from a few specialist companies such as Super Thin Saws Incorporated. These have a much narrower kerf than regular sawblades and are suitable for applications such as cross-cutting, cutting narrow grooves for weatherstrip slots, rebate (rabbet) sawing to split-off glazing beads, gang ripping on optimized saws, glue-joint ripping, single arbor splitting of louvre slats and double-arbor splitting of Lamellas. (See also Pages 212 & 213).

Lamella production is the process of splitting solid timber and then reforming it into wide sections for use as solid flooring, doors, kitchen cabinets, etc. By gluing the sawn lamellas, often in a re-assembling form to eliminate natural defects such as cup and warp, a high-quality, stable and defect-free wooden product results.

In addition to the huge reduction in saw-kerf loss - especially important when sawing narrow sections, ultra thin sawblades need less power to drive them, give minimal saw deviation and produce a superior finish suitable for immediate glueing or vastly reduced sanding for exposed surfaces.

Fig. 9.42

This shows a special machine for double-arbor splitting of lamells using ultra thin saws, Left. Top right shows the general arrangement, with thin saws for the centre cuts and a thick saw for the top cut. Bottom right shows a typical bossed-body saw. The following table is of this type of sawblade.

Source - Super Thin Saws Inc./Weinig

Diameter	Teeth	Rim		Kerf	
		Depth	Thickness	LP	MP
180 - 200	24 - 40	17.5	0.70	1.25	1.45
		27.0	0.75	1.30	1.50
		43.0	0.95	1.50	1.70
200 - 250	30 - 40	35.0	0.90	1.45	1.65
		43.0	1.00	1.55	1.75
		55.5	1.10	1.70	1.90

Compare the above to the table for regular saws on page 200. NOTE: LP refers to low pitch* material and MP to medium pitch* materials. *Resin content. Dimensions are in millimetres.

Ultra-thin sawblades are produced as a straight-plate saw (supported with special spacers) or as a bossed saw with an integral flange - both types preferably mounted on ETP hydraulic sleeves. Unlike regular saws, the kerf of ultra-thin bossed-body saws does not increase with sawblade diameter, but relates to the cutting depth of the sawblade beyond the bossed centre.

Ultra-thin sawblades are best tailor-made to suit specific applications, taking into account the machine type, sawblade requirements and the material to be sawn, but as a guide the table above gives typical details.

Noise pollution

Non-enclosed sawblades, such as on manually-fed panel and crosscut saws, produce an ear-piercing scream, especially when idling. The general attitude was once that an operator *"gets used to"* this noise and it then no longer troubles him, but it is now accepted that noise can disrupt concentration, and prolonged exposure can lead to partial deafness.

Noise comes from two sources on a rotating sawblade. One is the pressure waves thrown out by the gullets, and the other is natural resonance, or ringing, of the saw body.

With pressure waves, the highest pitch sounds are produced by fast-running sawblades with a large number of teeth. Unfortunately, the waves are distorted by the gullet shape to produce irritating harmonics rather than a more acceptable pure sound. The human ear is most sensitive to generated noise in the 3-4 kHz frequency bandwidth, so saw blades should preferably be manufactured to avoid this range. Regular saws when cutting particle board produce pressure peaks in the 2-4 kHz frequency bandwidth, and these are responsible for the irritating sound felt by the user.

Sound is alternating air waves of high and low pressure created by the saw gullets in a way similar to the mechanism used by wind instruments produce sound - so anything that disrupts this pattern also reduces the sound.

Fig. 9.43

This TopLine sawblade has laser-cut slots in place of the regular expansion slots which, it is claimed, reduces noise level by some 30%.

Source - Leuco

Noise from a crosscut saw ceases when in the cut, and then takes two or three seconds to build up again. The same sound dampening effect can also be partially achieved - even when idling - by fitting a hardwood block at the rear of the crosscut saw into which the teeth sink at least up to gullet depth when in the rest position.

On a panel saw a close-fitting fill-in piece can serve a

similar purpose. In both cases the hardwood should be replaced when the sawblade is changed, or as wear makes it ineffective. Another way is to deepen some gullets evenly around the rim of the sawblade, or to use sawblades with varying pitch. Variable pitch saws also reduce impact resonance - the source that stimulates and maintains body noise. Stopping material being sawn from vibrating by using a firm and properly-set hold-down also helps to dampen noise.

Fig. 9.44
The AKE Soundstar tct sawblade (left)combines traditional, plugged cooling slots with laser-cut slots, to provide an excellent sound-dampening effect on the saw body. The AS range of sawblades (right) have a visco-elastic layer covered by a steel foil on one side of the saw to dampen noise level up to 10dB(A).
Sources - AKE (left) Leitz (right)

The other sound source is the natural resonance of the sawblade - most have a characteristic ring when tapped. Although the pattern of resonance can be very complex, it has been found that altering the tension of the sawblade by rolling can reduce the noise level - but this is a rather hit and miss process - which could also affect the running of the sawblade of course.

An alternative way is to pressure some sort of steadying material agent the side of the sawblade body. The point where this is most effective depends upon a number of factors, and has to be found on individual sawblades by trying different points of contact *(radially)* until the most effective dampening position *(null point)* is located.

Fig. 9.45
Piano Plus sawblades have fine-tuned laser technology to give low vibration, quiet operation and an improved finish. The carbide tips used give maximum run-times.
Source - Guhdo

Many makers offer sawblades with expansion slot end holes plugged with metal or composition which act as a resonance-deadening feature.

Modern laser profiling of sawblades provide makers with a much simpler and lower-cost means of reducing body resonance simply by cutting two or more patterned laser cuts either within the saw body or into the saw body from the rim of the sawblade. These can reduce the level of the most damaging frequencies of between 3 & 6 KiloHertz - both when cutting and when idling.

Laser-cut slots alone have replaced conventional expansion slots on some sawblades, but other makers offer a belt-and-braces (belt-and-suspenders - US equivalent) solution to deaden resonance by providing both plugged expansion and plugged laser-cut slots.

Most dramatic results, however, are from sawblades with a sound-damping body.

One is the Leitz AS range of saws. These have a high-density saw body with a visco-elastic layer on one side protected by steel foil, with both securely bonded to the sawblade.

The Onci deSciebel is a composite saw body consisting of two steel bodies and a thin copper blade inserted between the two and assembled by spot welding. This technololgy transforms vibratory energy into heat, which both eliminates resonance when running free, and reduces the noise level, when cutting, on frequency ranges above 1 kHZ. By combining this type of body with a suitable tooth geometry, a noise level reduction of more than 5 dB(A) is possible when compared with a regular sawblade.

Fig. 9.46
The deSciebel low noise sawblade. A cross section is shown on the left, with a thin copper insert (B) sandwiched between steel bodies (A), (P) is the tungsten carbide tip
Source - Onci

Another similar type is the Gomex Minibel, with a body as two separate discs securely joined by a vibration-dampening insert. A noise level reduction of between 6 and 10 dB(A) is possible with this type, especially in the higher frequencies where the noise is at its most disturbing and harmful. This type of body is also claimed to be more stable and to require flattening only infrequently. To reduce the free-running sound level the Minibel can also be supplied with modified gullets to further reduce the general noise level by some 1-3 dB(A).

Fig. 9.47
For their Ultra-low-noise sawblades Freud laser-cuts question-mark shaped slots of different sizes as expansion slots, also complex sound-dampening slots. This reduces the noise from both air turbulance and body resonance.
Source - Freud

Chapter 10
Using and grinding carbide saws

TCT saws must be handled with care. Never let a saw come in contact with any hard surface - contact with metal, for example, can readily chip the teeth. Take particular care when removing and fitting carbide saws, perhaps even to the extent of placing protective wooden battens on the machine to prevent accidental contact with nearby metal parts.

Most carbide saws are provided with a box for safe storage and protected by rust-preventive. Always transport them in their boxes, and keep them there until actually needed - and after removal. When taken out at the machine, temporarily place them on a flat wooden support where they cannot be damaged by dropped wrenches, etc.

One protection material for new and repaired sawblades is a plastic coating applied in a hot bath. Another type is a U-shaped plastic strip cut to fit around the teeth. If practical, leave these protective covers in place while fitting the sawblade - and reapply the strip type before the sawblade is removed. As with all sharp tools use industrial gloves to avoid cuts to the hands.

Cleanliness is essential. Before fitting a sawblade make sure that the saw body and bore, the saw arbor and both saw collars are clean and free of dirt and grit, etc. If abrasive grit is trapped between the saw body and the collars the sawblade will not run true, and any grit remaining in the saw bore or on the saw arbor can scratch and wear both.

The sticky deposits that commonly build up on the sawblade while in use should be periodically removed with special solvents rather than scraping. TCT saws run for long periods, so there is considerable risk of resinous buildup on the sides of the sawblade which can cause overheating or even burning-up. Where lubrication systems are used with sawblades ensure these are working as intended.

Regularly examine sawblades for cracks, tension and flatness, and check that tips are securely held and sharp.

Safety checks

Before starting-up a sawblade make sure that it is free to rotate without danger of fouling any machine part when in operation. Check that all the safety devices are fitted, securely held and properly adjusted for the work in hand. Exhaust (or blower) systems should be connected and operating (portable blowers can add to the air pollution risk). Check the floor and working area around the machine to ensure that the machine can be worked safely - and without danger of material being processed fouling any standing stacks or nearby equipment.

Cleanliness is essential - on the saw table itself, on the machine proper and on floor around the machine. Keep the floor clear of sawdust and off-cut buildup that can make standing unsafe. Where a wooden floor has become polished and slippy over time, cover it with nonslip mats, or treat the floor with a nonslip surface.

When using a tct saw take particular care to any avoid any side pressure that could deflect it into contact metal parts of the machine - this can readily strip-off teeth. Side pressure can result from incorrect sawblade alignment with the rip fence or the traversing table guides on manually-fed machines, or with the traverse line of the saw unit on panel saws.

With modern machines the makers take care of such alignments and pin them, but some older machines allowed adjustments to be made - which can shift over time.

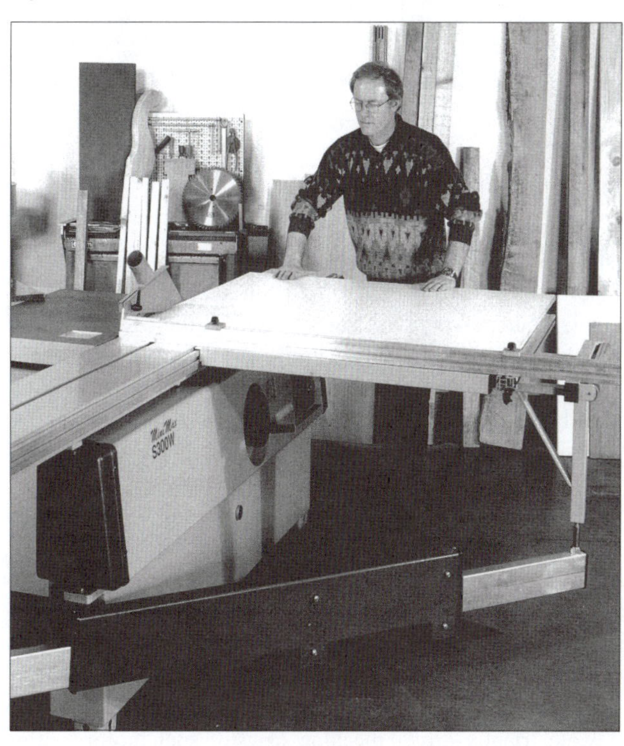

Fig. 10.1
Cleanliness in the work area is essential, with timber and tools stacked clear of machine operation, a clean floor and always with the machine properly guarded and used safely.
Source -SCM

It is also essential to correctly sharpen the sawblade to give either perfectly formed flat tops or equal-height points, otherwise the sawblade will deflect to one side when cutting - even when correctly aligned.

Wedge-shape off-cuts can jam between the machine table and the sawblade to force it over. To avoid this make sure that any table gaps are minimal and, using a push-stick, remove all off-cuts safely as soon as they form - or fit a deflector.

When manually ripsawing use a push stick or push block to clear material past the sawblade. Keep the push stick in a pocket on the saw for instant use.

Make sure that all material to be sawn lies firm and flat on the machine table before feeding it into the sawblade or starting the saw traverse movement - otherwise the sawblade could snatch it to twist and possibly cause damage.

For the same reason make sure material is properly seated, firmly against the fence and securely held when cross or mitre-cutting. Preferably use clamps to hold material being cross or mitre-cut - to keep hands clear of the sawblade.

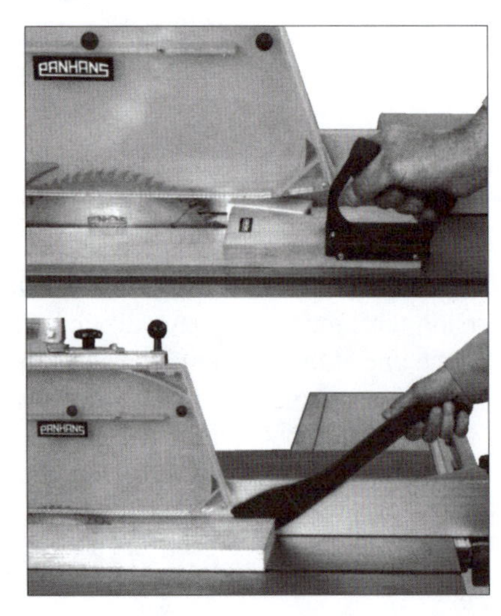

Fig. 10.2
Retain and use push sticks or push blocks when ripsawing natural timber and board material.
Source - Panhans

Machine braking

A new requirement under the Provision and Use of Work Equipment, requires woodworking machines in the UK to be fitted with brakes. This is to reduce injury should an accident occur, and to avoid injury from machinery which is switched-off electrically but remains running.

Older sawbenches, and in particular those with a direct-drive saw arbor, can take several minutes to finally stop after being switched-off. During this time the sawblade runs silently - but remains dangerous.

Mechanical brakes have been used in the past, but the easiest, most versatile and cost effective approach is to fit an electrical DC brake to those existing machines not originally fitted with brakes.

Add-on electrical units, which can be fitted to almost any electrically-driven machine, can stop such motors within seconds and so make them a good deal safer to operate.

They provide manual or remote triggering of an automatic braking cycle. Braking is achieved by disconnecting the electrical supply and injecting a DC current into the motor windings, which brings the motor to a rapid stop.

Adjustments are provided for setting braking torque and braking time, and braking can be initiated by pressing either a button on the unit or on a remote emergency stop button - or by an electrical trip caused through machine malfunction. The add-on units can be used for either direct-on-line or star/delta starters and usually are provided with motor control gear.

Fig. 10.3
The unit shown above, as all Allenwest Easy Brake Units, has Stand Still Detection. This protects a motor which has been stalled by ending the DC injection - other systems continue the DC injection for the set time and could overheat the motor. These units require little down-time to install, integrate with existing equipment and are maintainance-free.
Source - Allenwest

When to grind carbide saws

Tungsten carbide tipped saws retain their sharp cutting edges over a long period of time - with little change in their performance. Once the cutting edges do lose sharpness, though, the quality of the cut rapidly deteriorates. At the same time the noise of sawing becomes noticeable harsher, power consumption starts to rise and the dulling sawblade becomes dangerous tool.

In contrast, the cutting edges of steel saws progressively round-over, so the loss in cut quality and rise in power consumption continue at a steady rate - there is no rapid cut-off point.

Fig. 10.4
This automatic tct saw grinder, the HSN 600, is a highly versatile machine capable of re-grinding all regular tct sawblade types, including hollow fronts - in addition to back relief and side grinding.
Source - Stehle

Tungsten carbide saws should be re-sharpened before this rapid fall-off in performance is reached. If this is not done the edges will begin pounding to dislodge the fine tungsten carbide granules from the cobalt matrix to cause further edge breakdown - and at an increasingly faster rate. Thus, when the sawblade is finally re-sharpened, a lot of carbide has to be ground away in order to restore the cutting edges to their original condition.

Fig. 10.5
This Saturn HKS 600 is the latest in a long line of simple, manually-operated tct saw grinders made by this company. It is ideally suitable for the smaller shop wanting in-house saw maintenance.
Source - Saturn

This also takes far longer than would be taken in restoring the same sawblade before rapid breakdown takes place, costs more in saw grinding time and grinding wheel loss - and considerably shortens the overall life of the tips. Machine operators should be made aware of the high price of running tct saws beyond their optimum machine life.

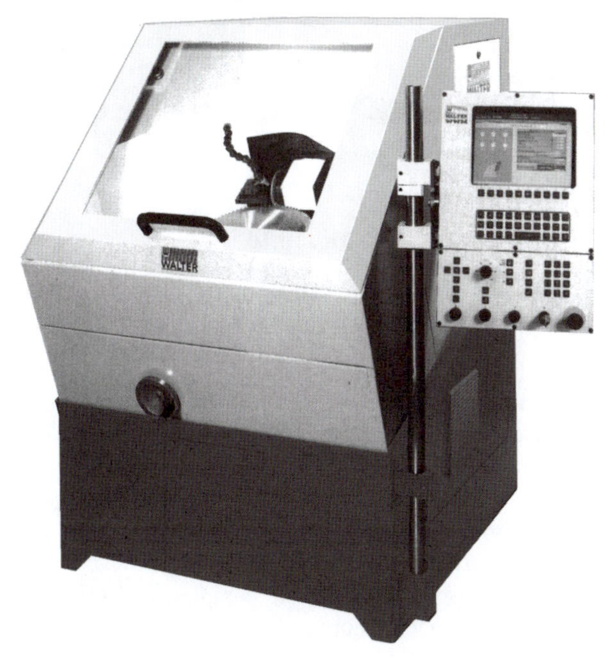

Fig. 10.6
The machine show here is the Woodtronic NC3, a reasonably-priced, numerically-controlled tct grinder with three NC axes and programs for all tooth geometries.
Source - Walter

Estimating run-times

When cutting the same material on a regular basis it is possible to predict fairly accurately when sawblades should be re-sharpened. This is done by taking careful records of the state of the sawblade, the quality of the cut and the power consumption (if relevant) at regular time intervals or at a specified linear footages.

By studying these records the point when this condition is reached can easily be found. An acceptable running time or linear production figure would be, say, 80% of this.

It is convenient to change sawblades at break times in the working schedule - even if this happens to be earlier than the estimated change-over point - in this way production is interrupted as little as possible.

If the precise figure is taken this may fall within a working period - and the temptation then would be to continue running the sawblade.

Sawblades run too long take more power and produce a poor quality of cut. To compensate for this a reduction in feed speed may be needed - and productivity will then also suffer. Changing sawblades sooner than later is by far the better option.

An overriding rule, of course, is that the machine operator should watch out for any loss in cut quality, a change in the noise of sawing, or any unexpected rise in power consumption. Any of these could indicate that the sawblades need replacing - regardless of whether or not the estimated change-over time or figure has actually been reached.

Carbide saw grinders

Fig. 10.7
The Akemat B automatic machine, for face grinding only of tct teeth, has an instantaneous head return movement to avoid crashes on damaged teeth. Straight, angled and hollow-front teeth can be ground with high stock removal rates and a short cycle time. The machine is used in conjugation with the Akemat U4 top grinder, See Fig. 10.67.
Source - Vollmer Dornhan

Grinding techniques for tct saws differ greatly from those for steel and Stellite saws, and to meet this need many specialized and dedicated tungsten carbide grinders have been introduced over the years.

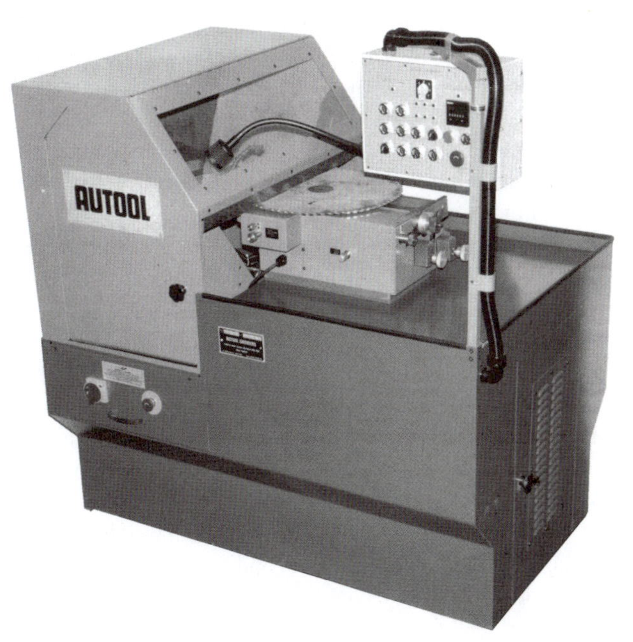

Fig. 10.8
This automatic grinder, the TCT/25, is a hydraulically-operated, plc controlled machine capable of either face or top grinding of alternate bevel and triple-chip teeth at a single setup.
Source - Autool

These range from small hand-operated machines and attachments for general purpose grinders *(intended for a single grinding pass at each setting)* to highly sophisticated plc, nc and cnc machines capable of fully grinding even the most complex tips at a single setup. Universal saw grinders are also available which are capable of grinding regular cv, Stellite and carbide-tipped saws.

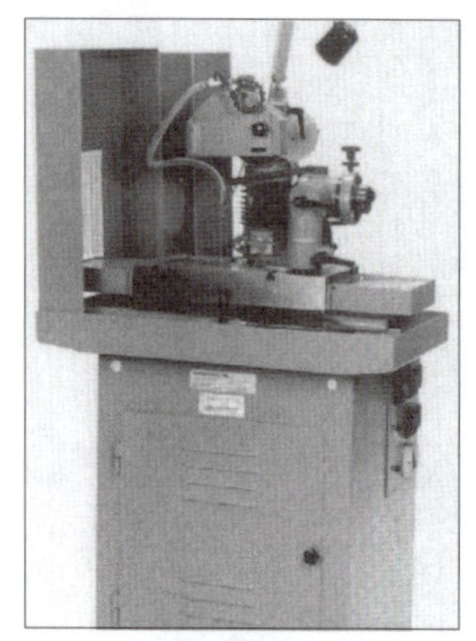

Fig. 10.9
This FKSC 450 universal grinder, shown here set for router cutters, can have an attachment for grinding tct saws.
Source - Saturn.

Any machine used for grinding tct saws must be sturdily built and capable of being adjusted to, and holding, the very fine tolerances that are essential if sawblades are to operate efficiently.

As grinding techniques vary according to the operating mode of individual machines, only general operating guides can be given - but there are certain basic techniques that apply commonly to all machines.

Safety precautions when grinding

Grinding tungsten carbide teeth *(and the tooth backing or gullet)* is like any other grinding operation for which certain precautions must be taken - and some materials are hazardous.

Always wear protective goggles or a face mask, and use a coolant wherever practical to keep dust to a minimum and to improve grinding efficiency.

On machines using coolant*, but which are not enclosed, adjust the coolant flow and direction so that spray and mist is kept to a minimum. *Use a non-leaching coolant

On enclosed machines keep the doors in good condition and fully closed during operation, and regularly clean and maintain any exhaust system provided on the machine.

Where coolant is not used, enclose the grinding area as much as practical and connect this to a separate and fireproof exhaust unit - not to the general plant exhaust or blower system. Grinding machines should be operated in well lit and well ventilated conditions, and the floor and working spaces should be kept clear of non-essential parts and clutter, and as clean as practicable.

Clean and lubricate all machines on a regular basis - and clear waste safely and in an environmentally-friendly way.

Preparations for grinding

The first operation is always to thoroughly clean each sawblade before grinding, possibly using one of the cleaning tanks now widely available. Gummy deposits remaining on the sawblade can inaccurately centralize the sawblade and give a jerky and unreliable indexing movement - both result in uneven tooth point heights or uneven side bevels.

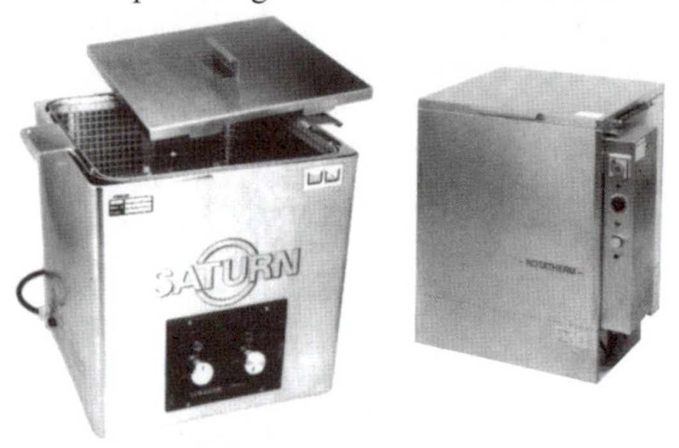

Fig. 10.10
Specialized equipment is available for cleaning saws and other tools - such as the ultrasound bath, left, and the industrial washing machine , right.
Source - Saturn

Also remove gummy deposits from all surfaces to be ground - otherwise these will tend to gum-up the grinding wheel. In particular the tooth tip and gullet must be cleaned down to the bare metal. It is against the tip or gullet that each tooth is indexed - so a buildup of gummy deposit here would give uneven face grind or an irregular tooth height.

Check also that all tips are complete and secure, and make good any that are not.

Grinding carbide-tipped saws involves a great deal more time and trouble than that needed for steel saws. Greater accuracy is needed, and the surface finish must be of a very high quality. The time and care needed in preparing tungsten carbide tipped saws, however, is fully rewarded by the longer running times and higher quality of finish that then result. Rushed or careless grinding gives a poor quality of finish and reduced running times.

Accuracy of tooth height

Fig. 10.11
To grind accurately sawblades must be indexed precisely and clamped securely, as shown on this CHC grinder. The tooth top is being ground using a saucer type diamond grinding wheel.
Source - Vollmer

Saw teeth which are concentric to their running centre *(i.e., the same relative height)* will share the work equally and will be at their most efficient - the sawblade will achieve the highest possible cut quality and the maximum running time between sharpening.

Sawblades which run eccentrically or wobble run inefficiently. The few teeth sharing the work dull prematurely and give a poor quality cut - however carefully grinding is carried out.

The possible causes of saw teeth running badly include an arbor on either the sawing machine or the grinder that is worn or runs badly, wear on the sawblade centre, a sawblade body that contains lumps or twists, or one which is run at a different speed than for which it is tensioned.

The condition of the saw arbor and sawblade fit has already been dealt with under 'Tungsten carbide saw applications'. It is assumed that these have been checked and corrected as necessary and, if so, then the role of the grinder is simply to grind the teeth accurately to the sawblade bore.

To do this the sawblade must be a good fit on a true-running saw-grinder arbor, the saw teeth must be accurately indexed and the saw body must be true and securely clamped during grinding.

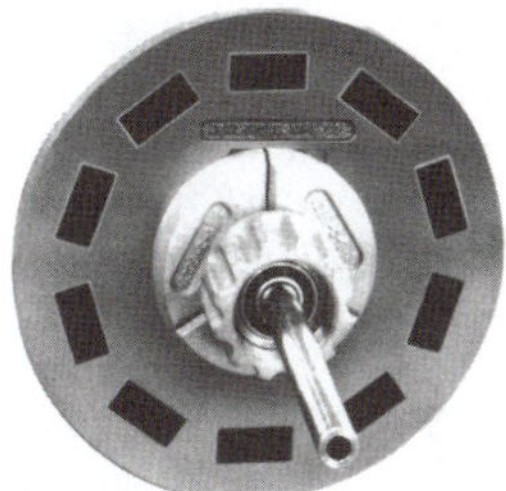

Fig. 10.12
This 12in. diameter, ball bearing backing plate has inlaid permanent magnets to grip and steady the sawblade, and a switchable, expanding three-jaw centre bushing to take up any variation in the sawblade bore - often present in the splined-arbor saws for which this type is intended.
Source - Armstrong

Fig. 10.13
The HSN 600 automatic saw grinder, shown here grinding a top bevel, uses a saw centre with a friction collar to grip the sawblade.
Source - Stehle

There are several methods of mounting the sawblade on a grinder. One is a fixed saw-centre, disc or cap, precisely machined to fit a specific sawblade bore. These interchange and are available in different styles and diameters.

Another is a cup and cone arrangement which accurately centres a range of sawblade bores. Both have been detailed under Chapter 6, Figs. 6.2, 3, 32 & 33. Those used when grinding carbide saws are very similar. More sophisticated methods of centering are also used, possibly incorporating a live, ball-bearing arbor, magnetic clamping flange, mechanical expanding centre or hydraulic bushing.

Before fitting a sawblade on the grinder, check that the saw body and bore, also the mounting device, show no signs of wear or damage and are clean and free of any dirt or grit. Lightly oil the surfaces and working parts.

Fig. 10.14
The TCT/25 automatic saw grinder mounts each sawblade on a saw centre and grips it via adjustable electromagnets built into the backplate. Shown face grinding.
Source - Autool

Fig. 10.15
A centering cup and cone is used on this TCT/2 manual grinder. An adjustable friction clamp above the backplate clamps the sawblade. Shown face grinding.
Source - Autool

Mounting the sawblade

After mounting the sawblade on a centre, a cone, or a mounting bush, adjust the assembly radially so the teeth are correctly positioned for grinding - supported close to but clear of the gullet base and with the sawblade firmly held against the rigid backplate. *(On machines with saws mounted horizontally the 'backplate' is the supporting plate beneath the sawblade close to the tip to be ground).*

Manual machines sometimes control the saw movement via a spring-loaded collar around the bore, but the regular method is to use a friction-faced clamp directly opposite the backplate which adjusts for pressure and possibly swings clear for easier loading.

On automatic machines friction clamps can be manual set for pressure or are powered by air or oil. An alternative is to use electromagnets fitted in the backplate. These power-up via a separate switch when setting the machine *(and automatically as the operation begins)*. They need no external mechanical clamp, so sawblades are quickly and easily loaded and unloaded.

Whatever method is used, though, the sawblade must index precisely and hold firmly in the indexed position throughout the grinding movement - otherwise grinding will be inaccurate. Clamping pressure is often made to reduce slightly when indexing a tooth on automatic machines, then increase to prevent any sawblade movement during the actual grind.

Fig. 10.16
The powerful clamp on the Akemat U4 fully automatic top grinder firmly holds the sawblade when grinding. The machine uses an edge-grinding wheel and is shown grinding the lower bevel of a triple-chip tooth.
Source - Vollmer Dornhan

Strob Saws

Fig. 10.17
When grinding strob saws the strob tips prevent proper seating if using a regular backplate and clamp. One solution is to sandwich the saw between two slotted side plates.

A problem may occur when grinding strob saws - sawblades with deep, carbide-faced body slots. With the regular clamp and solid backplate arrangement the strob slot facings will foul and could be damaged by them. In this case fit metal side plates on either side of the sawblade with slots into which the strob slot tips can protrude.

A simpler solution may be possible on machines fitted with small electromagnets in the backplate - or where small permanent magnets can replace the regular large-diameter electromagnets. If adjustable, these can be shifted around so they avoid the strob slot tips when this tooth, and those either side, are indexed. The strob slot tips slide over the magnets during the indexing movement so some magnet face wear is unavoidable - but the tips themselves suffer very little.

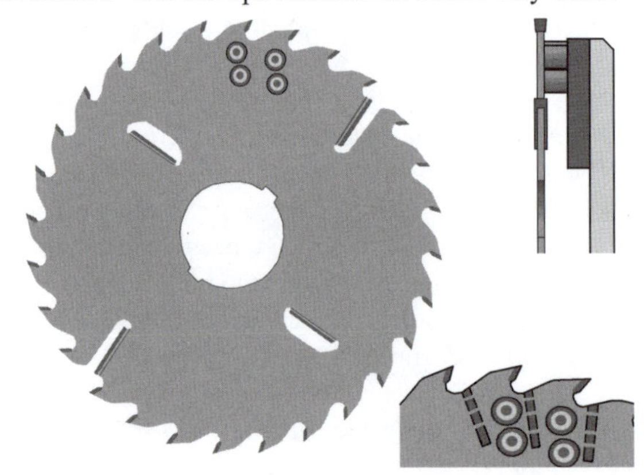

Fig. 10.18
Small diameter and adjustable magnets mounted on the backplate can usually be arranged to miss the strob slot carbide tips. The lower insert shows how these relate to the strob slot in different indexed positions - as shown in dotted outline.

Centralizing the sawblade

An essential part of most setups is to centralize the sawblade - *see also further details under setting top bevel angles on modern automatic machines.*

Machines using a saucer grinding wheel for top grinding commonly have a tilting head movement to set the top bevel angles. In this case the sawblade is centralized to the tilt centreline.

Some automatic machines instead use an edge-grinding wheel when top grinding - with a movement to offset the grinding wheel on either side of the back or support plate to form the top bevel angles. With these the sawblade is centralized to the centre-zero point of this movement.

Often the saw grinder backplate has a precision adjustment to accurately centralize the sawblade, although the method of setting varies.

The simplest arrangement has a loose pointer which mounts on the machine base to indicate where the centreline of the sawblade should be. Screw or eccentric centering devices instead have an engraved barrel or digital readout to show the setting.

The readout may indicate the actual distance of the backplate from the head tilt centreline - in which case the sawblade body thickness is measured and halved to calculate the actual setting needed.

More convenient are those where the readout relates to the actual saw body thickness. See more details under automatic saw grinders.

Note: All these devices centralize the sawblade body itself. This gives the right result in most cases but, if the tct tips do not project equally on each side of the sawblade, they will be ground with unequal alternate-bevel heights or uneven triple-chip bevels.

Saw indexing

Whatever the machine type, and whether grinding the face or the top, saw teeth are initially indexed *(positioned for grinding)* by some sort of finger or pawl acting against the tip face. On simple grinders the sawblade may be shifted manually against a pop-up pawl fitted at the point of grind - which is then retracted.

On most machines, though, the pawl first contacts the tip face, traverses it into the grinding position, and then withdraws well clear of the grinding wheel in readiness to pickup the following tooth. The movement is hand-operated on manual machines and power-operated on automatics.

Fig. 10.19
The index pawl on this Akemat U4 needs no manual setting - it has a sensor to automatically index any pitch of tooth between 8 & 125mm. When grinding sawblades with chip-limiters, the automatic indexing is set to skip the chip-limiter and pass onto the following tooth.
Source - Vollmer Dornhan

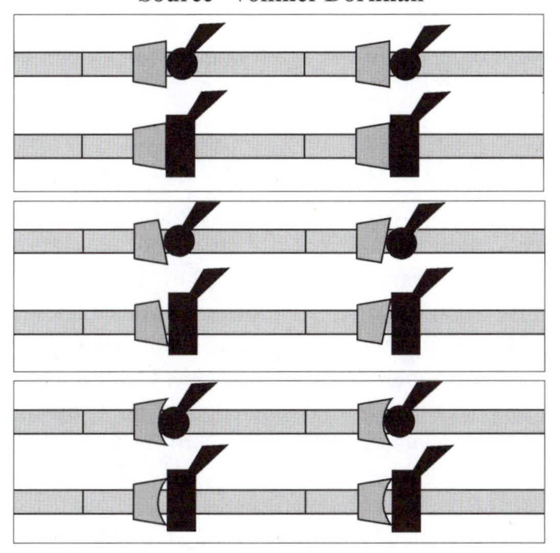

Fig. 10.20
The above sketches show the two most common forms of pawl contact faces, ball-shaped and straight. The ball-shaped type is universally suitable for all tooth face profiles. The straight-face is generally satisfactory, but may tend to chip the edges of alternate bevel and hollow-front teeth - and even straight-front teeth if badly re-ground.

Precise indexing of the tooth is absolutely essential to ensure accuracy of the grind, so the pawl must traverse to precisely the same position each time, and the clamping action must not allow any movement after the tooth has been indexed.

When operating a manual grinder it is all too easy to mis-index the tooth by attempting too rapid an indexing movement - and this can so easily crash the grinding wheel.

Because it is in repeated contact with a hard and sharp-edged tip, the pawl is subject to excessive wear and has itself to be made of a hard and tough material. Unfortunately, a wrongly-designed or badly re-ground pawl can possibly chip the carbide edges.

Some makers fit a pawl formed to a ball or point intended to contact the tip face clear of its sharp edges. This type is better for all tooth types, and is certainly better for indexing hollow-teeth saws and crosscut saws formed with an alternate front bevel.

Pawl movement

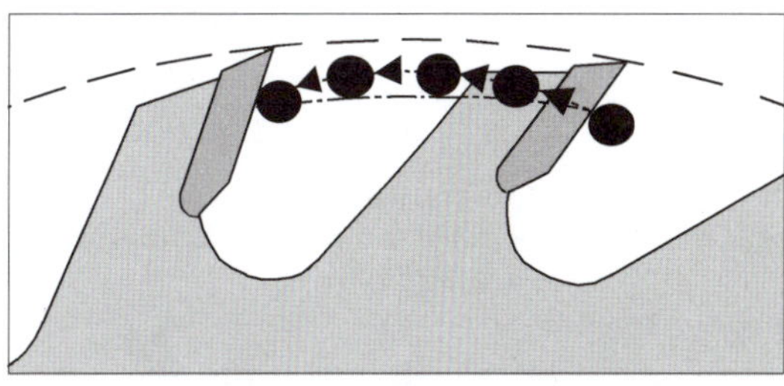

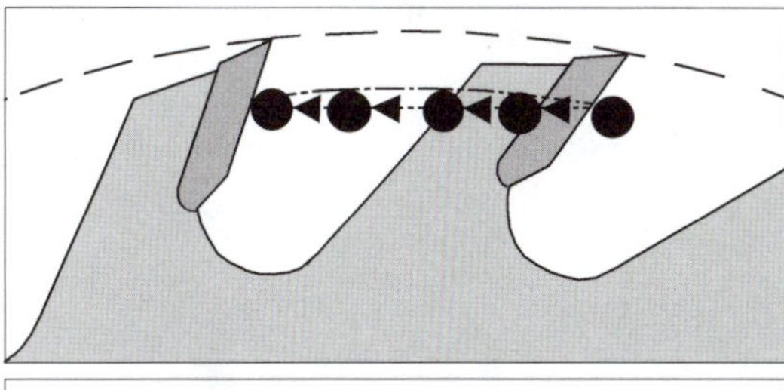

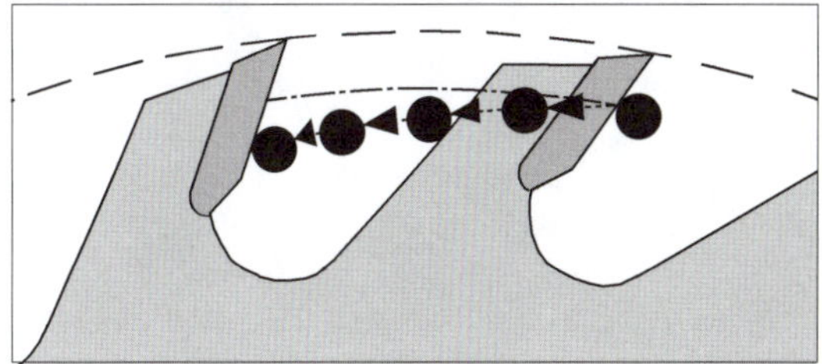

Fig. 10.21
The above sketches show three typical pawl traverse movements which create unnecessary wear because the pawl does not follow a true arc (shown as a chain-dot line).
In the top sketch the traverse movement has too small a radius, in the centre one the traverse is a straight line, and in the lower one the traverse arc is true but wrongly set.

Ideally the pawl should traverse in an arc centered on the saw bore. If it does not, then friction will be created because the pawl will move fractionally up and down the tip face during the indexing movement. This could lead to inaccurate indexing or possible tip damage - but in any event friction will wear the pawl prematurely.

On a few machines the pawl pivots on the saw mounting arbor to move in a true arc, or have a straight-line movement. Most, though, have some sort of arcuate movement either from a fixed pivot other than the saw centre, or by using a curved cam to guide the index arm.

On cam-type grinders, different cams can be switched to give an indexing movement roughly corresponding to whatever diameter of sawblade is fitted. Alternatively, some makers fit a composite cam profiled to match virtually any sawblade fitted within the machine's range.

Regardless of the actual arrangement, the pawl should be set to contact the same position on the tip face *(just below the point)* at both start and finish of the indexing movement - this keeps pawl wear to a minimum .

Many automatic grinders have a spring-loaded pawl which trips over the back and tip of the following tooth on its return movement - and this can possibly cause tip damage or a crash. To avoid this chance some makers provide instead some form of non-contact return movement of the pawl.

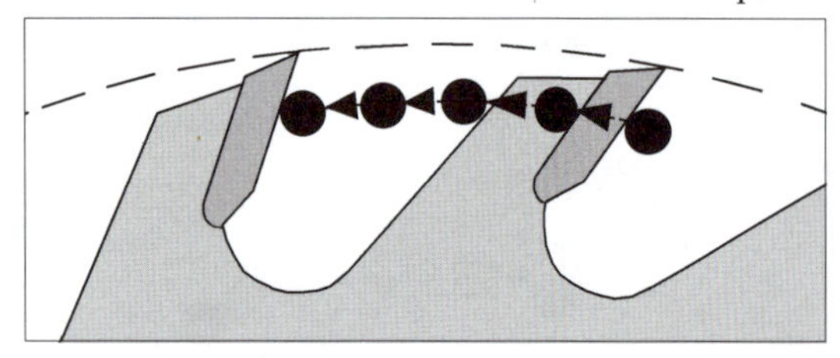

Fig. 10.22
In the sketch above both the arc of traverse and it's setting are both correct.

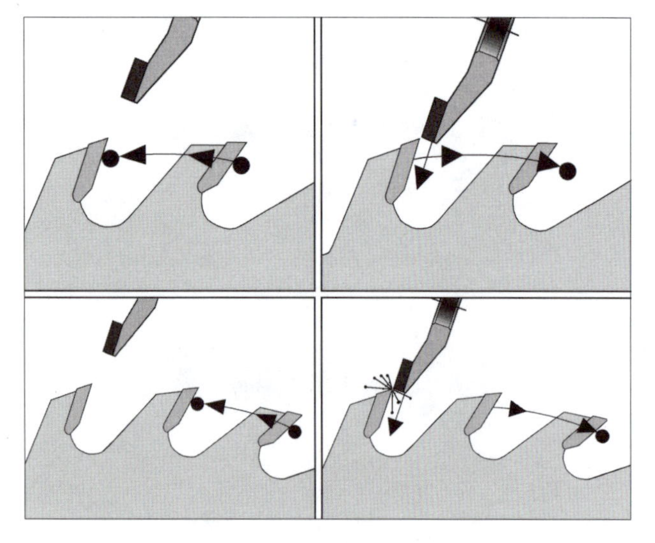

Fig. 10.23
When face grinding, the pawl should index the tooth to be ground to ensure that an even amount is removed from each tooth face (top left & right).
When indexing a tooth other than the one to be ground there is danger of a crash on a short-pitched tooth (bottom left & right) but in any event uneven grinding will take place.
Note: New carbide saws will have an even pitch - but any tip replacement will almost certainly alter this.

With all grinding operations, face, top and sides, it is essential that the pawl indexes the face of the tooth to be ground - or inaccuracy will result.

The reason is that some sawblades vary slightly in pitch - so indexing a tooth other than the one to be ground can give hit and miss contact when face grinding, uneven tooth heights when top grinding, and varying kerf widths when side grinding.

This will not be the case with new sawblades from reputable makers, but sawblades that have had teeth replaced are likely to have a irregular tooth pitches for this reason.

Some automatic machines have a pawl movement which automatically indexes teeth of any pitch. This type is ideal for group-teeth saws, for sawblades with a varying pitch*, and for sawblades with a suspect pitch. The device usually incorporates a sensor of some sort which detects when the pawl has passed across a tooth. Where chip-limiting saws are ground the chip-limiter could trigger the pawl, but on modern machines a skip program tells the machine to ignore these.

**Some saws are deliberately made with a varying pitch to avoid the quality-destroying vibration and irritating noise that can occur by the repeated impact of regularly-spaced teeth. This type needs particular care when setting some grinders to avoid miss-indexing.*

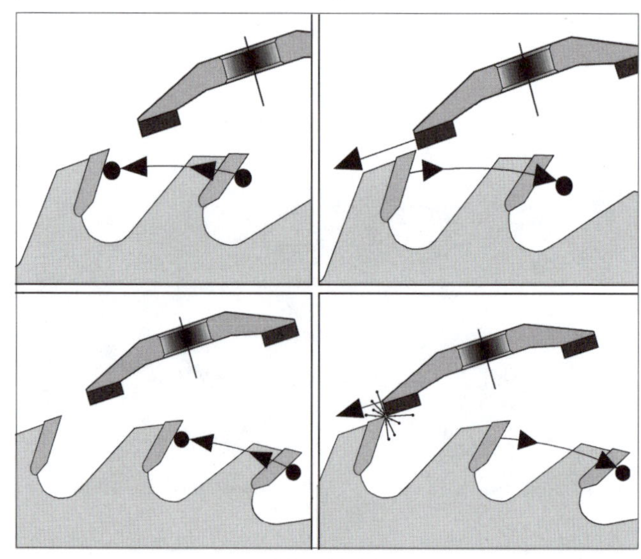

Fig. 10.24
Similarly, the pawl should index the tooth to be ground when top grinding, in which case even irregularly-pitched saws will be ground to the same height (top left & right). When indexing a tooth other than the one to be ground (on saws with an irregular pitch) teeth will be ground to an irregular height - with the outside chance of a crash with a long-pitched tooth (bottom left & right).

The grinding action

It goes almost without saying that the original grinding angles on each saw tip must be repeated absolutely.

The angles are precisely formed by the saw makers to match the intended application of the sawblade and to alter them - without good reason - will affect the cutting efficiency and power consumption of the sawblade. Unfortunately, it is all too easy to alter angles through poor grinding practice, or by taking shortcuts to speedup saw turn-round.

To work efficiently the cutting edges of newly-sharpened sawblades must be sharp, and with a consistent and high-quality finish free of lost or loosened carbide grains. This is best achieved by grinding slowly and carefully with the correct grit of grinding wheel, by providing ample flood coolant, and by using no more than the maximum recommended weight of grind *(a measure of the amount of carbide removed at a single pass)*.

Grinding tungsten carbide is similar to any other grinding action, but greater care is necessary because of its unusual nature - likened to that of a stone wall. Fine carbide granules are bonded together by cobalt, and these can be lost or loosened at the cutting edge by using too coarse a grinding wheel, by grinding too heavily, or by grinding too quickly.

The uneven edge formed by lost or loosened carbide granules might appear fine to the unaided eye, but when the sawblade is put back in service an edge-destroying pounding action sets in either immediately or at a much earlier stage than would otherwise be the case. Premature failure then results - leading to a short running time and a worsening quality of finish.

Grinding wheel types

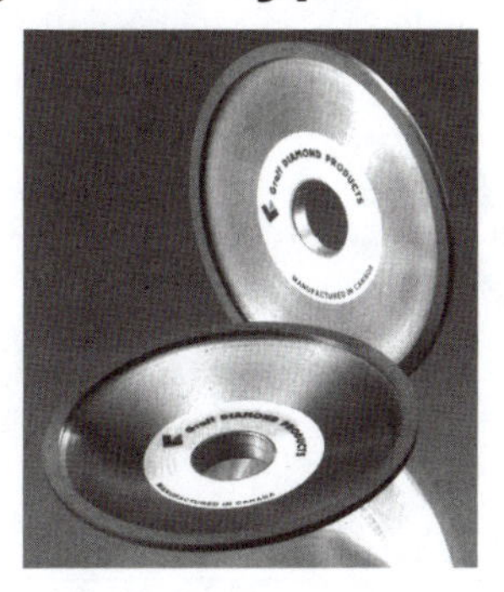

Fig. 10.25
Typical shallow and flat saucer grinding wheels as used in face grinding on all machines, and for top grinding on many machines.
Source - Graff

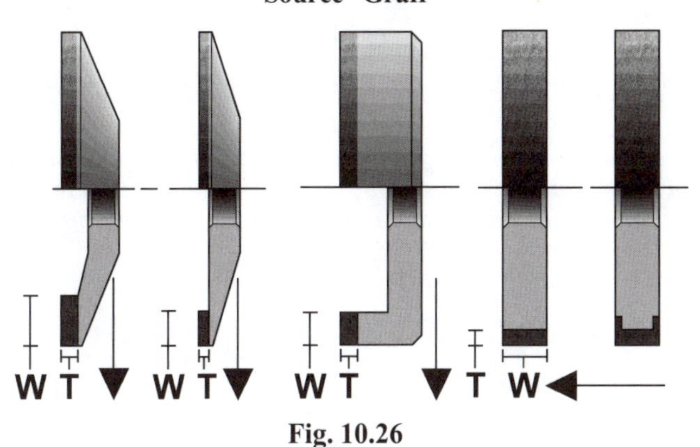

Fig. 10.26
In addition to different styles, the width (W) and thickness (T) of the diamond deposit can vary. The arrows show the direction of grind. The types shown are, (left to right) top grinding saucer, face grinding flat, top grinding cup, and two top grinding plain (edge-grinding) wheels.

At one time it was regular practice to use silicon carbide (green grit) grinding wheels to rough-grind tungsten carbide tools. This is not now recommended as these wheels rapidly loose their abrading ability and then tend to overheat and crack the carbide. Diamond wheels are the recommended type - and are used almost universally for grinding carbide tips.

Cup, saucer, dish or edge-grinding wheels can be used - depending upon the design of machine and the mode of operation.

They consist of a solid body faced or edged with a thin layer of diamond and should grind parallel with - not square too - the leading cutting edge of the tip.

Most machine makers recommend the use of resinoid-bonded wheels, with either resinoid-based or aluminium bodies - but individual makers may specify other types of grinding wheels, or grinding wheels of special design.

Solid aluminium-bodied grinding wheels are rigid and are ideal for holding tight grinding tolerances.

Resinoid-bodied grinding wheels can actually flex slightly in use. For this reason they give an excellent finish - especially if the grind is repeated without machine adjustment - but are not as precise when taking a heavy cut.

Grit size and concentration

Coarse-grit grinding wheels give a poor surface finish and are also tough on the carbide granules - they can detach or loosen them even with a light cut and a slow grind. However, they produce quicker results as they are capable of a heavier grind.

They are used solely when excessive grinding is needed - for example dressing down oversized, replaced tips, or when dealing with badly dulled teeth - but are never used for the final grind.

Fine grinding wheels give the best surface finish and are always used for finishing tips - but they grind much more slowly.

Extra-fine grit wheels cost more, take longer to re-grind and produce little improvement in performance. They are still preferred by perfectionists wanting to form a highly polished tip surface - though this is achievable by other means.

The diamond concentration *(the ratio of diamond-to-resin bond)* regularly used for manual grinding machines is 100 %, but a higher concentration is often recommended for automatic grinding machines.

A lower concentration grinding wheel can be used in either case - and is actually cheaper to purchase. However, the diamond concentration directly relates to the grinding wheel efficiency and its useful life - it is false economy to use grinding wheels with a lower concentration than recommended by the machine makers.

Abrasive dimensions

Cup, saucer, dish and edge-grinding wheels all have a narrow grinding face of securely-bonded diamond abrasive. These grinding faces must remain perfectly flat and parallel to the direction of the grinding feed movement if they are to operate efficiently. Fortunately, tct saw tip grinding is one of the few grinding operations where the grinding wheels can be self-dressing and so remain perfectly true throughout their useful life - provided the correct grinding wheel specification and grinding techniques are used.

Abrasive width

In general, use grinding wheels with a wide abrasive face - this gives the best and most even grinding action. How practical this is, though, depends on whether tooth tops or fronts are to be ground, and on the geometry of the tooth.

Tooth top grinding. Cup, saucer, dish and edge-grinding wheels all can be used for grinding tooth tops. They approach from clear of the tip and traverse in a perfectly straight line - preferably fully across the top so that the trailing edge of the wheel abrasive clears the back of the tooth tip.

In many cases the grinding wheel then returns in the same line to improve the finish by grinding a second time on its return stroke. When set to grind in this way the abrasive wears evenly and the grinding face remains perfectly flat and true.

With short tooth pitches, however, the trailing edge of the abrasive may not clear the back of the tooth - even when the grinding wheel is barely stopped short of the following tooth face. This can create uneven grinding wheel wear, but this can usually be made good if using the same grinding wheel between times to also grind wide-pitch sawblades.

If small-pitch sawblades only are ground, then use a grinding wheel with a narrow enough abrasive face to clear the tip trailing edge when grinding. Failure to do this creates uneven wheel wear which, if not corrected, can become so extreme as to radius-over the tip and so reduce the clearance angle.

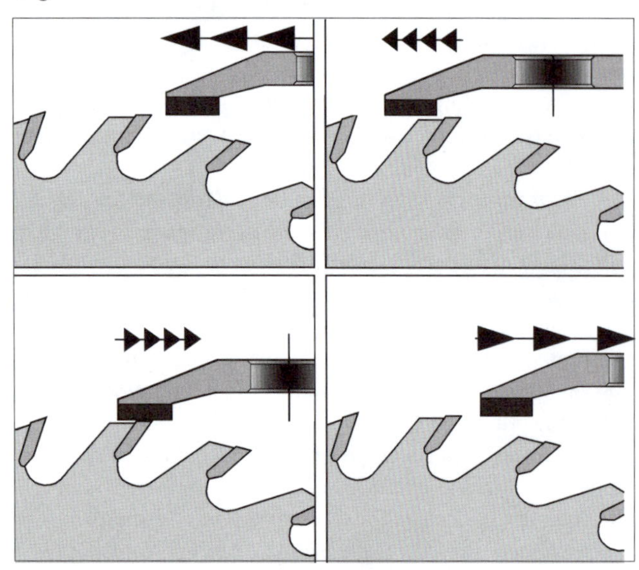

Fig. 10.27
One technique for grinding a tooth top is a fast approach (top left) a slow grind fully across the tip, top right, a slow return for a second grind (bottom left) then a fast clearance movement (bottom right).

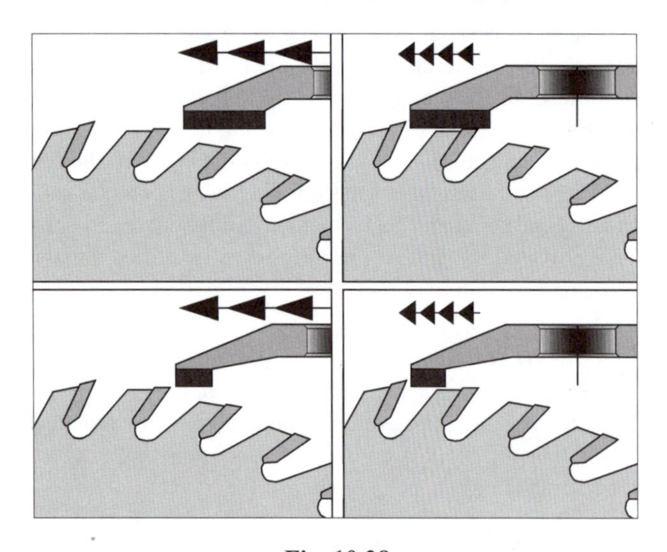

Fig. 10.28
The sketches above (top left & right) show the conditions which creates uneven abrasive wear - where the inside of the abrasive stops short of the leading edge of the tip. The lower sketches show the correct movement with a narrower abrasive to give consistent and even abrasive wear.

Tooth face grinding Only flat or shallow-dish grinding wheels can be used when grinding tip faces. Although carbide tips are bevelled to shorten the face to be ground, the downward grinding wheel movement is restricted by the depth of the gullet.

It is rare that the abrasive inside edge can actually clear the tip inside bevel - unless the grinding wheel has a very narrow grinding face. Where it does not, then a slight amount of uneven wheel wear can take place - but this is usually quite acceptable.

What must be avoided is when the inside edge of the wheel abrasive stops short of the tip - as would be the case when a wide-face wheel is used to grind a stubby tooth. This will create a step on the abrasive which will eventually radius-over the tooth point and effectively reduce the hook angle.

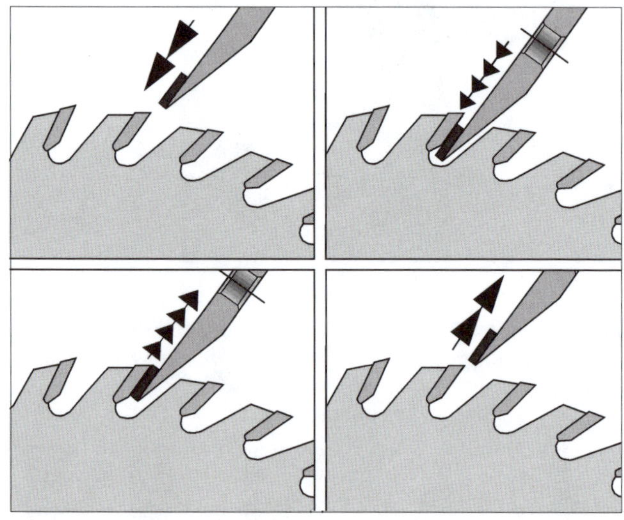

Fig. 10.29
Grinding a tooth face with a rapid approach (top left) then a slow grind into the gullet (top right) a slow return grind (bottom left) then finally a fast clearance (bottom right).

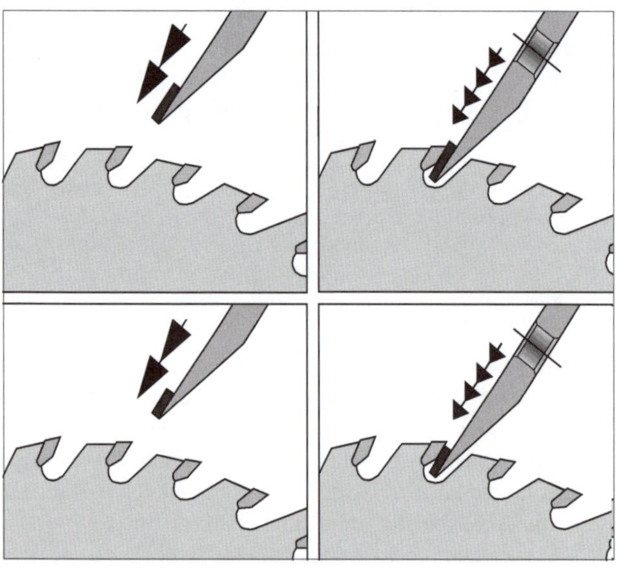

Fig. 10.30
When dealing with stubby teeth, the inside edge of the abrasive on a regular wheel may stop short of the tooth tip (top left & right). This creates uneven abrasive wear to eventually round-over the tooth face at the tip. To avoid this use a narrower abrasive (bottom left & right).

Basically, the grinding wheel abrasive width should relate to the tooth pitch when top grinding, and the tip depth when face grinding - bearing in mind always that the widest practical abrasive face will give the best result.

On some grinders the same grinding wheel is used for both face and top grinding. In this case switching between top and face grinding will normally correct any uneven wear caused when face grinding - so match the abrasive width to the top grinding operation.

Abrasive thickness

Grinding wheels are available with different thicknesses of abrasive. Those with the thickest abrasive *(usually)* contain most diamonds and offer the best value - but obviously cost more initially.

Abrasive wheel thickness is rarely restricted when top grinding. For face-grinding short-pitch sawblades, however, only flat or thin saucer grinding wheels can be used. For the narrowest of pitches the grinding wheel back has to be thinned down and the abrasive reduced in thickness to avoid the back fouling the adjacent tooth.

Because thin abrasive wheels are weaker and less cost-efficient than regular wheels, use these only for grinding the narrowest-pitch teeth. Never use them for face-grinding regular-pitch teeth, nor for top-grinding on machines that allow the same grinding wheel to be used for both operations. Rather than order thin-rim abrasive wheels for narrow-pitch saws, some use instead regular, partially-worn grinding wheels that have been turned thinner at the edge - most engineers can deal with this simple task.

Quality of finish

It is a well-documented fact that keen cutting edges produce superior results both in quality of production and run-time *(the length of time a sawblade continues to give satisfactory results)*. Spending extra time to produce the best possible edges is a practice that is well rewarded in the higher quality and greater productivity that this then brings.

Grinding tungsten carbide teeth is an abrading operation - diamond grains continuously pass across the tip surface to remove a mixture of carbide and cobalt. Although the surface of most wheels appear flat, some grains always project more than others to form shallow furrows in the direction of grind. *Of course there is not just a single diamond grain abrading the surface there are many, but the basic concept still applies.*

If left in this condition the furrows produce an irregular ground surface with a cutting edge that more readily breaks down than one where surfaces are ground perfectly flat. To get perfect flatness always use a fine-grit grinding wheel and use a slow traverse speed and the recommended grinding wheel rotary speed - or oscillate the wheel. Either ensures that these minute furrows disappear.

The quality of the ground surface finish can be affected by the grinding wheel rotational speed combined with its traverse speed - also the weight of grind - even though the main controlling factor remains the grit size and bond.

A slow wheel speed, a fast traverse speed and a heavy grind combine to give a faster re-grinding but a poorer surface finish. The grip of carbide granules at the cutting edges can also be weakened, perhaps with some being broken-away completely.

A fast wheel speed, a slow traverse speed and a light grind combine to give the best results - the preferred combination for quality finish-grinding.

On manual machines the grinding wheel speed is fixed by the machine makers, so the quality of finish can only be controlled by the grinding wheel specification and the speed at which the operator traverses the grinding wheel. Improvement to the surface finish is possible, however, by using two or more following grinding passes without adjusting the machine setting - a technique similar to sparking-out regular grinding wheels.

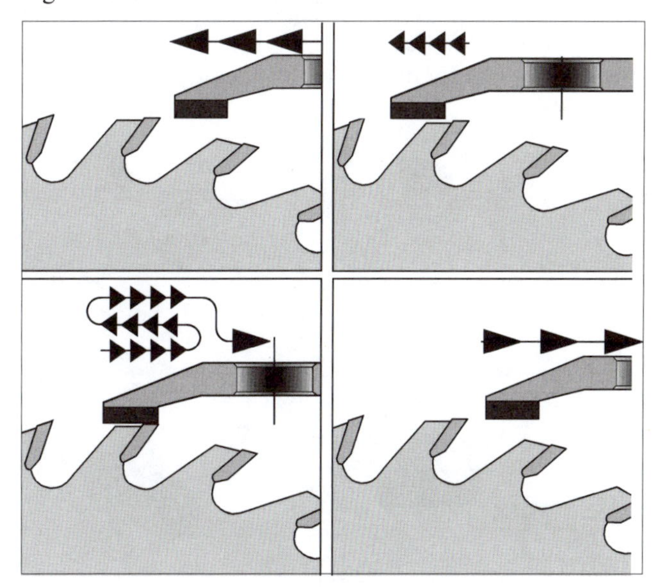

Fig. 10.31
Traversing several times across the tooth top or oscillating the grinding wheel - without adjusting the setting - improves the surface finish.

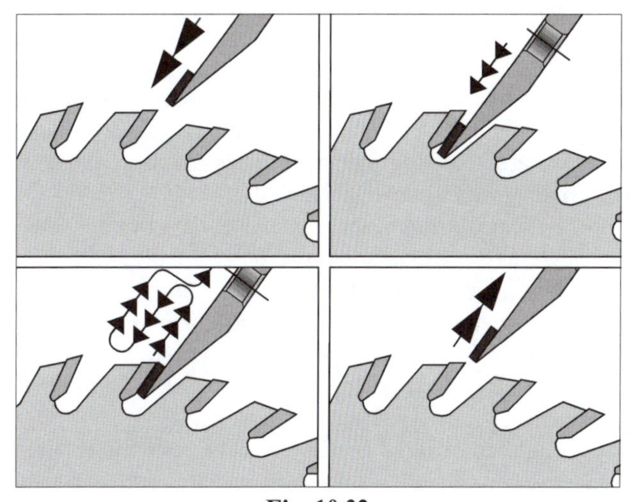

Fig. 10.32
The same technique applies when face grinding.

A keen and long-lasting cutting edge is formed by two meeting surfaces - the top and the face. Giving a good finish to one and a rushed or poorer finish to the other is bad practice - both faces should be ground to the best finish possible.

On automatic machines of the most advanced types the operator has much more control. Some machines allow a similar repeat grinding movement to that described for manual machines. Of these some may also incorporate a faster traverse for the initial, rough grind, and a slower traverse for the finish grind. Another possibility is a minute increase in the depth of grind between the rough and finish passes. As an alternative some machines provide the option of an automatic oscillating movement.

These refinements improve the surface finish to give longer sawblade run-time and a higher quality of sawing - and speed the grinding process at the same time to give the fastest possible sawblade turn-round.

Maintaining grinding wheels

A grinding wheel should not normally need truing - provided the above grinding recommendations are followed - but it should be dressed routinely using the type of abrasive and methods detailed earlier. In addition, dress the grinding wheel if it become glazed or filled, when it ceases to grind cleanly, or when it labours.

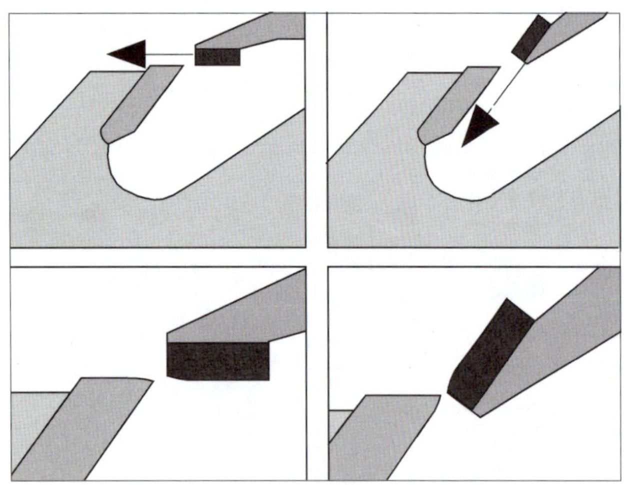

Fig. 10.33
A glazed grinding wheel can deflect slightly from the grind to become rounded-over, and then may tend to round-over the tungsten tip. This can occur both with face and top grinding.

On manual machines the 'feel' of the operation will change to warn the user of a changed condition, but it is not so obvious when using automatic grinding machines - so check and dress the grinding wheel at regular intervals.

With modern automatic machines the grinding wheel cut depth and traverse speed are precisely controlled and consistent. This, and the liberal use of coolant, creates ideal grinding conditions which so reduces the risk of glazing or filling that dressing is rarely needed. Nevertheless, still check the condition of the grinding wheel periodically and correct if necessary.

A glazed or filled grinding wheel - often the result of grinding teeth with resinous deposits - will not grind efficiently. If a heavy grind is attempted the wheel may either flex away after initially contacting the tip, or even shift the tooth slightly. In either case the effect will be the same, a rounding-over of both the grinding wheel and the tooth face or top. A badly rounded grinding wheel will steadily worsen if use is continued without correction.

The condition can be cured by returning the wheel to the makers for dressing (or brake-dressing by the user if suitable equipment is available) but the cure may not be cost-effective unless a reasonable amount of diamond remains after truing. However, slightly rounded wheels can be used - for rough-grinding replaced tips for example - provided any defect is made good by using an unworn wheel for the final grind.

Use of coolant

Unless there is a good reason to do otherwise, tungsten carbide tip grinding should be carried out with full flood coolant directed at the point of grind.

Most modern automatic machines now fully enclose the operation to prevent coolant spray contaminating the working environment - an increasing source of health concern. Keep enclosures in place and adjust the spray nozzle direction and flow to provide essential lubrication without forming excessive mist. Clean and replenish coolant on a regular basis.

If there is no choice but to grind dry - for which operation a different grinding wheel specification is needed - then make sure the workplace is well ventilated, wear goggles, and possibly a face mask, enclose the operation as much as practical and fit an independent exhaust or blower system.

Regularly clean down the machine to remove grit accumulation and so prevent this highly abrasive waste contaminating sensitive working parts. See also earlier notes on safety precautions.

Line of grind

To get the most life out of tungsten carbide tips follow a line of grind which keeps a ratio of top to face grind of roughly **5:1.** This maintains an acceptable kerf width through up to **25** grinds when removing **0.25mm**. *(0.20mm. minimum)* from the tooth top and a *(minimum)* of **0.05** from the tooth face - **Source - Leitz.**

There may be certain overriding reasons for not keeping to the recommended line of grind. For example, only the tooth face is ground where saw diameter loss has to kept to a minimum - perhaps when using sawblades in a gang - but this practice shortens the life of tips.

Another practice is to grind only the tooth tops - in an attempt to extend tip life* - but grinding only the tooth top on regular saws wears teeth down too quickly.

**Certain saw types are actually intended for the tooth top only to be ground, in particular DIY saws, and those with thin carbide tips sold as disposable.*

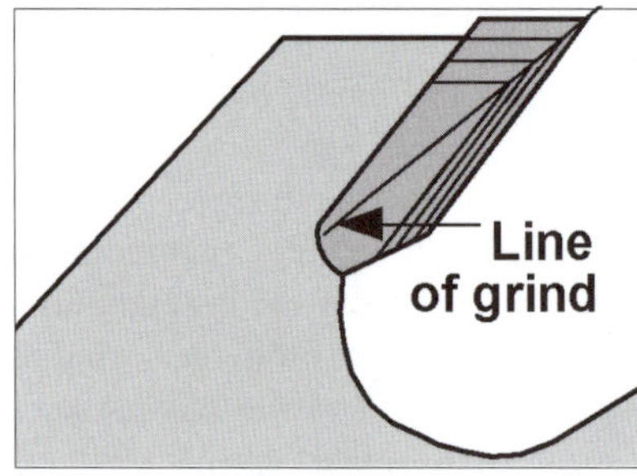

Fig. 10.34
Keeping to a suitable line of grind shown allows both face and top to be ground in a ratio of about 5:1. This minimizes kerf width loss.

Keeping to the recommended line of grind ensures the longest practical life for tips on most industrial saws and so reduces the frequency of sawblade replacement and the extra investment costs that go with it. In addition to reducing these costs it also gives other advantages:-

When tips are fully replaced on badly-worn sawblades it is necessary either to use short tips, or to wastefully grind-down regular tips. The tip seats alone could be ground deeper to take regular tips, but in this case the gullets should also be deepened to keep them in step - a major refurbishment task.

Replacing all tips may, in fact, not be cost-effective. New sawblades give a guaranteed performance, while in-

house repaired sawblades may not - and the cost of removing old tips, re-grinding the tip seats and gullets, brazing new tips and fully re-grinding them is extremely high in labour costs.

The recommended line of grind effectively means the removal of five times the amount of carbide from the tooth top than from the tooth face.

This has been misunderstood by some who believe that the tooth top should be ground up to five times to once only of the tooth face - this is not what is intended.

For good practice the tip face should be ground in advance of and each time the tip top is ground. Tips normally wear to a rounded point and, if ground only on the top, the amount needed to restore the cutting edge looses kerf width at a faster rate than if grinding both face and top.

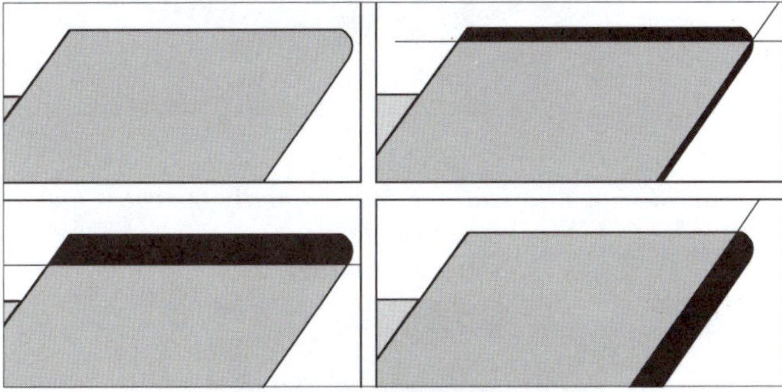

Fig. 10.35
With normally-worn tips (top left) carbide loss is minimized by keeping to the recommended line of grind, top right. Grinding only either the face or the top (bottom left and right) shortens tip life because too much carbide needs to be removed.

An all-too-common practice is to grind carbide tips excessively - in the mistaken belief that this will ensure that the cutting edges are properly restored. However, a saw tip ground excessively is no sharper than one where the cutting edge is barely restored - grinding beyond this point simply shortens the life of the tips for no good reason.

Saw-shop customers are sometimes concerned when they see what seems to be an unnecessarily heavy grind on their precious sawblades - and may query this. The commercial saw doctor or filer, however, has a valid point in taking the weight of grind he does. Repeating a grind on a sawblade initially ground too lightly adds extra cost to him which he cannot recover.

An experienced saw doctor or filer, however, can be trusted to get the balance right - not too light so that a second grind is needed, not too heavy so that sawblade life is wasted - it is all a question of experience.

Dealing with a heavy grind

If a heavy grind is needed, perhaps because the sawblade has been run too long or struck grit or a metal, then grind at two or more passes. Each time remove no more than the maximum recommended amount, and gauge the weight of the grinds to make the final pass a light one. By doing this a better finish will be possible, and the chance of loosening carbide granules during the grinding process will be very much reduced.

This recommendation mainly applies to older machines. Modern machines and technology now make heavy grinds possible - to give a fast saw turnaround - yet still produce an excellent cutting edge finish.

There are a few rare exceptions when the teeth wear to an irregular pattern, perhaps because of a difficult operating mode or due to converting unusual material.

Highly abrasive material tends to wear the tooth top excessively, in which case grind mainly the top. Softer material allows exceptionally long running periods, but friction then tends to wear a secondary angle on the tooth face at the tip - in which case grind mainly the face. In both cases doing otherwise quickly loses tip life.

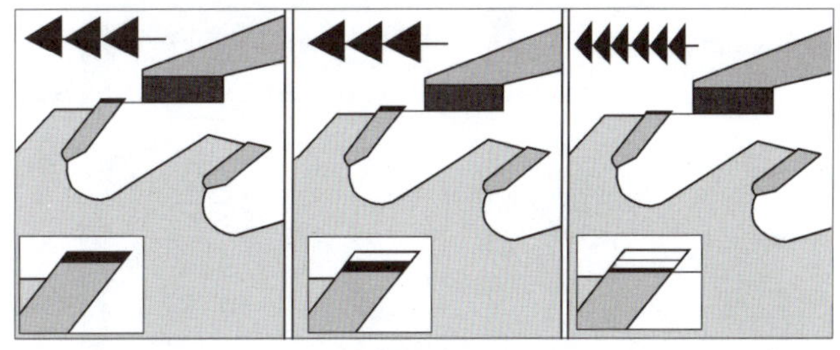

Fig. 10.36
For a damaged or badly worn tip grind in several stages. Make the final grind a light one and with a slower traverse movement.

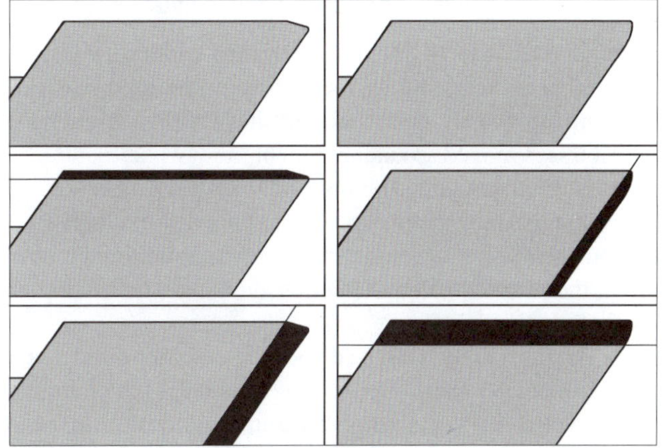

Fig. 10.37
A few carbide tips wear irregularly (on the top, left, or on the face, right). To keep carbide loss to a minimum grind more heavily on the worn face (centre left and right). Grinding on the non-worn face only (bottom left and right) rapidly looses tip life.

Checking tct edges

Check the face, top and edge of tct teeth - **before grinding** to check how much grinding is needed and **after grinding** to ensure that teeth are sharp - by using a bright light and some form a magnification.

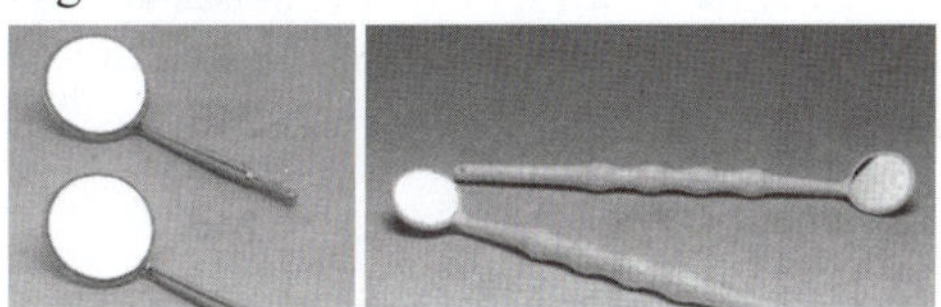

Fig. 10.38
Dentists mirrors are ideal for closely examining tungsten carbide teeth - as well as their intended use for human teeth. They are available in different styles, with magnifying types being the most suitable for the saw doctor and filer.
Source - Kent Dental

It is difficult to switch between the faces and edge of individual teeth when checking them, so use a dentist magnifying mirror in addition to a regular magnifier. Dentist magnifying mirrors are available from specialist stores which sell DIY dentist kits - but may not describe them as such for fear of scaring squeamish customers *(such as the writer)*.

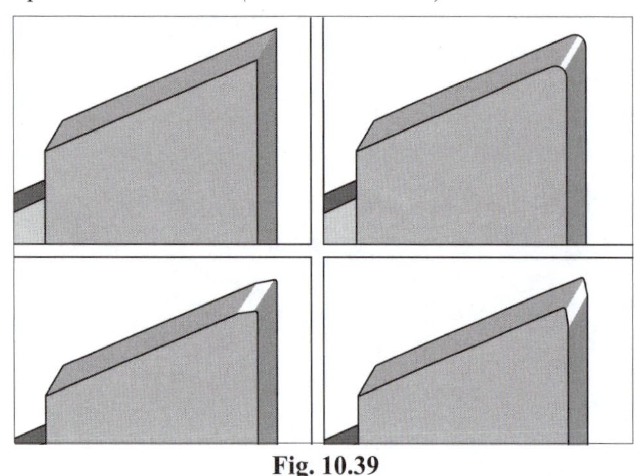

Fig. 10.39
Sharp edges, top left, do not reflect light - even though their faces do. The sketches show how light is reflected by normally dull tips (top right) by top-worn tips (bottom left) and face-worn tips (bottom right).

Wear patterns on tooth tips indicate by their width what weight of grind is needed, and by their position the type of wear. They are most easily seen in the way they reflect a bright light whilst this is moved around the rim of the sawblade.

Normally-rounded points tend to reflect a narrow band over a longish movement of the lamp. Tips with excessive top or front wear reflect a broader band over a shorter light movement, the position of the reflection showing whether top or face wear is present.

After grinding it can be difficult to determine when saw teeth are actually sharp. Unlike steel sawblades which form a burr when sharp - and so give positive indication that the cutting edge is restored - tungsten carbide teeth do not burr over.

The teeth are sharp only when all ground faces show a bright finish, and when wear patterns are completely removed along the full cutting edges of all teeth checked at random. Make sure that the faces *(and cutting edges particularly)* are smooth, continuous and free both of damage and missing carbide granules.

The more powerful the magnification and the easier it is to check the edges - but magnify the edge too much and a perfectly acceptable edge-finish looks terrible. A lot of judgment in this matter relies on experience - knowing what a good quality and long-lasting cutting edge looks like when magnified.

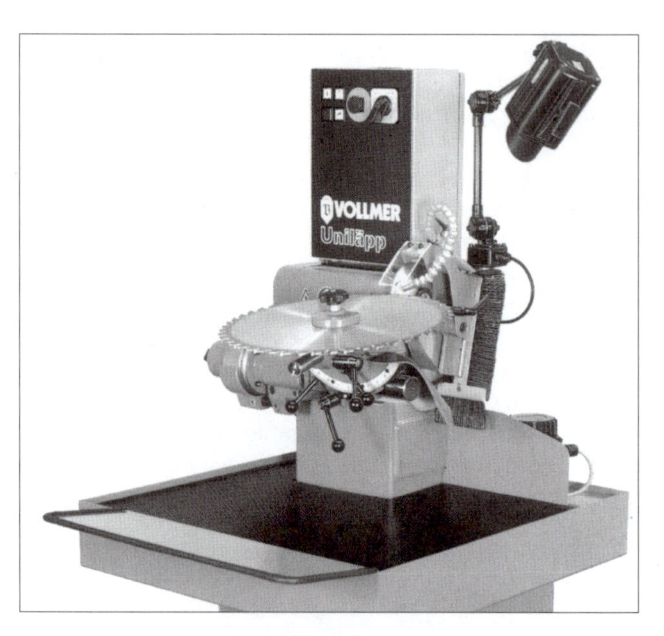

Fig. 10.40
General view of a Unilapp 600 manually-operated face and top grinder.
Source - Vollmer.

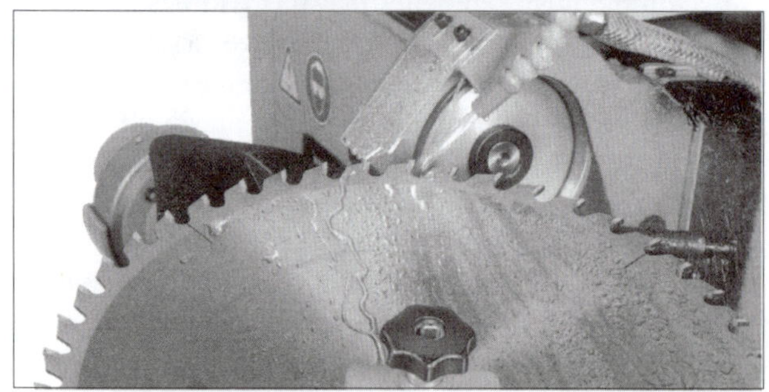

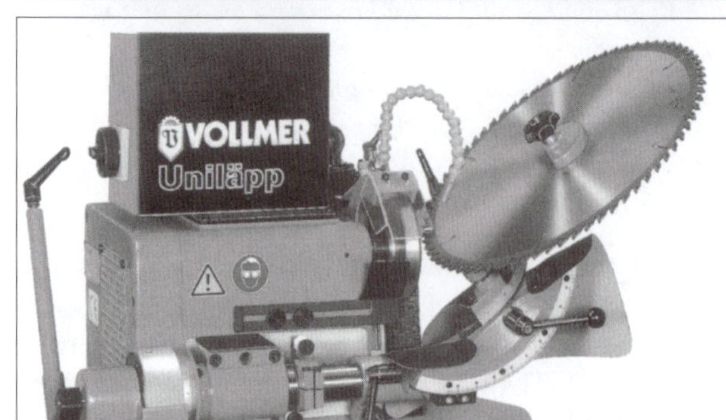

Fig. 10.41
Views of the Unilapp 600, shown face grinding (top) top grinding (center) and triple-chip bevel grinding (bottom).
Source - Vollmer.

Manually-operated grinders.

Operating sequence

These are quite simple machines to operate, but are capable only of grinding one face at a time. Five grinding setups are needed when servicing triple-chip teeth, one for the front face, one for the flat top, one for the triple-chip top and two for the triple-chip bevels. In the same way alternate bevel teeth need three setups.

Every second tooth needs to be indexed when top grinding either triple-chip teeth or alternate front and top bevel teeth. When wide-pitch teeth need to be ground - but only a short pawl movement is provided on the machine - instead index two teeth for each grinding pass.

On machines operated by lever handles, one to index the teeth and the second for the grinding movement, it is easy to adopt a rhythm of using these alternately - but get distracted and you can so easily crash the grinding wheel.

Make sure for each indexing movement that the pawl mechanism fully contacts the end stop in order to position the tooth accurately. Rushing this movement, or otherwise careless operation, will give inaccurate indexing and irregular tooth heights.

While it is convenient to simply plunge-grind *(straight in and out)* and then pass onto the next tooth, a better finish results if the initial plunge grind is followed by two or more slow passes. (See Figs. 10.31 & 10.32)

Grinding tooth faces.

In setting-up for a tooth front, first make sure that the grinding wheel *(usually a flat or shallow saucer type)* grinds square-on - all ripsaw teeth are square-faced. Accurately index the tooth to be ground, swivel the grinding head or saw support to the correct hook angle, i.e., parallel with the tooth face, then adjust the cut control *(for the weight of grind)* to bring the grinding wheel and the tip face into light contact.

Adjustment for the weight of grind is usually via an axial movement of the grinding wheel - with the cut-control setting indicated by a graduated barrel. On some machines the barrel is fixed permanently to the adjustment arbor, and on these the grinding cut out is calculated as the difference in readout between light and full grinding contact.

Fig. 10.42
Operating lever handles, as on this TCT/2 shown grinding carbide tip faces, makes manual operation both speedy and simple.
Source - Autool Grinders

On some machines the graduations are on a separate ring that can be reset to zero at any position - possibly with a click-in at zero setting for positive location. These are intended to be reset to zero after light contact so that the amount of cut is shown as a direct readout.

If it is not possible to set the hook angle using a scale, then first index a tooth and swivel the head *(or the saw support)* so that the grinding wheel can be moved into the gullet space. Then adjust both the hook angle and the grinding wheel cut control as needed to bring the grinding wheel parallel too and into light contact with the tooth face. Check the setting with the grinding wheel running. *(If the tooth face is first marked with engineers die or a spirit marker the parts where contact is made will be clearly indicated to show if and what correction is needed.)*

Set the forward movement so the grinding wheel stops just short of the gullet - this gives the best abrasive wear conditions for the grinding wheel. Set the rotation of the grinding wheel to grind towards the backplate for ripsaw teeth.

Adjust the cut control to give the recommended weight of grind *(the amount of carbide to be removed)*. This is shown by the noting the difference in the cut-control scale graduations between light and full grinding contact *(when using a fixed barrel)*, or by resetting to zero on light contact and then noting the precise amount as a direct readout *(on the adjustable type)*.

Refer to the makers instructions as to what weight of grind is recommended, and how the scale graduations relate to this. Failing this, check the amount of movement relative to the scale markings by using a dial indictor and follow previous guides.

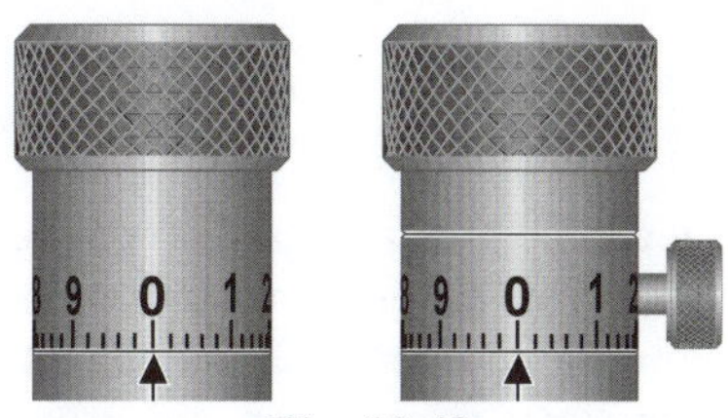

Fig. 10.43
Typical graduated barrel type scales as used to set the grinding cut. The barrel scale on the left is fixed, while the one on the right can be reset to zero at any position.

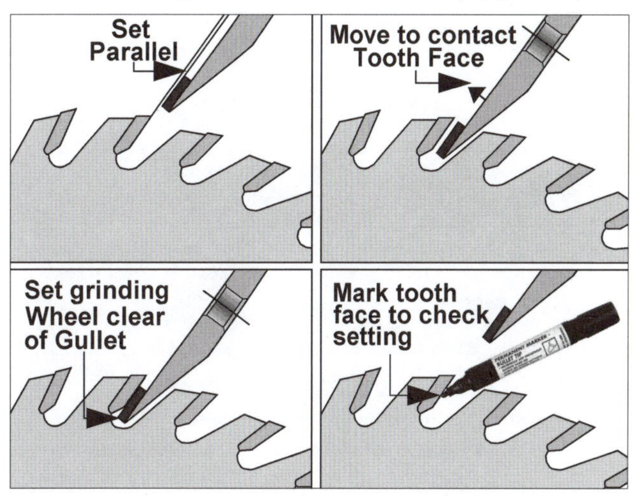

Fig. 10.44
Setting for face grinding

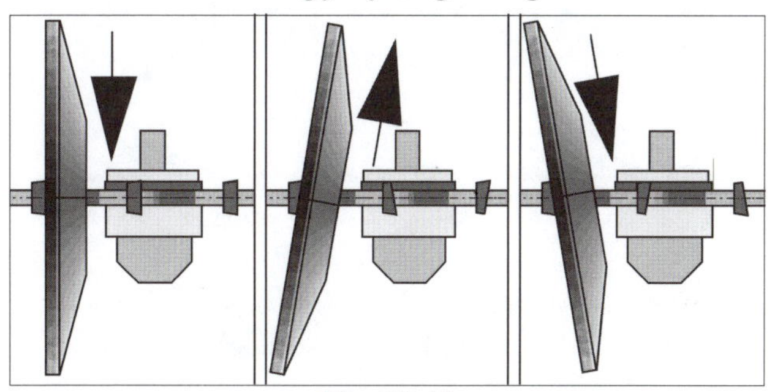

Fig. 10.45
When grinding tooth fronts the grinding wheel should grind towards the backplate when grinding square across, and towards the point when grinding alternate front-bevel teeth.

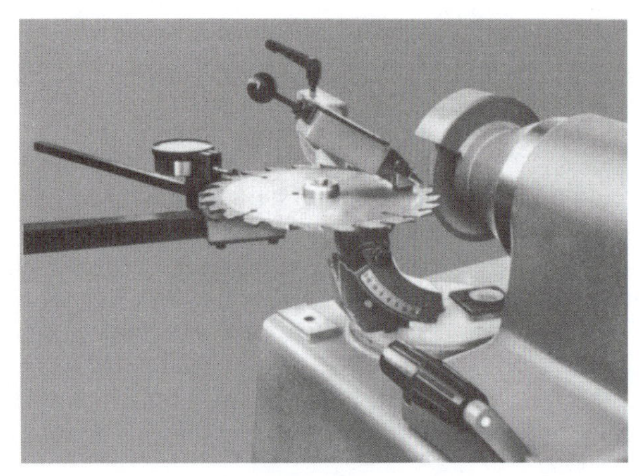

Fig. 10.46
This Saturn HKSC 600 H is shown face grinding.
Source - Saturn

Mark the first tooth, then grind fully around the sawblade, stopping when the marked tooth is again reached - otherwise face grinding will continue unnecessarily.

For ripsawing teeth the tooth face is ground square across.

For crosscut teeth the front faces are usually bevelled alternately, in which case tilt the grinding wheel to suit, and grind towards the point. Grind alternate teeth, then set to the opposite face bevel angle and grind the remainder, possibly adjusting the weight of grind between the two.

Grinding Hollow Teeth

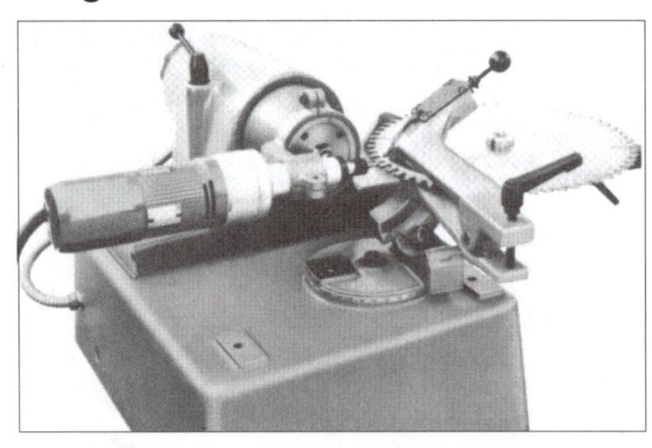

Fig. 10.47
Grinding hollow faces on an HKSC 600 grinder. The grinding wheel is a 6mm diamond barrel type mounted on a Bosch grinding head running at 26 000 revs/min.
Source - Saturn

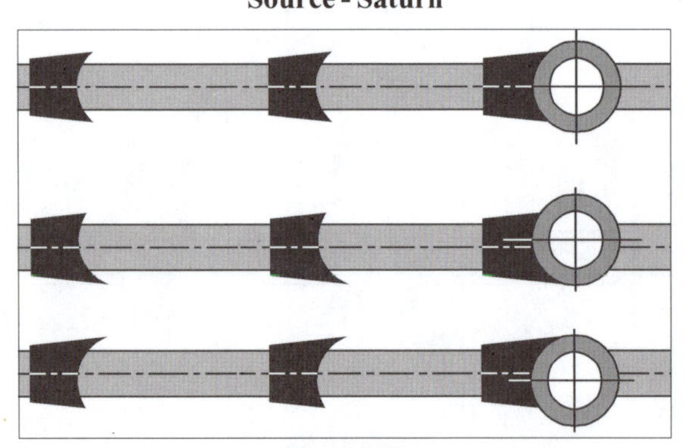

Fig. 10.48
Take care that the barrel grinding wheel is central to the kerf width (top) otherwise the saw will lead to one side (centre & bottom).

Only a few manual grinders have facility for grinding hollow tooth faces. Those that do use a separate high-speed grinding head with a small diameter barrel-type grinding wheel. Precise centering of the grinding wheel to the tip is essential - otherwise the sawblade will lead in the cut. Operation is similar to face grinding, using the regular movements of the pawl and grinding head or saw carriage. See further details under automatic grinders.

Grinding tooth tops

Grinding flat tops

Tooth tops are usually ground with dish or saucer grinding wheel. First accurately centralize the sawblade to the tilt line of the grinding head *(or the tilt line of the saw support)* and index a tooth.

Use the swivel angle scale to set the clearance angle, regularly 15 degrees, and set the tilt angle to zero. Adjust the cut control so the grinding wheel just lightly contacts the first tooth top. Run the grinding wheel to finally check the setting - after first marking the tip top.

If the setting scales are not accurate, then set by trial and error generally as described for face grinding.

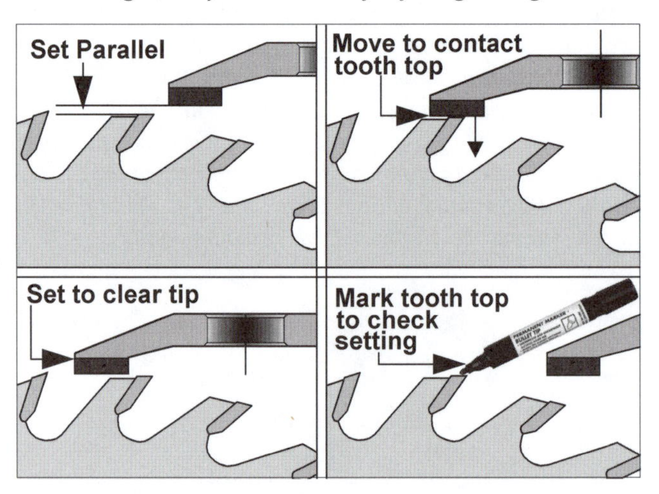

Fig. 10.49
Setting for grinding flat tops.

Set the grinding wheel to rotate against the saw support *(backplate)* and adjust the cut-control to give the recommended weight of grind. Continue to index and grind around the sawblade until all teeth are ground. It is not necessary to mark the first tooth as grinding will cease once this has again been reached.

Grinding alternate-bevel teeth

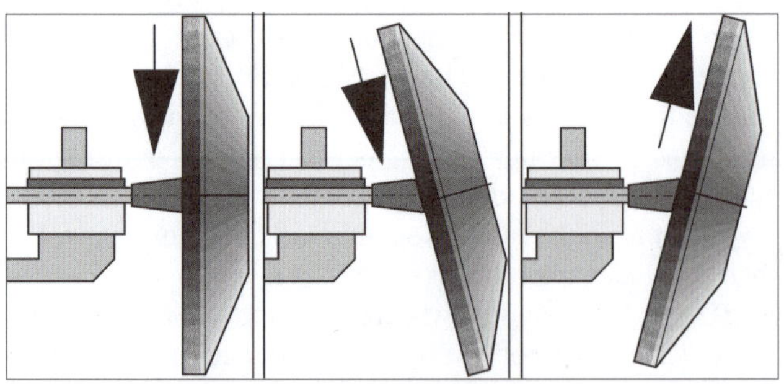

Fig. 10.50
The wheel should grind towards the point with alternate top-bevelled teeth.

Carefully centralize the sawblade, then swivel the grinding head to the correct clearance angle, and tilt the grinding head *(or saw support)* to the required top bevel angle for the first hand of teeth to be ground.

Set the grinding wheel to grind toward the tooth point - otherwise it will scream or start to vibrate the sawblade and destroy the quality of grind.

Set the pawl to index two teeth - or index two teeth between grinds. Grind all teeth of one hand, generally as described for flat top grinding by using the pawl and grinding wheel hand-levers alternately. On completion, reverse both the top bevel angle setting and the direction of rotation of the grinding wheel - then grind the remaining teeth as above.

This is the simplest and most convenient way of switching from one hand of teeth to the other - by making no cut control adjustment between the two top bevels.

However, unless the sawblade is accurately centralized in the first place, then practice this will almost certainly result in alternate teeth being unequal in height - a woodworking version of Sod's Law (UK) or Murphy's Law (US). If the means of centralizing the sawblade is unreliable then grind to a witness mark:-

Grinding to a witness mark

Set the grinding wheel to zero clearance angle and to zero tilt angle, i.e., square to the saw body. *(Make sure that the grinding wheel is absolutely square to the saw body - use an engineers square to check this if the tilt scale is unreliable.)*

Grind six or more following teeth to show a tiny witness mark on each. Use the pawl to index the teeth, and grind them as for flat-topped teeth.

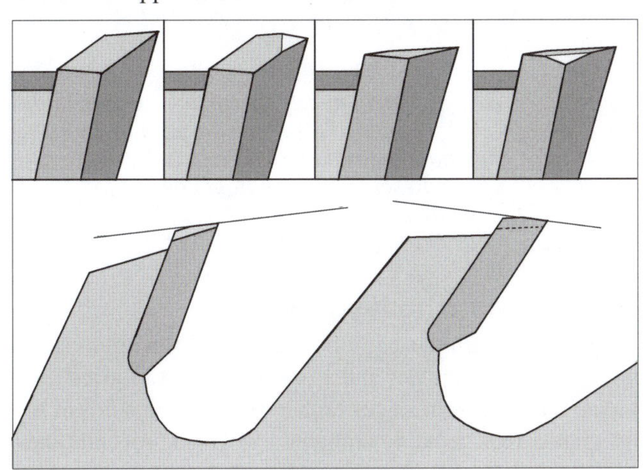

Fig. 10.51
These sketches show the effect of grinding a witness mark. The bottom sketch shows typical grinding paths for the witness marks, and those above show before and after witness grinding.

First grind any tip having a witness mark in order to set the correct weight of grind. Adjust the cut-control in small steps until the witness mark barely disappears. *(Possibly grind other teeth with a witness mark as a second check.)* Grind the remaining teeth of this hand at this setting.

Repeat the process for the opposite-hand teeth - after first switching the top bevel angle and grinding wheel rotation. This process ensures that alternate top-bevel teeth are all ground to the same height.

The amount ground-off as a witness mark, of course, predetermines the amount of carbide removed in grinding the top bevel angles. A little practice will quickly show what is needed according to the condition of each sawblade.

If the cutting edges have not been fully restored after grinding the tooth tops in this way, then re-grind using the grind-to-zero technique to keep both sets of teeth equal in height (see later notes).

Some operators grind to a witness mark only every third or fourth sharpening - this is an easier option - for the remainder they rely on the grind-to-zero technique to keep teeth in step.

Grinding triple-chip teeth.

Triple-chip teeth are ground at four settings:- Set first for the flat-topped teeth and grind these fully, noting the setting of the cut-control scale. Reset for the flat top of the triple-chip teeth, backing-off the cut-control by the tooth height difference recommended by the saw makers, and grind all these teeth. In both cases the grinding wheel should rotate towards the backplate.

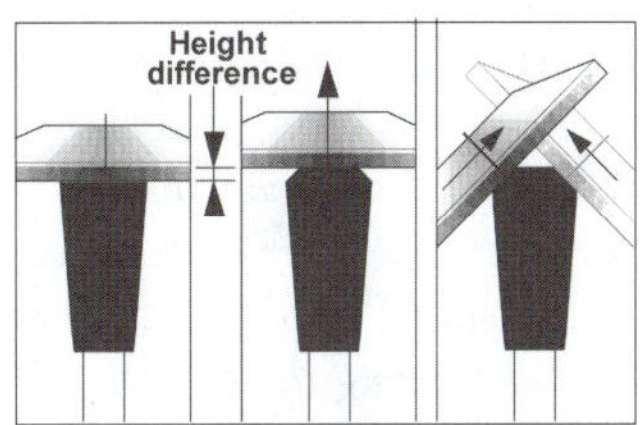

Fig. 10.52
The four settings for triple-chip teeth are the flat top tooth (left) the flat top of the triple-chip tooth adjusted for height difference (centre) and the two settings to grind the side bevels equal to one third of the kerf width (right). Note the direction of grinding wheel rotation for the latter.

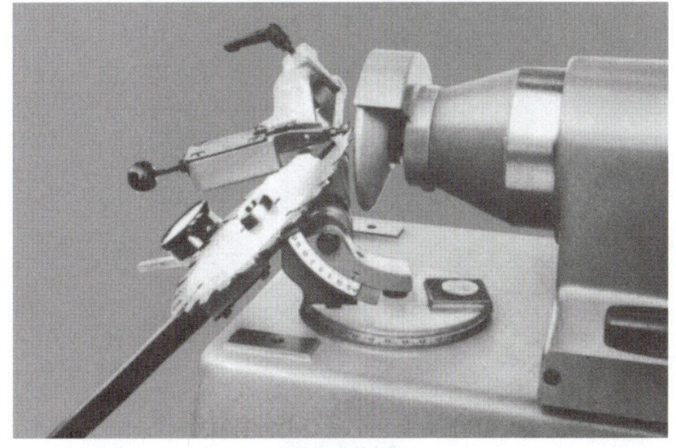

Fig. 10.53
In this illustration the HKSC 600 has been set for grinding the lower bevel of a triple-chip tooth.
Source - Saturn

Check the widths of the two bevels relative to the flat top to determine the weight of grind needed for each *(they all should finish equal)*. Tilt the head or the saw support to 45 degrees for the first of the two bevels, adjust the cut-control to grind this to the correct width, then grind the remaining teeth.

The second bevel can be ground after simply resetting both the tilt angle and the grinding wheel rotation - **provided the saw is accurately centralized**.

If the saw is not accurately centralized, then adjust the cut-control as needed to grind the second bevel to the same width as the previously-ground bevel.

When grinding the bevels on triple-chip teeth the grinding wheel must grind towards the centre flat, otherwise the grinding wheel will scream and tend to vibrate.

Grinding-to-Zero

The grind-to-zero technique is a practical way of re-grinding sawblades known to have teeth already of the correct relative height to one another. It ensures removal the same amount of carbide from each - and so to keeps them precisely in step*. It is also used when a heavy grind is needed, or when grinding replaced teeth to the same height or width as existing teeth.

** This method keeps teeth absolutely in step when ground to a flat top or to an alternate top bevel. With triple chip bevels, however, ignore the cut-control setting, instead adjust the grind to form the correct width of bevel.*

With inverted vee bevels it is not practical to simply grind the same amount from these as for the flat teeth - the two will eventually get out of step.

To avoid this happening reduce the weight of grind for these bevels (relative to the weight of grind for the flat teeth) according to the bevel angle, viz. - to 70 % for 45 degrees, to 76% for 40 degrees, to 81% for 35 degrees, to 86% for 30 degrees and to 90% for 25 degrees. In all cases periodically check that teeth remain the correct relative heights to one another by partially cutting into a piece of dense hardwood or mdf and closely examining the tooth-cut profile.

Accurate cut-control adjustment is essential. Some machines have a fixed cut-control scale, others allow the barrel scale to be re-set to read zero at any grinding wheel setting - and for this reason are easier to use. The best type has a click-in feature at zero setting which prevents adjustment beyond zero until further rotary pressure is applied.

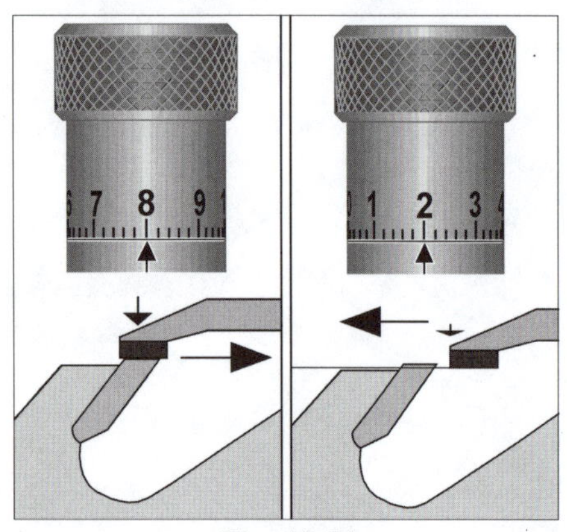

Fig. 10.54
This shows the method of using a fixed-barrel scale to grind precisely the same amount from teeth with different forms in the same sawblade. The technique can only be used on sawblades with teeth that are already the correct relative height to one another.

Note the scale settings when the grinding wheel is in light and then full grinding contact with the first tooth (e.g. 8 and 2,). Record the difference as the grinding cut (e.g.3, assuming the cut increases when rotating to the left) then grind all the teeth of this series.
With the other teeth, note the scale setting when in light contact with the first tooth, increase this by the recorded grinding cut (e.g. 3), then grind remaining teeth.

Machines with fixed-scales. Adjust the grinding wheel to lightly touch the first tooth top to be ground and note the scale setting *(choose the most worn or damaged tooth)*. Further adjust the cut-control and carefully grind to barely restore the cutting edge of this tip, then record the difference between the two scale settings. Grind all corresponding teeth at this setting.

For the other teeth again adjust the wheel to touch lightly, then increase the cut-control by the difference noted from the first grind *(or by the reduced amount previously noted for triple-chip and inverted vee teeth)* and grind all these teeth.

Machines with zero-setting cut-control. Adjust the wheel to lightly contact the first tooth top to be ground and re-set the scale to zero. Further adjust the cut-control to barely restore the cutting edge of this tip, and record this as a direct cut-depth measurement. Grind all these teeth of this hand.

Repeat for the remaining teeth, generally as described under machines with fixed scales, but in this case zero the scale each time on light contact, then reset to the recorded figure for the full grind.

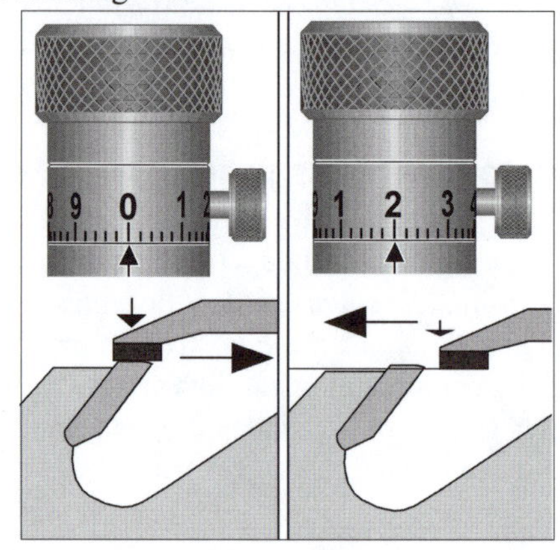

Fig. 10.55
Using a zero-setting barrel makes life easier. For the initial tooth of the first series of teeth reset to zero on light contact, then note the readout when in full grinding contact (e.g. 2). Reset to zero on light contact for the remaining teeth, then apply the same amount of grinding cut (e.g. 2) as a direct readout.

Duplicating a heavy grind.

When a heavy grind is required, for example on lightly-damaged teeth, it is quicker and more convenient to grind individual teeth down in stages (either to zero or to a preset figure) rather than grinding around the sawblade several times. First progressively grind down the most damaged tooth so that the edge is fully restored - Then reset to zero *(or note the scale setting)*. Back-off the cut-control, index the next tooth of this series and grind this down in stages to zero *(or the same scale setting)*. Repeat for the remaining teeth.

Grinding replaced teeth.

For each face to be ground *(front, top and sides)* bring the grinding wheel first into light contact with an existing tooth face of identical pattern and zero the scale *(or note the setting)*. Replaced teeth are then ground down to zero *(or the same scale setting)*.

Correcting setting scales

A wrongly-set swivel *(hook & clearance)* angle scale can

easily be corrected. First index any tooth and then, using a straight edge, swivel the grinding head *(or the saw support unit)* and simultaneously adjust the cut control to align the grinding wheel face, the indexed tooth point and the saw centre. Reset the swivel scale to read zero or 90 degrees - according to how it is graduated.

A wrongly-set tilt *(front or top bevel)* angle scale can also easily be corrected. Swivel the grinding head *(or saw support unit)* as though to grind a tooth top to zero clearance angle. Place an engineers square on any sawblade *(clear of the tips)* and adjust the grinding wheel *(or saw support unit)* until this is flush with the arm of the square. Zero the scale, or set to 90 degrees - according to how it is graduated. The best type of engineers square to use is the sliding type set as a tee to allow the body to rest on the sawblade and with the arm overhanging both above and below.

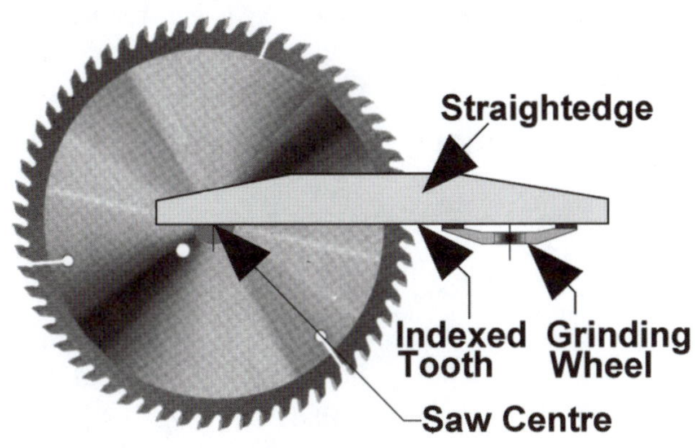

Fig. 10.56
One method of correcting the swivel (hook & clearance angle) scale. Use a straightedge to line up the saw centre, an indexed tooth point, and the grinding wheel face - then set the swivel angle scale to zero.

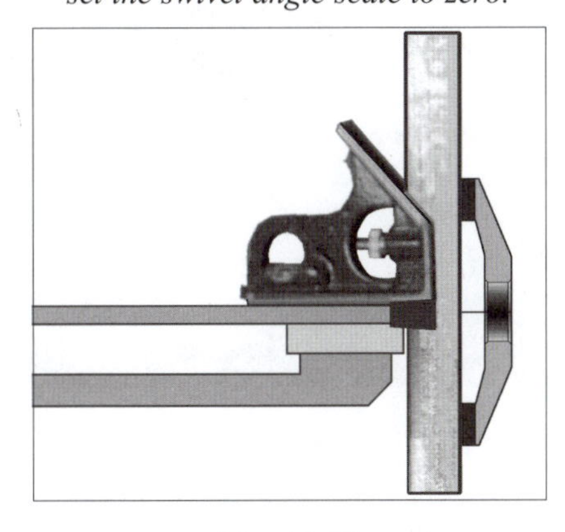

Fig. 10.57
To correct the tilt (bevel) angle scale place a sliding-arm engineers square on the sawblade clear of the tip and, with the grinding wheel set to zero clearance angle, tilt the wheel to align with the sliding arm, then set the tilt angle scale to zero.

Back relieving

New saws allow only a small projection of the carbide tip beyond the steel backing, so at some stage the tips become level with the backing. When at this stage the backing should be ground down using one of the methods described below:-

Note: Diamond grinding wheels should not be used to grind both the tip and the steel backing - steel tends both to fill the diamond wheel and wear it excessively. Whilst a combination abrasive of diamond and cbn is sometimes used to grind both tungsten carbide and steel at one pass on some other woodworking tools, this is at best a compromise solution. It is, in any event, an expensive and unsatisfactory method when compared to the back-relieving methods normally used on tct sawblades.

Conventional relief-grinding

The simplest way to relief-grind is to replace the diamond wheel with either a regular grit *(aluminium oxide)* or cbn *(cubic boron nitride)* grinding wheel, and grind **the backing only** at an angle of at least 5 degrees more than the regular tip clearance angle. Use the regular pawl and grinding wheel movements to give accurate and repeatable relief-grinding.

Grind as a series of small steps until the backing is ground back to the trailing point of the tip, then back-off the cut-control and repeat for all the remaining teeth. Use the grind-to-zero technique to keep the saw balance true.

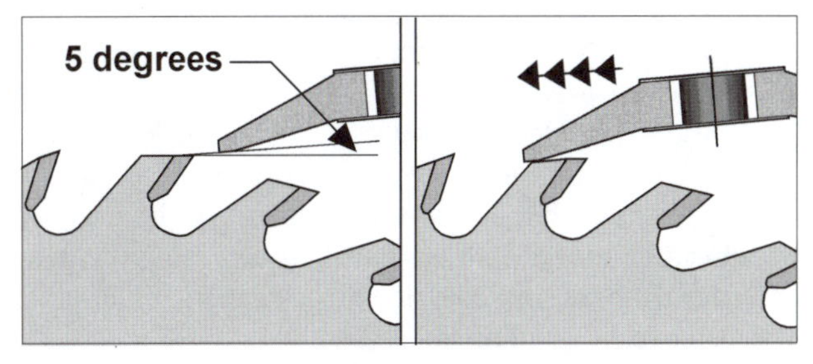

Fig. 10.58
The conventional way of grinding back relief on manual grinders. Use a regular grit or cbn grinding wheel set to the clearance angle plus 5 degrees. Index a tooth, then adjust the grinding cut so that the wheel barely clears the tungsten tip.

When the saw tips are next ground with the diamond wheel a little steel backing will also be removed, but the amount will be small enough to be ignored - at first. Eventually the amount of backing ground will slowly increase - so back-relieve in this way at regular intervals.

For the most effective back-relief for alternate-bevel teeth, grind these when set to the clearance angle (plus 5 degrees) and to the same the top bevel angle. For triple-chip teeth individually-set back-relief grinds should be made for all three faces. The direction of rotation of the grinding wheel is as for regular tooth top grinding.

Plunge relief grinding

This method - plunge-grinding using a wide edge-grinding wheel - grinds the back down well clear of the tip so the diamond grinding wheel never grinds steel.

Plunge-grinding is always square-on to the saw plate, with the edge-grinding face of the wheel set to the regular tip clearance angle and with the wheel set to rotate towards the backplate

With this method individual top-bevel relief grinds are neither needed nor practical.

Use the pawl to index the tooth, then plunge-grind just behind the tip. Use the grinding head control lever in a slow and careful grinding movement to avoid vibration.

This method is **only suitable** where the backing is relatively narrow, i.e., with short-pitch teeth. Regularly dress grit-grinding wheels square across to avoid uneven wear.

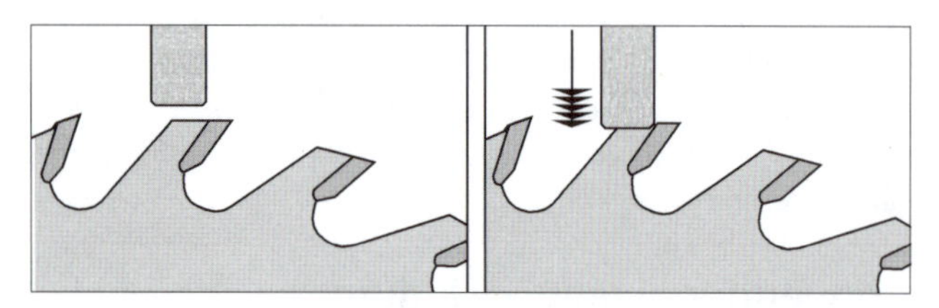

Fig. 10.59
Plunge relief grinding. Index the tooth then slowly and carefully plunge-in barely clear of the trailing tip point. Grind the back down by 0.2mm. (for saw kerfs less than 3.2mm) or 0.5mm (for saw kerfs more than 3.2mm).

Traverse relief grinding

For wider tooth backs the grinding wheel should preferably traverse-grind from the trailing point of the tooth top to just short of the tct tip. Few manual machines incorporate a suitable movement, but it may be possible to use the pawl movement instead. Use a cbn edge-grinding wheel only - *traverse movements apply an unacceptable side pressure to regular-grit grinding wheels.*

Set a wide edge-grinding wheel to the tooth clearance angle and the depth stop for a suitable weight of grind. Use the pawl to initially position the tooth so that the grinding wheel can be moved into the gullet space. Using the pawl movement, traverse the tooth slowly backwards to relief-grind it, then move the grinding wheel clear.

Possibly use a turnover block against the pawl-traverse stop to give a precise starting position, then flip it clear for the back-relief grind proper.

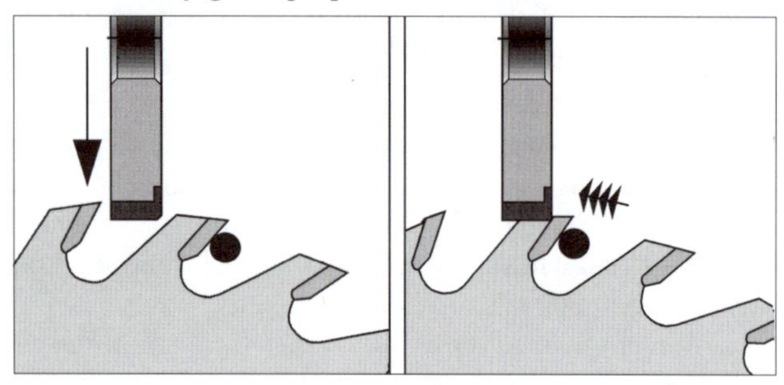

Fig. 10.60
Traverse relief grinding. Use a plain cbn grinding wheel with its edge-grinding face set to the tooth top clearance angle. Move it into the gullet gap, then use the pawl to traverse the tooth so that the grinding wheel stops just short of the trailing point of the tooth tip.

This method of grinding has a more controllable action than a straightforward plunge grind, but a slow and careful movement is still needed. The grinding wheel should be the same width as, or slightly wider than the tooth backing. If the tooth profile has a wide tooth back and a narrow gullet, initial setting has to be very precise. If the grinding wheel is wider than the gap, then initially position it to be clear of the following tooth to part plunge-grind before traversing back.

Maintaining chip-limiters

Saws used for hand feed normally have chip limiters *(anti kickback shoulders)* for safe operation. As the carbide tips wear down through re-grinding it is essential to grind down the chip limiters to retain the original height difference.

Whilst it isn't normally necessary to grind the chip - limiters each time the saw is top ground, it should nevertheless be a regular operation.

Use either a regular or cbn saucer abrasive grinding wheel and set this to zero clearance angle*. Use the pawl to index the preceding tooth so that the chip limiter is centrally placed, i.e., tangential to the line of grind.

Grind the chip-limiter down to restore the recommended height difference between the chip limiter and the tips, using the regular movements, with a dial indicator to check the settings. If the amount to be ground-off needs more than one pass, then use the grind-to-zero method to ensure all chip-limiters are the same height.

**Although this is the correct setting, it may not be practical to use a zero clearance angle as the grinding wheel may then foul the preceding tip. In this case reduce the angle, or instead use an edge-grinding wheel in a traversing movement - general as described under plunge and traverse relief grinding, but with the grinding wheel set square-on.*

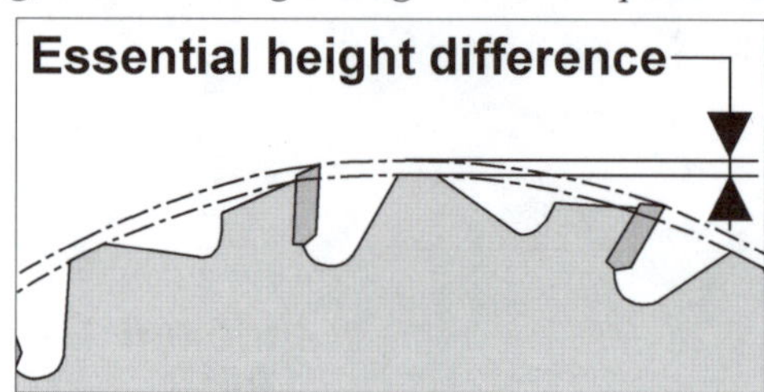

Fig. 10.61
As the tips wear down, the chip limiters must also be ground down to maintain the essential height difference. The sawblade maker will advise on the correct height difference.

Fig. 10.62
This shows a chip-limiter being ground down on an HSN 600 automatic tct grinder. Note the dial indicator used to check the chip-limiter height relative to the tips.
Source Stehle

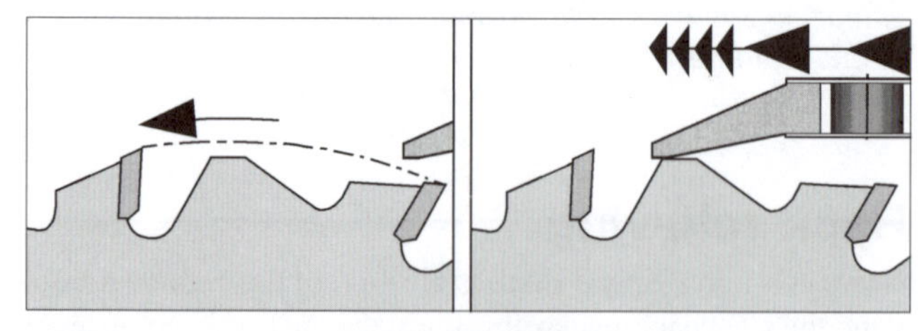

Fig. 10.63
When setting-up to grind chip limiters (left) make sure that the grinding wheel is far enough back to clear the teeth when indexing. Index to bring the chip limiter tangential to the line of grind. Grind carefully using the regular grinding wheel movement (right).

Automatic grinders

The term automatic saw grinder can sometimes be misleading as it covers a wide range of diverse machines - from simple machines intended to grind only one face at a time and which are little more than mechanized versions of manual machines, to complex, highly efficient and upmarket plc and cnc machines which can fully grind faces and tops at a single setup and handling.

In between are older automatic grinding machines which are not as speedy nor as controllable as their modern counterparts, but which operate with simple and dependable mechanisms.

Fig. 10.64
This HSN 600 automatic grinder, shown here face grinding, is a simple and economic mechanically-controlled machine for grinding one face at each setup.
Source - Stehle

The simplest automatics are set and operated generally as described for manual machines, and usually are driven mechanically.

They are capable of grinding only one face at a time at each setup, and the grinding traverse speed and cycle time *(the time to complete a single tooth)* is usually fixed - regardless of the actual grinding process. They need repeated setting and grinding operations to complete such as triple-chip saws, but they are simple to set and operate, their cost is low and they are easy to maintain in-house.

Virtually all automatics - even the simplest - have a tooth counter which stops the machine once a preset tooth count is reached.

More upmarket machines may be capable of grinding top bevels alternately, or the two bevels of a triple-chip tooth at one setup. Often they are hydraulic or air-powered and may be controlled by conventional electrical switches and/or electrical timers, or via a series of cams. They allow more control of the operating and cycle speeds, but still may need multiple passes for complex tooth patterns.

The most advanced plc and cnc automatic saw grinders, often called fully automatic grinders, have much more sophisticated operating systems which allow complete grinding of even the most complex teeth at a single setup.

Much of the sophistication goes in making these machines user-friendly. Advanced setting features make setup and operation far simpler than in their earlier counterparts - a combination that allows rapid machine setting and speedy turn-round of sawblades.

They may be driven wholly or partly via a self-contained hydraulic system, with the grinding program controlled by a plc unit, or by a built-in computer driving one or more numerically-controlled *(nc)* axes.

The ideal situation, of course, is to face and top grind tct teeth each time they are re-sharpened, and for this a huge choice of grinders are available.

Automatic Grinder Types

In rising order of sophistication and cost, fully automatic grinders are made for face grinding only (to be used in conjunction with a partner machine), for top grinding only, for top or face grinding at separate setting and operations, and for combined face and top grinding at a single setup.

There are excellent reasons for offering so many different machine types:- If financial constrictions limits purchase to a single grinder, then a machine which will face or top grind is the obvious choice. Where the capacity for two machines exists, then one face and one top grinder would be ideal - and would cost less than two face and top grinders.

Fig. 10.65
The fully automatic CHP tct grinder has a microprocessor and nc-controlled axes, allowing angle adjustments to be made from the control desk. All regular tooth configurations can be ground, including group and newly-tipped teeth, plus a facility to skip damaged teeth.
Source - Vollmer

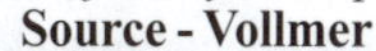

Fig. 10.66
The Unimat S, shown using a cup wheel to top grind, is a fully automatic grinding machine capable of face or top grinding of all standard tooth shapes. Classed as an entry level machine it has simple and convenient setting and operating controls.
Source - Vollmer

Where more than two grinders are warranted, then the addition of a face and top grinder would give more flexibility - unless a quick turn-round of some sawblades is a high priority - in which case a combined face and top grinder is better.

Machine manufacturers would be happy to detail the economics of all the machines they manufacture in relation to individual requirements - and in much more detail than the brief description above.

Ultimately, though, it remains for the user to decide which is best for him - he pays the money and makes the final choice. But, before finally deciding on what is a major investment, it is advisable to look closely at what the world market now offers - development has been rapid and still continues apace.

Fig. 10.67
The working partner to the Akemat B face grinder is the Akemat U4, an automatic top grinding machine for tct teeth - see Fig. 10.7. It uses an edge-grinding wheel to give exceptionally sharp cutting edges and economic carbide removal. It combines hydraulic and NC movements.
Source - Vollmer Dornhan

Modern Automatic Grinders

It will be appreciated that a saw grinder is working usefully only when it is actually grinding. All other operations - fitting and setting the sawblade, adjusting the machine for angles and bevels, indexing the sawblade, advancing the grinding wheel up to the tip and retracting it after grinding - are all nonproductive operations which must take a minimum amount of time for the machine to operate efficiently.

Modern grinders excel in this respect, with facilities provided for rapid sawblade loading and removal, and with precise angle and traverse setting scales. In some cases digital readouts or keyboard input provide virtually instant and precise settings - for sawblade centering, hook, clearance and bevel angles, cut-depth and speed of the grinding cut, the speed and traverse distance of nonproductive movements of the pawl and the grinding wheel, etc., etc.

The actual grind itself must still be carried relatively slowly and carefully to give the best possible results - but even here modern grinding wheel technology can speed processing. Modern abrasive wheels allow heavier and faster grinding than was previously possible - and modern machines are specifically designed and built to take full advantage of this huge gain.

Although modern automatics appear very complicated, setting them *(up to plc level)* is still basically as described for the simpler automatics.

Presetting checks for all automatics

Certain precautions need taking before starting-up an automatic grinder. Before fitting a sawblade check that all the carbide tips are complete and undamaged - a broken tip can cause a crash and write-off a grinding wheel in an instant*.

If a broken tip cannot be replaced, then use a secondary pawl to index any part of the preceding tooth that is undamaged. Take particular care that the secondary pawl does not override the main pawl on sound teeth - this can be a problem if the pitch of the saw varies.

**Some modern machines can be pre-programmed to skip damaged or missing teeth - easily the best way to avoid a crash. Alternatively, some face-grinding machines have a fail-safe system which prevents crashes for whatever reason - if an unusually heavy grind is encountered the grinding wheel withdraws instantly.*

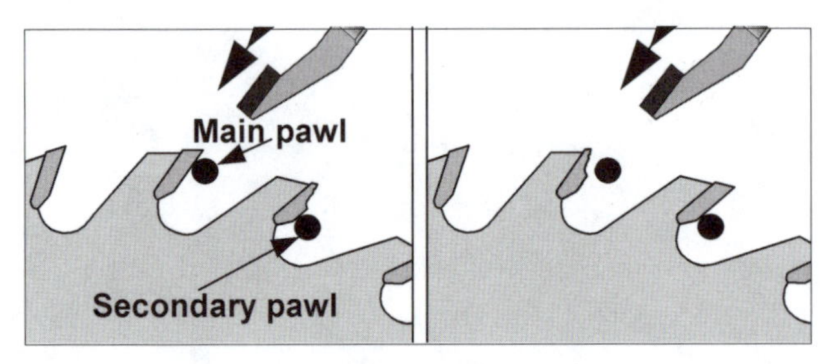

Fig. 10.68
Use a secondary pawl if some teeth are damaged. Adjust this to contact lower down the tip to avoid such damage, but ensure it does not override the main pawl. The two sketches show how a secondary pawl safely feeds a damaged tooth.

Setting automatic grinders

Automatic grinders, such as plc types, have a greatly improved speed and choice of operating cycle when compared to mechanically-controlled automatics. Different operating cycles can be chosen simply by re-setting one or more switches on a control panel.

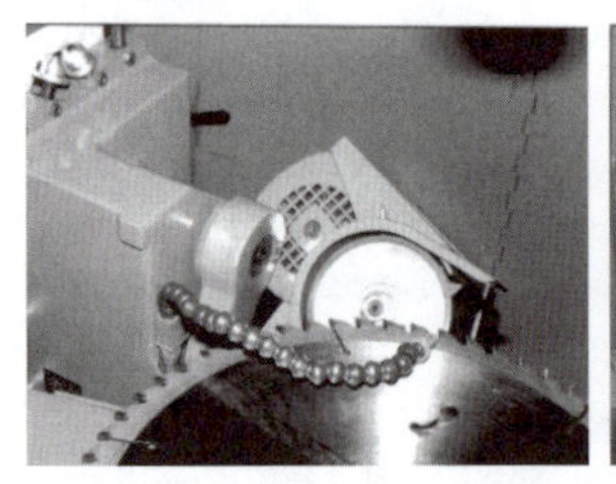

Fig. 10.69
The TCT 25 fully automatic grinder is a plc-controlled machine for either face or top grinding of all standard teeth - shown here grinding the two bevels of triple-chip teeth.
Source - Autool

Positioning and centralizing the sawblade, setting for the various angles and bevel widths of triple-chip teeth, adjusting the weight of grind and varying the grinding wheel traverse and return speeds, etc., are all manual adjustments made on the machine - which require resetting for each individual sawblade.

In most cases it is usual to tweak such adjustments once grinding starts and the first ground teeth can be examined. The process, a short trial and error run prior to full start-up, is no different to how grinders have traditionally been operated - home territory for experienced filers and grinders. Setting is made simpler, though, due to the more exacting control and the ability to step through the grinding program.

Centralizing the sawblade

With all fully automatic grinders it is essential to centralize the sawblade precisely to ensure that top bevel teeth are ground to the same height, and that the bevels on triple-chip teeth are ground equally.

Centralizing the sawblade means aligning the centreline of the saw body with a datum point, either the centreline of the tilting head - as used with saucer wheels, or the centre zero of the offset scale - sometimes used with edge-grinding wheels.

The adjustment can be a movement of the whole saw carriage, or merely a fine adjustment of the saw support backplate only. Centralizing is possible by using some form of scale giving a direct readout of the distance from backplate to the datum point - setting in this case to half the saw body thickness. An alternative is to show this as a scale or digital readout of the actual body thickness itself.

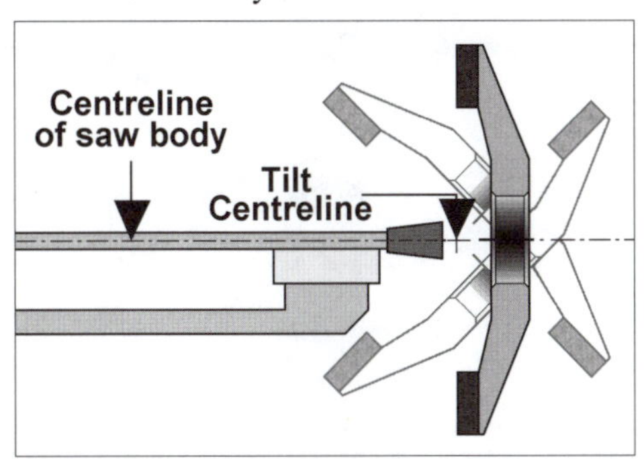

Fig. 10.70
The saw body must be centralized to the tilt line of the grinding head when using saucer wheels for top grinding.

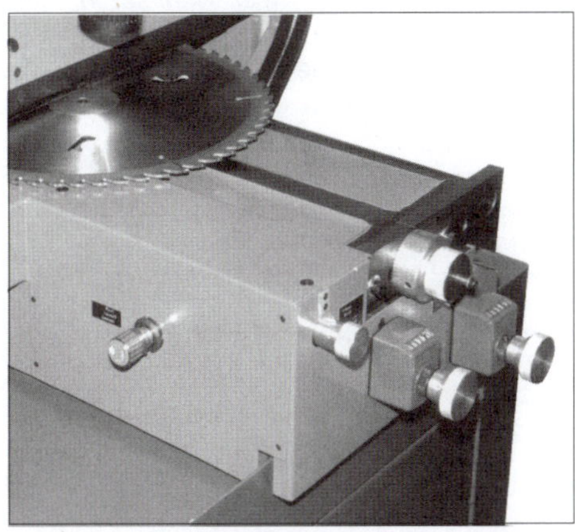

Fig. 10.71
The TCT/25 automatic grinder has a direct digital readout for sawblade centralization which is set to the saw body thickness, A second digital readout shown pre-sets the height difference between flat and triple-chip teeth.
Source - Autool

Centralization of the sawblade in this way is based on an assumption that the saw body is also central to the tips themselves. *If side projection of the tips is unequal, then alternate-bevel teeth and triple-chip bevels will still be ground unequally - however accurately centralization may have been carried out.*

Face grinding

The setting process for face grinding is little different to that for manual machines. Teeth for crosscutting saws can be ground to an alternate front bevel angle without problem on most machines - but ensure that the sawblade is centralized to equalize the weight of grind on both left and right-hand teeth.

Top grinding

This is also little different to that on manual machines for flat top and alternate top bevel teeth, but triple chip teeth need a more complex setup as the three grinds are made at following passes. In addition, tooth tops can be ground either with a saucer grinding wheel or an edge-grinding wheel - which the machine is designed to use determines the method of setting.

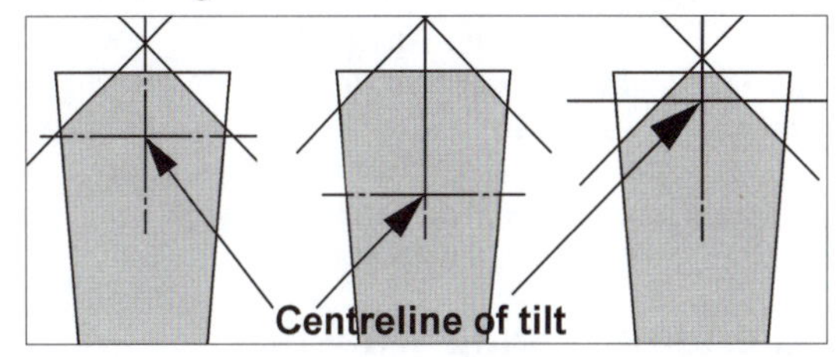

Fig. 10.72
A saucer grinding wheel is set square to the saw when grinding the flat top of triple-chip teeth, then angled alternately to 45 degrees to grind the two side bevels. By varying the distance of the grinding wheel face from its tilt axis the centre flat width can also be varied (centre and right). The correct setting (left) is when the two bevels are equal in width to one another and to the centre flat - at roughly one third of the saw kerf.

Using saucer grinding wheels

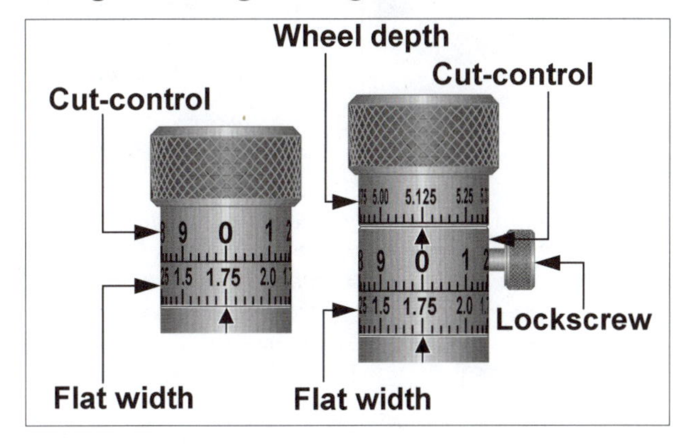

Fig. 10.73
The left-hand barrel scale shows both the cut-control setting and, with some calculation, the triple-chip flat width. The right-hand barrel scale does the same, but can be reset for the current wheel depth to then give a direct readout for triple-chip flat width. The same adjustment allows the cut-control scale to be used for resetting to zero.

A saucer grinding wheel tilts on its axis to form the top bevel angles, and adjusts axially via the regular cut-control scale to vary the relative widths of the flat and bevel faces on triple-chip teeth - making setting something of a trial and error procedure.

Some machines have a grinding wheel cut-control scale which also gives a direct readout for the flat top width. Although easier to use, this type needs correcting as the grinding wheel wears - by measuring this before fitting, then adjusting the scale by the amount of wheel wear from new.

Fig. 10.74
The cut control adjustment on this TCT/25 grinder has secondary scales to give a direct readout of the flat top width of triple-chip teeth. This includes one scale to register the amount of grinding wheel wear from new.
Source - Autool

Using edge-grinding wheels

Edge-grinding wheels offset to either side of the sawblade to form top bevel angles. On some machines the sawblade is centralized to the centre-zero point of the offset scale. Equal offsets then produce identical bevel angles of the correct width.

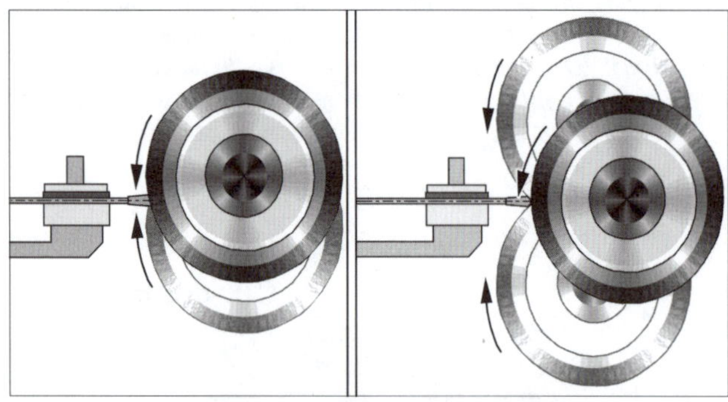

Fig. 10.75
Edge-grinding wheels are shifted to either side of the sawblade centreline when grinding top bevel angles. Arrows show the correct direction of rotation for the grinding wheel.

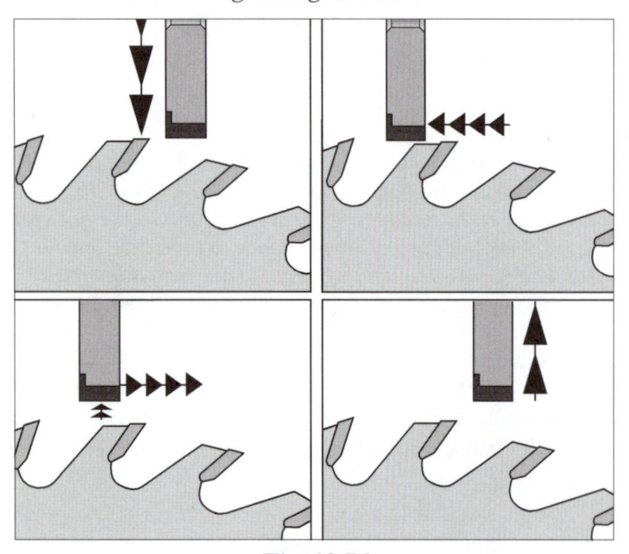

Fig. 10.76
This sketch shows the action of an edge-grinding wheel when grinding tooth tops. It advances rapidly into the grind position (top left) to grind slowly across the tooth top (right top). It then lifts clear (bottom left) and finally returns quickly to the rest position (bottom right).

In grinding alternate bevel angles the grinding wheel automatically offsets from the saw centreline whilst advancing radially. Bevels ground to the same offset but opposite advance settings produce equal height alternate-bevel teeth.

With triple-chip saw teeth the flat top is ground with the grinding wheel central to the sawblade. Bevels are ground by offsetting the grinding wheel and advancing further.

With edge-grinding wheels the bevel angle is determined both by the diameter of the grinding wheel and the amount of offset and advance. This can make setting a little difficult. If in doubt check against the following graphs.

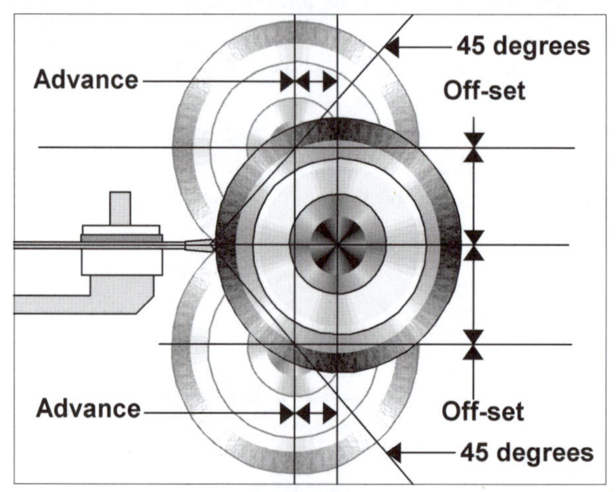

Fig. 10.77
The shows an edge-grinding wheel in position for grinding both the flat of a triple-chip tooth, and the two bevels (shown fainter).
A line from the tooth centre and at the required bevel angle (at 45° in this case), must pass through the centreline of the grinding wheel when set for bevelling. This position is determined by the offset distance (at right angles to the sawblade), and by the distance advanced beyond the set position when grinding the tooth top.

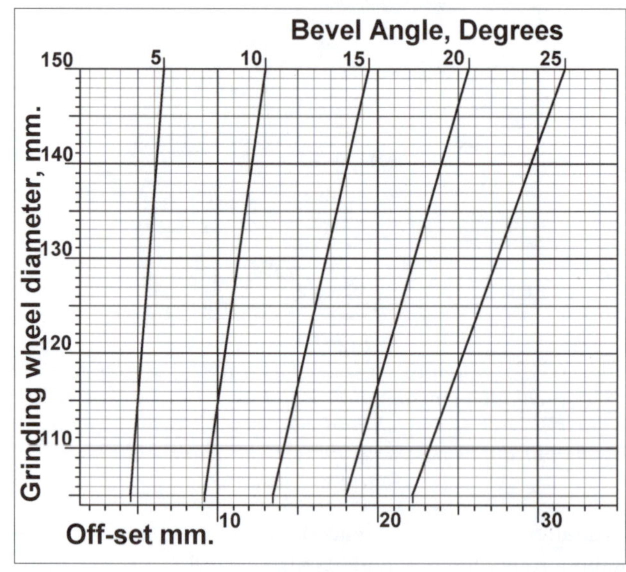

Fig. 10.78
Graph for calculating the amount of offset required. Trace vertically down from where the required bevel angle and grinding wheel diameter lines intersect to show on the bottom scale the amount of offset required from the

sawblade centreline.
This and following graphs are graduated in millimetres only.

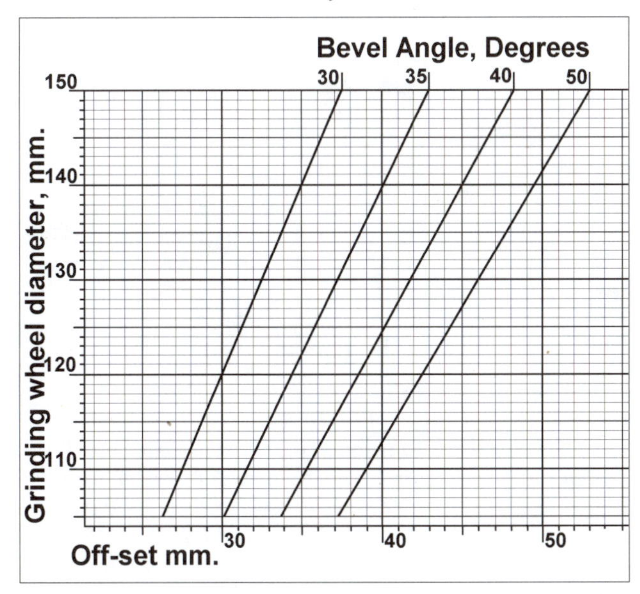

Fig. 10.79
As the graph above, but for bevel angles of 30-45 degrees

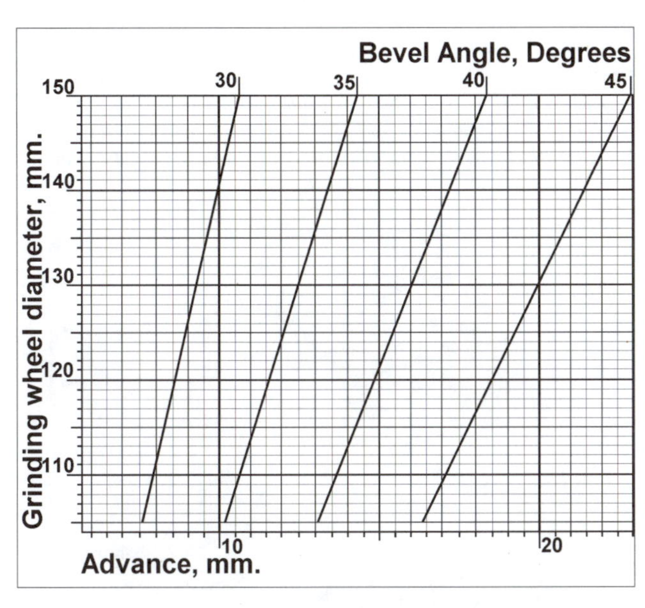

Fig. 10.80
Graph for calculating the extra advance setting needed to form the bevels of triple-chip teeth, i.e., the extra distance measured from the grinding wheel centreline when set for the flat top of a triple-chip tooth (based on equal widths of the flat and bevelled faces).

Starting up automatic grinders

The regular practice is to initially set the machine so that the grinding wheel lightly contacts the first face being ground.

The full operating sequence can then be stepped through to check that both the machine setting and the operating sequence are correct. Then set the tooth counter to the precise number of teeth in the saw and start the grinder.

Once the machine is up and running, increase the cut-control to give an acceptable weight of grind before tripping the tooth counter i.e., resetting it to the original tooth number count. This ensures that all teeth are fully ground - including those ground during the initial setup period.

Face grinding on all automatic machines will continue if the machine is not stopped when the first fully-ground tooth is again reached.

On machines with no tooth counter there is no warning when all teeth have been face ground - although the change in tempo to a more consistent grind may be noticeable. To avoid unwanted face grinding, mark the starting tooth and check the machine regularly.

There is no problem when grinding tooth tops as no further grind will take place once the first tooth is again reached, and the change in tempo will usually warn the operator when grinding is complete and so avoid continued idle machine operation.

CNC & NC Grinders

CNC (computer numerically-controlled) and NC (numerically-controlled) grinders are in quite another ballpark - far more advanced than even the most highly-developed plc types. Although the design and manufacture of cnc grinders is high-tech and complex, their setting and operation is simplicity itself - and they offer sophisticated and advanced facilities.

The reason for the high degree of sophistication is to make operation fast, precise and repeatable and, most importantly, to make these grinders simple to set and use by operators familiar only with manually-set machines.

CNC grinders have between one and four numerically-controlled *(nc)* primary axes* for setting or operation which, under the direction of the on-board computer, precisely control the main movements. The more numerical-controlled axes a machine has and the more versatile it is.

**The most versatile tct saw grinders are those with nc or similar control of at least four of the primary setting/operation axes. Some machines are part nc-controlled, with one of the other movements controlled hydraulically or via some other type of precise automatic control - but they all operate generally as described below. In addition, some machines may have more than four nc-axes to give further sophistication in machine setting or operation.*

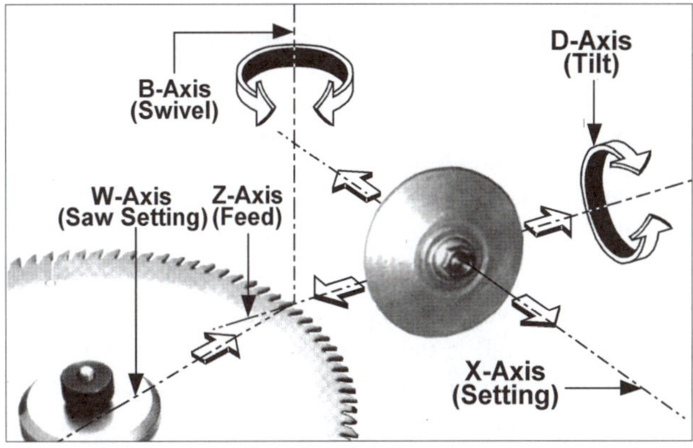

Fig. 10.81
This shows the arrangement of a four nc-axis machine when face grinding with a saucer grinding wheel. This machine is capable of grinding both square and angled face bevels at a single setup. The B-Axis sets the hook angle, the D-Axis the face bevel angle, the X-Axis the grinding cut and the Z-Axis the grinding feed. The additional W-Axis precisely positions the sawblade.

There are two main types of numerically-controlled axes, linear (straight-line) and rotary. Both are directly under the control of the computer and, on computer or manual command, move the linked machine part between any two points within the traverse movement - in either direction and possibly at any speed.

The speed of traverse and the start and stop positions are preset, and may alter between programs, or even within a single program at different points in the sequence..

The general terminology for the primary axes is D, Y, B, X and Z (the latter usually pronounced as Zee - US), possibly with additional axes for shifting and positioning the sawblade on transfer equipment.

The two main linear axes for the grinding head are the X-Axis and the Z-Axis - both of which traverse parallel to the sawblade. These primary movements often switch functions in different sections of the program - this is the unique advantage of nc axes - versatility and absolute control of distance and speed.

Where an nc movement shifts weighty machine parts (such as the grinding head unit itself) it usually starts up by traversing slowly, builds up to an optimum speed, then slows down when approaching the setting required - to prevent inertia overrunning the stop point.

The X-Axis moves the grinding head in a direction parallel to the grinding wheel arbor. It is used as the cut-control when face and top grinding with a saucer grinding wheel, and as the grinding feed when top grinding with an edge-grinding wheel.

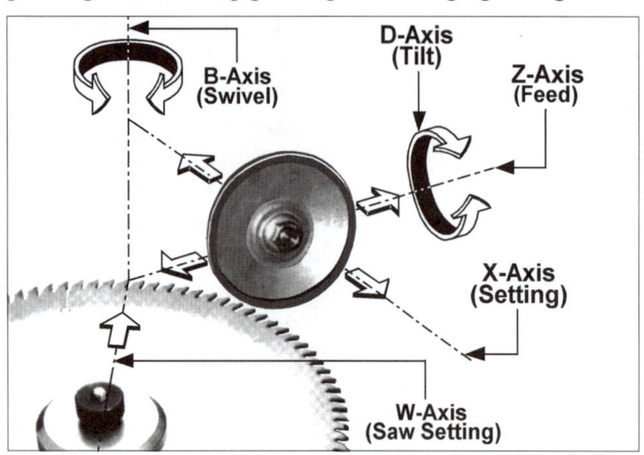

Fig. 10.82
This shows the arrangement of a four nc-axis machine when top grinding with a saucer wheel. The B-Axis sets the clearance angle, the D-Axis the top bevel angle, the X-Axis the grinding cut and the Z-Axis the grinding feed. The W-Axis precisely positions the sawblade.

The Z-Axis moves the grinding head at right angles to the grinding arbor and is the grinding feed when face and top grinding with a saucer grinding wheel, and the cut-control when top grinding with an edge-grinding wheel. Machines with only X and Z-Axes are suitable for straight face grinding, but can alternate-bevel grind at following operations after manually tilting the grinding head.

The D-Axis is the third primary axis. This tilts a saucer grinding wheel automatically, as part of the cnc program, to grind alternate-bevel tops or faces at separate settings and operations.

The corresponding movement to form alternate top bevel angles when using an edge-grinding wheel is Y-Axis which shifts the grinding wheel to either side of the sawblade.

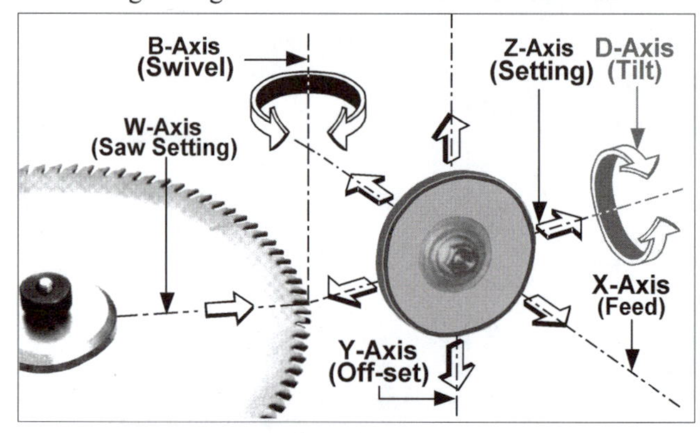

Fig. 10.83
This shows the arrangement of a four nc-axis machine when top grinding with an edge-grinding wheel. The B-Axis sets the clearance angle, the Y-Axis the top bevel angle, the Z-Axis the grinding cut and the X-Axis the grinding feed. The W-Axis precisely positions the sawblade. When this class of machine is switched between facing with a saucer wheel and topping with an edge-grinding wheel, the X and Z axes also switch functions. The D-Axis, shown lightly shaded, is a manual, nc, or cnc movement - an optional extra on machines designed to use an edge-grinding wheel for top grinding. It switches bevel angles when face grinding with a saucer wheel. It could be arranged to operate via a manual switch after face grinding every second tooth, or as a programmed-controlled movement to change automatically between teeth.

Fig. 10.84
The Woodtronic NC3 shown here is a reasonably-priced 3-axis nc grinder for either face or top grinding. The grinding head movements in this series of machines are totally isolated from the grinding section to avoid the chance of wear or damage from water, oil or grit.
Source - Walter

The B-Axis is the fourth primary movement. This automatically swivels the grinding head as a complete unit between the two primary settings - hook and clearance angle.

On some machines this movement is not under the control of the running program, but is used instead as a precise means of setting the hook and clearance angles from the control desk.

Machines with a B-Axis controlled by the computer program are termed combined face and top grinders in this book. The hook and clearance angles on these are set automatically during machine operation, so they are capable of grinding both faces and tops at a single setup and handling.

This operation is described in detail under Combined face and top grinding.

Programming nc grinders

Like plc grinders, nc and cnc grinders have a ready choice of different programs to control the machine operation but, unlike earlier machine types, few manual adjustments need to be made.

Virtually all setting parameters can be entered through a keyboard, either by typing-in the machine or sawblade data requested for new sawblades, or by selecting from data stored in the computer's memory for regularly-ground sawblades - as shown on alternative screen displays.

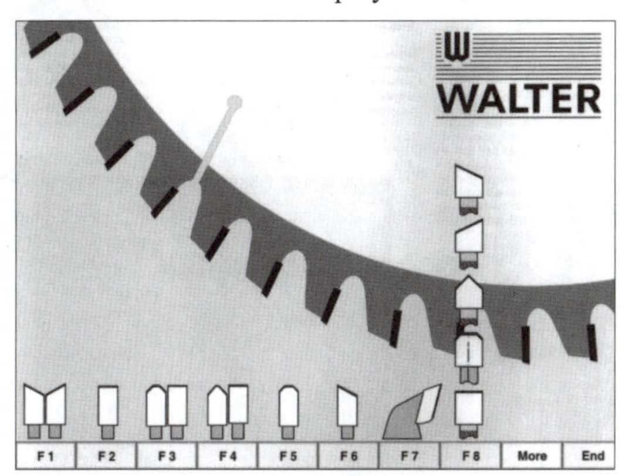

Fig. 10.85

The initial Woodtronic NC 3 screen shows alternative tooth profiles. Any one can be selected simply by activating the control button directly below. The 'More' program is to define a tooth profile not included in the regular programs.

Source - Walter AG

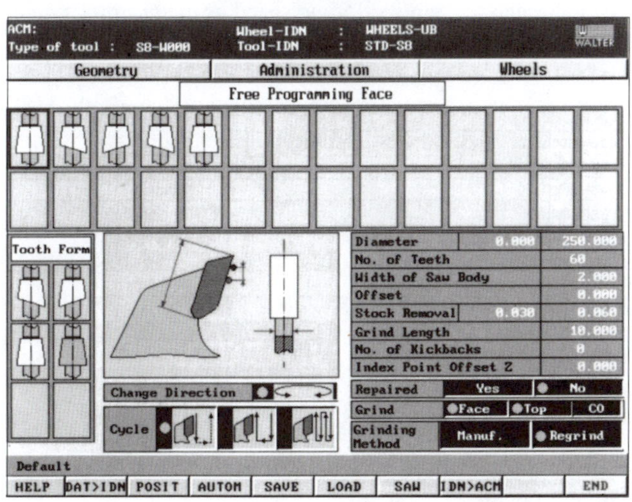

Fig. 10.86

The second and subsequent screens display complete details of the chosen tooth profile. Data for selection or alteration is entered via the integral, control desk keypad.

To make entry simple, the corresponding detail on the diagram for which data is required is clearly indicated as the lines are manually scanned-down. The screen shown is for free programming

Source - Walter

All nc movements are controlled directly by the computer, with manual input to the computer's memory from a keyboard, and digital commands from the computer to the various nc axes. This gives a degree of control unequalled by any other system of machine control.

Increase or decrease in distance and/or speed of all axes movements is possible in tiny, numerically-displayed increments, giving precise control - and absolute repeatable accuracy. Digital control also means that data can be recorded and stored for future use.

All nc machines have a screen to display different tooth profiles and the detailed information on them. To choose any one simply select the required tooth profile - flat-top, alternate bevel, triple-chip, etc. - from a multiple-choice listing on the screen or via individual buttons.

The computer then asks for specific details to be confirmed, usually by displaying another multiple-choice list covering only the tooth type selected.

This list includes such data as the sawblade thickness, weight of grind, grinding traverse speed, length of tip face and top, number of teeth, etc. etc.

The information is displayed in alphanumeric form *(a mixture of letters and numbers)* as a series of lines. By moving down the screen, line by line, any one item can either be accepted or modified.

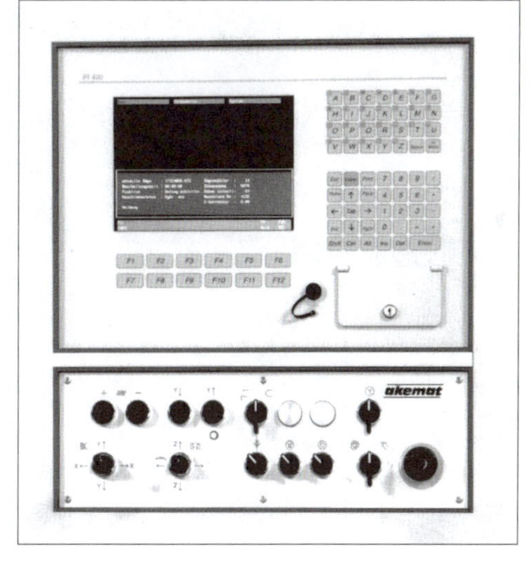

Fig. 10.87

The control panel of the Akemat B is easy to operate having a clearly-structured layout. The data for most common types of circular sawblades is pre-stored.

Source - Vollmer Dornhan

If incorrect data is displayed initially, then alternative data from another multi-choice list can be displayed.

The correct data can be substituted from this alternative list or, alternatively, completely new data can be entered via the keyboard number pad. When the displayed data is correct the information is finally entered into the computer's memory via the keyboard.

Once data has been entered the computer rapidly calculates the working parameters needed to preset the grinder for the chosen tooth profile.

An example of the ease of setting cnc grinders is when top grinding triple-chip teeth with an edge-grinding wheel.

Instead of needing manual calculations and exacting scale settings, the display panel merely asks for such data to be entered as the type of tooth, grinding wheel diameter, sawblade thickness and triple-chip tooth details, etc.

From this data the computer calculates the precise offset and advance settings needed for the grinding wheel to grind the correct angles and face widths of triple-chip teeth. To avoid inaccuracies which would otherwise occur through wear of the backplate or the magnetic support surface, a correction factor can be entered to take even this small error into account.

Fig. 10.88
This compact CX100 4-axis cnc face and top grinder is shown being programmed. Several different tooth shapes can be formed in a single cycle - including those with a differing hook angles - with the added option of different grinding speeds and path values.
Source - Vollmer

Once data is entered it only remains to fit and position the sawblade. This is different to earlier machine types in that the first tooth top has to be indexed and precisely positioned at the zero point of the top grinding cut*. Doing this ensures that the top grinding cut selected is precisely the amount that will be ground-off each tooth top.

**Such that the tooth top is just touched by the grinding wheel when this is advanced to the grinding position when set for zero cut.*

The pawl positions the tooth face at the zero *(or control)* point also for the face grinding cut - so ensuring that the amount ground-off each tooth face is precisely that pre-set.

On the most sophisticated machines the pawl always indexes to the precise 'control' position of the computers basic geometry. When the grinding head is swivelled on the 'B' axis to change the hook or clearance angle, it effectively swivels on the 'control' position. In this way the preset grinding cut is always correct irrespective of the hook or clearance angle setting. Through this basic premise the remaining tooth geometry can likewise be defined numerically - and guaranteed to be absolutely accurate.

Fitting and positioning the sawblade

Fitting and positioning the sawblade is simple enough. After fitting a suitable saw centre, place the sawblade on the grinder arbor, index a suitable tooth, and select the setup program. This moves the grinding wheel or a probe into the grind position at a cut-control setting of zero.

The sawblade is then advanced manually *(or automatically under power)* to bring the indexed tooth into light contact with the grinding wheel *(or the probe)* where it locks in place at the zero cut-control position.

Fig. 10.89
Top grinding, left, and face grinding, right, on a CHC grinder. The machine is controlled by pmc microprocessor accessed through a central control desk. The bevel grinding adjustment on this machine can be manual or automatic, and the grinding wheel can have a cnc-controlled grinding wheel infeed to compensate for wear.
Source - Vollmer

Accurate positioning of the sawblade is necessary not only when top grinding, but also when face grinding. The reason is that the computer controls both the change-over point from a fast approach to a slow grinding speed as the grinding wheel nears the tip - and the appropriate traverse distance down the tip face.

Once the sawblade has been fitted and correctly positioned, the machine can be started-up immediately. NC and CNC grinders are so precisely programmed through digital setting procedures that full grinding can begin immediately from first contact.

Tweaking machine settings, the normal practice following start-up with simpler machine types, is not normally required - but is still possible via the keyboard should it really be necessary.

Program memory

With all cnc grinders, the machine's memory bank stores grinding and operating setups for a whole range of regular tooth styles - and any configuration can be instantly recalled. Any custom style can be readily created from a copy of the nearest regular or pre-customized tooth configuration - and then itself recorded for future use.

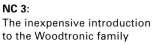

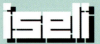

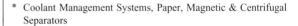

In this way instant recall of precise machine settings is possible both for sawblades ground on a regular basis - and for specials and one-offs. The system guarantees absolute consistency of sawblade grinding specifications - so combining rapid change-over with fast, highly efficient and dependable operation.

Fig. 10.90
Grinding the top bevel of a triple-chip tooth using an edge-grinding wheel on the Akemat U4. Machine setting for triple-chip teeth on this and other nc grinders does not require complicated graphs nor complex calculations. It is merely necessary to enter full details of the tooth profile, saw body thickness and grinding wheel data for the computer to work out the precise setting details. These are then stored in memory - and output automatically as grinding proceeds. Positioning the sawblade is a simple and fast process by using a setup program.
Source Vollmer Dornhan

Maintaining accuracy

CNC grinders are precision made, so care should be taken to keep them aligned and operating correctly. Most makers offer a regular maintenance service for these machines to guarantee their continued highly-efficient performance.

In particular any wear of the backplate, pawl and other critical parts must made good to maintain the integrity and accuracy of these machines.

Depending upon the part worn, this can either be replaced - or the wear taken into account through a correction factor entered via the keyboard.

Often makers provide jigs so the user can realigning machine settings that have become progressively out of step - but realignment must be precise! - and is best carried out by a qualified serviceman.

Grinding wheel wear.

Although long-lasting when used in the highly controlled conditions of an nc or cnc grinder, diamond wheels eventually wear, and this can affect both the cut control settings and the widths of triple-chip bevels, etc.

To take wear into account, the individual X or Z axis settings need re-calibrating from time to time. Whilst this can be undertaken manually as needed, some machines offer a program to do this automatically - and so are able to maintain absolute accuracy - indefinitely.

Prior to fitting a new grinding wheel this is carefully measured and the figure recorded. Then the machine is operated as normal. After grinding a set number of teeth (such grinders log a running total of the number of tips ground) the grinding wheel is again carefully measured to find the precise amount of wear on it. This figure is then divided by the number of teeth ground to determine the average grinding wheel wear-per-tooth.

The user then decides on an acceptable amount of grinding wheel wear before correction is needed. The number of teeth that can be ground to reach this degree of wear is then quite simply calculated from the above data - separately for topping and facing.

The final figures are entered into the computer memory to then trigger fully automatic grinding wheel wear correction as the calculated tooth number is reached.

The above system can get out-of-step if initial calculations are incorrect - or if grinding conditions alter to change the rate of grinding wheel wear - either of which could result in under or over compensation of grinding wheel wear.

To avoid this risk some makers can supply a system which automatically monitors and corrects for the precise amount of grinding wheel wear at regular, predetermined intervals. Such a system is recommended when grinders are fed by transfer equipment to operate continuously.

Pawl contact point

As already mentioned, the index position of the pawl has to be precise and absolutely repeatable - but it is possible to vary the contact point of the pawl to avoid damaged tooth points. Instead of altering the index position of the pawl, though, the grinding position of the sawblade is altered - under computer control, of course, and via the keypad.

Doing this also alters the top grinding cut and the hook and clearance angles (marginally) - but computers are so clever that these details are also taken into account and the machine settings automatically corrected.

Advance grinding techniques.

Many advanced grinding techniques are now possible on the more upmarket plc, nc and cnc grinding machines. These include dual-grit grinding, hollow-tooth grinding, relief and gullet grinding, combined face and top grinding, continuous-path profile grinding and fully automatic sawblade grinding using transfer robots.

Dual-grit grinding

Ideally, heavy grinding needs two grinding passes, a rough grind using a coarse-grit wheel, followed by a final grind using a fine-grit wheel. On simple and early automatic grinders this required two setups and two operations. This was considered impractical by most users who either allowed a fine wheel to grind sawblades twice around, tweaking the weight of grind between these, or fitted instead a medium-grit wheel to get a heavier and faster grind - but also a poorer result.

Modern machines with lift-off facility allow a third choice by using a dual-grit saucer or edge-grinding wheel to rough and finish-grind at a single pass. Dual-grit saucer wheels have an outer rim of medium or coarse-grit diamonds and an inner rim of fine-grit diamonds, both at the same level. The two grits may be butted one against the other as two annular sections, or as separate sections spaced slightly apart. Dual-grit edge-grinding wheels have a medium or coarse grit on the leading edge of the rim, followed by a fine grit at the trailing edge.

Both dual-grit saucer and edge-grinding wheels operate in the same way. The grinding wheel leads-in with the medium or coarse grit section to remove weight, followed immediately with the fine grit section to finish-grind - and is then lifted clear to return to the rest position.

If used on a machine without lift-off, a dual-grit grinding wheel would traverse back across the tip to grind a final time with the medium or rough wheel and so destroy the finish formed by the fine-grit section.

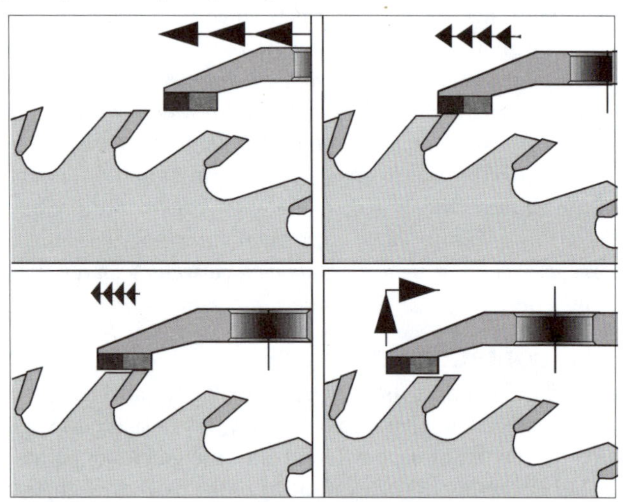

Fig. 10.91
Using a dual-grit saucer grinding wheel for top grinding. After a fast approach (top left) the outer, medium or rough grit, makes first contact (top right) followed immediately by the fine grit, possibly at a reduced traverse speed (bottom left). When the fine grit is clear of the tip the grinding wheel lifts away to return to the rest position (bottom right).

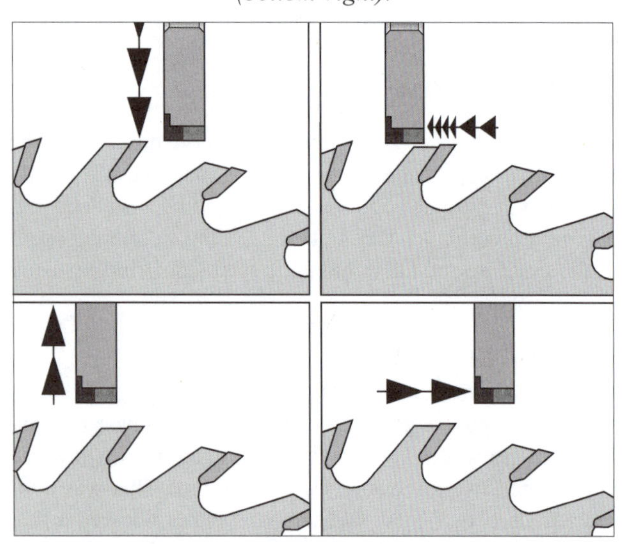

Fig. 10.92
A dual-grit edge-grinding wheel for top grinding has a similar form and grinding action, with a coarse or medium grit leading and a fine grit trailing,

In top grinding the grinding wheel gives the best finish and wear factor if traversed to clear the trailing edge of the tooth tip. The excellent finish this gives may be further improved if the traverse speed is slowed for the fine grit section.

Lift-off is in the opposite direction when face grinding, but in this case the amount of traverse into the gullet is so limited that the fine grit is unlikely to clear the inside of the tooth face.

However, it isn't necessary to grind down the whole face with the fine section of the grinding wheel - it is perfectly acceptable if the fine grind extends just a short distance down the tip.

For the reasons given earlier it is essential for the inside edge of the fine grit section to at least grind past the tip. When dual-grit face-grinding wheels are used with stubby teeth the abrasives need very narrow faces.

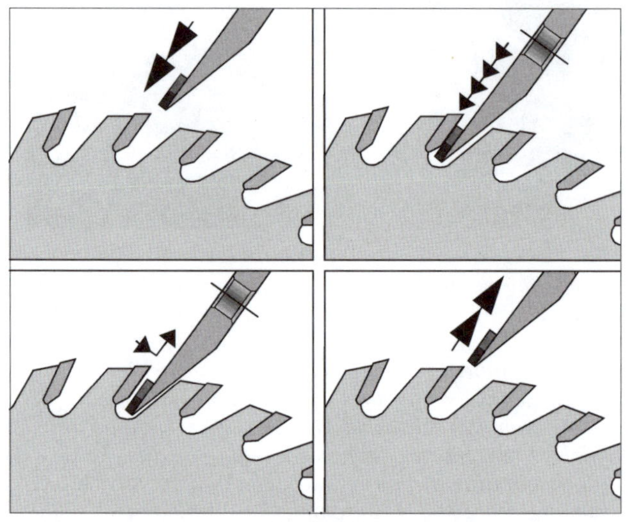

Fig. 10.93
Dual-grit face grinding using a saucer grinding wheel is also possible with small tooth pitches - but tricky. The grit arrangement and operation is basically the same as with saucer grinding wheels for top grinding.

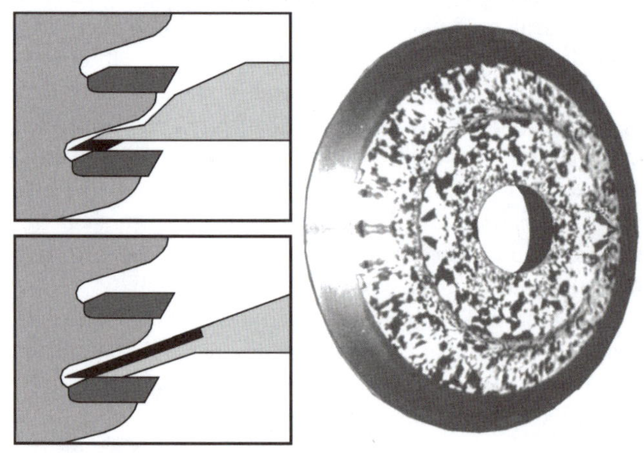

Fig. 10.94
For very short-pitched saws an alternative design of face-grinding wheel can be used on Vollmer automatic grinders, lower left and right. This, the K-plus, has diamond on the reverse face to normal, giving a narrow grinding face (less subject to uneven wear) which ensures that tooth faces remain flat. This wheel has an excellent resistance to breakage and a long wheel life made possible through natural paring by the saw teeth of the non-metalic body. A regular wheel is shown top left for comparison.
Source - Winter

Hollow tooth grinding

Grinding hollow teeth requires a small diameter, barrel-type grinding wheel with an axial movement for the grinding feed - but not all automatic grinders have a suitable movement.

In addition to setting to the correct hook angle and depth of grind, a barrel grinding wheel also needs precise centering to the tooth kerf. A high rotary speed is absolutely essential, with rotation of the grinding wheel towards the backplate. The operation is otherwise very similar to face grinding using the regular pawl and grinding head movements.

Fig. 10.95
This Akemat unit facilitates precise hollow-tooth grinding. The speed is steplessly adjustable up to 55,000 revs/min.
Source - Vollmer Dornhan

Fig. 10.96
This optional hollow-grinding attachment on the CNC 5 replaces the regular face-grinding wheel. At a single setup and in following processes it allows complete grinding of both hollow-face teeth and tooth tops - of any pattern. See also Fig. 10.116.
Source - Walter

Back-relieving

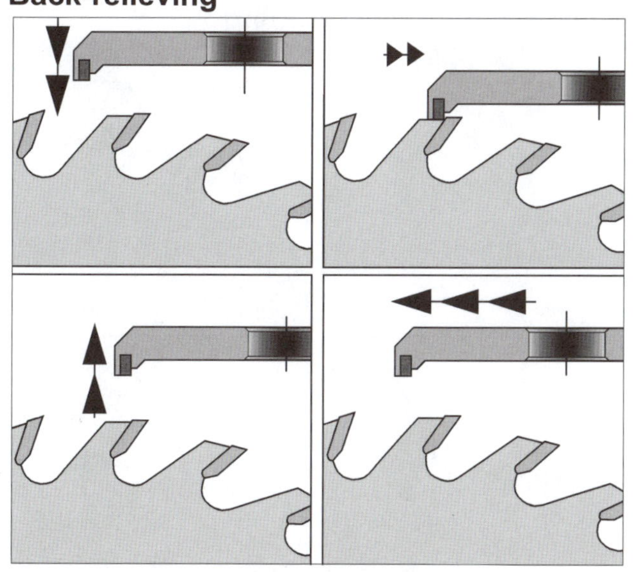

Fig. 10.97
Back-relieving on a purpose-made relief grinder, such as the Vollmer AS 810, is by means of a specially-made cbn grinding wheel and a program that allows the backing to be ground down below the tip in the sequence shown.

Back-relieving is possible on most automatic machines by using a regular-grit or cbn grinding wheel set to grind at the clearance angle, plus 5 degrees.

This operation is generally as described earlier for manual grinders, so a tiny but increasing amount of steel is ground with the diamond wheel as the tip height is reduced. An alternative and much better method incorporated in some cnc grinders uses a cbn wheel and a purpose-made cnc program to back-relieve.

Two methods are shown.

For processing large numbers of sawblades, however, it convenient and more efficient to use a dedicated back-relief grinding machine for this process. Such machines are simpler in design and operation, and cost less to buy and run than a dedicated type of saw grinder intended for primarily for tct tip grinding.

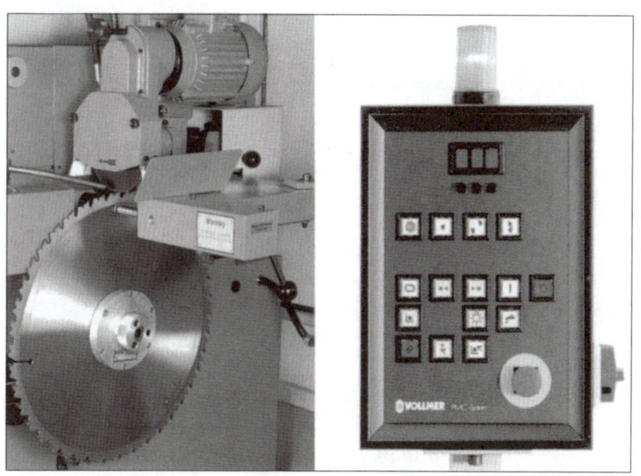

Fig. 10.98
The AS 810 machine is a dedicated back-relief grinding machine for tct saws. It is capable of quickly processing any regular type of tct saw - including bevel-angle relief grinding on alternate bevel and triple-chip teeth (to retain both maximum backing and adequate back clearance). Curved-back grinding, as needed for very wide-pitched sawblades, is also possible.
Source - Vollmer

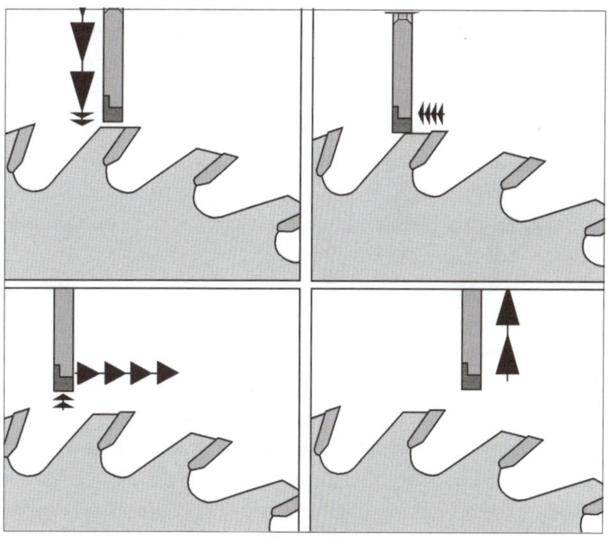

Fig. 10.99
An alternative method of back-relieving using an edge-grinding wheel. This plunge-grinds immediately behind

the tip (top left); traverse to the left beyond the tooth heel (top right); then lifts clear and returns to the rest position (bottom left and right).

Tip & relief grinding

When dealing **only with wide-pitched teeth** ground on cnc machines there is a possibility of relief grinding simultaneous with, or followed immediately by, tip grinding. If practical, it offers an ideal way of keeping the back relieved at a consistent, minimum level below the path of the diamond grinding wheel. These suggested processes, when practical, need add only a little increase in the cycle time.

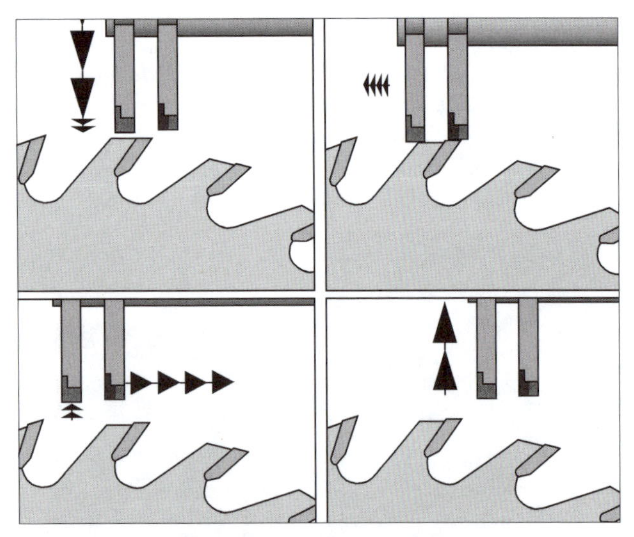

Fig. 10.100
This shows how twin, peripheral grinding wheels (with cbn/dual-grit diamond abrasives) can be used to simultaneously grind both the carbide tip and back-relieve. The cbn wheel slowly plunge-grinds into the backing steel (top left) then traverses clear. Simultaneously the dual-grit diamond wheel advances into the grind position to traverse across the tip only (top right). Both wheels then lift clear and return to the rest position (bottom left and right). This method is practical ***only*** *on wide-pitch sawblades.*

Saucer and edge-grinding wheels* can both be made to similar specifications for this purpose, with a narrow leading section of cbn, then a gap greater than the widest tip width, followed either by a fine grit or dual-grit diamond combination. The cbn grinding face projects marginally more than the following diamond face to grind away any steel that the diamond face would otherwise contact.

**Both the operating sequences shown can be undertaken with either saucer or peripheral grinding wheel types.*

When this becomes a regular operation, i.e., simultaneous grinding the back and tip each time the saw is ground, only a small amount is ground from the back. By using the sequence shown in Fig. 10.100 the grinding program would take only marginally longer than regular tip grinding .

When saws are not continuously relief-ground, perhaps due to being re-ground between times with a less-sophisticated program, then the more complex sequence shown in Fig. 10.101 would need to be used instead. The difference here is that any excess backing is ground away at an initial pass prior to the final grind - and the cycle takes longer.

It is appreciated that cbn wheels grinding steel and diamond wheels grinding tungsten normally operate at different depths of cut and traverse speeds.

For this reason cbn wheels of nonstandard grit and grade would be needed to operate in combination with regular diamond grinding wheels.

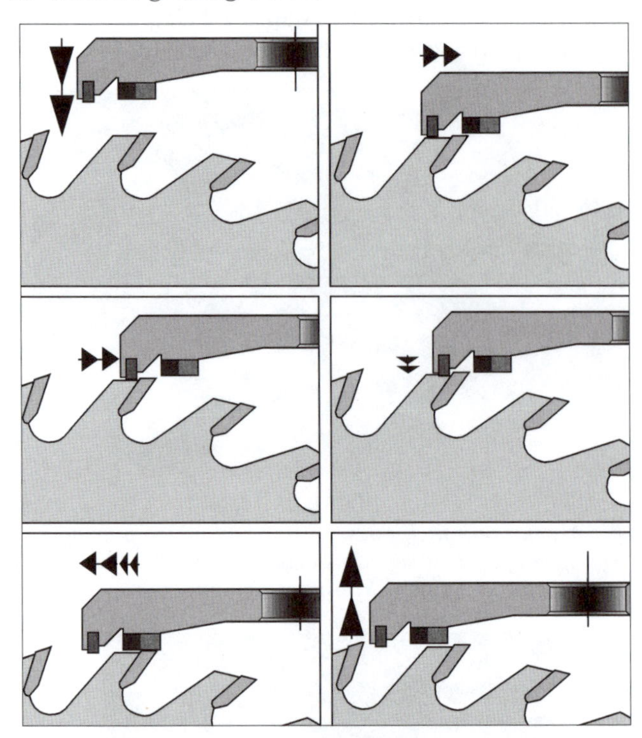

Fig. 10.101
A dual-grit diamond/back relief cbn grinding wheel combination could instead operate in the manner shown here when excessive back-relieving is needed. A rapid approach (top left) is followed by a slow relief-traverse from the left by the outer cbn section (top right) - with the grinding head set so that the diamond grinding wheel section barely clears the tip. When the cbn section reaches the back of the tip (middle left) the grinding head advances slowly for the cbn wheel to plunge-grinding to its final setting (middle right). This is followed by a slow traverse to grind the tip with the dual-grit diamond wheel - whilst also grinding a minimal amount from the backing (bottom left) - and finally rapid clearance, bottom right.

Gullet & back-relief grinding

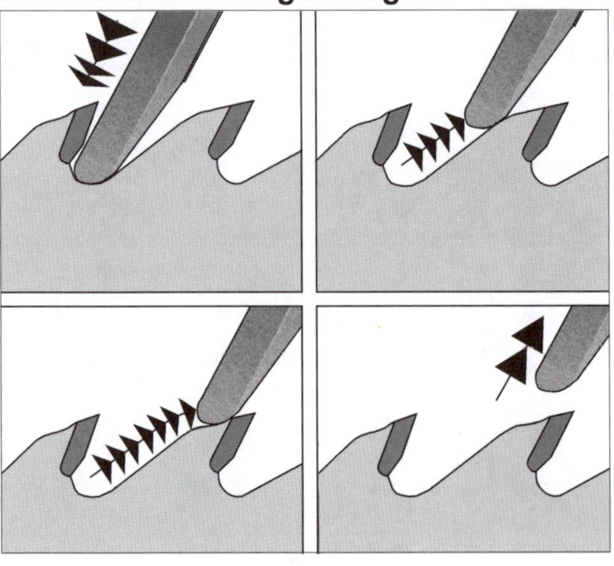

Fig. 10.102
Some universal saw grinders are capable of combined

gullet and relief grinding on wide-pitched, fast-feed ripsaws. The cbn grinding wheel grinds into the gullet, just clear of the tip (top left) then up the back (top right) onto the tooth top (bottom left), then finally clears the tooth just behind the tip (bottom right).

The base of carbide tipped saws tends to become worn and rounded-over due to the constant bombardment from fast moving waste. With slow feed speeds and a minimal cut it takes considerable time for such rounding-over to affect the performance of the sawblade - but with fast-feed ripsaws this can cause a serious problem over a much shorter period.

A rounded-over base, which should contain waste until clear of the timber and then discharge it, instead allows resinous dust to escape between the sawblade and the sawn timber. Such dust can build up as a resinous deposit on the sides of the sawblade to increase the power loading - with the outside chance of enough heat being generated through friction to badly blister or even wreck the sawblade.

To avoid this problem it is essential to periodically re-grind gullets of fast-feed ripsaws to restore them to their original square-edged condition - this gives the best chance of retaining waste within the gullet until discharged.

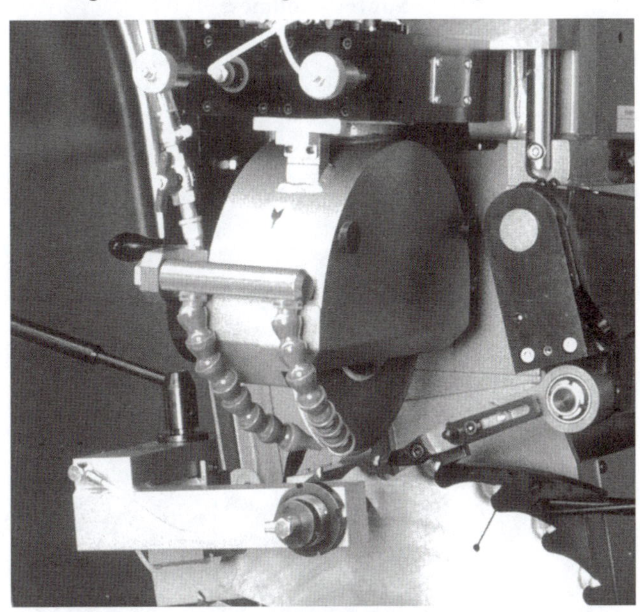

Fig. 10.103
The AS-2 universal saw grinder is suitable for gullet and relief grinding of tct saws, and for profile grinding of Stellite tipped saws - making it ideally suitable for maintaining all types of fast-feed ripsaws.
It is also capable of grinding carbide tips using a specially-profiled diamond grinding wheel. (see Fig.10.104)
Source - Iseli

In normal circumstances re-grinding of the gullet is only practical when all the teeth have been removed - simply because the tips themselves normally make profile grinding of the gullet impossible when using a regular cv saw grinder.

Some universal grinders, such as the Iseli AS-2 universal saw grinder can be programmed to re-grind the gullets of certain wide-pitch carbide saws - with all tips still in place.

To do this increase the hook angle to miss the tip but reach the gullet immediately below. In addition to this, a lift-off sequence stops the profile-grind barely short of the tip back - so the machine back-relief-grinds at the same time.

Face & top grinding.

Combined face and top grinding is possible both on universal saw grinders, and on cnc combined face and top grinders specifically designed for tungsten carbide tip grinding.

Universal saw grinders

Universal saw grinders are designed for profile-grinding cv and Stellite saws, also combined face and top grinding of tungsten tipped fast-feed ripsaws. The grind is carried out at two sequential passes following a program modified from that regularly used for cv and Stellite saws, i.e, to pass fully across the tip top from the back, then plunge-in to grind the face.

The profiled diamond grinding wheel is formed with a straight radial face for face grinding and with an angled and slightly rounded edge for top grinding. This allows the same grinding wheel to be used regardless of the hook and clearance angles of the tip being ground.

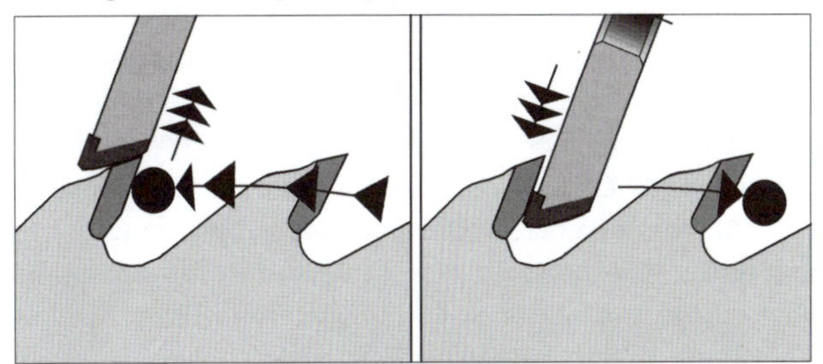

Fig. 10.104
These sketches show the form and operation of a diamond grinding wheel as used on a universal saw grinder for combined tct face and top grinding.

CNC combined face and top grinders

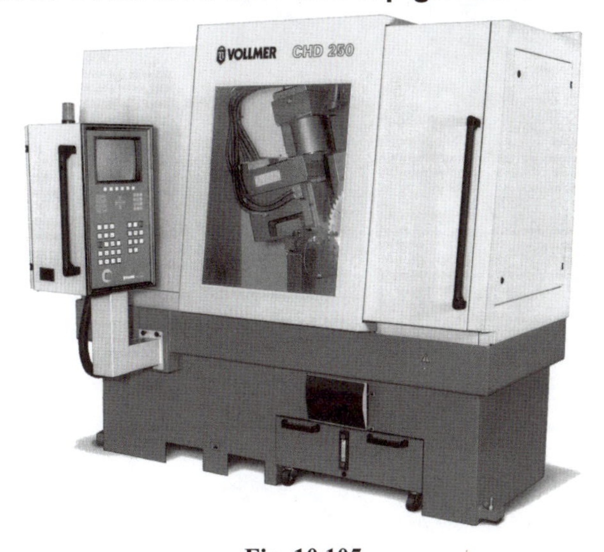

Fig. 10.105
The CHD is a dedicated tct, cnc combined face and top grinder operating with five nc-axes to grind tooth faces and tops at a single setup and sequential handling. It can be used with a loading robot for unmanned operation around the clock.
Source - Vollmer

In addition to specialist machines for tct face grinding, top grinding and face and top grinding at separate setups, several manufacturers also offer dedicated tct face and top grinders capable a completely facing and topping tungsten carbide-tipped saws one machine setting and a single handling.

Usually the top grinds follow complete face-grinding of

the sawblade - though the reverse sequence can be chosen if preferred..

These grinders give a faster turn-round than other types as the output is of completely finished saws. In addition, as sawblades are ground in precisely the same conditions for both facing and topping absolute accuracy of tooth heights is guaranteed.

On regular face and top grinders the machine has to be manually switched and fully re-set between facing and topping - in order to completely finish individual sawblades. For this reason it is more cost-effective to face-grind several sawblades before switching to topping.

In a workshop situation, where dull or damaged sawblades must be re-ground quickly to avoid production holdups, interrupting the schedule for a single sawblade is inconvenient and time-consuming - hence the need for combined face and top grinders.

For service companies the combined face and top grinder is also an obvious choice. Their business is dependent upon rapid turn-round of completely-finished sawblades - and these machines meet this need precisely - both as a stand-alone machine or combined with an automatic loader.

Stand-alone machines can be pre-programmed for the next and following sawblades while still in full operation - even though this may be of a completely different size and tooth style. When one sawblade is complete another can be loaded immediately - and the corresponding program started with minimum downtime.

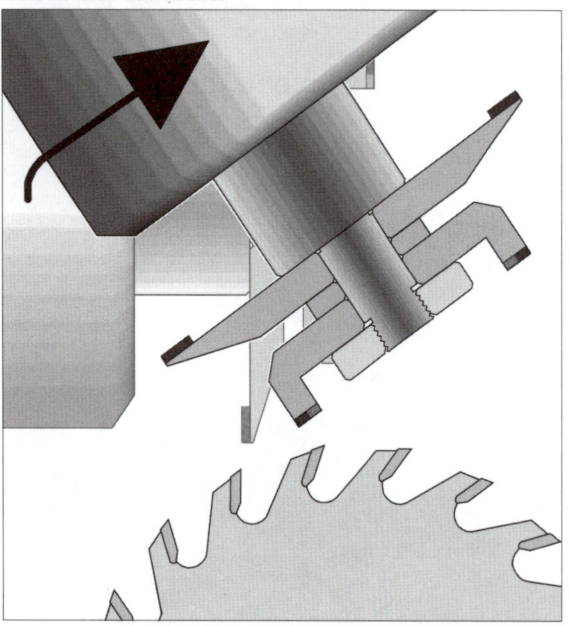

Fig. 10.106
The general arrangement of one type of cnc combined face and top grinder, using a flat cup wheel for topping and a saucer wheel for facing. The head swivels, as shown by the arrow, between facing (background) and topping (foreground). See also Figs. 10.107, 109 &110.

For scheduled grinding, sawblades can handled via an automatic loading system - which routinely fully grinds and transfers sawblades automatically while the operator sees to other urgent work *(a fuller description follows)*. But should a sawblade want instant attention, it is feasible to stop the automatic process to fully grind this sawblade - and then simply revert to the original program. Interruption to the grinding program can be made when the current sawblade is completely finished - but before the following sawblade is loaded - so downtime is reduced to the absolute minimum.

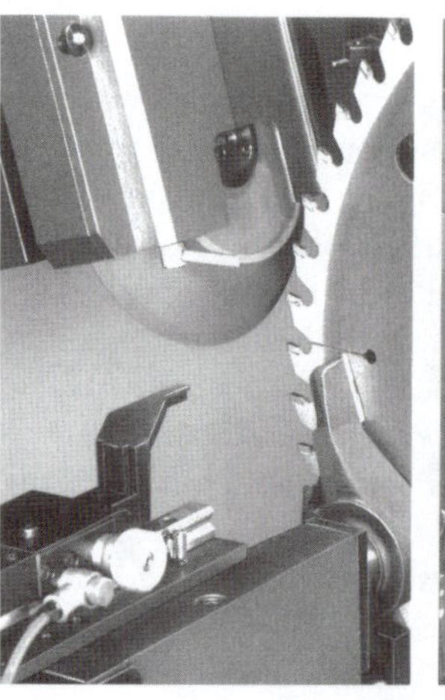

Fig. 10.107
The above shows the top grinding operation (left) and the face grinding operation (right) on the CHD combined face and top grinder.
Source - Vollmer

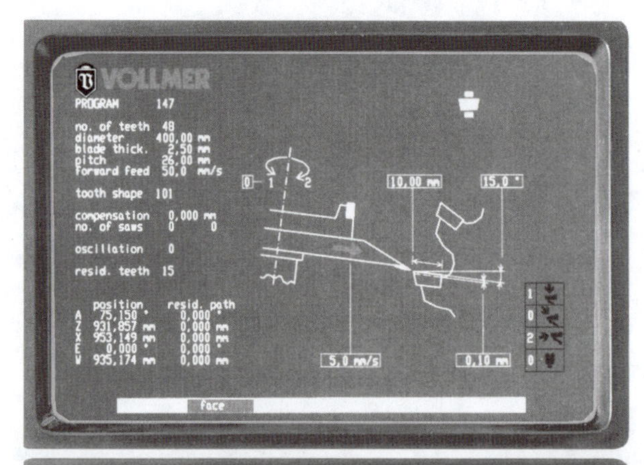

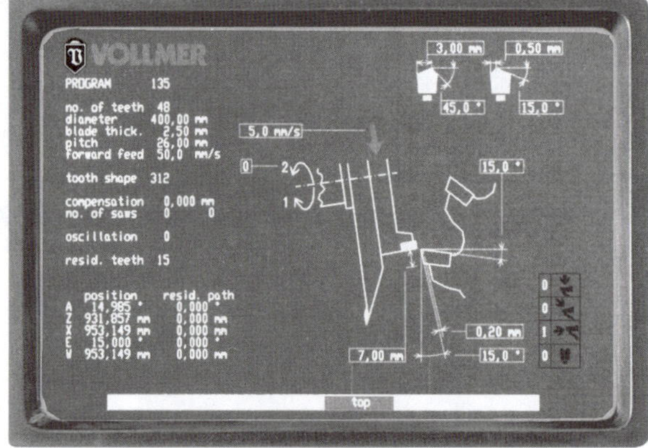

Fig. 10.108
Two of the machine setting screens for the CHD combined face and top grinder are shown here (top) for face grinding and (bottom) for top grinding. Both screens give a pictorial view of the operation - complete with full data - so making programming fast, user-friendly and simple.
Source - Vollmer

The modern cnc combined face and top grinder has face and top grinding wheels mounted on the same arbor and separated by a spacer, or separately mounted on either side of the grinding head unit. The grinding head unit is automatically switched between the face and top grinding setting (and simultaneously set to the precise hook or clearance angle required) by swivelling on the B-axis.

The 'B' axis movement on these machines is switched automatically as part of the operating sequence, then the three other axes take over. The D or Y-Axis sets the grinding head for the top or face bevel(s) as dictated by the program, leaving each grinding pass under the control of the X and Z-Axes operating as described previously.

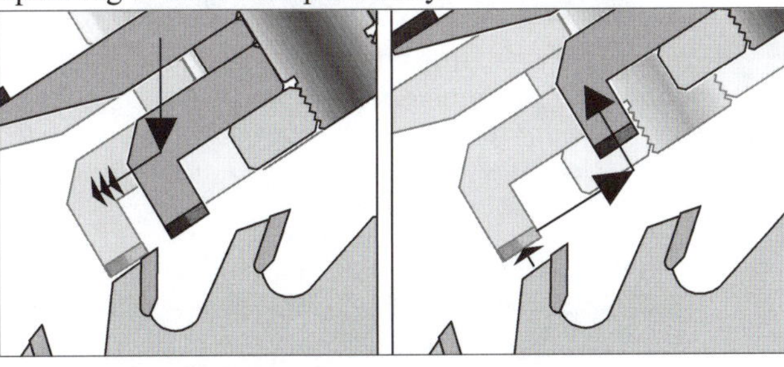

Fig. 10.109
Top grinding as on the combined face and top grinder arrangement shown in Fig. 10.107 when fitted with a dual-grit wheel and with the grinder programmed for lift-off. An alternative program can be used allowing a single-grit grinding wheel to oscillate and so further improve the finish.

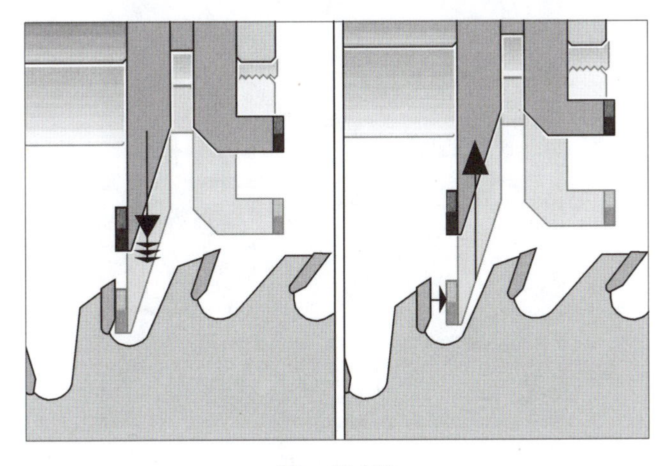

Fig. 10.110
When using a dual-grit wheel programmed for lift-off, face grinding on any of the combined face & top grinders is undertaken generally as shown here.

Ideally each tooth should be both faced and topped when indexed, but continually switching between facing and topping for each tooth is inefficient*. The recommended sequence is for the machine to be programmed to face grind all teeth, then switch automatically to topping - so completing the sawblade at two following passes automatically and with the absolute minimum of change-over time.

**Although switching between facing and topping only takes a few seconds, this is extra time taken for each tooth - which can add up to a considerable operating-time loss. The sequence may at first seem little different to switching sawblades manually between separate facing and topping machines. However, sawblades ground on combined machines remain in position, in the same grinding conditions for both operations, do not need operator intervention and production is a steady output of completely finished sawblades.*

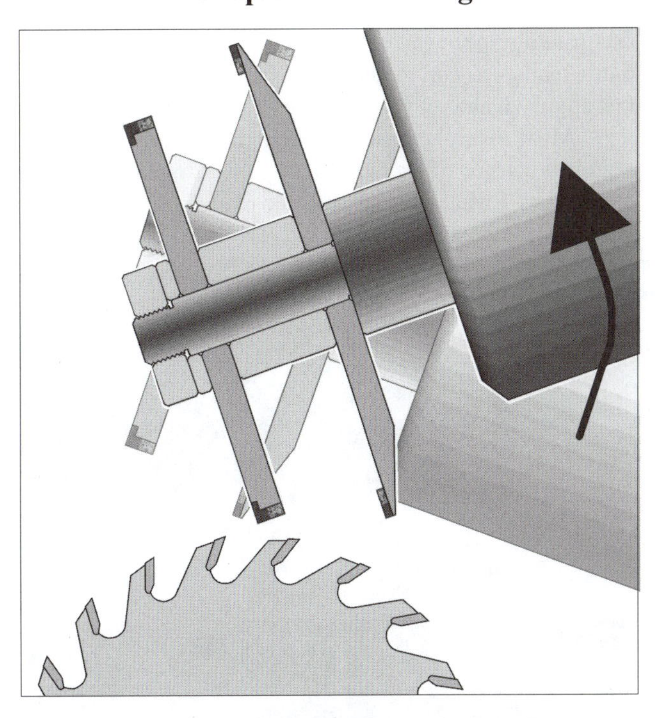

Fig. 10.111
The alternative method of two-wheel mounting uses a saucer wheel for facing and an edge-grinding wheel for topping. The grinding head unit swivels between facing (background) and topping (foreground) as indicated by the arrow.

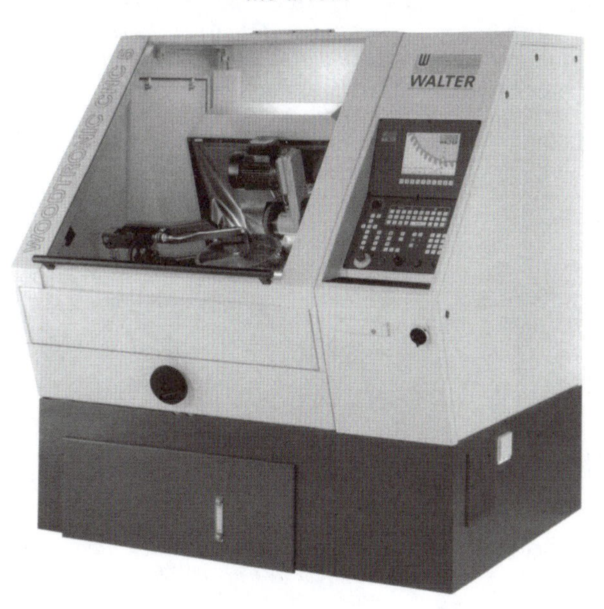

Fig. 10.112
The CNC 5 is a five-axis dedicated face and top grinder capable of grinding all currently-used types of saw teeth. In addition to the advantages listed under 'Combined face and top grinders', the CNC 5 also has continuous path control of both top and face grinding operations which allows it to form tooth profiles impractical with regular cnc grinders.
Source - Walter

Modern cnc combined face and top grinders incorporate far more control than earlier types - as regards quality and speed of grinding. The various cutting surfaces can be ground at different speeds, one to the other, each speed steplessly variable to match the individual requirements precisely.

This is the advantage of cnc grinders. The fast approach

and grinding speed can be varied infinitely - for each surface individually - as can the distances traversed both during the fast approach and retreat, and when actually grinding .

More than one pass can be made *(perhaps with a increase in weight of grind between passes)* or several quick oscillations - whatever the operator requires to best restore the sawblade being ground.

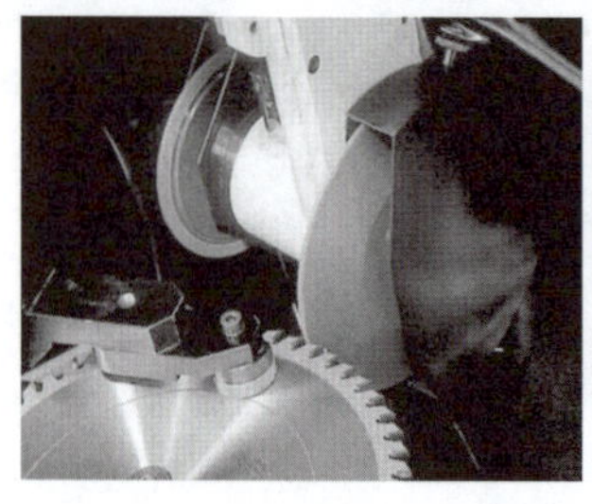

Fig. 10.113
Face on top grinding on the CNC 5.
Source - Walter

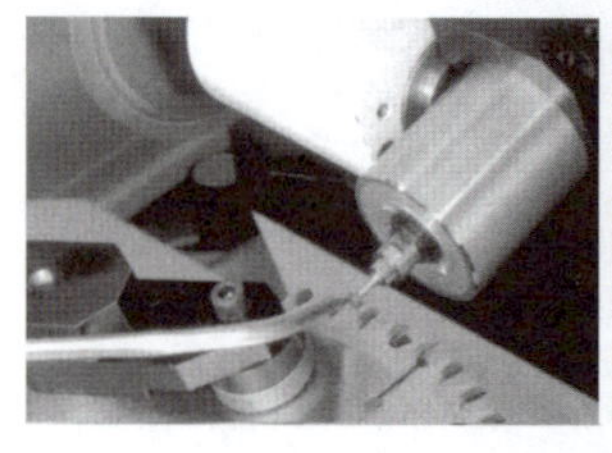

Fig. 10.114
The CNC 5 also allows top grinding and hollow front grinding to be perform as following operations - automatically and without operator involvement. The high-speed hollow-tooth grinding head is mounted in place of the regular face-grinding wheel.
Source - Walter

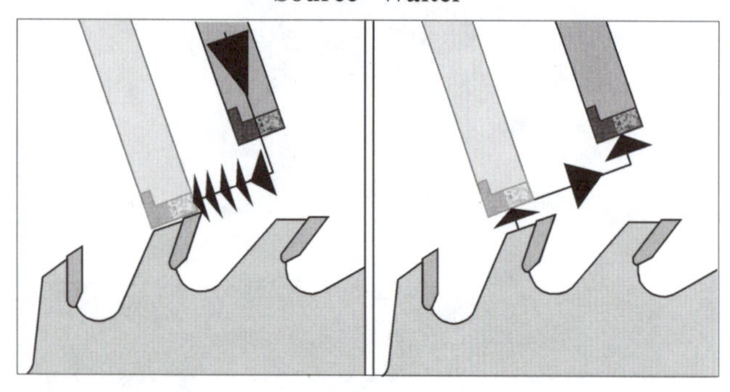

Fig. 10.115
Top grinding, with lift-off, as on the combined face and top grinder shown in Fig. 10.111

There is a growing realization that a break from traditional tooth profiles *(where the hook angle remains constant around a sawblade)* may have considerable merit - and the facility to grind different hook angles around a sawblade is possible with cnc grinders.

By doing this tooth configurations can be formed that previously were impractical - or at least difficult to produce and maintain in a workshop environment with traditional grinders. The advantage is that teeth for different functions *(for example scribing and cleaner teeth)* can each be given the hook angle most suitable for their particular application - fully automatically - during the machine's program

Although possible to produce with earlier grinder types, such a configuration needed at least two setups and two handlings.

Fig. 10.116
An optional, additional axis on the CNC 5 allows automatic alternate front-bevel face grinding to be carried out on cross-cut saws - also vee-shaped and hollow top grinding.
Source - Walter

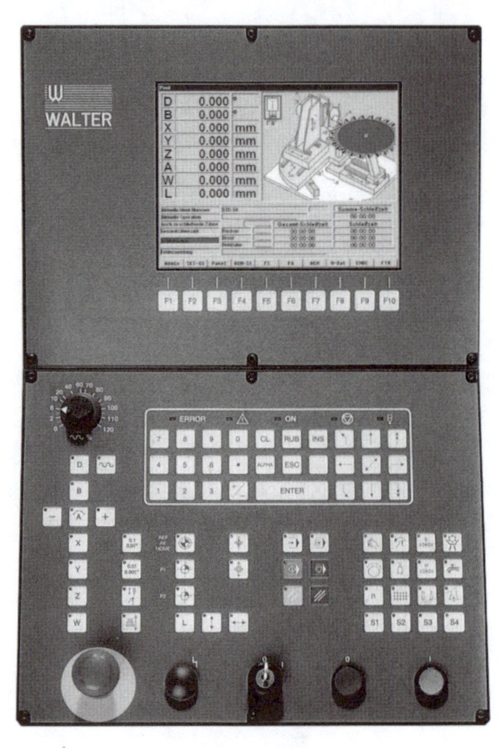

Fig. 10.117
All setup and operating functions of the CNC5 are controlled from this panel and colour display. The screen shows various machine or sawblade arrangements - in this case the basic machine movements
Source - Walter

Fig. 10.118
Face and top grinding on the NC 3 grinder. This particular machine is nc controlled and, though capable of grinding either tops or faces, change-over between the two requires manual movement of the grinding head. The CNC 5, shown earlier, is a combined face and top grinder which changes from face to top grinding automatically.
Source - Walter

Continuous-path grinding

Most regular tct cnc top and face grinders have one or more nc axes to set the grinding head for the tooth geometry required, but the grinding pass itself is usually powered as a single-axis, straight-line movement. It may seem that for most conventional sawblade teeth this is perfectly adequate - but tooth geometries are restricted to some extent because of this.

A more sophisticated system of grinding wheel control, continuous-path grinding, is an established method for grinding wide bandsaws and metal-cutting circular saws. The same system also has application for grinding tungsten-carbide tipped circular saws, as the traverse movements can follow a straight, tapered, curved or irregular path to produce tooth configurations that are impracticable with regular cnc grinders.

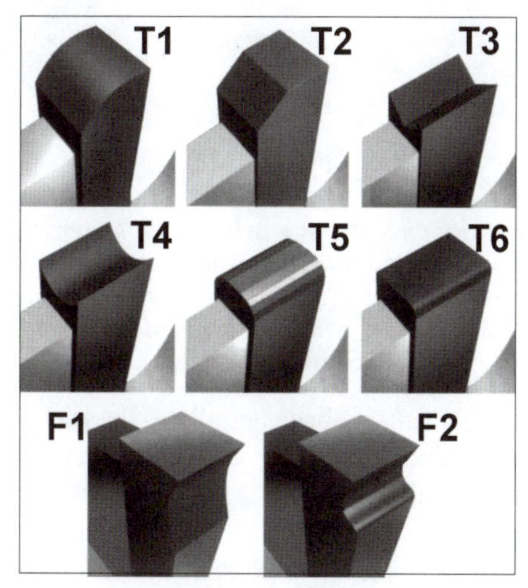

Fig. 10.119

Several nonstandard tooth profiles can routinely be produced on the Walter CNC 5 using the continuous-path grinding feature and the optional 'D' axis. Amongst others, these include top grinding profiles; T1, rounded; T2, doubled bevelled; T3, vee-shaped; T4, hollow; T5 fully rounded; and T6, edge-rounded; also face-grinding profiles; F1, softwood chip-deflector and; F2, cross-grained chip deflector. Any of the above, and indeed any regular form of tooth, can also be side-ground under continuous-path control on the Walter CNC 6F, see Chapter 12.

Source - Walter

To provide this feature, the grinding head traverse movement on the Walter CNC 5 grinder has two active axes which interpolate to give two-dimensional, continuous-path grinding.

The grinding paths can be precisely defined - under computer-control - and are absolutely repeatable. This can also be combined with other advanced cnc features - such as altering the hook angle between following teeth, variable grinding speeds and repeat grinding operations.

Continuous-path control is possible for both top and face grinding on the CNC 5, making it practical to produce some very innovative tooth profiles - especially when combined with the Walter continuous-path side grinder CNC 6F (see Chapter 12).

The use of continuous-path grinding for tct teeth opens up a whole new ball game - it allows sawblade teeth to be customized in a way that was previously impracticable.

The diagrams show some of the face and top profiles that continuous-path grinding makes possible. Combine these with continuous-path side grinding and the options are limited only by the saw doctor's or filer's imagination and his ingenuity to evolve special teeth to solve perennial problems and meet new challenges.

Automatic loading equipment

Automatic loading and discharging facilities are now offered for use with several makes of cnc saw grinders. The equipment picks sawblades from a stack and loads them individually onto a grinder for grinding fully automatically, then removes and places them on a second stack for further processing (or as finished sawblades) in a process that repeats until the supply is exhausted.

Equipment of this type was originally intended for sawblade manufacture only - and is still used for this purpose on conventional tct saw production lines.

Robot loading and unloading can also be used with a single cnc combined face and top grinder - to grind saws of any diameter and configuration.

Two basic systems offered - one intended for saw servicing shops *(for face and top grinding of saws in mixed sizes and types)* - and the second for saw manufacture *(where several machines are linked by a transfer system to fully grind new sawblades in production batches of the same or similar size and pattern).*

Saw servicing grinders

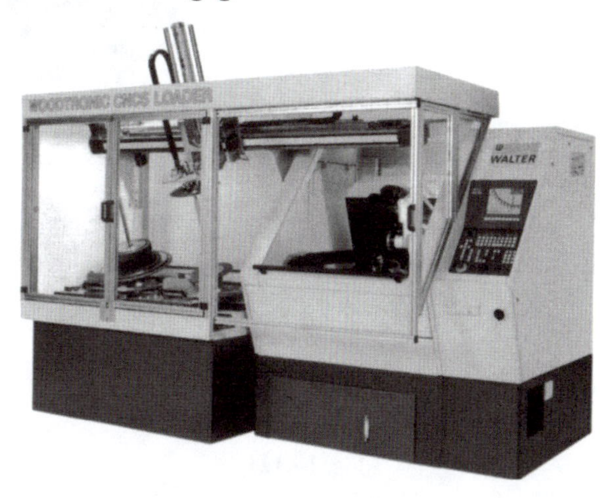

Fig. 10.120

The Woodtronic CNC 5 can be combined with a loader to automatic handle sawblades from 150 to 630mm diameter. It automatically processes sawblades - of any conventional tooth configuration and stacked in any order of tooth configuration and diameter - when used in conjunction with the Sawcheck.

The machine can be preset whilst working a minimal-manned day shift in preparation for fully automatic operation throughout the night.

Source - Walter

A saw servicing production line can consist of a cnc face grinder and a cnc top grinder linked by a transfer robot, or a single cnc combined face and top grinder served by a transfer robot.

Sawblades for re-grinding are first carefully measured for all the variable factors, and the data transferred to the memory of the grinder(s). At the same time, and in the same sequence, the sawblades are stacked at the infeed station in readiness for fully automatic processing.

It isn't necessary to pre-measure regularly-ground sawblades. Each newly-measured sawblade should be routinely given an identifying code - which can be etched or otherwise marked on the saw body.

This code is then the only data that needs to be entered into the program as each sawblade is loaded - so avoiding the need of re-measuring each time around. The only possible correction then required is to the sawblade diameter and tip face depth as the sawblade is worn down - although most equipment is capable of dealing with such variations automatically.

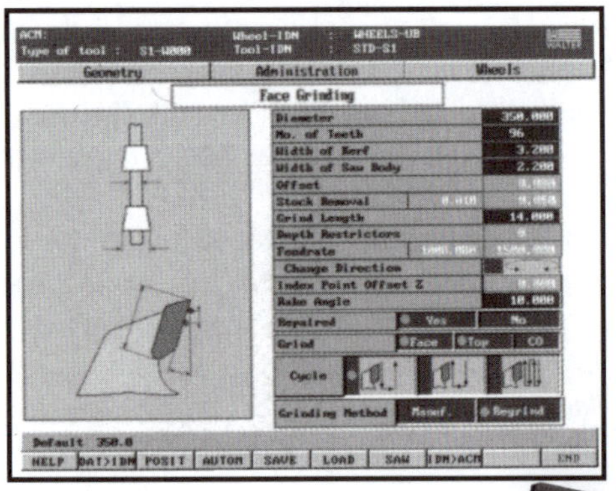

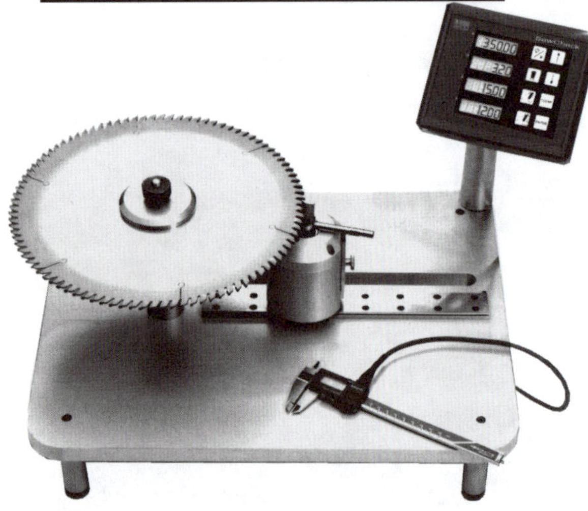

Fig. 10.121

The Woodtronic Sawcheck is used to pre-measure sawblades. Data is immediately transferred from the Sawcheck direct to the CNC 5 via an on-line link. As the data for individual sawblades is downloaded, they are also stacked at the machines infeed station in the same strict sequence.

Source- Walter

The grinding process is fully automatic. After picking-up the sawblade to be ground, this is placed on the saw grinding arbor, moved automatically into position, clamped and the nearest tooth indexed.

The machine then checks for the type of tooth by probing with a sensor in a sequence dictated by the computer program - to check that it corresponds to the downloaded data. Sawblades that do not correspond are rejected and the following sawblade loaded.

Alternate top bevel teeth are probed at two or three kerf-width positions to determined the angle and hand of the tooth, possibly with a second probe on the following tooth to confirm the configuration.

Fig. 10.122

The multi-probe on the CNC 5 has several functions. It automatically checks the sawblade diameter, tooth type and position, hook angle, clearance angle, tip face depth, compensation factor between face and top, saw body thickness, kerf width and tip side projection.

Source - Walter

Triple-chip teeth are probed centrally to the kerf width, with the following tooth checked to establish which is the higher *(triple-chip)* tooth.

Tooth faces are probed at two diametric positions to establish the hook angle, and possibly at two kerf-width positions to establish the face bevel angle - or confirm that the tooth is square-faced.

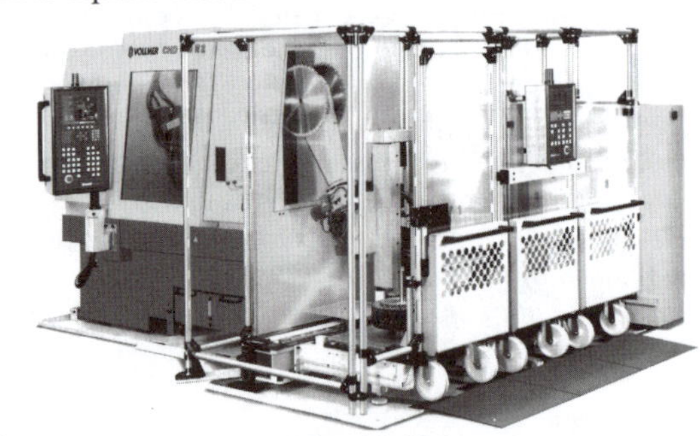

Fig. 10.123

This service centre consists of a CHD combined face and top grinder and a robot system suitable for fully automatic grinding of up to 100 sawblades of different outside and bore diameters and different tooth geometries. Saw data is recorded for each individual sawblade at a separate measuring point, possibly on memory cards.

Source - Vollmer

Following this the grinding program is started automatically - either at the indexed tooth or at the following tooth should be last-indexed tooth be the wrong one for the particular grinding sequence loaded.

When using an identifying code, sawblades are initially positioned according to their original measured diameter. This is followed by a cross-check via a probe to confirm that the correct sawblade is in place, then a forward movement to move the tip into the zero-grind position - this takes account of sawblade diameter loss through grinding.

In other cases the sawblade is measured each time for diameter and preset according to this, or simply moved forward under power until any indexed tooth contacts a sensor to stop the tip at the zero-grind position - regardless of its diameter.

In both cases a normal process is to index and probe the closest tooth to identify its type, then index a second time to check the overall tooth configuration. If the second indexed tooth is also the starting tooth for the particular tooth style, then automatic grinding starts immediately and automatically.

If this tooth is not the starting tooth for the program, then one or more further index and probe sequences may be needed to locate the correct tooth. Repeated indexing and probing naturally looses costly production time. This lost time can be turned into profitable production time if the starting tooth is correctly positioned for all sawblades as they are stacked for automatic loading.

Fig. 10.124
This optional laser guide, offered for use with the CNC 5, is designed to position the starting tooth correctly for each sawblade as it is stacked onto the CNC 5 loader. A thin laser line projected from above shows precisely where the point of the starting tooth for sawblades of any diameter should be positioned.
Source - Walter

The mix of sawblades that can be handled with equipment of this sort depends upon the degree of sophistication of the grinder and of the transfer equipment. Most can handle a surprisingly wide variation in sawblade diameter, bore size and tooth configuration.

Transferring sawblades

Transfer robots used for loading and unloading are carefully aligned to pick up each sawblade individually and place it precisely onto the grinder saw centre - which usually has a slight top taper to allow easier positioning. To avoid damage to the sawblade teeth an intermediate spacer ring is placed between each sawblade during initial loading.

Often these are of the top hat type, a combined spacer and saw centre bored to a common stack-arbor size. They are separately transferred to the finished stack as the corresponding sawblade is being processed - in readiness to receive the same sawblade when complete.

Sawblades of different bore can be processed in any sequence by entering full details into the computer program.

To cater for this a multi-diameter, stepped-cone type saw centre is provided on the grinder which resets to bring the next-required centre into place as the corresponding sawblade is being loaded.

Production line grinders

For continuous new sawblade production two or three automatic cnc grinders are connected by automatic transfer equipment. The top and face grinders in the production line may be separate machines, or combined top and face grinders, and production lines always include a side grinder.

Fig. 10.125
This shows an ND series robot of the type used for loading and unloading automatic grinders for unmanned operation around the clock.
With the next sawblade to be processed already held, the robot picks up the completed sawblade and quickly lifts, flips and lowers to substitute the new sawblade. By handling both the unground and ground sawblades simultaneously, reloading is extremely rapid.
Source - Vollmer

Fig. 10.126
This production combination consists of the CNC 6F and CNC 5 combined face and top grinder connected by a robot transfer system. Additional machines can be included in the line if the output required demands this.
Source - Walter

The grinding machines commonly used in these lines are adoptions of regular stand-alone grinders, and for this reason can also be used individually - independent of the transfer system.

To maintain accuracy on continuous production it is essential, of course, to program the grinders to compensate for grinding wheel wear, see previous notes.

Such a production line is produces sawblades in batches - and traditionally consists of a side grinder, a face grinder and a top grinder served by robots.

Alternatively a single combined top and face grinder can be linked to a side grinder or, to increase throughput, two combined face and top grinders can be linked to a single side grinder.

The latter arrangement gives more flexibility in that the two combined top and face grinders can be programmed to produce different sawblade types, but feed the single side grinder.

The number and types of machines in the production line can be varied according to sawblade specifications and production requirements. All the machines and the robot transfer system are programmed individually as production specification demands. In this way the operating sequence and final transfer position of the finished sawblades can be arranged to suit individual requirements.

With production lines of this type the unground sawblades are initially stacked onto transfer trolleys from where they are picked-up automatically by the transfer robot, with the finished sawblades stacked on empty trolleys by the same robot. By using trolleys the finished and unground sawblades are quickly shifted around to minimize downtime.

Dealing with broken and replaced teeth.

Replaced teeth are a problem on most automatic grinders because, to avoid a crash, they conventionally have first to be ground down to the level of the remaining teeth. Modern CNC grinders, such as the Vollmer CNC range, however, now offer special programmes to deal with this problem - and that of broken teeth:-

A) Feeding the replaced tooth but indexing past this without attempting to grind. This allows the replaced tooth to be dealt with separately - without interfering with the regular grinding programme.

B) Feeding and grinding only the replaced tooth. This allows the operator to take a slow, deep grind on these teeth. This reduces them sufficiently in height to allow a following regular grind of all teeth without danger of heavy grinding on the replaced teeth.

C) Feeding all teeth as normal (in order to maintain the correct tooth type sequence) but without grinding missing or damaged teeth.

D) Feeding and grinding all teeth in the normal manner, but with the replaced teeth only ground at several oscillating and gradually deepening passes in order to reach the correct height. This allows all teeth in such sawblades to be finished at a single handling.

The amount ground-off at each pass should be chosen according to the width and type of replaced tooth, and can be varied in increments of 0.1mm. By programming a small depth of grind per pass - between 0.1 and 0.5mm. and with a normal value of 0.3mm - excessive grinding wheel wear can be avoided*. In this way precise tooth accuracy can be guaranteed - even though the sawblade initially contains several high teeth. **If the full depth of grind required is removed at a single pass, the excessive grinding wheel wear this causes would affect the height of the following teeth.*

Looking to the future

The facility already exists for even further refinement in using cnc grinders for fully automatic grinding - by using bar codes to totally eliminate prior measurement, guarantee absolute repeatability and consistency of sawblade performance and eliminate the outside chance of machine-setting errors.

Grinders in this class can undoubtedly be adopted to read a bar code - and subsequently set the machine automatically and individually for each sawblade. If bar code data could be agreed between makers, then this could be laser-printed onto sawblades during manufacture - to permanently identity their precise type and grinding configuration. By doing this mistakes in grinding and subsequent unintended alteration of sawblades from their original pattern would be eliminated.

As an alternative, of course, stickers could either be applied to sawblades prior to processing for the grinder to read, or a bar-code sheet containing common configurations could be scanned as sawblades are loaded. Bar code data could be supplied by the machine maker for sawblades of common tooth configuration - suitably coded to suit their specific machines of course.

In this way sawblades of mixed diameter and tooth configuration could be processed fully automatically - without having to measure or enter their details prior to loading. This would make an ideal arrangement for saw servicing shops looking to hold labour costs to an absolute minimum - and ensure users that their sawblades could never be wrongly ground.

There is, of course, some variations that bar codes cannot take into account - as sawblades wear the saw diameter and length of tooth face alters. However, cnc combined face and top grinding machines already have the facility to take care of this problem - and to pre-select the starting tooth for the required configuration.

A second problem is the precise weight of grind needed to restore each sawblade - without under or over-grinding. This can be controlled in-house by carefully monitoring run-times to always remove sawblades before excessive wear occurs. In this case a specific weight of grind can be predetermined.

With saw service companies, however, sawblades can be received with a wide range of wear and damage, so that a predetermined weight of grind for all is impractical.

There is little doubt that equipment can be designed to scan sample teeth to establish a suitable weight of grind and then apply this. However, an experienced saw doctor or filer could do likewise - in the way he always has - and simply alter the program *(or scan the appropriate code from a selection of printed bar codes giving specific weights of grind between faces and tops)*.

An alternative and simpler method would be to segregate sawblades into different stacks with corresponding wear, then load these separately with the weight of grind adjusted to suit. This would, though, cramp flexibility in some degree, and make such equipment less appealing to service shops.

The process need not be confined to stacked sawblades fed automatically - although this would be the most efficient system.

A cnc grinder could be loaded manually, scanned for the bar code, a suitable grinding depth entered *(judged according to the degree of tip wear)* and then left to set itself to go automatically into a grinding mode.

The advantages of such a system would be obvious for saw service companies - speedy setup, absolute repeatability of specification, the ability to mix sawblade diameter and specifications, and rapid turn-round.

The trade is already very close to this ideal situation, and this Century will see even more sophistication than the exceptionally high state of automation already reached.

Chapter 11
Saw Maintenance

Maintaining a circular saw involves carefully checking the condition of the saw body and then levelling and tensioning it to make good any defect found. To do this properly it is essential to understand the stresses on a circular saw when rotating and cutting.

This chapter deals only with parallel plate saws, either alloy steel or tipped with Stellite, tungsten carbide or diamond. Swage and ground-off saws are a special case and difficult to maintain - but are rarely used now in any event.

When rotating and cutting, circular saws should keep to a true line, completely free from wobble and deviation. To achieve this absolutely was once practically impossible - especially with large diameter sawblades - but body steels and preparation techniques have so improved that new sawblades from reputable makers run as true as it is practical to make them. Problems mainly occur with sawblades damaged in use which then run badly - problems left for the saw doctor or filer to make good as best he can.

Running faults

Erratic running can have one or more causes :-
Badly prepared teeth.
Running at the wrong speed.
A lumpy or twisted saw body.
Overheating of the sawblade in operation.

Badly-prepared teeth can result in the sawblade running off the cutting line or snaking in the cut - and a host of other problems. Keep to the guidelines for alloy saws, and retain the original profile and tooth angles and heights with tipped saws.

When the tension is wrong for the running speed the sawblade could lean to one side, wobble when stationary, run off, or snake when cutting.

Running problems will almost certainly result if the sawblade is run at a different speed than that which it was prepared.

For this reason it is important, when ordering sawblades, to state the running speed as well as all other relevant details. Sawblades ordered without stating the operating speed will normally be tensioned to run at 50M/sec (10,000 ft./min) rim speed.

On multiple-speed machines make sure that the correct speed is chosen for the sawblade(s) fitted - note the running speed on them or check with the makers to avoid mistakes. Never use sawblades on machines with arbor speeds different from that for which the sawblade was originally ordered.

Tooth and body heating is caused through friction created by the cutting action. Some tooth heat is dissipated into the timber, some into surrounding air, some into the sawdust being generated, and the remainder into the sawblade body. The heat may spread up to the root of the tooth - or even into the rim where it can stretch this to create wobble. The amount of heat varies with the efficiency of the saw tooth, the nature of the cut and the material being sawn. To avoid excessive heat build-up make sure that the tooth profile, hook angle and the tooth bite are correct for the application. Check also that the number of teeth in the cut when deep sawing is not excessive. Too many teeth heat the sawblade - and absorb more power.

Body heating can also result from sticky buildup on the sawblade sides, or from defects in the saw body. This is a particular problem with multiple saws, so use strob saws which are effective in clearing waste before it has chance to stick - but also maintain a good dust extraction system. Ensure that the amount of set on alloy saws - and kerf width relative to body thickness on tipped saws - is correct for the material being sawn.

Lumpy and twisted sawblades give running defects. Lumps and twist occur in saw bodies for many reasons, but should always be removed by careful hammering to make the body level and correctly tensioned.

Tensioning sawblades

A sawblade expands fractionally due to centrifugal force when rotated, but returns to its original condition when stopped. However, heat created by the cutting action and during re-tipping or tooth grinding, mainly stretches the rim - which does not return to its original state when stopped and allowed to cool. Instead the amount of rim stretch progressively increases to make the sawblade 'stiff'.

The combined effect of these heat stresses and the centrifugal force can be effectively counteracted by first pre-stressing the sawblade. When this is done correctly the rim remains steady - regardless of the running stresses - to hold-up consistently and so maintain a perfectly straight cut.

In pre-stressing the sawblade no attempt is made to contract the rim in order to counteract these induced stresses - this is not possible - instead the saw body is stretched within a specific area by hammering or rolling. The amount of artificial stress put into a saw body by doing this should, however, only be sufficient to counteract these forces - **and no more**.

The process of pre-stressing the saw body by stretching it is termed tensioning. It is an essential part of circular sawblade preparation for all woodworking applications - but is of much greater importance for sawblades that are thin-bodied, large in diameter or working under a heavy load.

Tension shows up by the way the saw body sags when the sawblade is removed and supported at a slight angle - the method, in fact, of judging just how much tension there is in a sawblade, see Fig. 11.6.

Tensioning does not last the life of the sawblade, unfortunately, if it did it would save a lot of bench work. Tensioning is slowly lost because of the cutting action and the effects of sharpening, so all sawblades need re-tensioning from time to time. Some sawblades are claimed to need little if any tensioning - and certainly body steel and saw body treatment has greatly improved over the years to give some credence this claim. However, it is wise to treat such claims with caution by checking and correcting tension regularly.

Saw tensioning equipment

Levelling and tensioning hammers

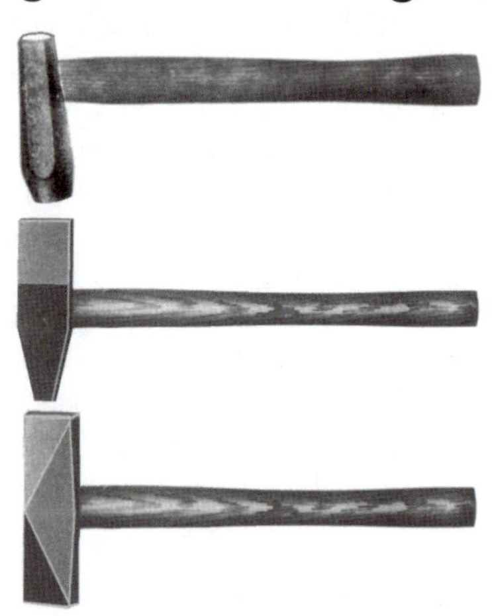

Fig. 11.1
Typical levelling hammers; dog head (top); cross-face (middle); and twist-face (bottom).

The dog head hammer gives a circular hammer blow which stretches steel equally in all directions, and is the type of hammer traditionally used for tensioning sawblades. Because the saw body really needs to be stretched more radially than tangentially when tensioning, however, the dog head hammer has less effect for the same effort than other types - it is actually the least efficient hammer for this purpose.

On the credit side there is less danger of a dog head hammer marking the saw body - and it does not need careful alignment relative to the sawblade as would other hammers. For these reasons it is easier to use when learning the art of saw hammering - and many continue to use only the dog head hammer for tensioning sawblades.

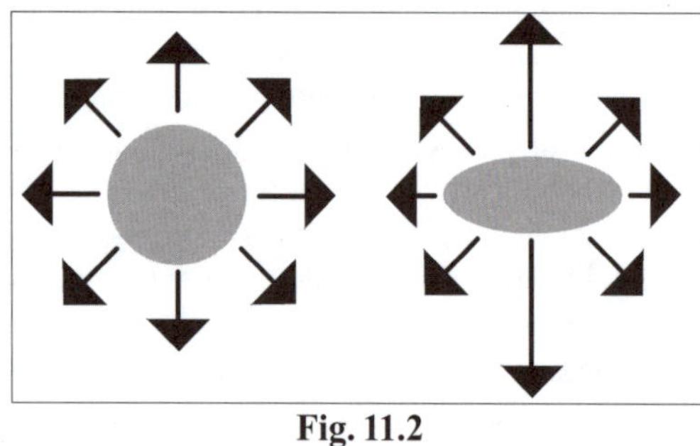

Fig. 11.2
A dog head hammer blow (left) and a cross-face or twist-face hammer blow (right). Arrows show the direction and magnitude of the forces applied.

The cross-face hammer gives elliptically-shaped hammer blows which stretches steel more in one direction than in the other. One of the two faces is at right angles to the hammer shaft and stretches steel more in line with the shaft, i.e., in line with the minor axis of the ellipse. The opposite face is in line with the hammer shaft and stretches steel more at right angles to the shaft. Both this and the twist-face hammer are traditionally used to remove lumps and twists.

The twist-face hammer has faces which give similar elliptically-shaped blows to the cross-face hammer, but these faces are both at 45 degrees to the hammer shaft. They stretch steel at 45 degrees to the line of the shaft outwards to the right with one face, and outwards to the left with the other.

Hammer weights and face sizes

For circular saws up to about 1200mm (48 in) diameter hammers weighing between 2kg (3-51bs.) are suitable. Heavier hammers are used for thick, large-diameter saws and lighter hammers for thin and small diameter saws. However, saw doctors and filers often have fixed ideas on these matters - and what suits one may not suit another.

The hammering surface of a dog-head should give a circular-shaped hammer-blow approximately 13mm (1/2in) diameter. Cross-face and twist face hammers should ideally give elliptically-shaped hammer-blows roughly 16 x 6mm (5/8 x 1/4in).

If the hammer surfaces are too flat - giving a large contact area - a much greater manual effort is needed by the saw doctor to stretch steel by the same amount - either by delivering more hammer blows, or by using a heavier hammer.

Sharper (more rounded) hammer surfaces give a smaller contact area, and expand steel more effectively - so less effort is needed by the user. However, sharp hammers more readily mark the saw body should the hammer be dropped badly. Marking the body of the saw should be avoided at all costs - a saw full of hammer marks shames a dedicated craftsman.

Many saw doctors freehand grind the surfaces of new cross-face and twist-face hammers to suit their own preference - or to restore hammers which have flattened through use. This is best done on an old type sandstone grinding wheel, grinding wet to keep heat to a minimum.

The radius of the curvature ground on a hammer face determines how sharp it is, so it is essential to grind them accurately - checking against templates of steel, aluminium, or plastic. These should be shaped to a predetermined radii, typically 230mm radius for the major axis and 40mm radius for the minor axis (9in. by 1 1/2in).

Test the hammers by dropping them on a lightly-oiled anvil to check that the shape and size of the hammer blow is correct. Each hammer should rebound upwards in the same arc as it is dropped - if it does not then the curvature has been ground off-centre.

No two craftsmen are alike, some prefer a flatter hammer and are prepared to use more effort, others prefer a sharper hammer and use lighter and more careful hammer blows.

Basically it is a compromise of hammer weights and shapes - to suit individual preferences. Many retain favourite hammers of known performance for many years - and are often loathe to loan them.

Anvil details

The steel-faced anvils used in circular saw maintenance are manufactured in various sizes and weights, typically 250 X 150mm (10 X 6in), weighing about 1 1/2 - 2cwts, and with a slightly crowned and highly polished face. Take care to avoid marking the anvil top as these marks transfer to the sawblade when hammering.

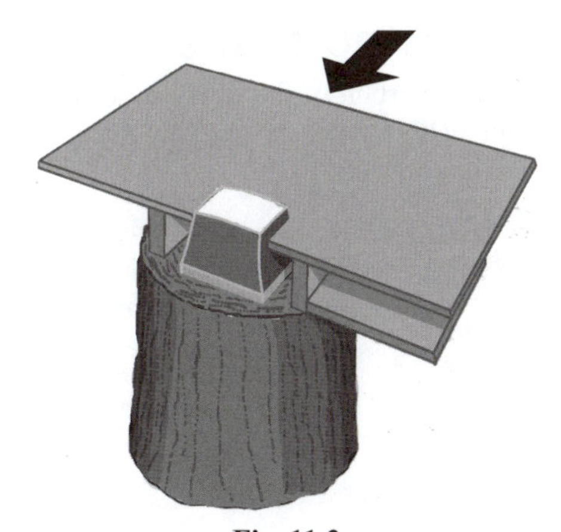

Fig. 11.3
This shows an anvil set into a workbench and supported (traditionally) on the butt of a large log. The direction of light should be as shown by the arrow.

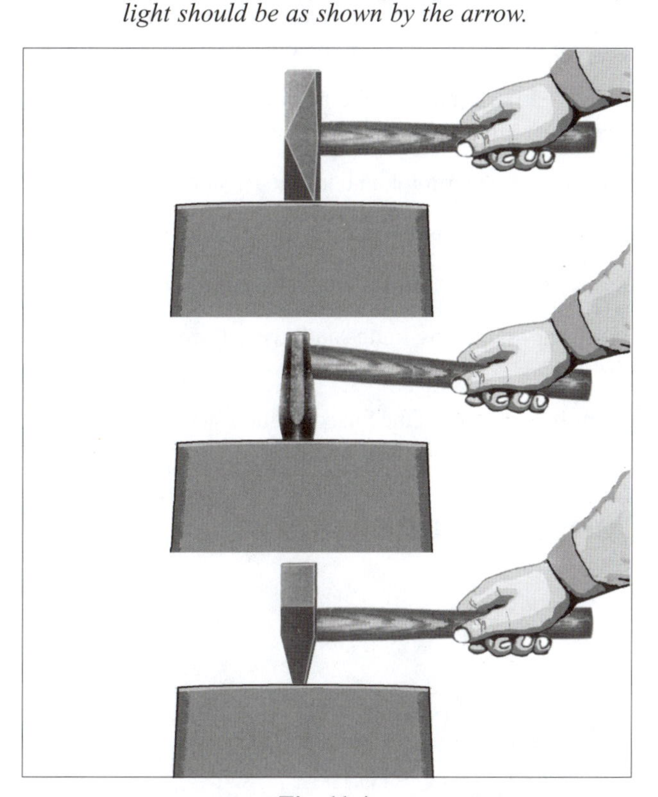

Fig. 11.4
The anvil should be the correct height, and the saw doctor or filer properly positioned to use it.
When each hammer is held in a comfortable grip on an extended arm the striking surface should lie flush with the anvil surface when directly on the crown.

Position the anvil in front of a window facing North or East - to avoid direct sunlight in northern countries (South and East in countries south of the equator). This should give enough light to allow proper examination of sawblades for tension when using straightedges, plainly showing any surface defects on the saw body in silhouette. Where daylight is not practical, a levelling bench with an illuminated back panel could be used, see Fig. 11.28

The anvil should be firmly seated, solid and free from vibration. Normally it is mounted either on a block of concrete or, more traditionally on the butt of a large log, to give a working height of approximately 840mm (2ft 9in).

The height really depends on the saw doctor or filer doing the hammering. The saw doctors or filers stance, whether levelling or tensioning and whatever hammer is used, should allow him to hammer comfortably and without strain. When he holds each hammer correctly with a fully outstretched arm, the hammering face of each should be in flush contact with the anvil directly on the crown. Note; there can be a difference in the alignment of hammer shafts - so a different grip might be needed for each.

The long axis on rectangular anvils should be parallel to the window wall. A level back-board made from hardwood should be placed between the anvil and the window, set roughly 40mm (1 1/2in) below its surface, or the anvil could be set into a bench top.

The remaining items of equipment needed are at least two steel straight edges, one corresponding to the average diameter of sawblade to be handled and the other to its radius. Useful sizes for sawblades of up to 760mm (30in) diameter are 305mm (12in) for **levelling and tensioning**, and 760mm (30in) for **plumbing-up**. Also required is an apron to prevent saw teeth catching on clothes, and a pair of heavy-duty industrial gloves.

Hammering practice

So much for the equipment needed, but the important aspect is in using them correctly. Craftsmanship cannot be learned from a book, but the hope is that this chapter will provide a reasonable basis for those wanting to take up the art of circular saw hammering.

The shaft of the hammer should not be held tightly as this restricts the rhythm needed in hammering - and could cause fatigue in the wrist. Grip the hammer more loosely at the moment of impact so that the rebound helps to raise the hammer for the next blow.

Constant practice on an old sawblade will help towards the complete mastery of the hammers. It is essential to get the right **feel** and sustain a **good rhythmic action** to stretch the steel **consistently**. Once these are learned it becomes easier to hammer sawblades of different thicknesses with the same set of hammers.

Hand coordination

Both when tensioning and when correcting defects, coordination of both hands is absolutely essential. The right hand* grips the hammer, while the left* grips and constantly moves the sawblade into position for the next hammer blow (it should be a gloved hand to avoid injury). See Figs. 11.7 - 12.

* *Vice versa for left-handed saw doctors and filers.*

Whatever a hammer is used it should always be aimed at the crown of the anvil. Striking off-centre gives a less solid blow and could distort or mark the sawblade. Between blows the sawblade has to be carefully repositioned so that the point for the next hammer blow is also directly over the anvil crown - so the hand two movements must be synchronized precisely something that requires considerable practice.

The anvil may not be visible - especially when hammering large diameter circular saws - so some saw doctors and filers fix a spotlight above the anvil pointing at the crown. This shows where to place sawblade for the next hammer blow, and the consistent point where the hammer should be aimed.

Although practice, patience and perseverance are all

required in learning the craft of saw hammering - it is essential to learn this thoroughly - it is needed repeatedly during the working life of every sawblade.

Straightedges and Tension Gauges

Unlike bandsaws, most saw doctors and filers use straightedges only for checking tension in circular saws - at least for saws up to 1M (40in) diameter. Two straightedges are needed for fully checking each sawblade, a long one for general examination and for plumbing-up the sawblade, and a short one for checking tension in more detail, examining the collar area, and checking for lumps and ridges.

The long straightedge should be slightly shorter than the sawblade it is to examine, and the short straightedge should be slightly shorter than the distance from the centre hole to the gullet bases. For this reason different lengths of long and short straightedges are ideally needed for different sawblade diameters, although most use standard straightedges for a range of sawblades.

Fig. 11.5

This adjustable tension gauge incorporates a wedge ***'B'*** *to vary the amount the gauging strip* ***'A'*** *is deflected into a convex outer edge to match the tension required in the saw being checked. Thumbscrew* ***'C'*** *locks the wedge in place.*

Source - Hanchett

A tension gauge is a short 'straightedge' with a slightly convex contact edge to match the saw body profile precisely *(when testing for tension)* and is used only for testing for tension - not for checking lumps or ridges.

Tension gauges can be made by the saw doctor or filer from an existing straightedge, but are not available as such from equipment makers, at least as far as the writer is aware. In any event such a tension gauge would be suitable only for a specific diameter of sawblade to run at a given speed.

A more useful, variable, tension gauge is available from Hanchett. This infinitely adjusts between straight and convex, and is for use with sawblades from 1M12 (44in.) to 1M50 (60in.) diameter.

The standard gauge is 610mm (24in) long and consists of an outer, rigid frame supporting an inner tension strip - the curvature of which is controlled by an adjustable wedge locked in place by a thumbscrew.

Non-standard gauges of from 460mm (18in) to 840mm (33in) are also available to order. However, the standard gauge is suitable for wide a range of saw diameters if used as described later.

The adjustable tension gauge provides a much more precise check of the condition of a sawblade than the regular method of estimating the gap between the saw body and a straightedge.

Being adjustable it can take into account the factors which vary the degree of tension required in sawblades - diameter, running speed, saw body thickness and the feed speed/ depth of cut. See notes under 'Tensioning large saws'.

Saw doctoring terms

Fast or Tight - a portion of the sawblade that contains no tension, traditionally marked by chalking a pair of brackets with a straight line between (-).

Loose - a portion of the sawblade that is over-tensioned, usually indicated by a pair of brackets with a circle between (O). The steel in the *"loose"* area has been stretched too much.

Lump - a distorted portion of the sawblade showing concave on one side and convex on the other. Lumps may be circular or ridged *(elongated in one direction)*, with the ridge running in any direction.

Twist - overstretched sections of the sawblade that distort the rim.

Dish - a condition when a sawblade is over-tensioned on one side and under-tensioned on the other.

Blister - a burn mark on the body of the sawblade, usually indicating a circular lump.

Stiff - a general term for an under-tensioned sawblade.

Opening the sawblade - putting tension into the sawblade.

Drawing the sawblade out - reducing excessive tension.

Levelling - removing lumps and twists from the body of the sawblade.

Plumbing-up - one of the final operations with the sawblade is stood up vertically to find defects.

Tension examination

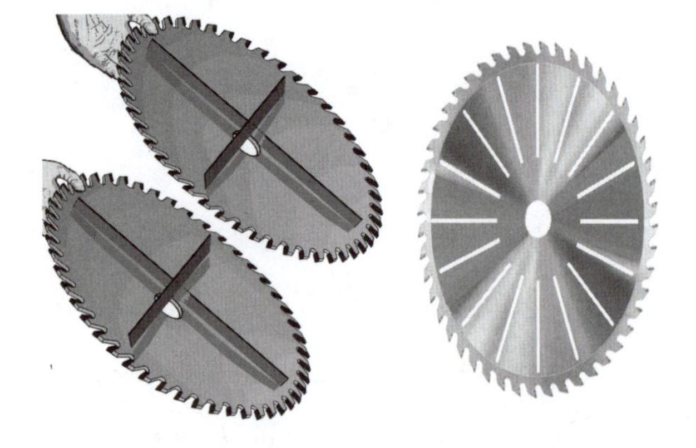

Fig. 11.6

Left: A tensioned sawblade conforms to a dish when supported at an angle (top) while an untensioned sawblade becomes roughly saddle-shaped (bottom). The straightedges show how both sawblades drop virtually by the same amount when a straightedge is placed between the hand support and the back-board, but quite differently where the straightedge is placed across the sawblade at right angles.
Right: Tension is checked using a straight-edge placed at equally-spaced radial positions. To aid this draw chalk marks to show where checks should be made.

A tensioned sawblade is one that has been artificially stressed by hammering to stretch the body of the saw to counteract running and cutting stresses.

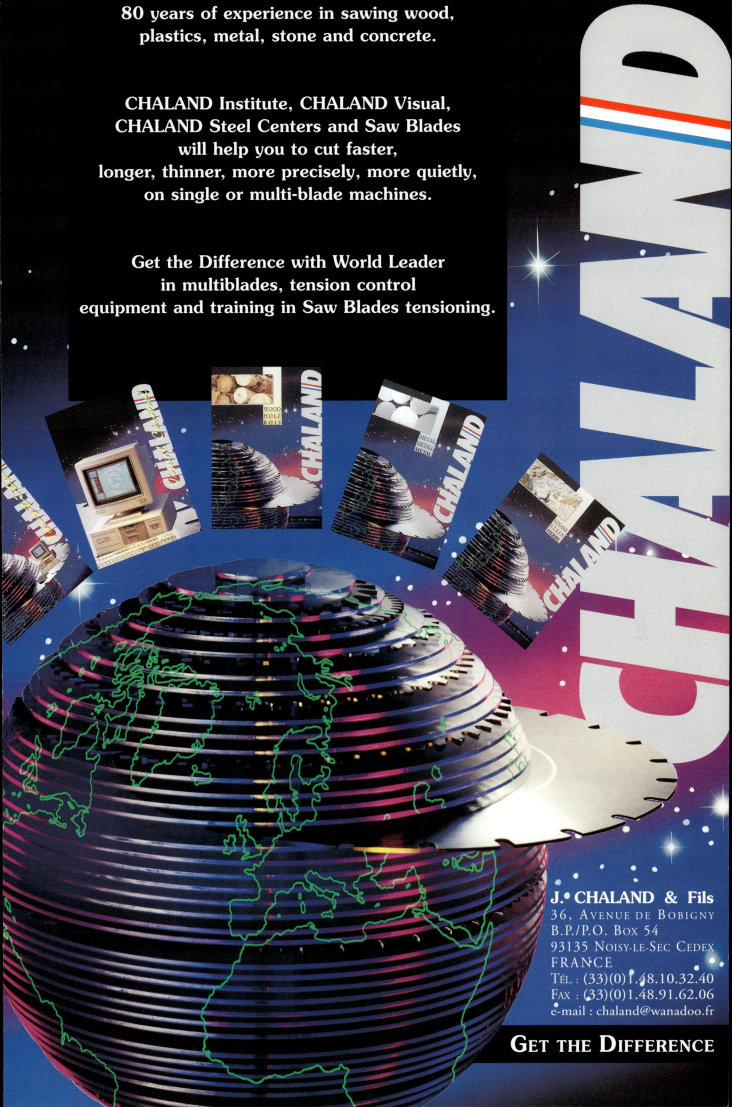

Didac Limited

DIDAC

Didac Limited

Didac Limited employs skilled staff, ensuring that all services are to a very high standard, ensuring complete customer satisfaction. All services are tailored to the individual company needs. Didac is committed to a policy of providing equal opportunity.

Didac Limited is a recognised training provider, delivering training and other services to the Wood and Associate Industries.

The company is accredited with City & Guilds, Edexcel, RTITB (Road Transport Industry Training Board), CIEH (Chartered Institute of Environmental Health), IMWoodT, FFINTO (Furniture, Furnishings and Interiors National Training Organisation) and the BWF (British Woodworking Federation).

Services include :

- **CNC Training**
- **Computer Aided Design and Computer Aided Manufacture (CAD/CAM)**
- **Classical Machine Woodworking Training for new and existing staff**
- **Abrasive Wheel Training**
- **Health a Safety**
  - ***basic Health & Safety Course, certificated by CIEH***
  - **manual Handling Training, tailored made course or certificated by CIEH**
  - **machine audits in line with PUWER 98 and ACOP**
  - **Safe use of Woodworking Machinery**
  - **general advice on policies, risk assessments, safety audits, legislation updates**
- **Fork Lift Truck Training National Vocational Qualifications (NVQ's) - machine woodworking, saw milling, hand crafted furniture, furniture production, performing manufacturing operations.**
- **Apprentice Training within the Wood and Associate Industries.**

For further information, please contact Jon Gibson.

Didac Limited

Kingswood House,
South Road,
Kingswood,
Bristol BS 15 8JF

Tel 0117 9049377
Fax 0117 9150009
Email email@didac co uk
Web www didac co uk

City & Guilds

FFINTO
Corporate Member

The Institute of Machine Woodworking Technology

The Institute was founded in 1952 to support and protect the status and interests of suitably qualified persons in the machine woodworking industry, and to promote serious study and good practice. Membership grades include Student, Associate, Member, Fellow and Affiliate.

The institute is involved in many activities, including meetings, lectures, works visits, SkillBuild and liaison with many safety and training organizations. It also produces a twice-per-year journal, newsletters at regular intervals, and has a website on the Internet giving the latest news and information from the Institute.

The Institute has a stand at the Woodmex Exhibition where apprentices display their skills in manufacturing wooden articles sold for Charities.

A register is maintained by the Hon. Secretary of Trainers and Assessors for both machine and hand woodworking - also for most industries associated with wood. Our registered membership have between them many years of sound experience in the many specialized areas of machine woodworking, and cover all parts of the country.

Those requiring advice or training in any of these tields should first contact the Hon. Secretary at Website:http://www.imwoodt.org.uk.

Hon. Secretary John Fryer
Website:http://www.imwoodt.org.uk

WMSA

WOODWORKING MACHINERY SUPPLIERS ASSOCIATION

WOODWORKING MACHINERY SUPPLIERS ASSOCIATION

The WMSA is the leading UK Trade Association representing suppliers of woodworking machinery, both engaged in manufacture and import of products. WMSA members are established suppliers of woodworking machinery, tooling and dust extraction equipment and aim to provide a high standard of service and reliability.

Woodmex Exhibition:

Owned and sponsored by the WMSA, Woodmex takes place every two years at the NEC, Birmingham. To find out more please visit - www.woodmex.com

WMSA Directory:

For further information and to receive your free copy of the WMSA Directory of members please contact the WMSA:

WMSA

The Counting House,
Sir Richard Arkwright's
Cromford Mill, Mill Lane,
Cromford, Derbyshire,
DE4 3RQ

Tel: +44(0) 1629 826998
Fax: +44(0) 1629 826997
Email: info@wmsa.org.uk

This is always the first operation before attempting to hammer - fully examining the sawblade to find the degree of tension in it and the position of any lumps or twists.

While the saw doctor or filer is checking the sawblade all the defects should be clearly marked and the degree of tension noted. In this way he can see the overall condition of the sawblade and judge where to hammer both to correct the defects and tension the sawblade at the same time.

Before starting, rub a little oil both on the surfaces of the sawblade and on the anvil. This will make it easier to move the sawblade across the anvil while examining and hammering it. The oil also helps to cushion the blows from the hammer and allows defects to be marked on the saw body using a finger rather than a piece of chalk.

Theory of saw tension

Tensioning profile

Many saw doctors and filers believe a properly tensioned sawblade should conform more or less to a regular curve from collar to rim, but the best results are from sawblades that are basically flat in the centre section and then bend upwards either side into a slow curve towards the rim.

For sawblades to run and cut successfully the saw body should conform as closely as practical to the ideal sawblade profile shown in Fig. 11.14.

Checking the saw body for tension

Fig. 11.7
A tensioned sawblade drops in the centre when checked with a long straightedge placed at as shown.
NOTE: the daylight gap in this and following sketches is purposely exaggerated to show the effect more clearly. The actual gap, between the straightedge and its reflection from the body of the saw, is really very small.

As a general check for overall tension, support the sawblade with the left hand at an angle of about 30 degrees to horizontal, with the opposite teeth resting on the rear back-board. Place a long straightedge fully across the sawblade at right angles to an imaginary line joining the left hand and the back-board. Repeat this check at several radial positions.

A tensioned sawblade will show a daylight gap (drop) between the straightedge in the centre section, while a flat, untensioned sawblade will show a daylight gap at both sides, see Fig. 11.6.

Large diameter sawblades show this effect clearly, but correctly-tensioned sawblades of under 750mm (30in) diameter may not readily show up in this way - and might then be wrongly diagnosed as **fast**. With these, put slight hand-pressure on the saw centre to show the gap that indicates correct tension at this point, see Fig. 11.10.

Fig. 11.8
A sawblade without tension drops towards both edges when checked with a long straightedge.

Turn the sawblade over and check again. A properly tensioned sawblade will show the same amount of drop in the centre. If the drop is different then the sawblade is slightly dished. If the sawblade shows the rim dropping away on one side, but the centre dropping away on the other side, then the saw is **permanently** dished and cannot be used in this state *(permanent, that is, until corrected)*.

To check tension in detail support the sawblade in the same manner, but use a short straightedge between the saw centre and just below the gullets. Hold the short straightedge in the same line as described for the long straightedge - and square to the face of the saw body *(holding it at an angle will give a misleading result)*.

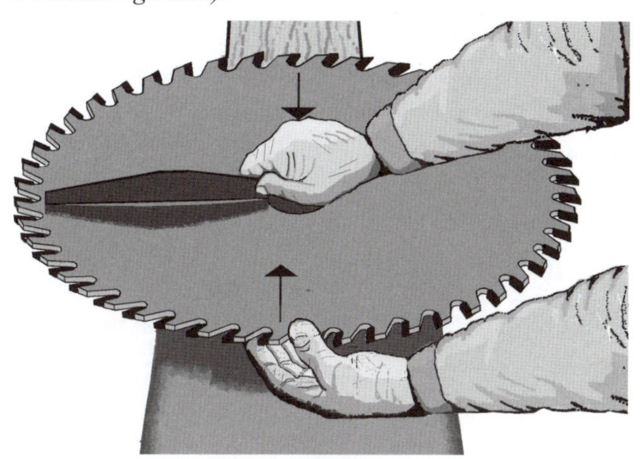

Fig. 11.9
To tension-test a large diameter sawblade simply support this with the left hand.
For medium diameter sawblades press up with the left hand and resist this pressure with the palm of the right hand.
In both cases a tensioned sawblade should show a slight amount of daylight between the straightedge and the saw.

Check for tension at a number of radial positions around the sawblade, then turn the sawblade over and repeat the process. The daylight gap, seen roughly midway between the saw centre and the teeth, should be consistent at all points checked.

For sawblades between 760mm (30in) and 450 mm (18in) diameter place the palm of the right hand on the collar region of the sawblade and close to the centre hole, holding the short straightedge between thumb and forefinger. Apply an upward pressure with the left hand while resisting this pressure with the right hand.

A correctly tensioned sawblade will show a daylight gap when pressure is applied - but may not do otherwise. If hand pressure is released and re-applied, the daylight will likely **blink** - appear and disappear. If no daylight is seen then the section of sawblade being examined is **fast,** i.e. it requires tensioning.

Fig. 11.10
Hold small sawblades by their edge as shown, and apply a twisting pressure between the thumb and fingers of the left hand.

To learn the correct pressure to apply to small diameter saws *(and what daylight should be seen)* practice by tension-checking several new and correctly-tensioned sawblades in the manner described above.

Mentally record the position and amount of daylight and the pressure needed to show it.

Fig. 11.11
A sawblade without tension drops away towards the edge.

For sawblades under 450mm (18in) diameter. Place the right hand as above, but with the four fingers of the left hand on top of the sawblade and the thumb only underneath. Exert a downward pressure with the four fingers and a counteracting upward pressure with the thumb to make the saw can be made to **blink**.

A properly tensioned sawblade needs only slight pressure to show daylight - undue pressure can show a thin-bodied sawblade as being correctly tensioned when it is fast. The skill is to apply just the right amount of pressure - and know the right degree of daylight that should be seen as a result. These factors all vary with the diameter and running speed of the sawblade, its plate thickness and hardness and quality of steel.

Fig. 11.12
An over-tensioned sawblade shows too much daylight.

Similarly check new, larger-diameter saws *(which need no pressure to check tension)* and again memorize in the same way. Practising this will give an excellent guide when tensioning sawblades of identical specification, as well as giving a reasonable guide for others. As a guide for how much tension should be in a sawblade, see the table below.

Table of typical figures for natural drop

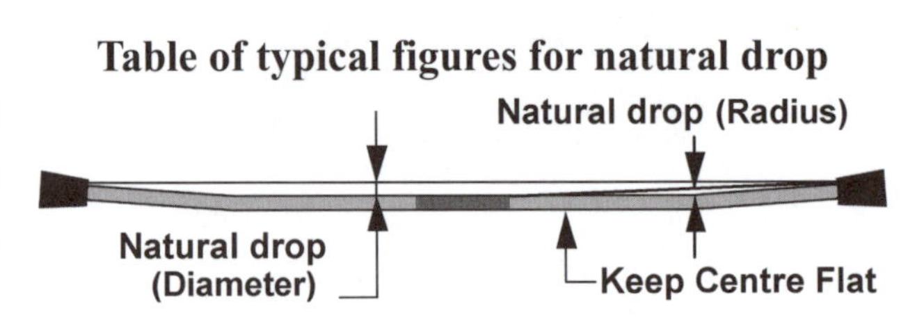

Figures for natural drop *(without pressure)* interpolated from sample sawblades tensioned to run at 50M/sec. (10,000 revs/min) are as follows :-

Saw diameter, mm. **Saw diameter, in.**	**750** **30**	**500** **20**	**300** **12**
Diameter drop, mm.	0.015	0.010	0.006
Diameter drop, in.	0.0006	0.0004	0.00025
Radial drop, mm.	0.005	0.0035	0.002
Radial drop, in.	0.0002	0.00013	0.00008

The radial drop amounts are very small - hardly measurable in fact. However, if the figures for diameter drop correspond, and the all drops tested radially with a short straightedge show consistent, then tension should certainly be in the ballpark.

These figures do not apply to all circular saws, but the author at least hopes they give some sort of guide for sawblades of average gauge and application.

Tensioning large diameter sawblades

For regular, small diameter sawblades the 'daylight' method of checking tension is acceptable.

For large diameter sawblades, however, tension is much more critical - and this method may not be good enough unless the saw doctor or filer is highly experienced. More precise and dependable results can be obtained by using an adjustable tension gauge, see Fig. 11.5.

In practice the amount of tension needed for any sawblade is dependent primarily upon the speed the sawblade is run at - the above figures relate to sawblades intended to run at 50m/sec. (10,000ft./min). Certain other factors also affect the amount of tension needed.

Basically more tension is needed if :-
A sawblade is run at a higher feed speed
Using thinner sawblade
The sawblade is to be used for harder or more dense timber
The feed speed or depth of cut is above normal.

The amount of tension needed is also partially dependent upon the quality and temper of the saw body - harder sawblades need less tension.

Specific instructions cannot be given in a book such as this because of the many variable factors. It is up to the sa doctor or filer to make his own judgement based on ex perience, observation and common sense. What is of paramou-nt importance is that tension in a sawblade is even all ar und and equal on both sides. Even if the amount of tensio-n is not correct, this will give a better results than from a sa blade of generally correct, but uneven, tension.

At first tension a new sawblade by the minimum amount, then gradually increase this - carefully noting the affect this has on the way the sawblade runs and performs - until the best running condition is realized. At this point note the precise amount of tension and use this as a starting point for similar sawblades.

Using an adjustable tension gauge

This type of gauge precisely fits the saw body when checking for tension.

The amount of convex curvature of the tension-checking edge can be varied in small increments simply by adjusting a wedge. This allows the tension gauge to be precisely matched to any sawblade once the correct amount of tension has been established.

Once set for a particular sawblade, of course, the setting can be quickly repeated by following the graduations on the wedge.

The comments below assume the use of a straightedge in checking tension. When using an adjustable tension gauge the same general remarks apply - but the sawblade should contact the gauge evenly - no daylight should be seen.

Noting the results of tension checking

In checking each section of the sawblade for tension, carefully note the position and amount of daylight gap between the straightedge and the saw body and mark accordingly.

If the daylight gap is correct for the particular sawblade type and remains consistent for both sides and all radial positions, then the sawblade is correctly tensioned and requires neither marking nor further treatment.

If the surface of the sawblade falls away to show daylight at the rim but contacts the saw body elsewhere - or does not fall at any point - then that section is **fast** - so the body of the saw requires hammering. These sections should be bracketed using chalk with a dash between (-).

If the daylight gap seen is too large in comparison to what it should be, then that section of the saw is **loose** - it has been stretched too much and requires hammering near the rim at this point. Bracket all these sections with a circle between (0).

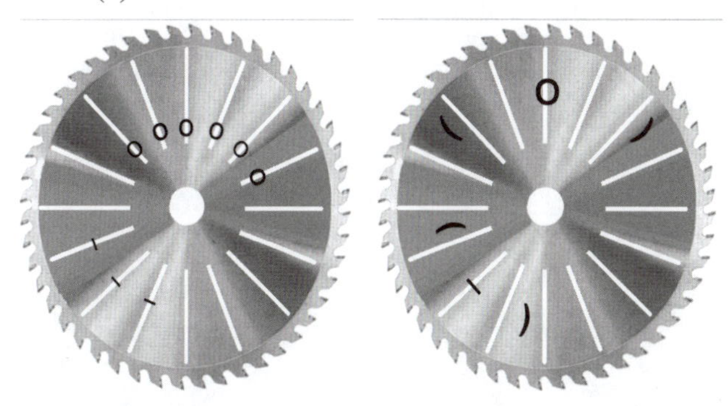

Fig 11.13
Left: Initially a sawblade may be marked as shown here, with some 'loose' sections opposite 'fast' sections. Right; These can be resolved into bracketted 'fast' and 'loose' sections.

When the sawblade shows a consistent **fast** or **loose** condition at both sides and all radial positions, then the sawblade needs hammering evenly all round - a **fast** saw in the body and a **loose** saw near the rim. See Figs 11.14-16.

If the sawblade shows even tension all round on each side, but the amount of tension is unbalanced between the two sides - a slight dish - then light hammering should be carried out as follows:-

*Carefully examine both sides of the sawblade for tension, and note the side which most closely matches the recommended tension profile. If the opposite side appears slightly **fast**, then lightly hammer in the **body area** on this side. If the opposite side appears slightly **loose** then hammer lightly in the **rim area** on this side.*

Hammering techniques

When learning the art of hammering it helps to mark the sawblade with both radial lines and concentric circles at equal spacings - both sides the same - to show where hammer blows should fall. It isn't necessary to show a concentric circle for each hammer blow, but include at least four concentric circles as a general guide.

Hammer along these radial lines, on or between the concentric circles, using the same number and weight of hammer blows at each radial line and on both sides of the sawblade. If one side of the sawblade is hammered more than the other, the daylight seen when checking tension will be unequal. In extreme cases the saw may become temporarily dished, i.e., concave on one side and convex on the other.

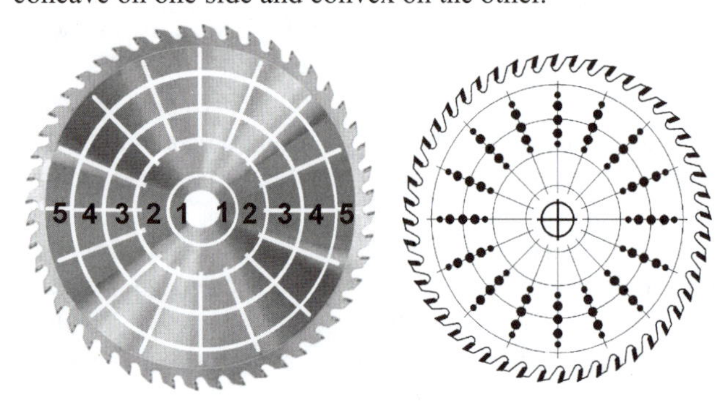

Fig. 11.14
*Left: Before hammering, mark the saw radially and with concentric circles to show where hammerblows should fall. Right: To make sawblade movements simpler when tensioning a **fast** sawblade, hammer along any radial line moving inwards, then hammer on the adjacent radial line moving outwards, and so on. Hammer more heavily within the saw body (as shown by large circles) and lighter towards the rim and centre (as shown by small circles).*

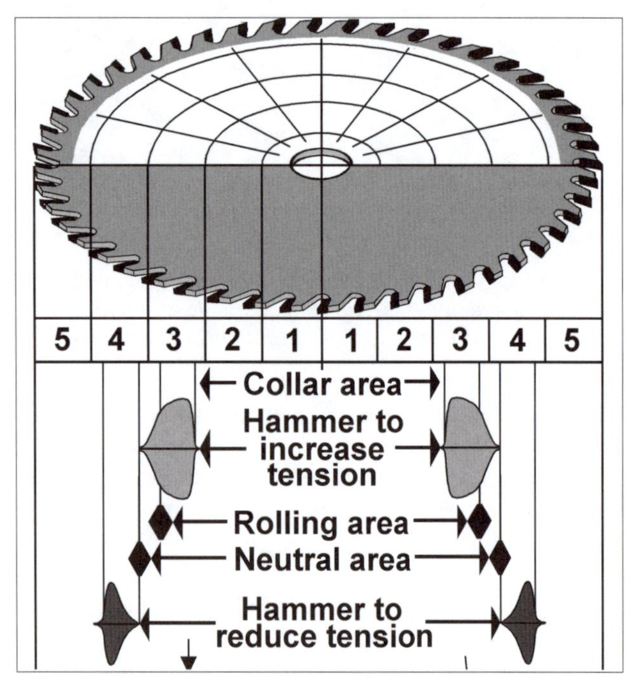

Fig. 11.15
This shows how a sawblade is divided by four concentric circles into 5 sections to identify areas where hammering should take place.
The width of the enclosed areas (lower sketch) indicates the differing affect of equal hammer blows in the amount by which the saw is stretched - lesser width shows little effect, greater width shows more effect.
Around the neutral area hammering has least effect, outwards towards the rim it reduces tension, and inwards towards the centre it increases tension.

Which hammer to use ?

The traditional hammer for tensioning saws is the dog head - cross-face and twist-face hammers are normally used only for correcting defects in the saw body.

*Because cross-face and twist-face hammers stretch steel more in one direction than the other, they are more effective in tensioning than the dog-head type - but they need more skill in use to stretch the saw body **only** in a radial direction.*

It is more difficult to ensure that each hammer blow falls correctly when working quickly - in the right place and *(in the case of cross and twist-face hammers)* with the hammer correctly aligned.

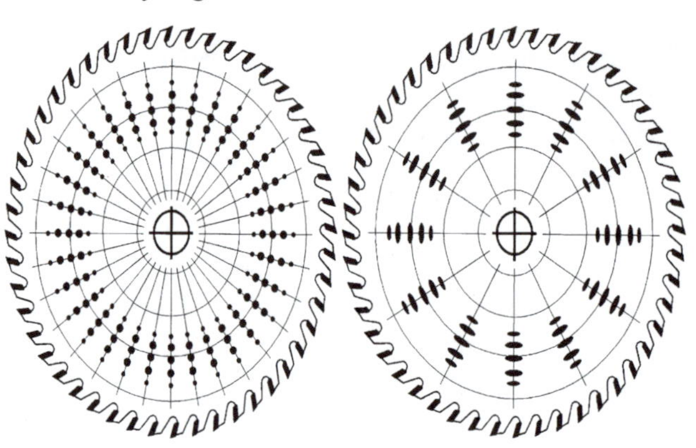

Fig. 11.16
Left: To give more even tension on large diameter sawblades increase the number of radial lines and, in so doing, the number of hammer-blows.
Right:Both cross-face and twist-face hammers can be used for tensioning as their blows are more effective.
Take care, though, to align the minor axis of the blows along a radial line.

The stance taken up by the saw doctor or filer - and which hammer and which face he uses - are critical when using cross-face and twist-face hammers for tensioning sawblades.

It is essential that the minor axis of the elliptically-shaped hammer blows always fall radially - to stretch the steel only in this direction.

Mis-aligning the hammer blows will twist the sawblade - almost certainly cause it to wobble in running and make levelling and tensioning much more difficult.

Hammering fast sawblades

Because sawblades normally become stretched in the rim area, adding tension to the saw body is the most regular operation in saw maintenance. The sawblade first becomes **stiff** when some tension is lost then, if the condition is not corrected, the sawblade progressively becomes **fast**, so that the rim tends to wander in the cut.

Hammer-blows for a fast sawblade should be as shown on Figs. 11.15-16. Divide the distance from the saw centre to the rim roughly into five segments for hammering as follows :- none within the inner two segments *(the collar area)* most within the outer half of the third segment and the inner half of the fourth segment, and none in the fifth segment *(the rim area)*.

The intensity of the hammer blows are shown as larger circles or ellipses for heavy blows, and smaller ones for lighter blows.

The best hammering technique is to complete all the hammer blows in a single radial line, moving inwards, then rotate the sawblade to position it for the next radial line to hammer along this line moving outwards, and so on.

To do this the saw doctor or filers right hand delivers the hammer blows, while the left hand controls the movement of the sawblade.

With large and heavy sawblades the saw doctor or filer

can use his right thigh to take the weight of the sawblade and use it to assist the movement - standing with the right foot slightly in front of the left.

Hammering loose sawblades

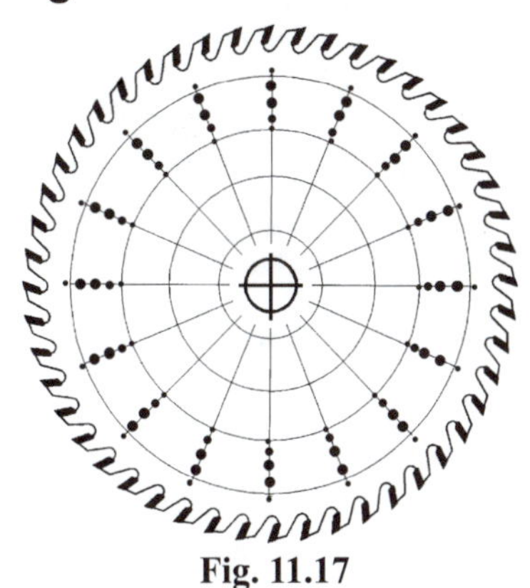

Fig. 11.17
Hammering to reduce tension in a loose sawblade.

Loose sawblades are those which have too much tension through overzealous hammering, or as a result of overheating in the saw body. In both cases hammering mainly takes place closer to the rim, mostly in the fourth segment.

Checking the collar area

It is important to keep the collar area flat using minimal hammering. In fact there is an old saw doctoring saying worth remembering *"Keep of the eye (centre bore) - never get it loose"*. If the collar area does become **loose** then the whole sawblade will wobble and wander from its true sawing line - or it may even become **dished**.

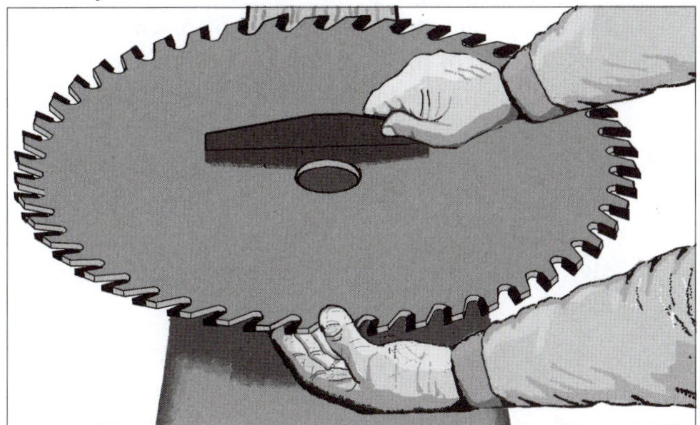

Fig. 11.18
The collar area should be kept flat, as shown here, with no daylight visible.

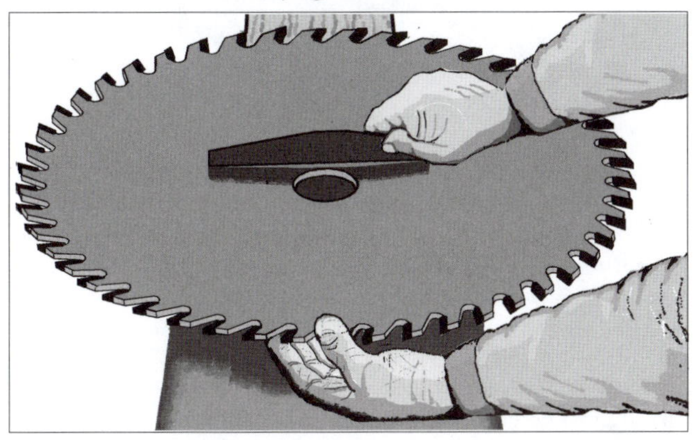

Fig. 11.19
If the saw collar area becomes loose, then hammer from the collar area outwards to correct this condition.

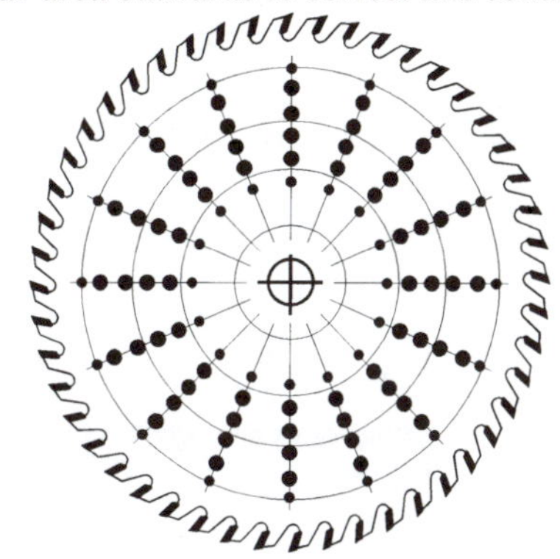

Fig. 11.20
Hammer as shown here to cure a loose eye or a slight dish.

To examine the collar area hold the short straight edge as in Figs. 11.18-19. If correct this should show virtually no daylight.

Apply the usual slight pressures if checking small diameter sawblades - which should show just the merest amount of **blinking.**

When a sawblade has a loose eye carefully check the amount of tension in the body of the saw. If the only defect is a loose eye, then hammer blows should of equal intensity along the radial lines. If not, then make good any tension defects by hammering more intensely in the body area if the saw is fast, or more intensely near the rim if the saw is loose.

Correcting combined defects.

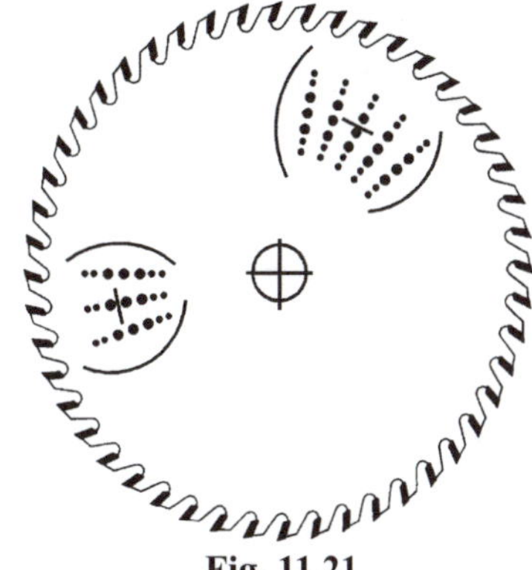

Fig. 11.21
Hammering a sawblade that is fast in only a few places.

In only a few sawblades will the remaining tension be even throughout, in which case correction is as described above.

Fig. 11.22
Normally use a dog head hammer for removing fast places.

In most cases tension will have become irregular, for example, altering from acceptable to either loose or fast at different points around the sawblade. In this case bracket the fast sections (-) and the loose sections (0), and treat as follows:-.

When just one or two small sections are fast, then hammer these sections only. If several small, fast places show up, first check that a lump or ridge is not causing this defect - if so make this good by levelling before tensioning - see later notes.

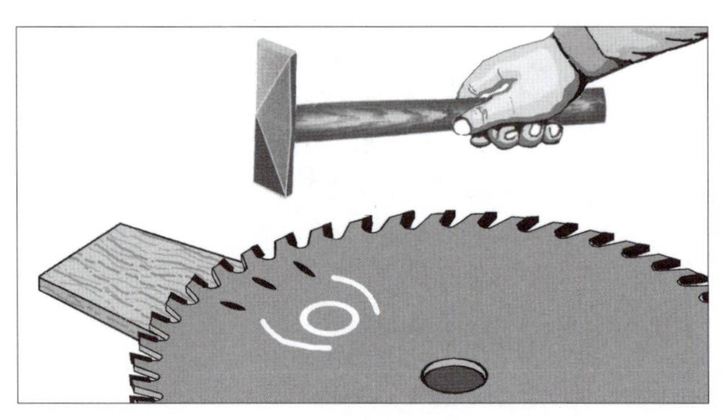

Fig. 11.23
Removing narrow loose places inward of the rim using the twist-face hammer. Take care to align the long face to the radial lines - or chance twisting the sawblade.

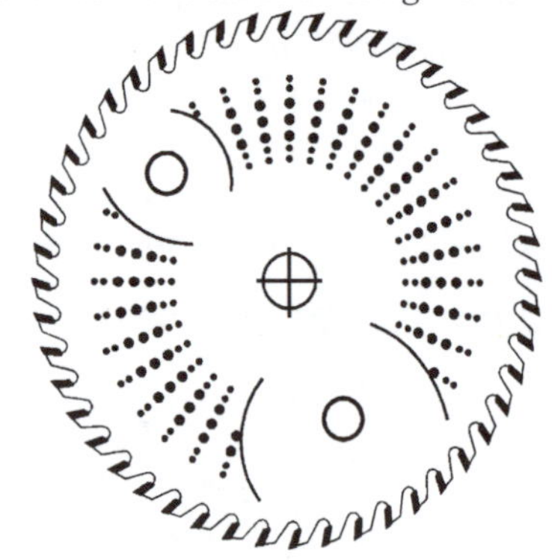

Fig. 11.24
Treatment of a sawblade that is loose in a few narrow sections but fast elsewhere.

When the sawblade shows a loose section inward of the rim in one or two narrow sections but correct elsewhere, then stretch the rim just at these points, possibly using a twist-face hammer for maximum effect.

If the sawblade is loose in the rim in one or two places but fast elsewhere, than hammer to increase tension everywhere but in the places marked as loose. If the sawblade is extremely loose in wider sections then ridges in the rim or a twist may be the cause, so treat these first by levelling.

If the sawblade has both fast and loose places, then hammer all but the loose sections initially, then further hammer the fast sections if the intial hammering does not also make these good.

Levelling the sawblade

Levelling the sawblade is the process of examining the saw body in detail, marking any defects, and then hammering it accordingly to make the defects good.

In levelling the saw doctor or filer first finds and outlines lumps, ridges and twists, and brackets fast and loose places, marking accordingly.

By defect marking the sawblade in this way its overall condition can be clearly seen - so that the most effective hammering can take place - hammering-out defects found individually can actually worsen the problem!

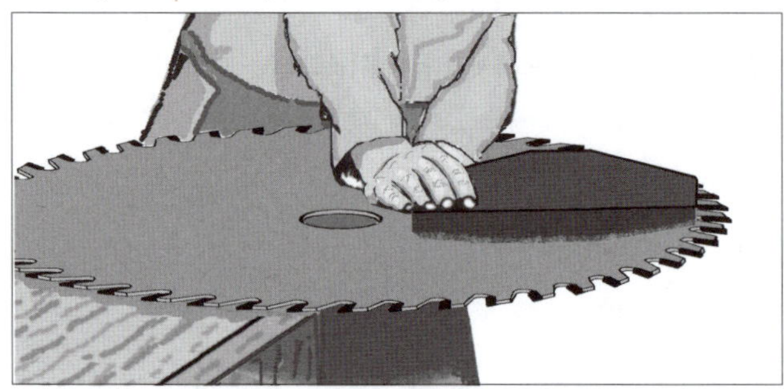

Fig. 11.25
Checking for level. With the sawblade supported by the anvil and the back-board, move the short straightedge around the saw body to check for lumps.

For example, a lump may show the area around it to be fast - but this could be restored to correct tension when the lump is hammered out. In other cases lumps or ridges can be 'drawn out' by hammering close to them rather than on them - it all depends upon the overall condition of the sawblade.

For this reason levelling and tensioning can never be treated as separate process - it is impossible to level a circular saw without regard to its state of tension - the two operations must be combined.

Use the short straightedge to check the condition of the sawblade both radially and tangentially. Ring the raised section of any defect found.

These regularly take the form of burns, lumps, ridges and twists, showing up as hollow on one side and raised on the opposite side.

The defects are most easily found by placing the short straightedge on top of the lump where daylight can be seen between the straightedge and the saw body on either side.

These defects can occur on one or both sides of the saw body, so turn the sawblade over to check it completely.

To determine the size and shape of lumps and ridges, place the sawblade flat on the surface of the anvil, then move it slightly away so that the teeth furthest away rest on the back-board or bench top.

Fig. 11.26
To check the size and position of a lump, rotate a straightedge on it. When the gap shows equally all round then the straightedge is directly on the crown.

Check the upper surface of the sawblade by moving short straight *(held at right angles to the saw body)* completely over the area to be checked - while slowly moving the sawblade around with the left hand.

When a lump is found, move the straightedge around

this, at various angles, to find the high spot. A round lump shows the same degree of daylight all round when the straightedge is turned on the high spot.

A ridge will show the maximum amount of daylight on either side when at right angles to the line of the ridge, but little, if any, when placed along the ridge.

The line of a ridge is best found by noting the direction the straight edge has to be moved to keep its mid point in contact with the ridge. Once the rough direction has been noted, hold the straightedge at right angles to the ridge line to show it more clearly. Then carefully outline the ridge to show its shape, size and direction.

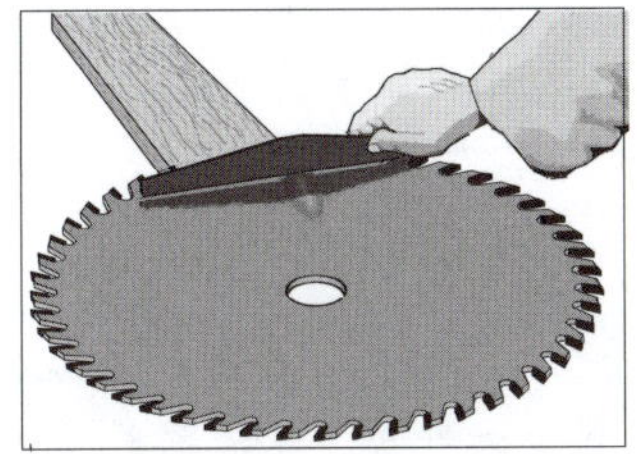

Fig. 11.27
Left: Finding the line of a tangential ridge by moving the short straightedge along it.
Right: Finding the line of a radial ridge.

Fig. 11.28
Although levelling is regularly carried out on a crowned anvil, it is easier to use a flat, round levelling block. This is especially so for tipped saws as care has to be taken when hammering near the rim to avoid damaging the tips. This levelling bench has switchable round blocks of different sizes, and an illuminated back panel. A replaceable UHMW, heavy-duty plastic work surface allows sawblades of all types to be rested or stood up on it without danger of damage to the teeth.
Source - Armstrong

Removing defects

Burns, lumps and ridges are regularly removed by hammering around the centre of the raised side. Doing this, though, may also stretch the sawblade in the immediate area - and so upset the tension.

To prevent this happening pad the anvil surface with leather, thin card or brown paper to cushion the effect - and vary the impact point slightly for each successive blow. Alternatively use an oval-section wooden block instead of a hammer. Doing either tends to flatten the lump without either knocking it through or altering the tension. After levelling, though, check and correct tension.

Burn marks or black eyes

A burn mark on the surface of the sawblade, sometimes called a **blister** or a **black eye**, is often a circular lump. It is usually caused by excessive local heat through friction between the saw body and the timber - often aggravated by a buildup of resinous deposits. It expands the saw body to show as high on one side of the sawblade and as hollow on the other side.

Burn marks show up clearly because of the distinctive blue and purple colouring - the result of surface oxidation of the steel - mostly seen on the hollow side of the 'blister'. To remove this colour, pour a few drops of well-diluted hydrochloric acid on it, then spread with a hand rag until the colours disappear.

To neutralize the acid rub a chalk stick against a file to deposit chalk dust on the mark, then go over the whole area with an oily rag.

Do not attempt to remove oxidation with an abrasive cloth as this leaves scratches on the surface which encourage sticky deposits. Removing the burn mark, though, does not remove the lump. This must be levelled or the effect will worsen - and may even start a crack in the centre if the sawblade is run with the lump untreated.

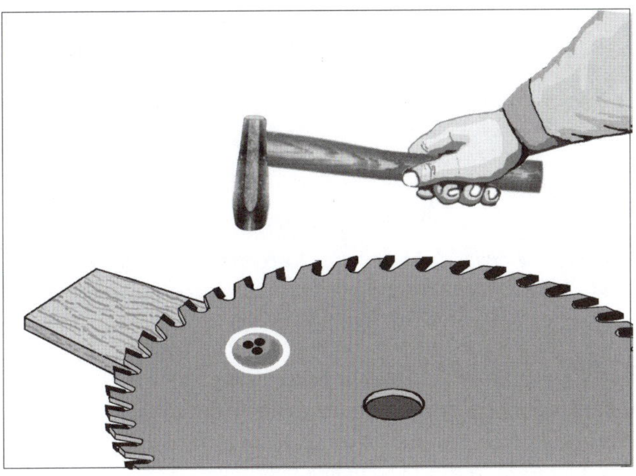

Fig. 11.29
Removing the lump associated with a burn mark using a dog-head hammer.

For small lumps, of up to 1in (25mm) in diameter, hammer directly on the lump using the doghead hammer *(this is practically the only time the doghead hammer is used for levelling purposes).*

A lump can actually be hammered through to the other side - in which case turn the sawblade over and again level. According to some experts doing this can actually help to balance-up the stresses set up by the initial heating-up of the sawblade.

Note: Most burn marks start as a small lump with a shiny centre resulting from friction between the sawblade and the timber.

When checking sawblades look out for such spots and check these carefully for level - correct those that show as lumps before they turn into burn marks.

Lumps and ridges

A lump is more or less circular, but called a ridge when stretched in one direction. Lumps are treated as in the same way as burn marks by hammering as described above.

Ridges, which can run any direction, are hammered-out in a similar way, but by using a twist or cross-face hammer directly on the ridge. Whichever hammer is used it must be held so that the major axis lies along the ridge. Failure to do this will put a twist in the sawblade that is hard to make good. The most convenient hammer to use rather depends on the line of the ridge relative to the stance of the saw doctor or filer.

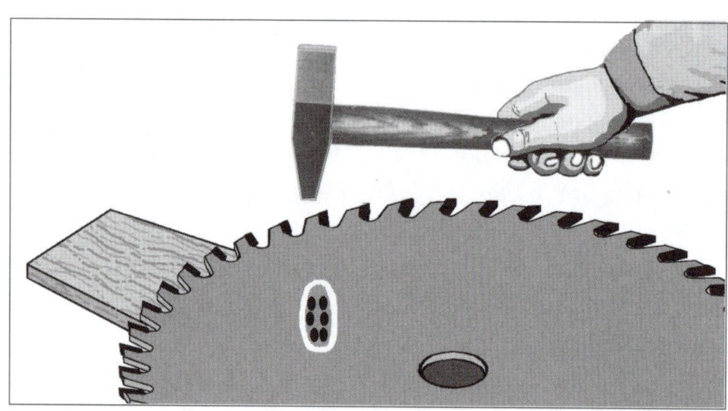

Fig. 11.30
Using the cross-face hammer to remove diagonal ridges. Take care that the long axis of the hammer-blow lies along the line of the ridge.

For ridges running radially or tangentially the twist-face hammer is probably the most convenient. Whichever face is used, hold it with the hammer shaft, wrist and forearm in more or less a straight line, and so that the shaft is at 45 degrees to the line of the lump. This gives a clear view of the hammering area *(a cross-face hammer the shaft tends to hide its in-line hammering face)* and allows the hammer to be gripped comfortably.

Ridges must be hammered with the long axis of the hammer precisely aligned to the long axis of the ridge - if they are misaligned this will twist the sawblade.

The cross-face hammer is better with diagonal lumps as the line of the shaft, wrist, and forearm can be straighter than if attempting to use the twist-face hammer.

Hammering out lumps and ridges can affect the sawblade tension in different ways. If large, both these, and part of the surrounding area, may need hammering to restore tension in the saw body.

Radial ridge hammering tends to reduce tension - which may then need hammering lightly within the body. Tangential ridge hammering tends to increase tension - which may then need correcting by hammering lightly under the rim. By bearing this in mind the saw doctor or filer can often correct two defects with fewer hammer blows than might otherwise be needed.

The effect or hammering out a ridge depends upon where this lies. If in doubt hammer out the ridge first, then check and correct tension. Only experienced saw doctors and filers have the skill to know exactly what is needed - and get it right first time!.

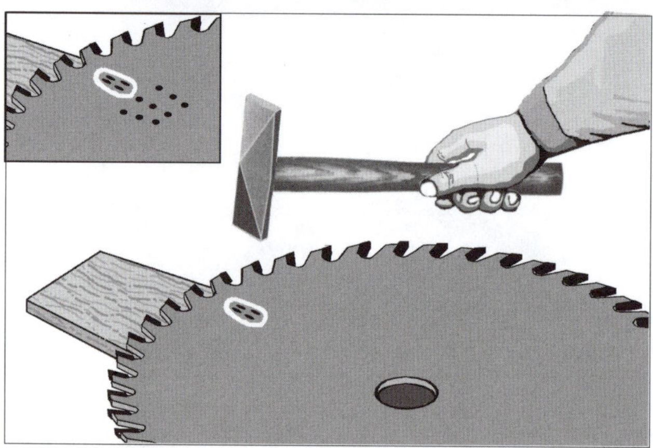

Fig. 11.31
Using the twist-face hammer for radial and tangential ridge removal. The additional light Hammer-blows (inset) are to make good tension lost through hammering out the ridge.

Twists:

Twists distort the rim of the sawblade to cause it to wobble. Unfortunately they are also one of the most difficult defects to diagnose - and the easiest to put into the sawblade by incorrect hammering.

A twist deflects the sawblade rim slightly in one direction at one point, and in the opposite direction elsewhere. This effect is often caused by opposing pairs of ridges, one ridge running radially from centre to rim, complemented by a similar ridge roughly at right angles to this, but on the opposite side of the sawblade. There may be more than one pair of ridges in a badly twisted sawblade, and the ridges may not run exactly radially or tangentially - in which case they are classed as long-face or cross-face twists.

Check for twists with the sawblade resting on the anvil

and back-board in the regular levelling position. Move the straight edge across the sawblade to find the line of the ridge, where it will show as a high point when the straightedge is square to it. If the straight edge is then turned through 90 degrees *(parallel with the line of the ridge)* the straight edge will show either a reduced high portion - or even slightly concave with an excessive twist.

In removing the twist it is not necessary to keep turning the sawblade from one side to the other to hammer the two complementary ridges alternately. Treat both ridges individually as described earlier. Reduce the intensity or number of hammer blows on the ridge when nearing the rim of the sawblade.

Plumbing the sawblade

Fig. 11.32
Plumbing the sawblade means examining it while held upright on a non-marking block, and checking both with a long and short straightedge.

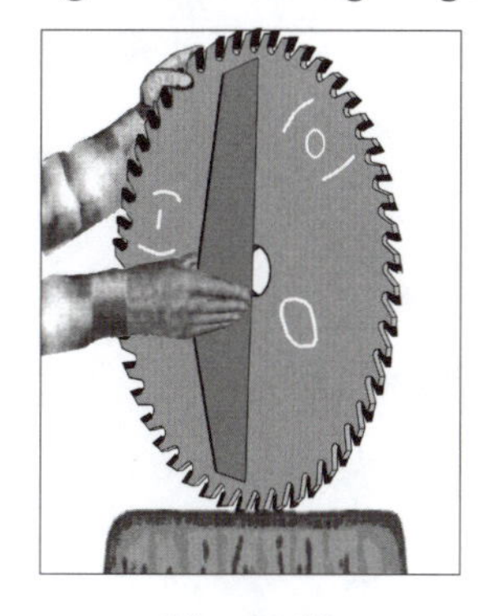

Fig. 11.33
Examine the sawblade completely on both sides, first using the long straightedge as an overall check, then switch to the short straightedge for more critical examination of any defect found. Mark defects as they are found.

After levelling and tensioning the sawblade as described above, the final test is **plumbing-up,** allowing a more critical overall inspection. Stand the sawblade vertically the bench top close to the anvil, resting the teeth on a piece of wood or similar which will not damage them but which will allow the saw both to be rotated and spun around. Take care that the support does not slip.

Hold the sawblade vertically with the left hand, and use the long straightedge in the right hand. Check both sides of the sawblade overall, generally as described previously, to find and mark lumps, ridges and twists. Use the short straightedge to check in greater detail any defect found. Finally make good any defects and re-check again.

The Try Mandrel

The try mandrel supports the sawblade vertically, as in a sawbench, where it can be checked for both run-out and wobble - and many of the other defects described earlier. In skilled hands it is used as a final means of fully checking the sawblade.

Fig. 11.34
A basic try mandrel consists of a rotating arbor to hold the sawblade vertical, and a dial indicator to show lumps or rim run-out.

A simple try mandrel can be made by the saw doctor or filer himself. It is simply a rotating horizontal arbor *(preferably mounted on opposing thrust bearings to eliminate end-play)* with the back saw collar permanently fixed to the arbor and trued-up while rotating on its bearings. Use some form of centering device or interchangeable saw centres to mount the sawblade. Arrange a firm support for the dial indicator needed to probe the saw body.

Alternatively, various floor and bench-standing try mandrels are available from Petschauer, and possibly other makers.

The method of examination for run-out, lumps and twists is straightforward. With the sawblade mounted on the arbor, set the dial indicator against the side of the saw body to give a reading, then twist the bezel to read zero.

Rotate the sawblade a few degrees and note the new reading, rotate again and again note the reading, and so on, until the full circumferences had been covered. Make the first check near the rim, then move the dial indicator down to a smaller radius and repeat, and so on, until the full area of the saw body has been covered.

It may be necessary to adjust the position of the dial indicator should a long gap occur without reading. However, it isn't the actual reading that matters, but the relative readings around one complete circle and between concentric circles.

By noting the change in readings it is easily possible to

mark the saw body run-off and for some other defects. A higher reading shows a lump or ridge on the side being measured, while a lower reading shows a depression - a high spot on the opposite side.

As an alternative to full examination and marking, it is practical to simply rotate the sawblade until a high or low spot is encountered. Then move on either side of this to plot the extent of the defect, ringing all lumps and ridges.

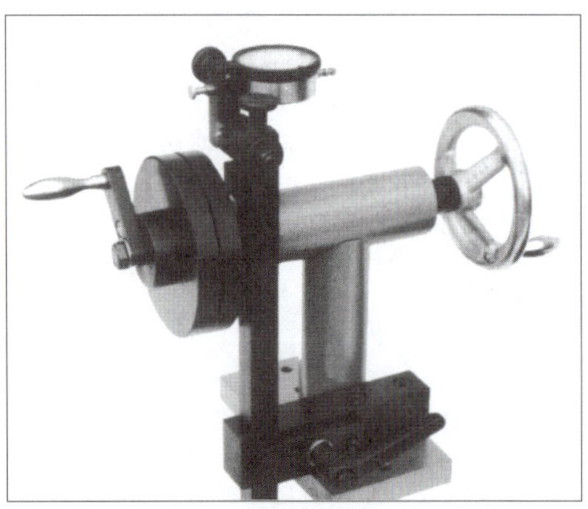

Fig. 11.35

This bench-mounted try mandrel is a simple piece of equipment ideally suited to even the smallest saw doctors or filers room.

Source - Petschauer

Repeat the process on the opposite side of the sawblade. This should show up defects that are opposite to, and the reverse of, those on the side first tested - so confirming the initial marking. To do this reverse the sawblade, or mount the dial indicator on the opposite side. Carefully marking a sawblade in this way gives an excellent all-over view of many defects. It also clearly shows if there is run-out at the rim - something that other methods of examination do not show. Run-out, as previously mentioned, is usually caused by ridges or twist in the saw body.

After fully examining and marking the sawblade for defects, remove and hammer as needed to correct them. Take particular care to reduce run-out as much as practical - this is where accuracy is most important.

Body marking using a try mandrel

This method clearly shows saw body defects. Hold a chisel-edged marking pen rigidly against the saw body while this is slowly rotated through a full circle. Then set the marker at a different radial point and repeat, and so on, until the whole saw body shows a series of concentric circles of varying line width.

A chisel-edge spirit marker shows lines of different width according to how hard it contacts the saw body, a broader line when contacting a lump, and a narrower line when contacting the same lump as a hollow from the opposite side. Run-out of the rim show in the same way. To get the best effect clean the saw fully, trim the marker to a point, and experiment by holding it at different angles. Even if the marker misses the saw body in places, this will still show where a defect lies.

Marking in this way gives an overall view of the condition of the saw body - from which the saw doctor or filer can more easily decide where hammering is needed. To automate the process, the fixture holding the marker could be mounted on a slide parallel with the saw body and traversed radially by a screw mechanically linked to the sawblade rotary movement. By driving both with a small motor the sawblade would be quickly and fully marked in a continuous spiral path.

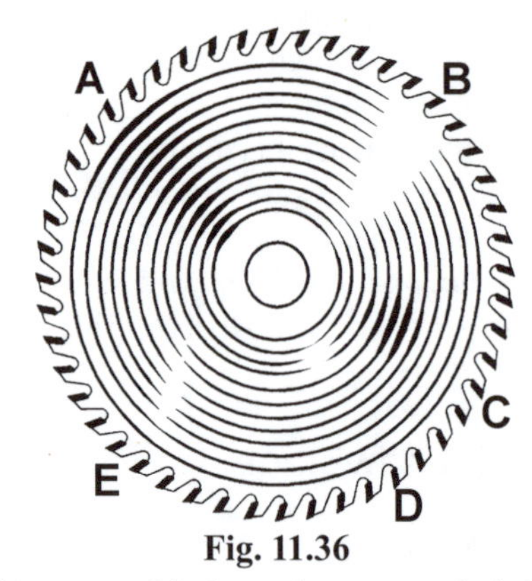

Fig. 11.36

*Spirit-marking a sawblade on the try mandrel shows defects quite clearly as; **A**, radial ridge; **B**, radial ridge on the opposite side (probably a pair creating twist on the saw); **C**, A round lump; **D** another round lump, but on the opposite side; **E**, a diagonal ridge on the opposite side.*

Interpretation of try mandrel examination

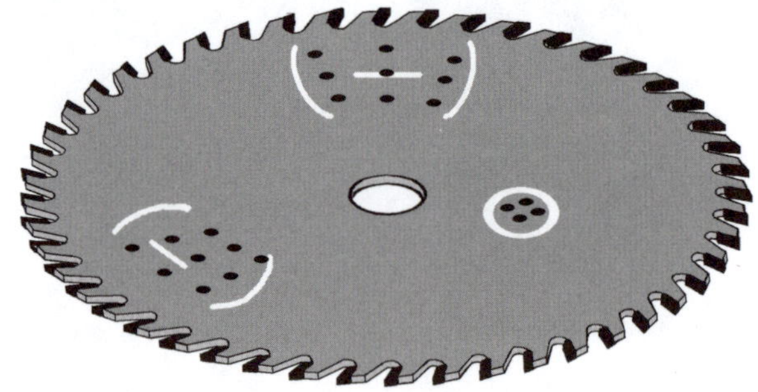

Fig. 11.37

This shows a fairly simple combination of two fast sections and a small round lump - all on the same side. Hammering as shown should remove these.

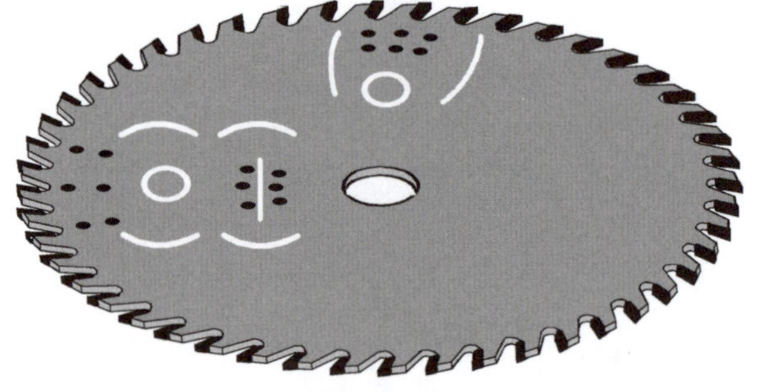

Fig. 11.38

For combined defects such as these first open up the inner fast section (left), then hammer near the rim to correct the loose section - which hammering might have worsened. The loose section near the rim (top) is treated separately.

Once a sawblade is fully marked up during try mandrel examination, take appropriate action as previously described.

Where several defects are present, hammering one may alter the condition of the others. Preferably remove large defects first, then re-examine the sawblade to check how this has affected the remaining defects and the overall tension of the sawblade.

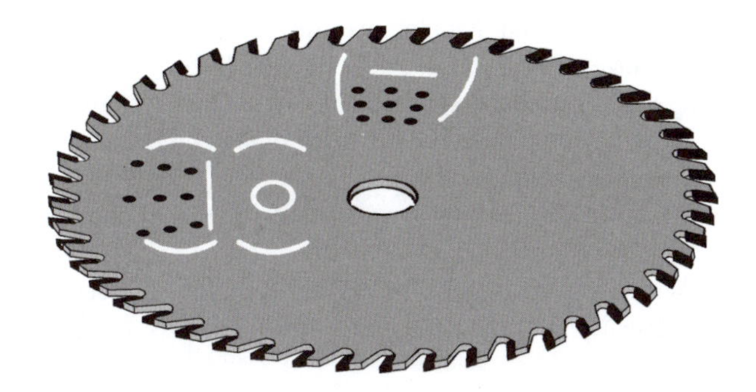

Fig. 11.39
In this case hammer to cure the fast section near the rim (left) - which could also make good the loose section nearer the centre by drawing out the excessive tension.

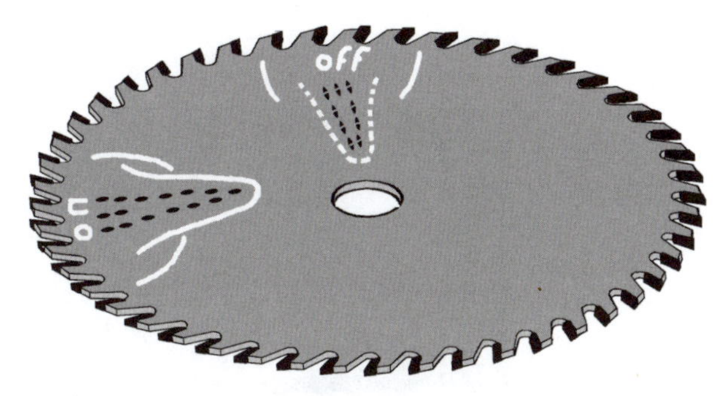

Fig. 11.40
*These show typical radial ridges which deflect the saw rim in opposite directions in a classical twist. The ridges are on opposite sides of the sawblade, and the rim is deflected in opposite directions because of them. Treat the two independently. The dotted ridge adjacent to the **off** marking is on the opposite side of the sawblade, and is hammered from the opposite side.*

Try mandrel tension examination

The Petschauer range of try mandrels includes the SP models to check both for sawblade tension and regular try mandrel testing.

They test for tension in a way similar to that used by the saw doctor or filer, but by mounting the sawblade vertically and flexing it mechanically.

The sawblade is mounted on a suitable centre, but rigidly held by collars identical to those in the machine using the sawblade.

Two adjustable pressure points, which should be positioned just under the gullets, are provided diametrically opposite one another on the operating side of the machine. They incorporate variable-pressure pneumatic cylinders to flex the sawblade when checking for tension..

By flexing the sawblade, a reading can be taken from a dial indicator mounted at right angles to the line joining the cylinder pressure points, and roughly midway between the saw centre and its. rim*. The amount of deflection this causes is in direct proportion to the amount of tension the sawblade contains. The support cylinders are triggered by a foot pedal. On the P version, the outer saw collar pressure cylinder retracts to permit rapid sawblade change.

**As can be seen from manual examination with a short straightedge, Fig. 11.9-12 , it is at this point that maximum deflection takes place on correctly tensioned sawblades.*

Two readings are taken, one static without cylinder pressure, and one dynamic with cylinder pressure. The same measurements are taken at several radial positions around the sawblade - after which the sawblade is reversed to check the opposite side in the same way for equality of tension. The required reading is the actual deflection - the difference between static and dynamic readings.

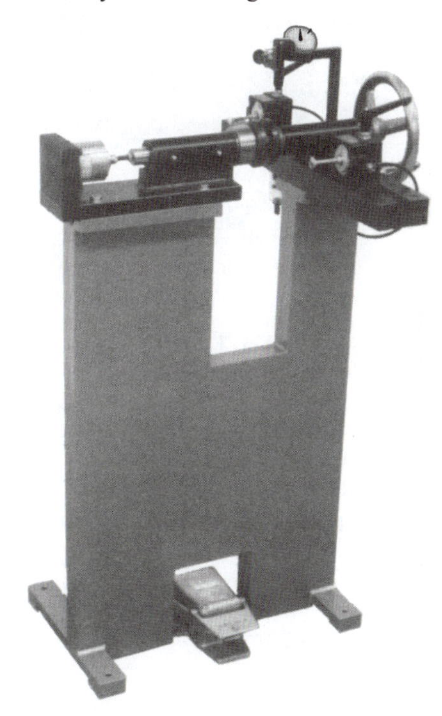

Fig. 11.41
This universal machine operates both as a try mandrel and for tension-testing sawblades.
Source - Petschauer

Although tension is not measured in the way previously described for manual measurement - so this guide does not apply - but the general method of calibrating the machine is very similar.

First check a number of correctly-tensioned sawblades of different thickness and diameter, and record all their details in a comparison table for future tension testing. The tabled test results should show the sawblade diameter, thickness, sawblade deflection and cylinder pressure needed (the latter is indicated on a gauge on the device).

The cylinder pressures are adjustable, and should be more when a thicker or a smaller sawblade is being tested and less for a thinner or larger sawblade. The pressure is correct when repeated tension checking of a test sawblade gives consistent and measurable results. Obviously some experimental testing is first needed to establish suitable cylinder pressures for each diameter and thickness of sawblade to be tested.

As wide a range of correctly-tensioned sawblades should be tested in this way, including sawblades of the same diameter but of different thicknesses. This will then provide a guide as to how both thickness and diameter effect what cylinder pressures are needed, and what deflection readings then apply. Once a range of results are obtained it is practical to interpolate data for sawblades of different diameters and thicknesses than those actually tested.

The cylinder pressures and dial indicator positions need adjusting each time a sawblade of different diameter is to be tested, but keep to these precise settings or tension-testing results will be unreliable.

Fig. 11.42
Two diametrically-opposed cylinders deflect the saw body at a pre-set pressure while the dial indicator at right angles to these shows the deflection this causes.
Source - Petschauer

Automatic tension and run-out measuring

Fig. 11.43
The Chaland Jack Equipment is a simple-to-use control bench designed specifically for Sawmills and Saw Sharpening and Servicing Centres. It has an electronic control system and a screen which displays, as a diagram, the side run-out and tension of any sawblade up to 1M30 diameter. The diagram can, if required, be printed-out. The inset view shows the computer in more detail.
Source - Chaland

J. Chaland & Fils manufacture equipment in different models for measuring the flatness, side run-out and tension of all types of circular saws.

The equipment comprises basically of a substantial frame with a free-running arbor and flanges for mounting the sawblade, an angular detector near the arbor, a linear detector for measuring deviation, a pneumatic deflector for tension testing, and an associated computer and control system.

The detector and deflector are mounted on linear slides at right angles to one another and can be precisely positioned, via scales, at the settings appropriate for the diameter of sawblade being tested.

Once the sawblade is placed on the arbor it is gripped pneumatically, and then manually rotated slowly using the crank handle.

As it is turned the linear detector measures any deviation as side run-out and, in conjunction with the angular positioning sensor, records the findings on the computer. At a second testing the sawblade is flexed mechanically and a second set of readings taken as the measurement of tension.

Fig. 11.44
The paths for run-out and tension, and the working tolerances between which the side run-out should lie, are clearly shown on this Visu 2 computer screen.
Source - Chaland

Both readings are then displayed on the computer screen as roughly circular paths, shown in red and blue to readily distinguish them. Their deviation from a true circle indicates by what amount the sawblade either runs-out or has irregular tension.

The saw doctor or filer can then easily and quickly correct any irregularities in side run-out or tension in the sawblade - after studying the computer screen.

A more sophisticated version the Visu 1 displays a tolerance value for side run-out - both on the screen and on print-outs - as two concentric circles between which the side run-out path should lie. Users of the system can modify the program to display their individual preferences for side run-out tolerance - which varies with sawblade diameter.

The Visu 1 also records this data, along with a customers details, etc, on 3½ in computer discs for future reference or customer information. Individual diagrams can be printed out for each tested sawblade. If a batch of saws are tested, a collective data listing can be printed-out instead of individual diagrams.

The option of recording the side run out and tension parameters for each sawblade checked fully meets the ISO 2000 in respect of traceability of manufactured products.

The company, in addition to manufacturing sawblades,

runs training courses of from 2 to 5 days on the theory and practice of sawblade straightening and tension control. The courses cover all aspects of saw preparation and can include the theory and practice of brazing carbide tips.

Fig. 11.45
This shows part of the training courses which run at Chaland for between 2 and 5 days.
The illustrations show hammering practice (above), and the theory of saw tension (below).
Source - Chaland

Hammering hints

Clean off all the resin and gum from the teeth and the sawblade before starting.

Rub the surface of the sawblade and the top of the anvil with a slightly oily rag. This allows the sawblade to slide easily over the anvil, cushions hammering and allows the saw doctor or filer to mark defects with his finger alone.

Note that levelling can increase or decrease tension in the saw body.

No two circular sawblades are alike - treat each on its own defects.

Always finish levelling on the face (outer) side of the sawblade.

Keep off the collar area of the sawblade - except to correct other defects.

Be methodical; place hammers and straight edges neatly in their positions close to the anvil.
Never lay a hammer on a straight edge, it may bruise it.
Practice dropping the cross and twist-face hammers so as to avoid twisting the sawblade.

Try not to mark the surface of the sawblade.

Remove any chalk marks after hammering - too many old marks mislead.

Use the appropriate hammer when removing ridges.

Do not use the doghead to remove twists - it won't.

Cross-face and twist-face hammers add and remove tension more efficiently than a doghead hammer - but can mark and twist the sawblade if used badly.

Take a pride in your work and do not allow others to use your hammers.

Rolling circular sawblades

Fig. 11.46
Stretching rolls are a standard piece of equipment in modern saw shops, they are faster than hammering sawblades for tension, and leave no hammer-marks.
Source - Armstrong

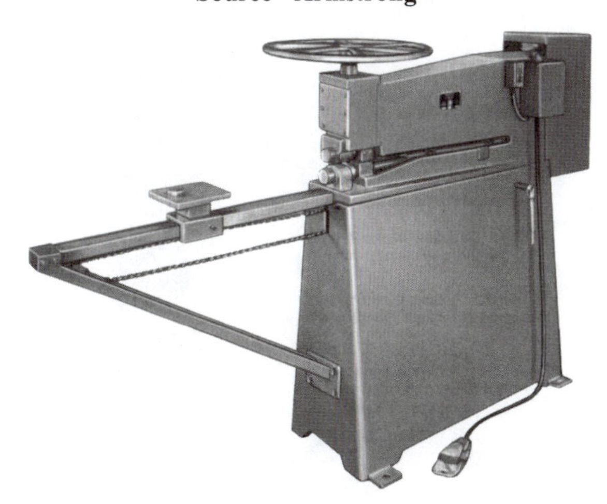

Fig. 11.47
This model 104-C-72 is a heavy-duty stretcher suitable for circular sawblades from 200mm - 1M83 (8 - 72in) diameter.
It has a reversible motor for operating from either side.
When rolling a marked section only, the best way is to first position the sawblade at the start point with the drive motor stopped and apply pressure, then start-up the motor and stop it again at the final mark.
Source - Hanchett

Circular saws were traditionally tensioned using hammers, but the current practice is to use a stretcher roll.

Similar machines, with upper and lower power-driven rolls, are widely used for tensioning wide bandsaws. The ones originally used were, in fact, wide bandsaw stretcher rolls modified for circular saws, but machines are now made specifically for this purpose. The sawblade is mounted on some form of vertical arbor directly in line with the roll arbor centreline. This allows the sawblade to rotate, driven by the rolls, so that they roll in a full and true circle.

The saw body is squeezed between them to stretch it - as would a series of closely-spaced hammer blows - but in this case both faces are affected simultaneously and by the same degree.

The rolls are very precisely radiused to a crown of roughly 100mm (4in) radius. They stretch the body by a specific amount - according to pressure applied. A sawblade can be tensioned using a stretcher roll more accurately and more consistently than by hammering.

But this is not the only advantage, roller tensioning is very much faster than hammer tensioning, perhaps ten to twenty times as fast, and there is no danger of marking the sawblade in the way that misplaced hammer blows do.

When dealing with new sawblades, or when re-tensioning sawblades which have lost tension without developing other defects, stretching rolls alone are needed.

Saw rolling is of less benefit when treating sawblades with lumps and twists* - these first have to be made good before attempting to roll.

**Some stretcher rolls can be supplied with specialized equipment to remove lumps, twists and dishes. Certain types of sawblade, for example the splined-bore types used on some double and multiple edgers, require only levelling.*

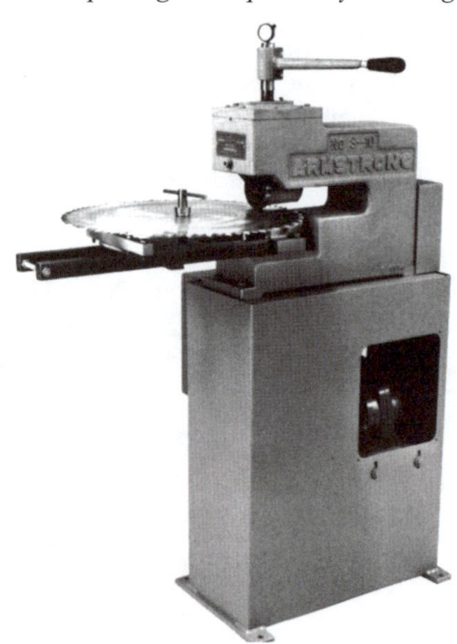

Fig. 11.48
This 3-10-C stretcher roll is suitable for sawblades up to 760mm (30in.) diameter.
It can have a dial indicator to show roll pressure, also an eccentric saw arbor for spiral rolling.
Source - Armstrong

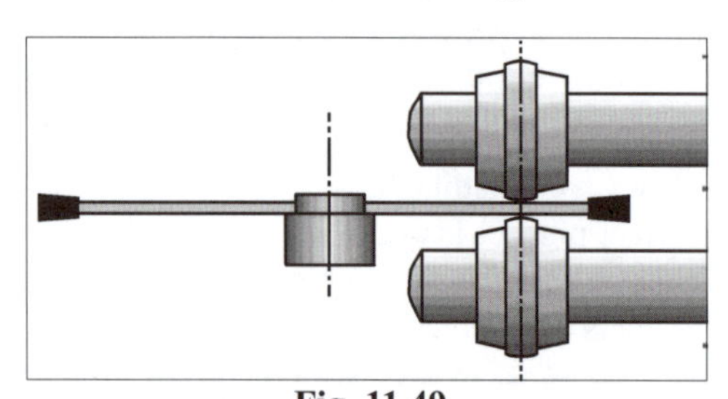

Fig. 11.49
The rolls must be directly opposite one another and positioned so that tha they are directly above and below the rolling line. Normally the saw centre adjusts.

Most saw makers apply just a single, heavy roll in sawblade manufacture - this tensions new sawblades to a commercially-acceptable level. However, many saw doctors and filers prefer to control tension more precisely by using less pressure and making several concentric and closely-spaced rolls. To do this the saw arbor must be reset precisely between the rolls to give the closely-spaced rolls this technique demands.

Fig. 11.50
A single roll is regularly applied in both sawblade manufacture and repair. The off-set distance saw centre to the roll line is critical. See Fig.11.14 for details. As a guide the off-set should be around 143mm for a 500mm diam. sawblade, 115mm for a 400mm. diam. sawblade and 86mm. for a 300mm. diam. sawblade. Bear in mind, though, that sawblades vary in their hardness, thickness and application, etc., which can affect this figure.

Fig. 11.51
Left: Closely-spaced rolls give a more even tension to sawblades. After the first roll at the off-set distance quoted above, further rolls are made by shifting either the rolls or the sawblade inboard and outboard of this first roll.
Right: To roll in a spiral form, off-set the sawblade centreline from the roll centreline. The amount of centreline off-set needed is approximately 1.6mm for 10mm roll spacing, 2.4mm for 15mm roll spacing and 3.2mm for 20mm roll spacing.- regardless of sawblade diameter.

To save adjusting the arbor position between rolls, and to provide a more controlled rolling process, Armstrong offer an eccentric saw centre mounting which automatically rolls the sawblade in a spiral, or snail-cam, form, effectively giving two or more virtually concentric rolls at a single setting.

The eccentric movement adjusts to give closer or more widely-spaced rolls as preferred, and can be set to roll towards or away from the saw centre.

Some treatment is also possible by rolling only certain sections of the sawblade - parts that are under-tensioned, for example.

Doing this can be quite tricky as the rolling action has to be started and stopped very precisely, but foot control makes this practical. Some users also claim lumps can be rolled-out - but this takes much more skill.

The rolling track recommended for tensioning a sawblade is critical. It should correspond basically to Fig. 11.14. at slightly outboard of a point midway between the saw centre and the rim. All rolling for tension should take place at or close to this point.

Rolling further out towards the rim gives less effect for the same roll pressure, and rolling further in towards the saw centre initially gives more effect, but if taken too far tends to dish the sawblade.

To avoid positional errors being made when rolling, fit an expanded scale for setting the saw arbor position.

This would be calibrated for the saw diameter - but would actually set the saw arbor in the ballpark rolling position for the diameter indicated.

Another critical point is the actual pressure applied - this should be constant to give consistent and repeatable results - but many makers simply fit a screw and pressure lever and rely on the user to apply his skill.

Some makers can supply a dial indicator which shows the precise amount of pressure applied by the lever handle. With hydraulic roller stretchers, roll pressure is controlled by a hydraulic valve with actual pressure shown by an indicator. In both cases knowing the applied pressure is a great advantage in achieving reliable and consistent results - especially for learners.

One basic feature with stretcher rolls is that precisely the same pressure is applied to both sides of the sawblade at the same time, so unequal tension - a slight dish - cannot be cured using regular stretcher rolls.

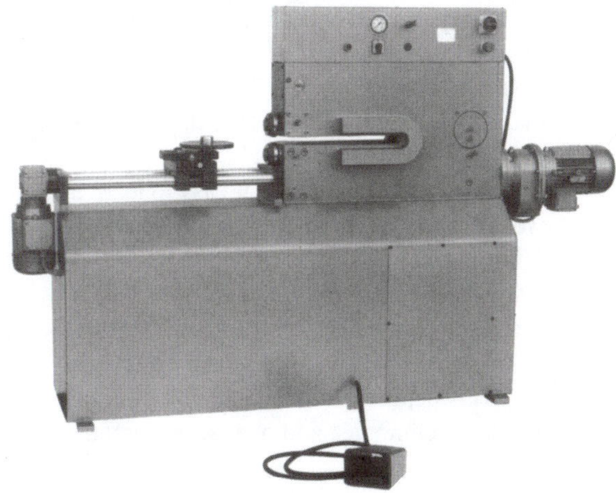

Fig. 11.52
This stretcher roll, type SW, has hydraulic variable pressure and externally-mounted rolls for easier monitoring of the rolling operation.
Source - Petschauer

Fig. 11.53
The SW stretcher roll, when required for volume production, can have this automatic transfer system fitted - which is suitable for processing both saw blanks and finished sawblades.
Source - Petschauer

Stretcher rolls are primarily used for tensioning sawblades by rolling fully around them, but in skilled hands it is possible to roll partially around the sawblade to correct isolated fast and loose sections. This is done by first positioning the sawblade with the rolls at the start of the defect, then apply pressure and start-up to rolls to run to the defect end, and again stop.

Fig. 11.54
Start-stop rolling to correct isolated loose sections. Fast sections are treated in a similar way by rolling nearer the rim - generally as shown for hammering.

Fig. 11.55
A levelling attachment can be supplied with Armstrong stretcher rolls for dealing with local defects (top) also special levelling rolls for correcting dished saws (bottom).

Armstrong offer special rolls for correcting dished sawblades, a regular bottom stretcher roll and a notched or concave levelling top roll. Used together they can make good dished sawblades by rolling them concave side up.

Also available is a levelling attachment consisting of two small-diameter hardened and crowned rolls which are positioned on either side of the regular bottom roll, and set slightly above its level. This attachment can remove local defects without affecting the overall tension already in a sawblade.
Source - Armstrong

Automatic Sawblade Levelling

Levelling, making the sawblade flat and free of lumps and twists, is a process that often precedes tensioning - although a skilled saw doctor or filer can often combine the two operations.

A few types of sawblade require little or no tension, for example, small diameter, thick-gauge sawblades, and sawblades used on certain types of timber conversion equipment.

Regular gang saws and edgers for rough mill conversion have splined-centre sawblades that allow sawblades to traverse freely along a, large-diameter, splined saw arbor.

The saw guides, one on each side, support and steady the sawblade and prevent it deviating in the cut. They also traverse parallel to the saw arbor, either under manual or machine control, and in doing so move the sawblades with them.

In this way sawcuts can be preset to give the best conversion factor for each individual piece of rough timber before it is fed into the machine. (See also Chapter 16).

These machines are capable of roughly following the grain of the timber. This is done to produce high-strength construction timber that contains virtually no cross grain, and to provide a much high conversion factor than regular straight-line sawing. When drying curved-sawn timber there is also less distortion than with regular conversion of timber and less down-grading.

Splined saws used on these machines require to be commercially flat, and whilst they can be run without tension, a minimal amount of tension certainly improves their performance.

This is because the guides serve as supports in place of the regular saw collars and, being close to the cut, do not subject the body of the sawblade to same the stresses in cutting.

Equipment for automatically flattening splined sawblades initially tests each sawblade by rotating it at least once to map the irregularities before attempting to correct this by rolling or hammering.

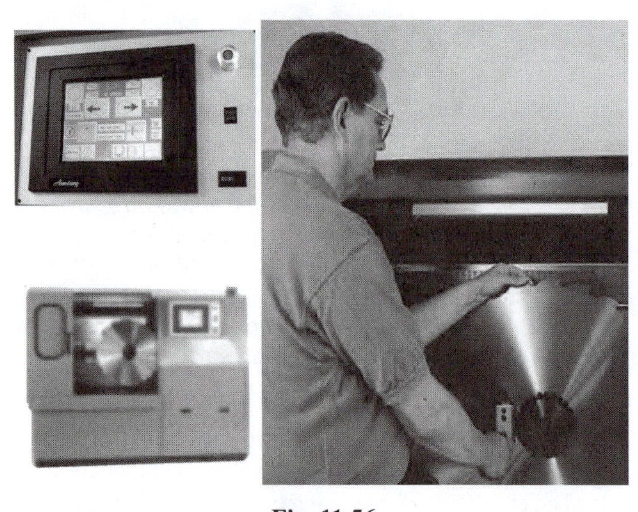

Fig. 11.56
The Armstrong automatic leveller shown being loaded, main illustration, with the touch-screen upper left and the general view lower left.
Source - Armstrong

The Armstrong circular saw leveller is computer-controlled machine specifically intended for flattening spline-arbor type guided sawblades of this type and of any diameter between 400mm and 815mm (16in and 32in).

There are two modes of operation - which are chosen simply by pressing a single button on the touch screen. The ***'Standard'*** mode allows the saw plate to be mapped and leveled simultaneously in just two passes - a process made possible by using a non-contact sensor and a superior sawblade stabilizing system.

This ***'Standard'*** mode is suitable for most sawblades and makes the entire process of flattening extremely fast - in fact a 22" (550mm) sawblade can be levelled in less than 8 minutes. More severely out-of-level sawblades are ***'mapped'*** first in the second mode, allowing the biggest lumps to be pre-levelled before automatically completing the levelling process by two ***'standard mode'*** levelling passes.

The saw is loaded vertically onto the machine arbor. This method offsets the effects of gravity and makes the machine easier and safer to load. As the saws *'float'* on the centering arbor - just as they do on splined saw arbors - this exclude the chance of outside pressures distorting process of flattening.

Each sawblade is clamped between a pair of three-piece levelling rolls which automatically center the sawblade and simultaneously level at each pass.

The amount of roller pressure needed to correct a distorted sawblade varies according to its hardness. Hard sawblades need more pressure to correct their condition, softer sawblades need less. To ensure that the correct roller pressure is always applied, a hardness gauge is provided. This allows the precise hardness of each sawblade to be accurately measured and the machine preset accordingly.

By doing this, working time is never lost in having to correct sawblades ineffectively treated through initially using the wrong pressure.

The color touch screen makes it easy to learn and operate the Leveller. The screen reports the sawblades condition before and during leveling so the operator immediately knows if one or more additional passes are needed.

The levelling parameters for up to six different sawblade types are stored - so every sawblade can be levelled to the head filer's precise specifications. This combination of features makes the machine fast and precise in operation.

And for the future...

A lot of progress in automating bandsaw levelling and tensioning has been made, and it seems likely that similar progress will in the future apply to circular saw tensioning.

One possibility is a try-mandrel linked to a computer to memorize vertically-mounted sawblade defects while rotating it, giving a readout on completion of the defects found. The data could then be fed into an automatic tension and levelling machine to which it is linked. Another possibility is a combined defect-measuring, tension and levelling machine. In both cases the levelling section would incorporate levelling rolls and/or hammers.

Mechanical probes or sensors would detect defects in the saw body as it is rotated, spiralling inwards from the rim to check the complete sawblade. The rolls for tensioning and levelling could possibly be three-position switchable between regular rolls, flat rolls and concave rolls - or hammers switching between flat and regular - to cure whatever defect is encountered.

The sawblade could be rotated primarily by a driven arbor or by contact rubber rolls driven to give the same surface speed as the regular rolls, the latter switching and engaging or disengaging as dictated by the measuring program.

Sawblade loading or unloading would, of course, be incorporated to fully process finished sawblades or saw bodies in a production environment.

Chapter 12
Maintaining Carbide Saws

Because tungsten carbide tipped saws are made to be long-lasting, the tips have to be hard and brittle, and for this reason they are all too readily damaged or broken-off entirely.

Where finish is not critical, for example in fast-feed timber conversion, sawblades with a few slightly damaged teeth or one or two teeth missing are often kept in use until sufficiently dull to be removed for sharpening, and only then repaired.

Unlike the rough and tumble of lumber conversion, damage is less likely when converting manufactured boards or dimension-sawing natural timber - but damaged or missing teeth in their cases can downgrade the quality of finish. For this reason immediate repair is recommended - and routine servicing by a saw repair shop may be inconvenient.

This Chapter deals with how such urgent repairs can be carried out in an average tool room, in-house, with simple devices or inexpensive manufactured equipment. It also details the more upmarket equipment used by larger companies and saw sharpening and repair shops.

Brazing and sharpening teeth to repair tct saws needs certain skills - but these are well within the scope of any saw doctor or filer - with the right equipment. This need not be expensive, but must include oxyacetylene equipment to provide the heat for brazing, cleaning and de-greasing materials, and a saw grinder suitable for both regular grit and diamond grinding wheels.

The problem thermal expansion

A difficulty with brazing tc tips is that the saw body, the brazing bond and the tungsten carbide tips all have different coefficients of thermal expansion. Steel expands on heating and contracts on cooling at twice the rate of the carbide tip.

While being heated they can expand freely - and so are bonded together in this expanded condition. On cooling, however, contraction stresses can be created which may result in crazing, or shear-type cracks developing near the tip face which lead to sections splintering off.

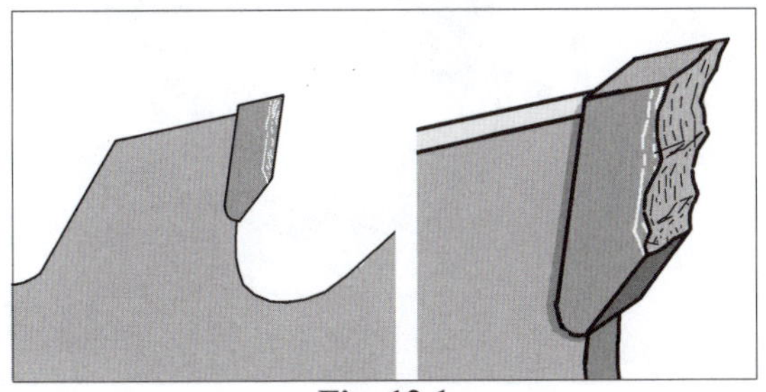

Fig. 12.1
The different expansion and contraction rates of tungsten carbide and steel compresses the tip seat on cooling and stresses the face - possibly to shear-off sections of carbide near the tip face.

Thermal stresses increase with temperature rise, so use a low melting-point solder to keep brazing heat to an absolute minimum - especially when brazing manually.

Silver solder, the regular bonding medium, is sold in strip or sheet form 0075-025mm thick (0003in-0010in). Preferably use thicker strips as these act as more of a cushion and guarantee a more secure grip.

A flux, compatible with the silver solder, is needed to coat the silver solder, the tip back and the tip seating to prevent oxides forming prior to the solder starting to flow *(oxides prevent the tip and seating bonding reliably.)*

Silver solder paste, now regularly used instead of strip silver solder and flux, is more convenient than the latter for manual brazing. It has silver solder granules suspended in a flux-type paste, and is packaged in a tube - but has a short shelf life. The granules are kept free of oxides in the paste, so allowing this to be applied direct to clean tips and seating without further preparation.

Machine- based and fully automatic brazing equipment allows much more precise control of heating - so in their cases follow the makers advice on bonding materials.

Treatment of damaged tips

Tips that are completely missing obviously need replacing. Also replace badly damaged tips - rather than grinding the remainder down to their level.

In conversion sawing, sawblades are often run with one or two tips damaged or missing. Provided the quality of sawing or productivity is not affected this is acceptable - but when grinding these sawblades use a double pawl to avoid a crash. Periodically examine each sawblade to replace missing or damaged teeth - or clearly mark them to warn the saw doctor or filer of their condition.

Sawblades with many missing or badly damaged tips should be scrapped rather repaired. Service companies recommend this course - not merely because they want to make a sale *(of course they do!)* - but for continued customer satisfaction. A new sawblade has a guaranteed performance and long life expectancy - a repaired sawblade may not. The same reasoning applies to sawblades where the side clearance is so reduced that rubbing takes place - the cost of fully re-toothing may actually be higher than that for a new sawblade.

Choosing the right tip

Tungsten carbide tips vary in size, proportion and classification, and should be carefully matched both to the sawblade and the sawing application.

Tip thickness should be equal to, or more than, that of existing tips.

Tip depth should be only slightly more than that of existing tips - to avoid creating an unbalance in the repaired sawblade, and so the same depth of tip seating can be used.

Tip width is depends primarily by the body thickness of the sawblade. To this is added the side clearance required and a small allowance for side grinding.

For example a 3.4mm bodied fast-feed ripsaw needing teeth with a 0.7mm. side clearance on either side, plus an allowance for sideways inaccuracy of, say, 0·1mm, would need a tip width of 5.0mm.

The grinding allowance has to be very carefully calculated - too much and a lot of time and diamond abrasive will be wasted - too little and the tip may finish under-size.

Width-wise positioning of the tip on the saw body for brazing is also critical -place them with great accuracy - otherwise they may need heavy grinding on one side and fail to make the width on the opposite side.

The grade of tip must be matched to the sawing application - hard materials need a softer tip to avoid chipping, and a harder tip is suitable for easily-sawn materials.

Most makers and agents offer tips in many size, type and shape combinations.

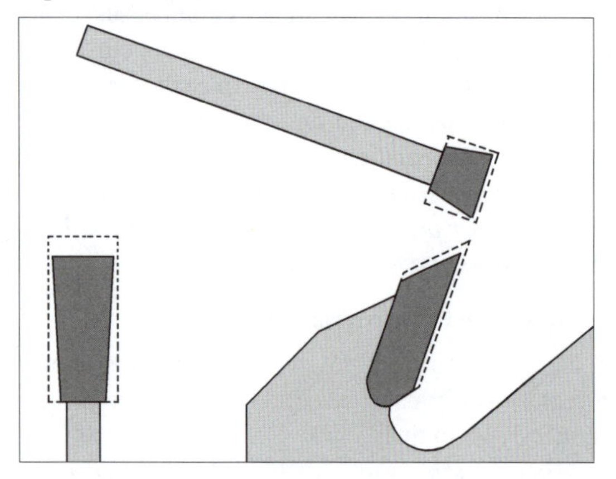

Fig. 12.2
New tips must be oversize when compared to existing tips to allow for sizing. Dotted outlines show the new tip after brazing. Solid outlines show the finished tip.

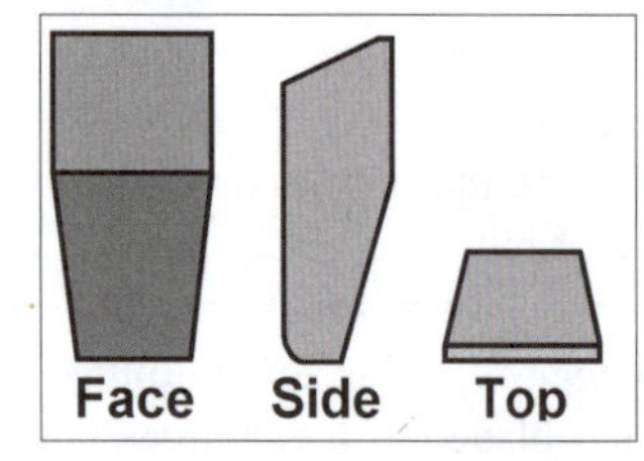

Fig. 12.3
Pre-shaped tips, with such as the profile shown here, need less grinding when finally sizing them.

Fig. 12.4
Typical carbide tips of various sizes and types.
Source - Dymet Alloys

Pre-shaped tips reduce the amount of grinding needed to finish-size them as they have clearance angles already part-formed - plus a lower bevel to give a short face depth (this lessens the chance of uneven grinding wheel wear and allows wider abrasive to be used.

Pre-tinned tips have a thin coating of solder applied to give a better and easier bond - but cost more.

This type is convenient for occasional tip replacement, but may be uneconomical for regular tip replacement.

Basic Equipment needed

The equipment needed for manually removing and brazing tips is:- oxyacetylene cylinders, regulators, hose, and welding torch fitted with a No.0 or No.1 size nozzle, a saw holding fixture, silver solder* and flux or brazing paste**, a pair of tweezers, de-greasing fluid, a dial indicator, wire brush, and a pocket microscope. *Or a braze alloy, alternative types are being deveoped. **A mix used in place of silver solder and flux. It contains flux and silver solder particles in a squeezable tube. This is convenient for certain brazing operations, but can have a short shelf life.

CAUTION - Most cleaning materials give of vapours which can injure health. Likewise, the brazing fluxes and pastes produce health-hazardous gases when heated. Keep the area well ventilated during both the cleaning and brazing operations.

Wear a suitable filter mask if the operations are intensive, or prove to be of high risk. Refer to the makers instructions and follow their recommendations. Clearly display a notice with these details in the repairing area. Wear suitable eye protection when grinding and brazing.

Tip removal. An old tip can be easily removed after melting the silver solder using the oxyacetylene flame to apply heat to the tip itself. A gentle tap with the tang of a file will then detach it.

Any solder remaining may cause the pawl to mis-index when tip-seating grinding. To remove this, gently reheat the tooth and, as the solder liquefies, rapidly brush towards the rear with a wire brush.

Tip seat grinding.

Fig. 12.5
Use a saw grinder to clean-up the tip seats, for example this HSN 600 low-cost automatic saw grinder.
Source - Stehle

The tip seating must be cleaned-up and shaped to suit the replacement tip, preferably by grinding. Form it to the same hook angle and to the same depth as the existing tips.

While hand filing could be used, this can give an im-

perfect seating for the tip and create unreliable bond. However, a light stroke with a file after tip grinding, according to some, can produce a better bonding surface.

The tip seating can be ground freehand, but for precision results use a saw grinder. Fit a regular grit wheel that is shaped to match the back of the tip - usually straight a section ending in a small radius. Where the volume of repairs warrants it, use instead a pre-shaped cbn grinding wheel.

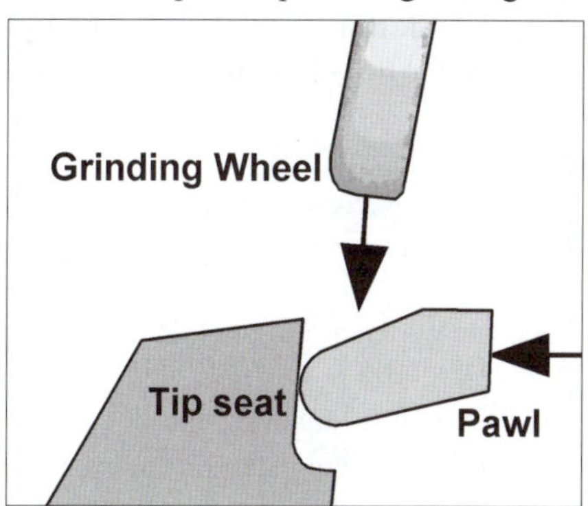

Fig. 12.6
Use the pawl to index the tooth for tip seat grinding. Grind with a shaped regular grit or cbn grinding wheel.

Set the saw grinder feed pawl to index each seating prior to grinding - basically as face grinding - contacting the actual seating itself - although with deep gullets it may be possible to index the pawl against the gullet below the tip seating. Use the regular stop to gauge the depth of the seating.

If several teeth are missing use an automatic saw grinder - but with care! Set the machine as described above. Use the slowest indexing speed, and manually raise the grinding wheel clear of the saw when complete tips are indexed. Lower the head only when indexing empty tip seats - or shift the saw around manually between these with the grinding wheel raised.

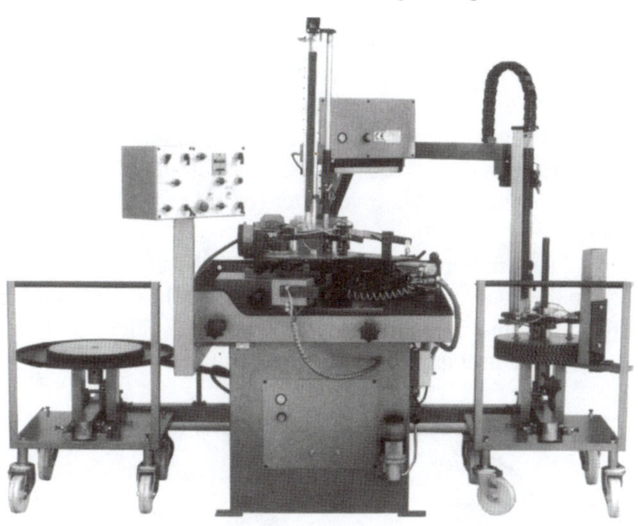

Fig. 12.7
This Petschauer PLS 800 MB, designed for production tip-seat grinding, is shown in the fully automatic version - capable of continuous operation on sawblades of between 100 and 800mm diameter.
Source - Petschauer

Production tip seat grinding. When manufacturing from new, or if completely re-tipping sawblades, use a fully automatic saw grinder - most modern automatics have a program for this.

Machines made specifically for tip-seat grinding are available either as stand-alone units for manual loading - capable of tip seat grinding of up to four saw bodies simultaneously - or with automatic loading and discharge facilities.

After tip seat grinding

Fig. 12.8
After tip seat grinding, and prior to brazing, carefully check the sawblade for run-out - and correct.
This illustration shows part of the high-quality line producing AKE sawblades.
Source AKE

To ensure that tips will be accurately centralized on the saw body check the sawblade for lumps, tensioning defects and side deviation - and correct as necessary. Failure to do this could result in tips being brazed off-centre on equipment that relies on the saw collars for centralizing the saw body.

Fig. 12.9
Use a device such as this RP850 for checking run-out or side deviation of the sawblade. The sawblade is mounted on a free-running arbor, and a dial indicator registers against the rim just below the teeth.
Source - Saturn

The brazing process

Brazing fixtures. When brazing teeth the saw body must be firmly supported or clamped. The tip must be located centrally to the body, and some means is needed to press the new tip into its seating as the solder cools and solidifies.

The saw doctor or filer can make a suitable jig or fixture. One simple type is merely a flat steel plate larger in diameter than the sawblade, a circular spacing piece smaller in diameter than the sawblade to raise it from the plate, and a weight to hold everything firmly in place.

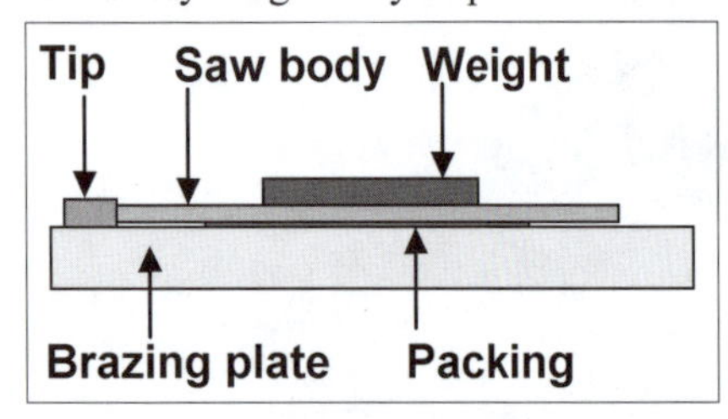

Fig. 12. 10
A simple tip brazing device consisting of a flat welding plate, a spacer and a weight.

The tip under-edge rests on the steel plate while being brazed, so the thickness of the spacer must correspond exactly to the amount of side clearance required - plus grinding allowance. *(The sawblade cannot rest on the existing saw tips because the new tips must initially project more.)* Alternatively, fit set screws into the base in place of the spacer. During the brazing process hold the tip in place by such as the tang of a file.

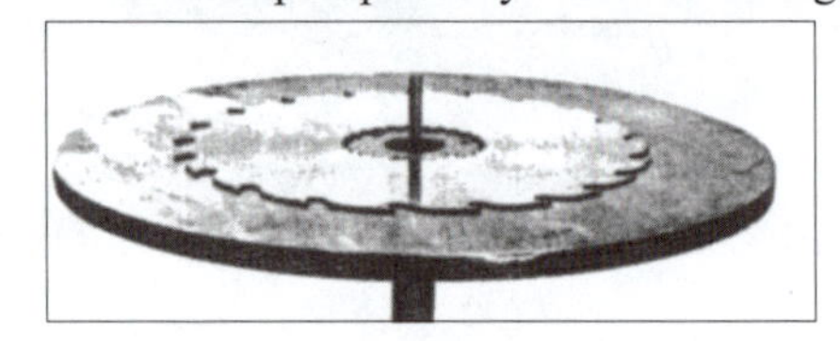

Fig. 12. 11
A table similar to this can be used for brazing a wide range of sawblades.

Fig. 12.12
The Petschauer ML 600 is a simple manual device for removing and brazing tct tips using a welding torch.
Source - Petschauer

Petschauer offer a simple brazing attachment *(for use with a welding torch)* suitable for both for tip removal and replacement. The sawblade is clamped magnetically on an arbor which quickly adjusts both for sawblade diameter and hook angle.

Damaged tips are removed by heating and tapping them off. New tips are placed on a plate, then clamped and centered by precision screw adjustments. During the brazing process manual pressure is applied via the saw body to seat the tip correctly.

Brazing procedure: The tip seat and tip must be thoroughly cleaned and de-greased before brazing - as any remaining impurities *(grease or oxide)* would create an unreliable bond. Use a solvent such as carbon tetrachloride or trichloroethylene.

Preferably clean the back contact face of the tip by rubbing it on a diamond wheel prior to de-greasing *(pre-tinned tips merely require regular cleaning)*. Immediately apply flux to the contact faces of the tips and the seatings to prevent re-oxidation. Clean the silver solder by using a fine piece of abrasive cloth, and cut it into small rectangular pieces each slightly larger than the tip. Place them in the flux until ready for use

Some brazing machines use a coil of silver solder. On these pre-cutting is not necessary as the action of brazing itself actually parts it automatically, this also gives minimal wastage.

Use tweezers to place silver solder into the tip seating - where the flux will hold it in position. Similarly place the tip firmly against the silver solder and edge-on the welding plate or in the tip recess of a brazing stand. *(Use tweezers to avoid contamination with natural grease from hands and fingers.)*

When using a brazing paste clean as above, then immediately coat the tip seating with sufficient silver-solder paste and simply snug the tip in position.

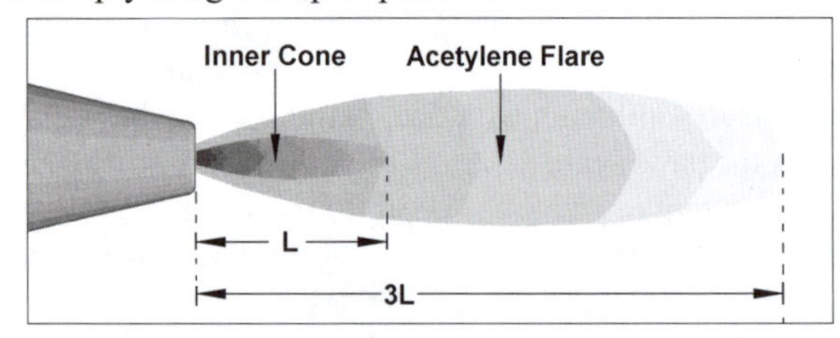

Fig. 12.13
The type of oxyacetylene flame suitable for tip removal and brazing; a. Inner cone; b. Carburizing feather (inner flame); c. Outer flame.

The correct oxyacetylene flame for this type of work is that of a carburizing flame, the cooler of the three flames obtainable with oxygen and acetylene.

A regular neutral flame, using equal amounts of oxygen and acetylene, is too hot. Use a carburizing flame *(with a lower oxygen level)* to give better control of the heating process.

It is important that tip heating is by conduction only from the body of the sawblade. In this way heating is kept to the absolute minimum needed to melt the solder - so minimizing the problem of contraction faults.

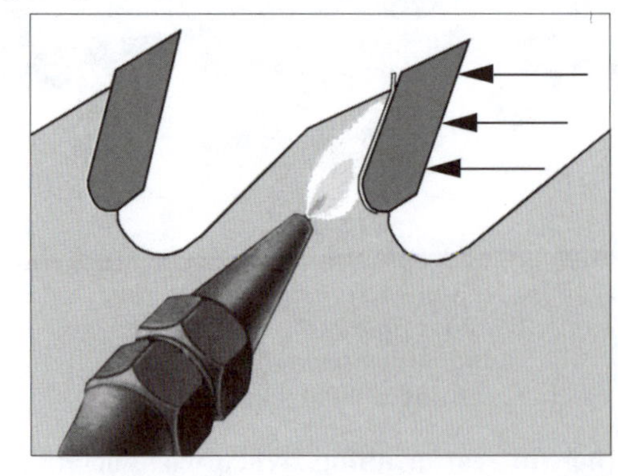

Fig. 12.14
This shows the how the outer, cooler envelope of the flame should be directed onto the saw tooth - not the tip itself.

Apply and maintain light pressure, using whatever means are available, to keep the tip snugged into the tip seating. As the silver solder melts - usually as the saw steel reaches a dull red colour - apply further slight pressure so that an even joint thickness results* *(an uneven solder bed may later fail)*. To reduce thermal stresses to a minimum avoid further temperature increase at this point. Continue to apply light seating pressure as the silver solder liquefies and spreads around both the tip and its seating. Gradually withdraw the flame to reduce the temperature as the silver solder begins to solidify - slow cooling is absolutely essential!.

Both overheating and rapid cooling should be avoided. Doing either will tend to leave the tooth hard and brittle - and increase stress problems. When several consecutive teeth are replaced, direct and maintain heat from each following brazing operation towards the heel of the previously-brazed tooth to further extend the slow cooling sequence.

When all tips have been brazed, clean the glassy, vitrified flux around them using a hand or rotating wire brush. To test the strength of the joint tap the top of the tip lightly and squarely, using a small aluminium hammer or wooden mallet. Do not to hammer the point itself as this could actually damage the tip.

**Tips are more securely held and absorb impacts better when the braze is even in thickness and around 0.08mm (0.003in.) thick. A thicker bed is less secure, and a thinner one can lose tips through operational impact. The braze alloy substantially increase the impact resistance of carbide tips, and this is further increased by forming an even fillet around the finished joint. The fillet should extend virtually to the edge of the tip, and a similar distance into the body, and have a convex appearance - this also indicates an efficient brazing process, see Fig 12.15..*

Brazing machines

Brazing tips using an oxyacetylene flame is fine for a small shop, but not when tips are replaced on a regular basis.

Two basic types of machine are suitable. One uses electrical resistance for the heating process and the other, which gives better results but costs more, uses high frequency generation. Using either machine is faster and more reliable than manual brazing - though needing less skill. Both machines allow rapid adjustment for sawblade diameter and hook angle, and precisely locate and hold new tips.

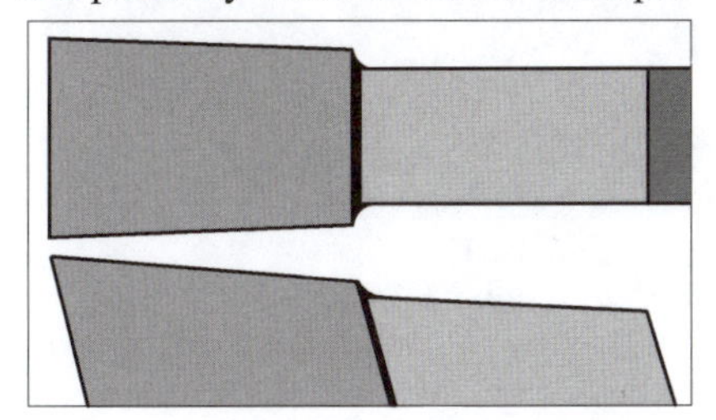

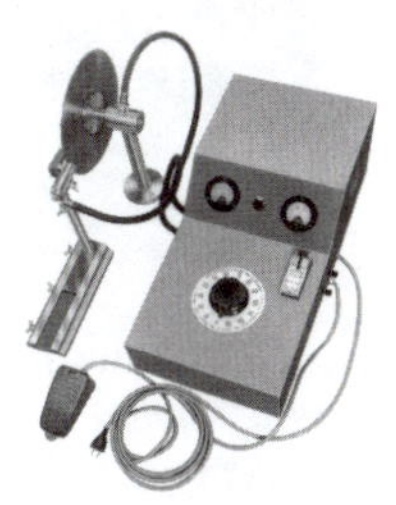

Fig. 12.15
Left, top and side views of a brazed tip with good fillets. Right: This small, bench-mounting resistance brazer shows the component parts - electric lead and foot control, transformer unit, saw support arbor and brazing fixture.
Source - Matsushita

Resistance heating.

With this type the saw tip is held in contact with one electrode of the electrical circuit, while a second electrode is attached either to the arbor on which the sawblade is mounted, or to clamps on both sides of the sawblade close to the tooth being treated. A large transformer and rectifier unit creates a low voltage*, high-flow current between one electrode and the other.

**This heats the joint area only because, where a low voltage current flows in a good electrical conductor (the body of the sawblade) it creates a minimum amount of heat, but where electrical conduction is poor (in the joint area) it creates the greatest amount of heat.*

Fig. 12.16
This is a typical resistance brazing machine, the LG 21H.
Source - Vollmer

When de-tipping, the broken tip is located in a recess of the tip electrode and the current applied. Immediately the solder melts, the broken tip and saw body are separated by quickly shifting the sawblade sideways to leave the tip in the electrode recess.

As this action sometimes releases the still-hot broken tip entirely, take care that it does not contact clothing or drop onto footwear - which should in any event be of an industrial safety pattern.

When tipping saws first clean and flux the new tip, silver solder and tip seating as described previously. *Brazing paste is actually more convenient as tips and seatings can be pasted well in advance of brazing.*

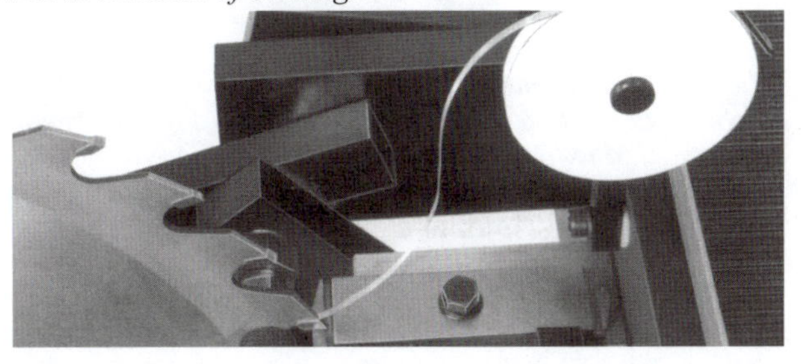

Fig. 12.17
This close-up of the LG 21H clearly shows the tip electrode, and partially shows the clamps acting both as the second electrode and to accurately position and hold the saw body. Silver solder in strip form is fed from a roll.
Source - Vollmer

Place the tip in the recess of the tip electrode and centralized it to the saw body via a fine screw adjustment. *(With some machines the tip needs only placing roughly in position to be gripped and positioned automatically when the machine is activated.)* Once the tip is in place the brazing process is carried out, generally as described for torch brazing but using the foot pedal to control heat input.

The amount of current flow is matched to the tip dimensions and the thickness of the saw body using either a switch with a range of current-flow settings, or by stepless adjustment via a rotary switch. Setting the current flow is critical - if too little heat is applied the joint may be imperfect, if too much heat is applied a 'dry' joint could result.

Resistance-brazing machines are simple in design and build and are low in cost - but require some expertise to operate them satisfactorily.

High-frequency brazing

This type of machine has a heating process which is fast, easy to set and more precisely controllable - the preferred type when cost is not an issue.

The machine has a high frequency generator connected to a double coil *(antenna)* sitting astride the tip to be brazed or removed - so arranged that the high frequency output is concentrated only where heat is essential to the process. When switched on, the tip is almost instantly heated to allow rapid tip removal or replacement.

Fig. 12.18
The ML 100/2 is one of a range of high frequency brazing machines in the Petschauer range. Two brazing fixtures are provided, one for regular tips and the second for wiper slots tips and similar.
Source - Petschauer

The generator works in a similar way to modern microwave ovens, except that the output is concentrated into the tip area rather than being deliberately spread more widely into food on a microwave dish. Heat is generated by eddy currents *(high-frequency alternating electric currents)* induced within the tooth and tip area.

The term 'high frequency' refers to speed at which the electrical current reverses in the heating coils. This it does much faster than in a regular mains ac supply - which operates at only 50 or 60 Hz.(cycles per second).

All rapidly reversing electrical currents produce electromagnetic waves - even overhead power cables. With higher frequencies the waves extend over a far wider area, so these are used for broadcasting sound and vision from radio and television stations, sending data to mobile phones, etc. - and for heating within a confined area.

The term wavelength is another way of describing the output of a high frequency generator. In this case the figure quoted is the distance between the electromagnetic waves as they travel outwards at the speed of light. For example, radio frequency short waves can be up to a maximum of 30 MHz. (30,000,000 cycles per second) and with a wavelength of 10 metres (30ft.). Microwaves operate at frequencies of up to 30,000 MHz. with a wavelength as short as 10mm. ($^{3}/_{8}$in.)

The amount of energy produced depends upon the wattage of the high frequency generator. For sawblade tipping these can range from 2 kW to 10kW. The choice of generator depends upon the size of the largest tips to be brazed. As a general rule a higher wattage is needed for large tips. Too-low a wattage generator may provide insufficient heat to guarantee a good bond on these tips - or heat them so slowly that heat spreads to create thermal stress problems.

High-frequency heating is powerful, extremely rapid and readily controlled. It must be set to input only the optimum amount of heat needed. To verify this some machines can be provided with a heat sensor and control device to allow fine-tuning of the setting. Otherwise the process is generally similar to that described for resistance brazing.

NOTE: Both with resistance and high-frequency equipment the amount of heat generated should be as rapid as practical, no more than necessary to fully liquefy the solder, and maintained for no longer than absolutely necessary. By controlling heat in this way the chance of imperfect brazing and body problems are both reduced to the absolute minimum.

Fig. 12.19
This shows the high frequency heating coil, the fine adjustments for both the saw body and the tip to ensure accurate sideways alignment, and the magnetic saw mounting flange. This equipment is capable of brazing up to 400 teeth per hour.
Source - Petschauer

Tip sizing.

After brazing, replaced tips must be ground down to the height, width, hook and clearance angles as corresponding tips. The general method of grinding carbide tips is fully described in Chapter 10.

Tip sizing sequence
Fully grind tooth faces and tops before side grinding - only when teeth are finally profiled should tips be side-ground to the correct kerf width. Failure to follow this sequence will result in unnecessary loss of kerf width.

When face-grinding the tips use the pawl to position the tooth in the regular way, and set the grinder merely to fully clean-up the tip face. Although the replaced tips may be thicker than existing tips this is not a problem - there is no need to grind them down to a common thickness. Widely differing tip thicknesses will alter the tooth pitch slightly, but this will not effect the performance of the saw - and modern grinders can handle slight pitch variations without difficulty.

Once the face is cleaned-up, the tip top and sides can be ground, again indexing each tooth tip by using the pawl in the conventional way. When dealing with individually replaced tips use the grind-to-zero method.

Grinding-to-Zero

A single-wheel side-grinder, or an automatic grinder under manual control, are perfectly suitable for this purpose. The method of grinding-to-zero is as follows:-.

Trigger the pawl movement to index any suitable existing tip. This should be the same hand and type as replaced tip selected for grinding. Choose the face to be ground first, then carefully set the grinding wheel to brush lightly against the full width and depth of this face on the tip indexed.

It helps to ensure proper contact with an existing saw tip face if this is first coloured with engineers blue. Light contact shows as either an overall reduction in the intensity of the blue, or as irregular light scoring of the surface.

When light contact has been made, reset the grinding cut-indicator scale to zero *(or note its actual setting)* and index the replaced tip. Back-off the grinding wheel, then make several grinding passes - each time increase the cut by a small amount - until the cut-indicator scale again reads zero *(or the previously-noted setting)*.

Repeat the process for the same faces of other replaced tips of similar hand and type, then treat the remaining faces in the same way. Repeat the process for any remaining teeth of different hand or type.

As a final check to avoid any inaccuracy that might result from grinding wheel wear when needing to grind heavily, repeat the above sequence on all faces.

Side grinding.

Side grinding is only needed during manufacture of sawblades and when sizing replaced tips. It is not part of the regular resharpening procedure of dull carbide tipped saws - routine side grinding unnecessarily reduces the kerf or cutting width and so shortens the life expectancy of the tips.

Replaced tips should be ground to the same side projection, kerf width and clearance angles as existing tips.

Completely re-toothed sawblades need to be side-ground with equal tip projection at either side *(side clearance)*, to suitable radial and back clearance angles, and to an appropriate kerf width to match the body and diameter of the sawblade and its intended application.

Single-wheel grinders have a solid support for the saw body close to the tip seating. Automatic grinders usually have two opposing grinding wheels to side-grind both sides at single pass. Sometimes they have two substantial clamps which open wide for saw insertion, then securely clamp and accurately centralize the saw plate to the grinding wheels.

Basic machines require the tooth kerf width and clearance angles to be set manually using dials and gauges, but with cnc grinders the kerf width required, the saw body thickness and all other relevant details are simply entered into the program - and the machine then sets itself.

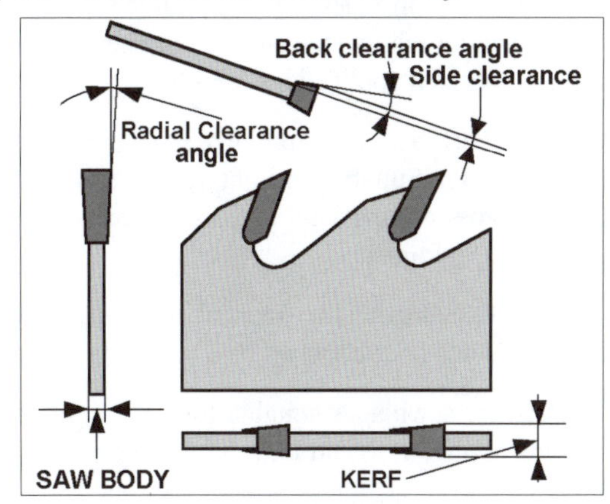

Fig. 12.20
Typical tct ripsaw tooth, showing the side clearance, and the back and radial clearance angles.

Side clearance is the amount the tips project each side beyond the body of the sawblade.

With too little projection the sawblade may bind in the cut - because waste is then more readily trapped between the saw body and the sides of the sawn material. This encourages waste to build up on the saw body as a sticky deposit - which then absorbs excessive power to drive the sawblade and could lead to burning and blistering.

If the kerf width is too wide for the saw body the sawblade may lack strength at the rim for the width of cut - to vibrate, run in the cut *(lead to one side or the other)*, snake *(lead first to one side and then the other)* or dodge *(deviate to one side or the other on meeting knots and other hard sections)*.

Leading, snaking and dodging can also result from using too-thin a saw body for the nature of the work undertaken - even when other factors are correct. Do not, however, attribute these faults only to a saw body being too thin - there are other possible causes as detailed elsewhere.

The amount of side clearance should increase in step with the saw body thickness. This itself should increase in step with sawblade diameter and increasing difficulty of sawing - so choose a saw body thickness from a table that relates to both sawblade diameter and sawing application.

For example, sawblades for manual cross-cutting or manual feed when sawing joinery components and single manufactured boards can have a thinner body and narrower kerf than sawblades of similar diameter for panel saw conversion of stacked boards or fast-feed sawing of wet timber.

Saw body thickness also depends upon the diameter of the saw collars relative to the saw diameter - these vary slightly between different machines.

For example, a machine with large diameter saw collars is suitable for thinner-bodied sawblades than a similar machine using the same sawblade diameters but fitted with smaller diameter collars.

Some sawblades are made with a thick body that is ground-off thinner outwards from the collar area to allow for a very thin kerf, but these are suitable only for a limited depth of cut. Some of the saws in the T2 range in the table below should be of this type.

Body thickness and kerf width table
The following table provides a general guide to body thickness and kerf width relative to sawblade diameter and application. The suggested applications are:-

T2 - Very thin kerf sawblades, possible with a reduced body thickness at the rim. Suitable for light manual sawing of thin material and cross-cutting light and small sections such as mirror and picture frames.

T1 - Thin kerf sawblades for manual feed of joinery and similar components on table saws.

R - Regular-kerf sawblades suitable for most solid timber, panel sawing applications and light-duty gang ripsawing.

H1 - Thick bodied sawblades for heavier-duty and faster feed speeds than above.

H2 - Thicker bodied sawblades for converting deep stacked boards on panel saws, for fast-feed ripsawing on one and two-arbor multi-ripsaws, and for the outer saws on hogger sets.

H3 - The thickest bodied sawblades for fast feed converting of wet timber on double edgers and as the outer saws on one and two-arbor multi-ripsaws.

Table of body thicknesses & kerf widths

Diameter	T2	T1	R	H1	H2	H3
150	**1.4**	**1.6**	**2.2**	-	-	-
	2.0	2.4	3.2	-	-	-
200	**1.6**	**1.6**	**2.2**	**2.8**	-	-
	2.2	2.4	3.2	4.0	-	-
250	**1.6**	**1.8**	**2.2**	**2.8**	**3.0**	**3.0**
	2.2	2.6	3.2	4.0	4.4	4.8
300	**1.8**	**2.0**	**2.2**	**2.8**	**3.0**	**3.2**
	2.4	2.8	3.2	4.0	4.4	5.0
350	**2.0**	**2.2**	**2.5**	**3.0**	**3.2**	**3.2**
	2.6	3.0	3.5	4.2	4.6	5.0
400		**2.2**	**2.5**	**3.0**	**3.2**	**3.4**
	-	3.0	3.5	4.2	4.6	5.2
450	-	-	**2.8**	**3.0**	**3.4**	**3.4**
	-	-	3.8	4.2	4.8	5.2
500	-	-	**2.8**	**3.2**	**3.4**	**3.4**
	-	-	3.8	4.4	4.8	5.2

Bold figures give the body thickness in millimetres.
Light figures directly beneath give the kerf width in millimetres.

Checking side projection

After a trial side-grind carefully check the amount of side projection of the tips at their outer point using some form of precision measuring instrument - at both sides relative to the saw body - and correct as necessary. Check individual replaced tips against existing tips *(after facing and topping, of course)*. Check new and re-toothed saw tips against the recommendations above - or against any saw already used for a similar application.

Consistent and accurate side projection of the tips will guarantee precision cutting and a good finish - provided, of course, that the saw body also runs true. (This should have been checked before replacement tips are brazed-on.) It is advisable, at this final stage, to also check the overall side alignment of the tips using a test stand similar to that previously shown.

Fig 12.21
After test side-grinding, check side projection of the tips using a precision dial indicator similar to the type shown here. This particular model has a magnetic base and clamps firmly to the saw body.
Source - Matsushita

Back or tangential clearance is measured as the angle between two lines drawn from the outer points of the tip and square to the hook angle, one line parallel with the saw body and the other along the tip itself. The tip should taper in from the cutting edge towards the following tooth at an angle of up to 5 degrees.

Radial clearance is measured as the angle between two lines drawn from the outer points of the tooth tip and in line with the hook angle, one line parallel with the saw body and the other along the tip itself. The tip should taper-in from the cutting edge towards the saw centre at an angle of around 1 - 2 degrees *(most side-grinders are capable of forming a radial clearance angle of up to a maximum of 5 degrees).*

With small clearance angles the saw gives a clean cut. Also the tip width reduces at a slower rate through regular grinding - and so has a longer life before replacement becomes necessary. If too-small a clearance angle is used, however, there is greater risk of friction between the tips and the sides of the cut material. This may cause burning of these surfaces, need excessive power to drive the sawblade, and result in faster dulling of the radial edges.

With large clearance angles the cut is an easy one and so takes least power to drive the sawblade. Clearance angles which are too large, however, lead to faster loss of kerf width through regular face and top grinding - and so require much earlier replacement than would otherwise be the case.

Both clearance angles should be the least practical to produce and maintain a clean and low-power cutting action, and so that kerf width is only slowly lost through grinding. A careful balancing act between all the conflicting factors is needed to achieve this.

As a general guide, more side clearance and greater radial and back clearance angles are needed for fast ripsawing of wet softwood, and the least are needed for dimension sawing of seasoned timber and manufactured boards.

When side-grinding replaced tips the existing tips will be already ground to the correct side projection and side clearance angles, so carefully repeat these for the new tips. Use the grind-to-zero method to ensure that side projection is the same between new and existing tips.

When side-grinding a completely re-toothed sawblade, set-up to the side clearance and clearance angles of any similar, new sawblade that gives satisfactory results for your particular use - but be sure to check a new and not a part-worn sawblade - or use the above table.

When side-grinding a new or re-toothed sawblade for other than a regular application, set initially for small clearance angles, then gradually increase them (if necessary) after carefully noting the affect on quality of finish and power consumption.

The Side Grinding Process

The operation is very similar to top and face grinding. The new tip is first indexed by the pawl to position it for side grinding, side-ground, then the next tooth is indexed and side-ground, and so on.

A simple, single-wheel, general-purpose manual grinder or side grinding attachment is perfectly adequate for side grinding occasionally-replaced carbide tips, but for completely re-tipped and newly manufactured sawblades use a semi-automatic or a fully automatic two-wheel side grinder.

Some two-wheel side grinders are capable of fully side-grinding at a single pass (perhaps incorporating an oscillating movement) or permit rough and finish-grind passes as part of the regular grinding cycle. Use these features as necessary according to the weight of grind needed and/or the class of finish demanded.

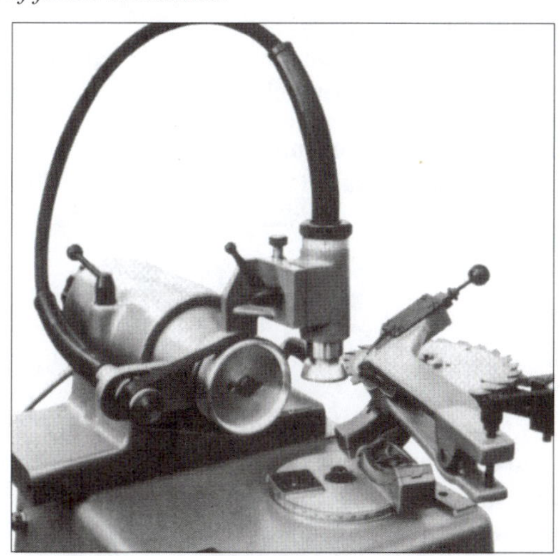

Fig. 12.22
This shows the optional flexible drive to a side grinding attachment - fitted with a saucer grinding wheel - as used on a Saturn HKSC 600 tct saw grinder.
Source - Saturn

All side grinders use similar grinding techniques and similar diamond grinding wheels - either saucer face-grinding wheels, or plain edge-grinding wheels.

Saucer face-grinding diamond wheels. These have diamond mainly on one face of the grinding wheel. They effectively grind parallel to the radial cutting edges of the tip, and so conform to the recommended method for grinding carbide. For technical reasons the abrasive has to be relatively narrow, as explained later.

Plain edge-grinding diamond wheels. These have diamond formed on the periphery only of the grinding wheel (although this may be keyed at the edge) and they grind tangentially to the radial cutting edges of the tip - which is technically wrong*. They form a minute hollow grind on the tip *(but which is too shallow to have any detrimental effect)*. Because of the nature of the grind the edge abrasive can be relatively wide - and produces a particularly good finish. There is another advantage in using a edge-grinding grinding wheel - a larger clearance angle is possible when grinding close-in.

**Although grinding tangentially (at right angles to the radial cutting edge) is contrary to normal carbide-tip grinding practice, other grinding conditions are such that excellent results are produced.*

Both types of grinding wheel wear evenly - provided of course that the traverse movement is correct set.

Side grinder setting

Setting the back clearance angle

The method of setting for the back clearance angle depends upon the type of grinding wheel used.

Using saucer grinding wheels

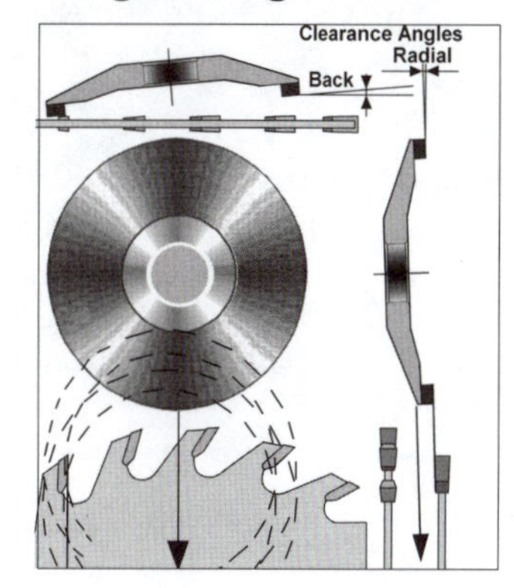

Fig. 12.23
Using a saucer grinding wheel for grinding side clearance. Both this and Fig. 12.24 show the traditional grinding method - traversing parallel to the hook line, while tapering-in to the centre of the sawblade to give the required radial clearance angle.

Saucer grinding wheels are set to grind only with the extreme left-hand edge. They are set to traverse parallel to the radial cutting edge of the tip, while also closing-in towards the saw centre *(toeing-in)*. To give the required back clearance angle the wheel is tilted slightly.

Take particular care that the inner edge of the diamond face overlaps at least the leading edge of the tip. By so doing the grinding wheel will wear evenly - failure to do this may result in uneven diamond wheel wear and possible rounding-over of the cutting edge. The abrasive on saucer grinding wheels has to be relatively narrow for this reason.

Using edge-grinding wheels

Edge-grinding wheels are off-set to the rear of the hook line so that the curvature of the wheel forms the required back clearance angle.

A zero clearance angle is formed when the grinding wheel centreline is set both square-on and central to the tip. As the grinding wheel is shifted towards the tooth heel the clearance angle increases. Refer to the makers instructions for precise setting details, or note the following table.

Off-set value table

Wheel diam.	Back clearance angle		
in mm.	5°	4°	3°
200	10mm	8mm	6mm
175	9mm	7mm	5.5mm
150	8mm	6mm	5mm
125	7mm	6mm	4.5mm
100	6mm	5mm	4mm
75	5mm	4mm	3mm

Fig. 12.24
The grinding head on the HSN 600 automatic side grinder tilts a few degree off vertical when setting for the back clearance angle to use a regular saucer grinding wheel.
Source - Stehle

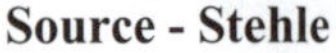

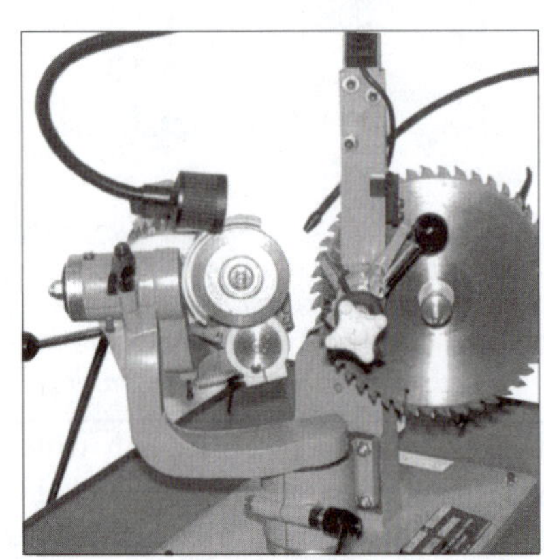

Fig. 12.25
When side-grinding on the TCT/2 manual saw grinder, use a regular saucer grinding wheel together with an optional right-angle bracket to mount the sawblade.
Source - Autool

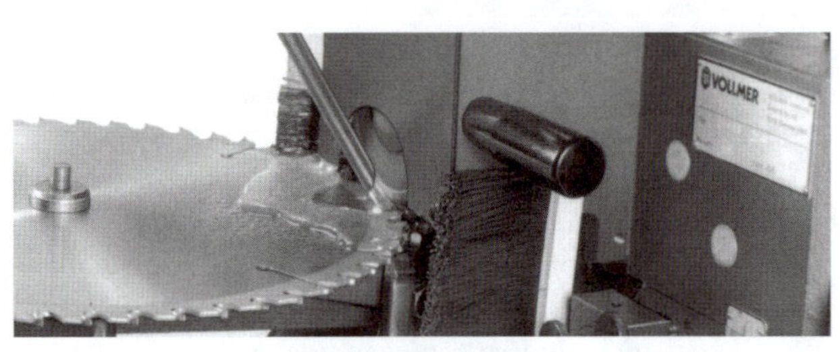

Fig. 12.26
The CHHF 21H manually-operated grinder uses an edge-grinding grinding wheel.
Source - Vollmer

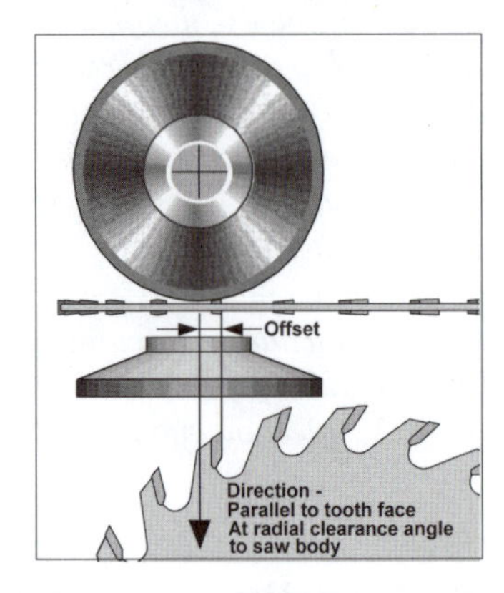

Fig. 12.27
Using an edge-grinding wheel for grinding side clearance. In this case the grinding wheel traverses parallel to the hook angle and toes-in radially.

Setting the radial clearance angle

This is possible in one of two ways, either by tilting the traverse movement slightly relative to the saw body (toeing-in) or slewing the traverse movement relative to the hook angle of the teeth.

Tilting the traverse movement

This is the regular and most direct way of setting for the radial clearance angle. Use the scale to tilt the traverse movement relative to the saw body, i.e., to toe-in towards the sawblade centre in the case of regular tips, or to toe-out for reverse taper tips.

In this case the grinding wheel *(or the sawblade itself)* is always set to traverse parallel to the hook angle of the saw.

Slewing the traverse movement.

On a few machines the radial traverse movement of the grinding wheel *(or that of the sawblade itself)* remains parallel with the saw body.

With these it is necessary to slew the traverse movement of the grinding wheel *(or that of the sawblade)* to a specified angle slightly less than that of the hook angle of the sawblade teeth.

By traversing at a slew angle, the radial clearance angle for the tip is formed through contact with a lower part of the grinding wheel as it traverses towards the sawblade centre. The actual slew angle depends upon the back clearance angle setting and type of grinding wheel used:-

Using saucer grinding wheels

Using a saucer grinding wheel gives a straight radial edge and a consistent back clearance angle. The slew angles given in the table below can be taken as a general guide only:

Slew angles for saucer grinding wheels

Back clearance angle		5°	4°	3°
Radial clearance angle	2.0°	20°	22°	26°
	1.5°	15°	18°	21°
	1.0°	10°	11°	13°

Using edge-grinding wheels

The amount of slew required when using an edge-grinding wheel varies according to the grinding wheel diameter and the back clearance angle to which it is set.

The figures given in the table below should be taken as a general guide. This is based on forming a specified radial clearance angle only at the tip. *(For this reason, set for the back clearance angle only at the outer point of the carbide tip.)*

Slew angles for edge-grinding wheels

G.W. Diam in mm	Back clearance angle 5°			4°			3°		
200	20°	**15°**	10°	22°	**17°**	12°	26°	**20°**	14°
175	18°	**14°**	10°	20°	**15°**	10°	26°	**20°**	14°
150	17°	**13°**	9°	20°	**15°**	10°	26°	**20°**	14°
125	17°	**13°**	9°	20°	**15°**	10°	25°	**19°**	13°
100	16°	**12°**	8°	19°	**14°**	9°	23°	**18°**	12°
75	15	**11°**	7°	18°	**13°**	8°	21°	**16°**	11°

The table above gives the slew angle required relative both to the back clearance angle setting and the grinding wheel diameter. The three figures under each back clearance angle column refer to the three radial clearance angles of 2.0°, 1.5° and 1.0° respectively, in that order, with the 1.5° column printed bold.

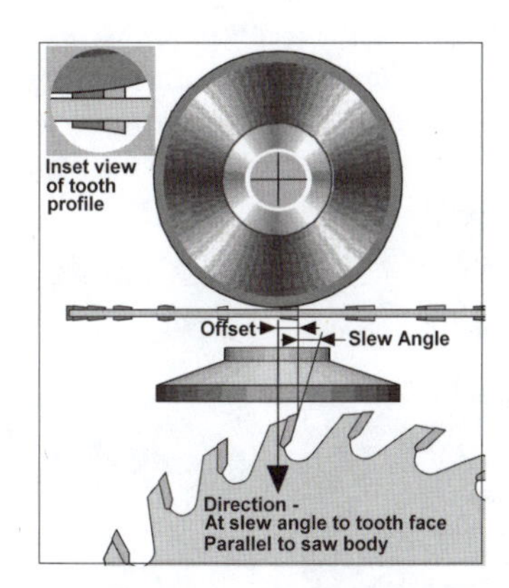

Fig. 12.28
By slewing the traverse movement of an edge-grinding grinding wheel relative to the hook angle, a radial clearance angle can also be formed - even though traversing parallel with the saw body.

Using a plain, edge-grinding wheel gives a slightly curved radial edge, and a back clearance angle which reduces slightly towards the saw centre.

This effect is greater with a small diameter grinding wheel and less with a large diameter grinding wheel. The inset view in Fig. 12.28 shows this clearly. The greatest amount of clearance is always at the tip - for which some claim a technical advantage.

NOTE: *Reverse-taper tips should only be ground by tilting the grinding wheel traverse movement. Attempting to grind clearance angles on these is impractical when grinding parallel to the saw body and slewing the traverse movement - except in certain exceptional cases.*

Setting the traverse movement.

The grinding wheel must draw back enough to clear both the tip and the pawl while the next tooth is indexed.

The amount of forward traverse down the tip for both plain and saucer grinding wheels should be such that they effectively clear the innermost edge of the tip. This guarantees an even and regular amount wheel wear - provided that this is also set as recommended above.

Take care that the grinding wheel does not foul the saw body at the end of its stroke as this can fill the grinding wheel - so needing regular dressing - as well as making the saw look untidy and unprofessional.

In this respect edge-grinding wheels are better as they permit a shorter working stroke and so are less likely to foul the saw body.

Setting the pawl

Index the individual tooth to be ground in the same way and with the same care as when top and face grinding. Set the pawl to make contact only a little way down the tip, and set the arcuate traverse movement to maintain a consistent point of contact throughout the stroke.

Make sure that the grinding wheel does not foul the pawl either during the full movement of the pawl or during the grinding operation. In some cases excessive clearance movements are needed for both grinding wheel and pawl.

With some machines the pawl intentionally remains in contact with the tooth face during the side grinding operation. This provides a stabilizing effect - a distinct advantage when processing thin-bodied sawblades.

Fig. 12.29
The CHHF 21H is typical of the type of small, manually-operated side grinders suitable for single-sided grinding.
Source - Vollmer

Automatic side grinder operation

Automatic two-wheel side grinders can side-grind with a single grind or multiple passes - depending on the control features offered.

In basic versions of this type the grinding wheels usually make full contact with the tip on both the inward and return movements.

For this reason two or more setups may be necessary when processing new or fully re-tipped saws - unless the grinder is heavily built, with powerful motors and infinite control of both the grinding wheel and traverse speeds.

Fig. 12 30
The True Sizer is a rugged and powerful automatic two-wheel side grinder suitable for Stellite or tct saws, but which also incorporates a manual saw tooth indexing system for grinding single teeth.
Source - Armstrong

Fig. 12.31
The handwheel shown here is the means of instantly setting the True Sizer side-grinder to one of five preset radial clearance angles.
Source - Armstrong

Modern automatic side grinders offer refined control of the grinding operation - the most advanced permitting an instant choice of one of several grinding sequences, see Fig. 12.37:-
Grind on the **forward** movement only, with lift-off and a rapid return movement.
Grind on the **return** movement only, with a rapid forward traverse clear of the tip.
Grind on both the **forward and return** movements.
Grind with oscillating movements on both, or either, the forward and return movements.

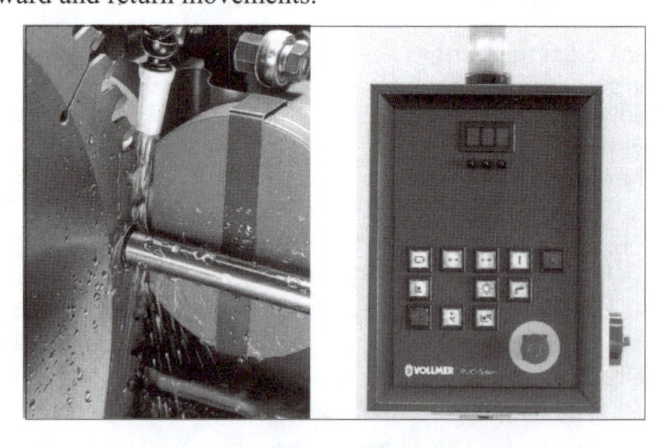

Fig. 12.32
The FS 2A is a low-cost and uncomplicated two-wheel automatic side grinder intended for small and medium-size tct saws. Dial gauges are provided for accurately setting. Left; the ample cooling system of the FS2A and, Right; the central control desk.
Source - Vollmer

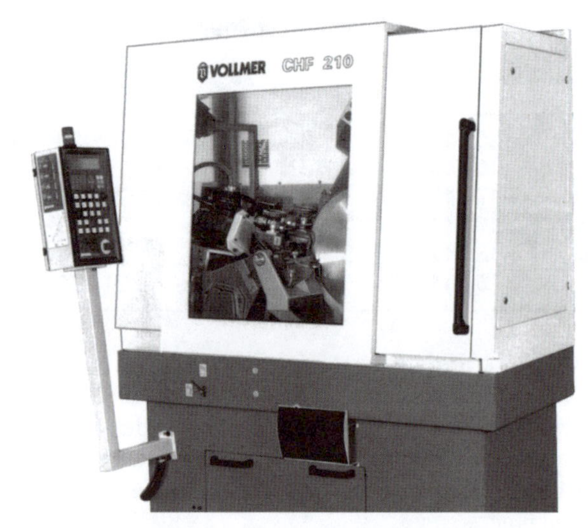

Fig. 12.33
The CHF (shown above) & CEF automatic side grinders are both cnc controlled and offer a wide range of grinding programs - each designed to give short cycle times.
Source - Vollmer

Fig. 12.34
Left; immediately below the grinding wheels can be seen the wide-opening sawblade clamp, and above the optional devices to measure grinding wheel loss and so provide automatic loss compensation.
Right; the side-grinding operation.
Source - Vollmer

Fig. 12.35
The Akemat F3 automatic, two-wheel side grinder compliments the Akemat top and face grinders. With a transfer system this machine can be incorporated into a production line together with Akemat face and top grinders.
Source - Vollmer Dornhan

Fig. 12.36
The Akemat F series of automatic side grinders incorporates infinite control of both grinding wheel rotary speeds and traverse speeds. The radial clearance angle control readily adjusts to any angle between -5° and $+3^{o}$
Source - Vollmer Dornhan

Continuous-path grinding

Fig. 12.37
The CNC 6F automatic side grinder has CNC control and six dynamic axes. These are arranged to give automatic head positioning, continuous-path control of the side-grinding movement and arcuate movement of the pawl. The control panel clearly displays - in an easy-to-follow form - both the operating process and input required to set the machine. It requests input for the main dimensions and grinding parameters both for the sawblade and grinding wheels etc., and from these computes the settings required in readiness for implementation on command.
Source - Walter

Walter make a side grinder, CNC6F, which has continuous-path control of both side-grinding wheels.

For the two grinding wheels there are single-axis movements to initially set both for the required hook angle and the appropriate amount of offset for the back clearance angle. The Z-axis, which powers the traverse movement along the hook angle, operates in conjunction with two Y-axes, one for each grinding wheel. The Y axes operate independent of one another and control the movement of the grinding wheels towards and away from the saw body as they traverse along the tip.

The Z and Y axes operate simultaneously, interpolating their movements to control tip side projection, kerf width, radial clearance angle and the radial tip-face profile. *(The back clearance angle is controlled by the relative position of the grinding wheel centrelines to the tooth face - as previously described - and is set automatically from the data input.)*

It is not necessary to calculate the precise machine settings needed. The display screen merely calls up for such sawblade details as diameter, body thickness, number of teeth, back and radial clearance angles, kerf width, etc. to be entered.

The computer then undertakes the necessary calculations and sets the machine accordingly - there is no need either to work out such details as the pawl arc movement, the amount of off-set for the peripheral-grinding wheels, or the interpolation of the Z and Y-axes - the computer does all this automatically.

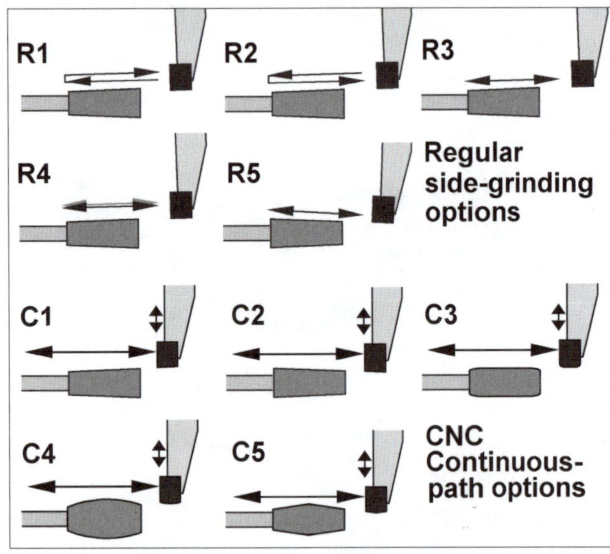

Fig. 12.38
This shows typical side grinding options for regular and cnc controlled grinders.

R series, regular side-grinding profiles.
R1 - Slow grind in, lift-off and fast return.
R2 - Fast movement in clear of the tip, slow grinding return movement.
R3 - Slow grinding movements in and out.
R4 - Rough and finish grinding passes in sequence.
R5 - Reverse taper tip - any of the above options.

C series - continuous-path computer-controlled movements
C1 - Regular tip, any combination of R1-R4 movements
C2 - Reverse taper, any combination of R1 - R4 movements.
C3 - Flat sided with rounded tips and tails.
C4 - Spoon profile.
C5 - Pentaform profile.
C3 - C5 are planer-type profiles.

Programming a grinder in this way is simplicity itself - only the actual sawblade data is needed - not the machine settings needed to achieve it. As a result, machine setting for regular, reverse-taper and radial-profile grinding is both fast and precise.

Machine settings for standard tooth forms can be instantly called-up from the computer's memory. Customized settings for new sawblades are readily created by modifying standard settings and allocating separate and individual identifications - then memorized for future use.

Simple, rapid and precise machine setting is not the only advantage of continuous-path grinding. Tooth profiles which are impractical with conventional side grinders can also be produced routinely using the software provided. In addition to the regular and reverse side-grinding profiles, this also allows round-edge, spoon-shape and pentaform radial profiles to be produced simply by data entry.

The round-edge, spoon-shape and pentaform are side-point-free shapes which produce a planed finish on the sides of the material being sawn and gives the saw doctor and filer immense scope to develop new profiles to suit particular applications - something which was not previously available to him.

A companion to the continuous-path side grinder is the continuous-path face and top grinder, as shown, the Walter CNC6F and the Walter CNC5. These together give immense scope for innovation - for example inter-spacing planer teeth with hollow-face scriber teeth to prevent side-chipping ahead of the cut in brittle and difficult-to-cut materials. The facility for development of new and more advantageous saw teeth is as wide as the user wants to make it!.

Certainly the future prospect for controlling and developing tooth geometries with such equipment is a challenging and exciting one and I regret that retirement came before such highly sophisticated equipment was readily available.

Quality of grind and run-time

Regular straight-line grinding of faces, tops and sides on CNC grinders provides full contact between the abrasive and tooth face to give the best possible grinding conditions. This provides a high quality of grind and minimal wear of the grinding wheel.

Continuous-path control, by its very nature, does not necessarily give full contact between the abrasive and the tooth face being ground.

For example, side grinding wheels have a grinding face formed parallel to the grinding arbor, perhaps with a slight angled relief at the leading and trailing edges. This is progressively dressed to a more practical profile as grinding proceeds, varying according to the type of grind:-

For regular side clearance (C1) the grinding wheel faces wear to an angle corresponding to that of the side clearance of the teeth being ground - to then give the full contact and ideal grinding conditions. If the side clearance angle is altered slightly (by varying the cross to the main traverse movement ratio) the grinding wheels will quickly bed-in to the new angle.

For reverse-taper grinding (C2) the same applies.

For flat-sided grinding with rounded tips and tails (C3) the initial wheel profile is ideal for the flat sides, but may wear irregularly for the rounded tips and tails until a compromise wheel profile is generated with matching rounded edges. This will not affect the quality of the grind.

For spoon profiles (C4) the abrasive face wears to a corresponding convex shape, but line contact only takes place - and a different grinding wheel specification or grinding condition may be needed to provide a high quality of ground surface.

For pentaform profiles (C5) the grinding wheel will quickly wear to vee profile to match the two angles, and will then give full face contact - but of only half the width.

For these reasons a perfectionist would switch grinding wheels when changing from one type of side clearance to another - although the type of grinding wheels and grinding programmes used with these machines do allow switching between the above operations - without need of switching wheels. Profile CP grinding has similar characteristics.

Chapter 13

Diamond Saws and Carbide tools

Diamond saws

Diamond (polycrystalline diamond, or PCD) saws have become a firmly-established tool for converting faced and unfaced particle board and similar - materials for which carbide saws were once exclusively used.

Diamond tipped saws look little different to regular carbide saws - but their performance is vastly superior. This is due to the difference in their physical properties, as the table below clearly shows:

Physical properties MCD, CVD diamond, PCD and TCT

Property	Monodite*	CVD dia.	PCD	TCT (K20)
Density (g/cc)	3.52	3.51	4.10	15.00
Compressive strength (GPa)	9.0	16.0	7.4	5.0
Fracture toughness (MPa $m^{0.5}$)	3.4	5.5	9.0	11.0
Knoop hardness (GPa)	50-100	85-100	50-75	18
Young's Modulus (GPa)	1050	1180	800	600
Coefficient of Thermal expansion (x 10^{-6}/K)	2.0-5.0	3.7	4.0	5.4
Thermal conductivity (W/mK)	1000-2000	750-1500	500	100

*Synthetic, single-crystal diamond (MCD) manufactured by De Beers Industrial Diamond, see later details.

Source - De Beers Industrial Diamond

The impressive performance of PCD saws is directly related to their physical properties, mainly their hardness and wear coefficient. But the two factors alone would be of little use unless the other properties were also in line - which fortunately they are.

In addition to retaining their cutting edges for far longer periods than carbide saws, diamond-tipped saws do this without the rapid cut-off in performance that carbide saws suffer from. In practical terms this gives an effective saw life of somewhere between 100 and 300 times that of tungsten carbide-tipped saws.

Diamond saw applications

While there was some initial concerns that diamond saws would be too-readily prone to accidental damage, doubts about this - and the suitability of diamond for wood-based materials in general - have largely been resolved.

Where concern is still felt that manufactured boards could contain highly damaging tramp metal, some makers recommend the use of a metal detector to scan boards prior to conversion. Perhaps a more viable solution, though, is to always purchase boards from a reputable source who can guarantee that their product is tramp-metal-free.

The same general advice *(given under carbide saws, Chapter 10)* regarding the need to carefully handle these sawblades and the importance of the sawing machines being in prime condition, applies equally well to diamond saws - in fact doubly so.

Although they can be damaged with rough or careless handling, in the right application and with care on the users part they give excellent service.

Tooth geometry

There was some initial doubt about whether diamond saws could be given a suitable hook angle for woodworking applications. They were at first restricted to a minimum included angle of 75° *(a figure originally stipulated by the diamond-tip manufacturers.- see Fig. 7.12 page 80.)* This apparently limited diamond sawing only to materials for which small hook angles were also suitable. Fortunately, small hook angles were also the norm for the first uses to which diamond saws were so successfully put - converting the difficult materials that rapidly dulled carbide saws.

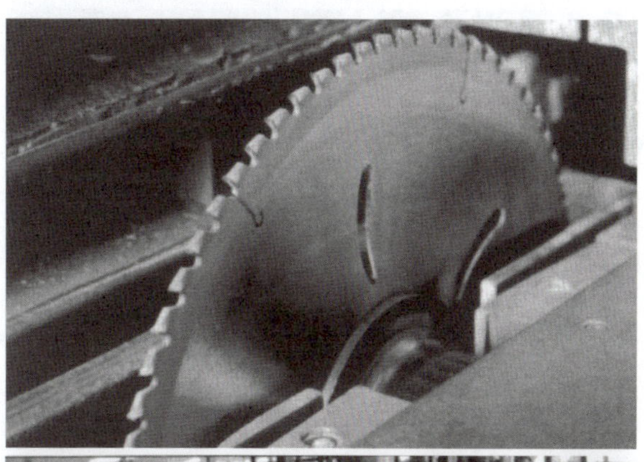

Fig. 13.1

This is a typical PCD success story, where a 600mm. diameter pcd saw (above) used on a Holzma panel saw (below) gave spectacular results when compared to the tct saws used previously.

In converting MDF, plywood, plain and laminated chipboard, the tool life increased by 300%, - yet allowed a 10% increase in the cutting speed while giving the same quality of edge finish. Taking all this into account, plus the more consistent results and decreased downtime, the pcd saw paid for itself within 3 months.

Source - Supreme/Holzma

This included-angle stipulation was probably too over-cautious - at least for wood-based materials. Through their accumulated experience, leading diamond saw makers realized that rapid breakdown and costly edge-damage *(forecast for using a smaller included angle)* did not necessarily occur. Hence they were able to adopt carbide tooth geometry to a much larger degree than had initially been thought practical - so greatly widened the scope for diamond saws within the woodworking industry.

Diamond saw tooth geometry, although mainly based on corresponding carbide types, does not include their larger softwood hook angles. This does not automatically exclude diamond from sawing the softer materials for which large hook angles are traditionally needed. Experience in other woodworking applications has shown that, if a cutting edge is exceptionally sharp and remains so for long periods, then this alone will ensure reasonably clean and low-power sawing - even with too-small a hook angle.

Fig. 13.2
The PCD hogger unit sawblade (above) replaced a corresponding carbide saw on the IMA edgebander (below). The diamond-tipped saw lasts nearly 2 months - compared with carbide saws which had to be changed every two days. This gave a welcome 30-fold increase in tool life and huge reduction in downtime.
Source - Supreme/IMA

PCD saws are not normally used for softer materials - not because they cannot be cut with diamond - they can - but because the biggest run-time gains are made with materials that are difficult to cut. With easier-to-cut materials the run-time gain is considerably less - and below a certain critical level tungsten carbide becomes more cost-effective than diamond.

For this reason diamond saws are still mainly used for materials that give problems for carbide saws. But there is a move away from the traditional woodworking approach to more of an engineering viewpoint - of fitting and forgetting tools - expecting sawblades to remain in service for months rather than days. Diamond tipped saws are ideal for this - right in the ballpark.

The only visual difference between carbide and diamond saws, without close examination, is that diamond saw tips are noticeably shorter, at only 4, 5 or 6 mm. deep.

The reason for this is to economize on the cost of individual tips - to the benefit of both maker and user. Although this requires a much more reliable bond between the saw tip seating and the carbide substrate of the diamond tips - the brazing methods used by leading makers ensure that the bond is as secure as it can be made.

A diamond saw is a high-cost investment when compared to carbide -but the investment cost can be reclaimed within a relatively short space of time - in the right circumstances.

When purchasing diamond saws rely on the expertise of specialist companies who lead the field in PCD saw manufacture and who have the necessary background and experience to produce the right tool for your application. Buying an unsuitable diamond tool is a highly costly mistake.

Sharpening diamond tipped saws

Sharpening is carried out on an EDM *(electro-discharge erosion machine)* specifically designed and manufactured for diamond tools. Never attempt to sharpen PCD except on the as shown in Fig.7.16 *(across, not on, the diamond face)* and with any machine other than an EDM type. Trying to grind PCD on a conventional saw grinder with regular or super-abrasives will be slow, costly and inefficient - and certain to produce a poor edge that rapidly breaks down.

Note previous details on PCD tips in Chapter 7.

Fig. 13.3
The QW Universal Electrical Discharge Machine can machine PCD tools for both the woodworking and metal industries. It is just one of a series of specialist erosion machines manufactured by Vollmer in a wide range of specifications and price brackets.
Source - Vollmer

Modern erosion machines are highly developed pieces of equipment at the very cutting edge of technology. Although complex, much of the technology is aimed at making the machines operator-friendly, simple to set and use, and fully automatic in operation.

Naturally this is reflected in their high investment cost - but they have the capability of manufacturing and re-sharpening high-cost diamond tooling of the type that in likely to be used in greater numbers - particularly when considering the newer and more difficult-to-process materials.

However, for those who can find the capital and justify

machine purchase for in-house re-grinding, there are eroding machines with fewer nc axes intended for PCD sawblades alone which are less complex and lower in cost.

For all but the few large woodworking manufacturers, however, manufacture and re-grinding of diamond saws has to remain with specialist diamond saw makers and repairers - and no doubt this will be the case for some time to come.

Fortunately there is a growing number of servicing companies who are able to re-grind these sawblades, and more and more saw makers are offering them - though not necessarily making them. The increase in the use of diamond-tipped saws is likely to continue apace, with the likelihood of PCD taking over from carbide saws in an ever-widening range of applications within the woodworking industry.

Erosion machines for PCD tools

There are two basic methods of spark erosion - either by using relatively inexpensive rotating graphite or copper/tungsten electrodes, or by using a fine wire electrode.

Rotation spark eroders.

The basic technique is not one of grinding but of spark erosion - an erosion process created by a pulsed electrical discharge between an electrode and the tool. *(This requires the tips to be electrically conductive. While diamond itself is not conductive, the cobalt matrix bonding the individual particles of diamond in PCD is - so these tips can be machined in this way.)*

In the erosion process an electrical discharge occurs within the spark gap that separates the rotating electrode and the tool. The generated electrical energy is transformed into thermal energy which serves to melt/vaporize the cobalt and PCD micro-particles. A continuously-fed dielectric liquid is needed to encapsulate the process in a similar manner to coolant in regular grinding machines.

The rotating graphite and/or copper/tungsten electrodes are formed to patterns generally similar to those of regular carbide grinding wheels, and erode the surface in an action that appears similar to a grinding cut.

They are manufactured both as edge-eroding discs *(similar to plain edge-grinding wheels)* for machining tooth sides *(flanks)*, and as face-eroding discs *(similar to cup and saucer grinding wheels)* for machining tooth tops.

Edge-profiled discs are also available for making and sharpening shaped router cutters and similar.

Fig. 13.4
Left - during erosion the area is flooded with dielectric fluid to promote the process.
Right - a typical diamond-tipped saw.
Sources - Vollmer (left), Guhdo (right).

As the electrical discharge progressively degrades the rotating electrodes, these are dressed at pre-set intervals in order to maintain accuracy and quality of finish. This function is controlled by the grinding machine, which then automatically re-sets the rotating electrode to its original position.

Fig. 13.5
The tungsten/copper alloy electrodes shown here are re-sharpening a diamond hogger (left) and a diamond sawblade (right) on a QW spark erosion machine. The electrodes are of the face-eroding type similar to cup grinding wheels.
Source - Vollmer

Fig. 13.6
Here the PCD tips are ready for side-dressing using twin edge-erosion graphite discs on the QF 20 P.
Source - Vollmer

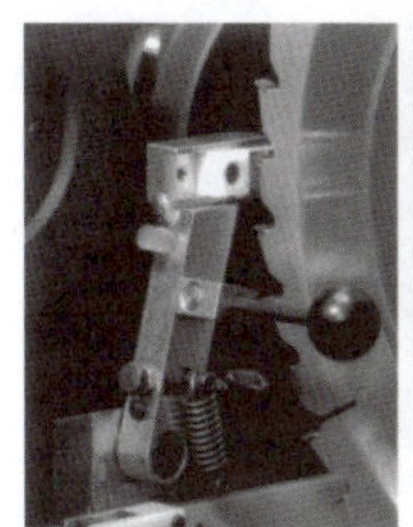

Fig. 13.7
When side-eroding on the QF 20 P the pcd sawblade is pre-set for position by moving it against a permanently-fixed swivel-mounted stop (left).
After a pre-selected number of teeth have been machined, an automatic sequence is started to dress and then precisely to re-set the graphite eroding discs to their original position and eroding condition (right). This indefinitely maintains both quality and accuracy of the erosion process.
Source - Vollmer

The whole operation is controlled by a multi-processor system. Amongst other functions it gauges each tool before and after eroding, maintains a precise spark gap, and ensures continued machining accuracy and tool quality. The computer also regulates and monitors the automatic erosion sequence to ensure that this continues to operate at the optimum working speed.

The machining speed and depth of erosion control the quality of the machined face and the cutting edge. The multiprocessor can vary these conditions throughout the process of eroding each individual tip - to produce the finest practical finish at highly efficient stock removal rates.

The process can start with a fast stock removal which leaves a poorer finish, through a medium cut with a better finish to a final light pass giving the best finish. All these conditions can be changed via the keyboard - according to the type and condition of the tool being treated - to make these machines universally adaptable to a whole range of tools.

Wire eroders.

Fig. 13.8
General view of the operating section and control panel of a the QWD 76 P wire eroding machine. As with all eroding machines absolute stability is essential, and for this purpose polymer concrete is used in its construction. The tool and wire eroder are under five-axis control.
Source - Vollmer

With wire eroders a fine electrode wire is stretched taught between guides on either side of the workpiece, and erodes the tool in a similar way to rotating discs. In this case, though, the wire can either cut through the tips to slice-off a section, or erode-away at pre-determined depth - according to how the machine is programmed. This allows these machines to remove greater depths than disc eroders.

Wire eroding machines are, of course, capable of eroding a wide range of metal & woodworking tools. The latest multi-axes machines are suitable for even the most difficult types, and are even capable of forming complex profiles with appropriate back and side clearance angles.

Examples are profile and step tools used in metalworking and router cutters for faced shipboard and mdf - in addition to regular milling cutters, woodworking hoggers and circular sawblades.

The tension on the wire is automatically maintained - and does not deflect during processing as there is no physical contact between the wire and the tool - the spark gap is maintained automatically as the tool is eroded away. The spacing between the guides can readily be varied according to the tool type and shape.

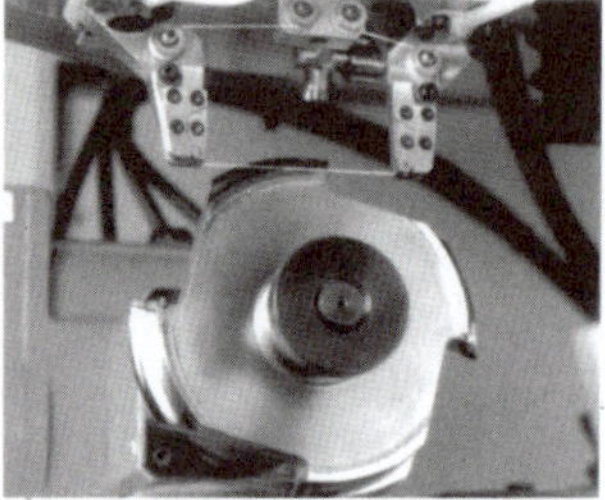

Fig. 13.9
The spacing of the wire electrode guides can be adjusted, so allowing the wire to be placed as close to the tool as its contour permits so yet remain absolutely stable.
Source - Vollmer

The wire degrades in the same way as the rotary discs, but fresh wire is continuously fed between the guides from spools within the machine - and so accuracy and quality of finish is continuously maintained without the need to interrupt the erosion process.

Grinding machines for MCD

Monocrystalline diamond is non-conductive and cannot be spark-eroded. Tips faced with MCD have to be pressure-ground on high-tech, specialized grinders.

On these the degree of contact pressure between the special diamond grinding wheel and the tool can be steplessly adjusted - and multiple passes can be undertaken as described for disc eroders. The diamond grinding wheels used in this process naturally wear and loose their grinding efficiency as grinding progresses. To correct this they are dressed at regular intervals with a ceramic grinding wheel. This is immediately followed by adjustment to compensate for the amount of wear in the diamond grinding wheel *(the diamond wheel is accurately and automatically measured for wear and appropriate compensation applied automatically).*

Combination grinders/eroders.

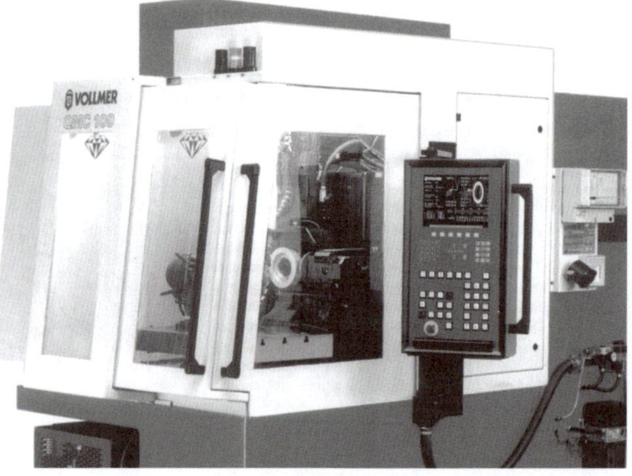

Fig. 13.10
The QMC 100 is a machine capable of automatically grinding MCD, CVD diamond and PCD tools, also for automatically eroding CVD diamond and PCD tools. A common fluid can be used for both processes.
Source - Vollmer

Machines are now available for both grinding and eroding diamond tools of all types.

When grinding, the entire process runs automatically, with the tools measured, ground and re-measured in a single clamping process. The degree of essential contact pressure between the grinding wheel and the tool is infinitely adjustable, and the grinding wheel is dressed and reset automatically for wear at pre-set intervals. In addition to tip grinding, the machine can also relief-grind the tool body - a process normally needing another machine.

Fig. 13.11
Dressing the grinding wheel (left). Grinding an MCD hogger (right) on the QMC 100
Source - Vollmer

The diamond grinding wheel can be quickly switched for a disc eroding wheel - and the machine can then be re-set to become a fully automatic eroding machine *(to operate generally as described above - complete with fully automatic disc trimming and re-setting, etc).*

All operations in both operating modes are fully computer-controlled and self-regulating - to give the most efficient working conditions. In this way the fastest possible turnround of diamond tools is made possible.

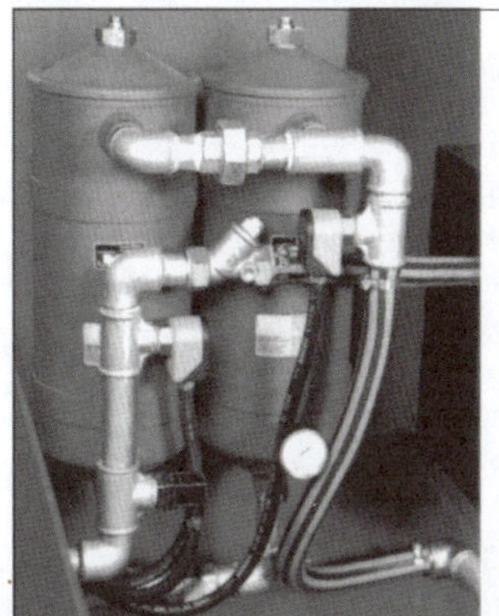

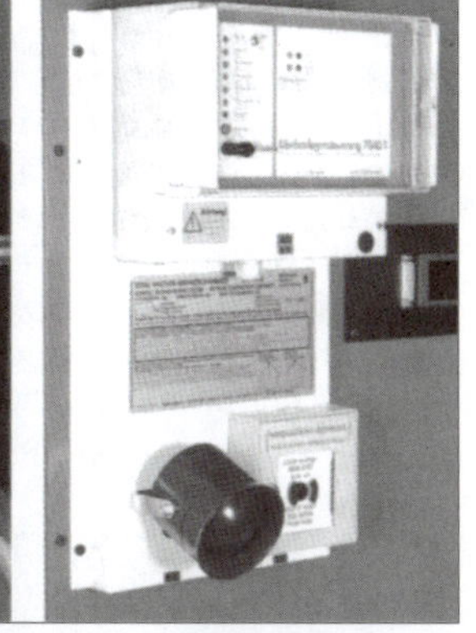

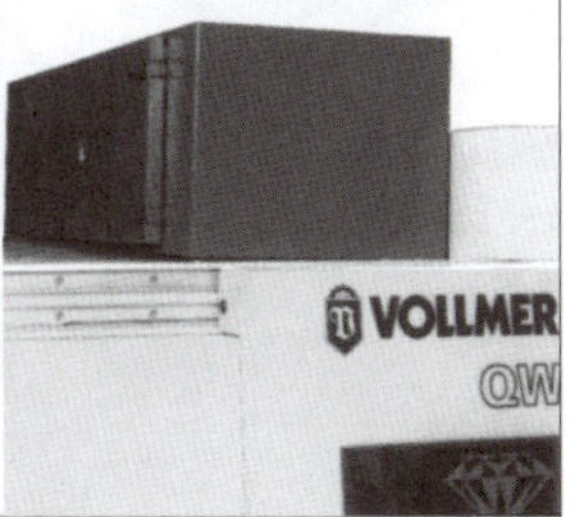

Fig. 13.12
The above are provided, or are an optional fitment, for all Vollmer diamond grinders/eroders. Filtration unit for the dielectric fluid to ensure uniform properties (top left). Cooling device to maintain dielectric fluid temperature (bottom left). Automatic fire extinguisher using environmentally-friendly CO_2 (top right). Extraction unit to maintain friendly working conditions (bottom right).
Source - Vollmer

Other saw-like tools

Splitter & Planer Saws

Tipped splitter and planer saws are used mainly on moulders. They are usually fitted on the final or beading head of the moulder for parting the planed and/or moulded output into two or more pieces - so multiplying the throughput and raising the efficiency of the machine. Another application for planer saws is on cnc routers or workstations, where work once carried out on several traditional woodworking machines is now undertaken on a single, high-tech machine.

Regular plate saws were originally used for splitting, but these gave a sawn finish which was not always acceptable. Hollow-ground saws gave a better finish but were prone to overheating - mainly due to the small amount of side clearance provided with this type of saw body.

Two types of *'saw'* were, however, developed in the mid sixties and these, the reverse taper and the Woodsaw, both split timber efficiently - and give an *'as planed'* finish. Both are tipped sawblade types, with tips of high speed steel, Stellite or tungsten carbide

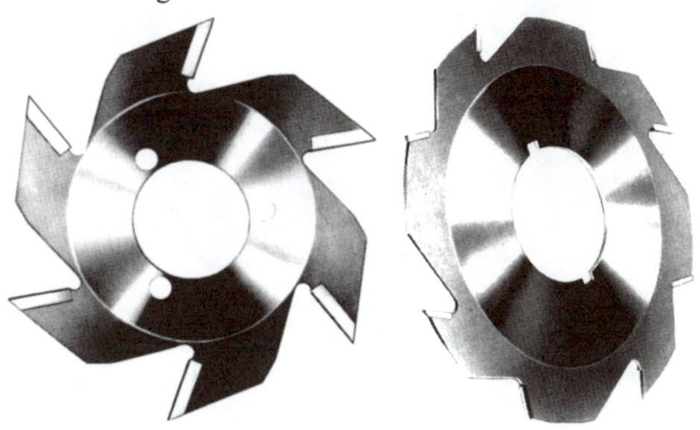

Fig. 13.13
This shows two planer-type splitting saws; left a reverse taper saw and; right, the Woodsaw splitter saw.
Source - Wadkin

The reverse taper planing saw, as its name suggests, is more like a multi-point groover with long tips that taper outwards from the rim towards the saw arbor by roughly 1^0 - 1.5^0. The planing action of the reverse taper saw, however, is only good for the depth of tip fitted - splitting deeper than this still gives an unacceptable sawn finish on the lower section.

Being a reverse taper saw the sides of the timber are formed slightly out-of-square - an unacceptable feature for some applications.

The front face only is ground to restore the cutting edges, and this must maintain the original hook angle - usually around 25^0. This type of sawblade runs sweetly at lower spindle speeds and, while it produces a good finish on interlocked-grain hardwoods, it tends to produce a woolly finish on stringy softwoods.

The Woodsaw - so called after its inventor, ***Geoff. Woods,*** is a *'saw'* with four rip teeth and two pairs of planing teeth. The rip teeth ,which remove the bulk of the kerf, have a positive hook angle of around 20^0- 25^0 and are square-fronted and square-topped. The planer teeth have a negative hook angle of around 25^0, and an alternate face bevel of around 25^0.

The planer teeth project sideways fractionally more than the rip teeth and traverse the timber almost as regular planing cutters - effectively along the grain. This is due to their negative cutting angle and height setting *(as recommended for all planer saws)* with the planing teeth protruding above the timber by less than a millimetre.

In sharpening, the front faces only should be ground, removing an equal amount from both ripping and planing teeth - measured peripherally. *(This is essential to keep the relative widths of the teeth in step with one another.)* Like planing cutters themselves, the planing teeth can be hand-honed with an oilstone on the front face to form a keen cutting edge. Preferably the outer and inner corners of the tips should be slightly rounded-over from the **sides** - **not** the **front** - to prevent these scoring the finished surface.

Using splitter saws

Certain precautions must be taken in order to produce a high quality of splitting. The regular steel bed plates cannot support the underside of the timber being split satisfactorily - and certainly not at the point of cut - so some spelching of the underside is likely if regular bed plates only are used.

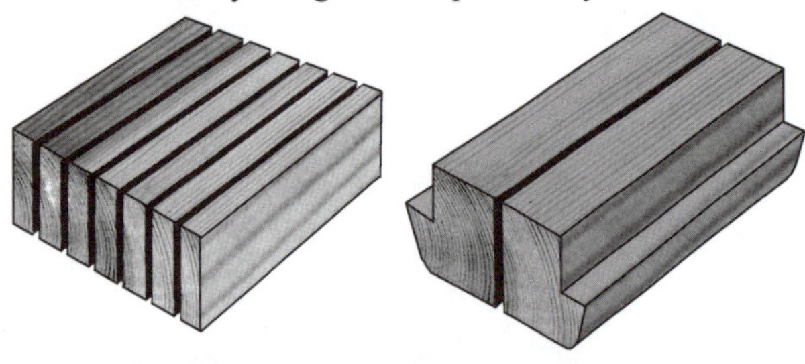

Fig. 13.14
The traditional use of splitter saws on a moulder is to produce multiple squared-up pieces at a single pass through (left). It is also practical to produce two moulded window sections, for example, split by a single saw (right).

To avoid splintering and break-out on the underside, securely fit a one-piece hardwood insert in place of the plates on either side of the splitting head. With the sawblades running, slowly wind up the head so they break through the insert - but only sufficient for the sawblades to project fractionally above the timber to be split. *(Before doing this check that the sawblades do not foul any metal when at the correct operating height.)*

The insert will support the exit point of the sawblades and prevent splintering on the underside of timber for some time - but will eventually wear and become less effective. If at all possible build-in some means of periodically shifting the insert slightly in the direction of the feed, with the sawblades wound down, then break through again to restore the chipbreaking action.

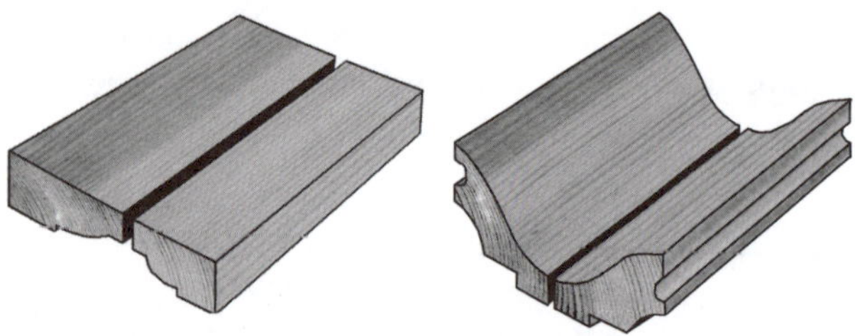

Fig. 13.15
Moulded sections can be split as here from the top (left), or from underneath (right).

It is essential to fit a wooden pressure above the timber being split to hold it firmly down and steady the cut.

Riving knives *(splitters)* should also be fitted immediately behind the sawblades to prevent their up-cutting action at the rear spoiling the finish.

If a single sawblade only is being used, then the riving knife can be fractionally thicker than the saw kerf *(by something less than a millimetre)* to open-up the timber and positively prevent up-cutting contact. In this case the outfeed fence behind the sawblades needs off-setting by at least half this amount.

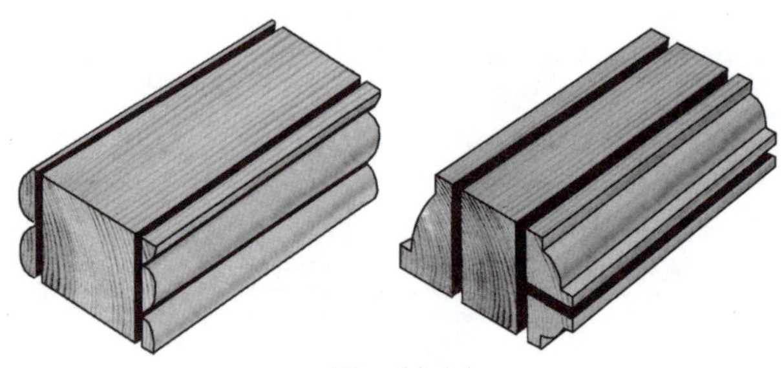

Fig. 13.16
Another application is to produce a square section, three half-round beads and two quarter beads (left) or a square section and three ovolo sections (right) - in the latter case using an additional splitter saw on a vertical arbor.

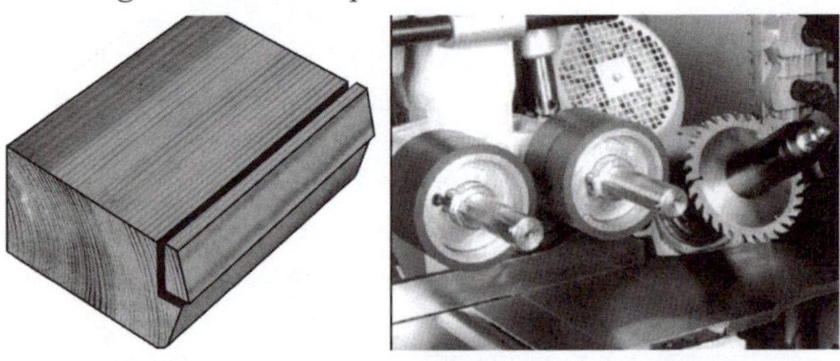

Fig. 13.17
It is also practical to split a glazing bead from the glass rebate on door stock (left) using a splitter saws on both horizontal and vertical arbors. This keeps the two parts to the same count, and also retains the grain and colour match when stock is intended for clear varnishing - provided these are stocked together, of course.
The right illustration shows a regular carbide saw on a vertical arbor of a moulder which, in this machine can be tilted to any angle.
Source - Weinig

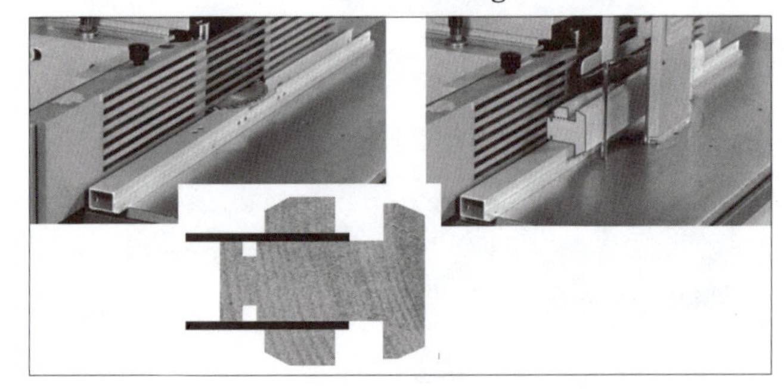

Fig. 13.18
Where the glass rebate dimensions do not allow a planing head to be combined with a splitter saw on the moulder, an alternative is to form the section, a mullion in this case, to include the glass bead, and with a groove to correspond with the glass rebate.
The final cuts are then made on a spindle moulder (shaper) using the fence attachment shown here. The splitter saw is in line with the top surface of the 'Glassledger' and the separated glass bead passes through the hollow section.
Source - Aigner

When using more than one sawblade, then the mid riving knives or splitters can be thicker than the body of the sawblade, but must be fractionally less than the saw kerf - otherwise they will jam. In all cases accurate alignment between the sawblades and their riving knives is absolutely essential. It is also essential that the moulder side guides are absolutely square to the splitter saw arbor.

Planing saws are thicker than equivalent regular sawblades and waste more timber in the kerf. Because they plane the sides, though, subsequent planing operations are unnecessary. In addition, the total timber loss in producing PAR. is less in comparing the overall processes *(planer-splitting as against sawing and double planing)*. Consequently, planer-splitting is far more production-efficient.

Thinner sawblades can be used if large diameter collars are fitted. But, in any event, collars should always be large enough to give the maximum possible support the sawblade. They should be as close to the bed insert as practical - if necessary slightly relieving the underside of the insert.

Fig. 13. 19
This shows the regular arrangement for multiple splitting on a moulder. Note the bed insert, the top pressure (right) and riving knives behind the saws (left). The trailing fingers in front of the saw prevent kick-back.
Source - Wadkin

Splitting saws today

Splitter saws of the regular pattern are still widely-used, and with modern manufacturing methods they are better than ever.

The introduction of better methods of saw grinding has raised the quality and life-expectancy carbide tips. This has vastly improved the performance of tct saws of regular patterns - to the extent that they are now widely used for splitting. The scoring effect they still produce is much less noticeable than with earlier plate saws, and is commercially acceptable in many instances.

An interesting development is the viability of continuous-path side-grinding of carbide teeth (see Chapter 12). This, combined with the option of different hook angles and form-grinding of tooth tops (see Chapter 10), gives far more scope to tungsten-tipped saw types than was previously the case.

For example, it is now possible to produce fast-feed multiple-point carbide saws of the Woodsaw pattern routinely - with absolute repeatability - and with the absolute minimum handling.

Continuous-path side-grinding gives the option of spoon-shaped, corner-rounded or pentaform tooth front profiles - any of which is capable of producing a planed finish. The use of modern cnc grinders also allows hollow fronts and vee-tops to be formed without problem.

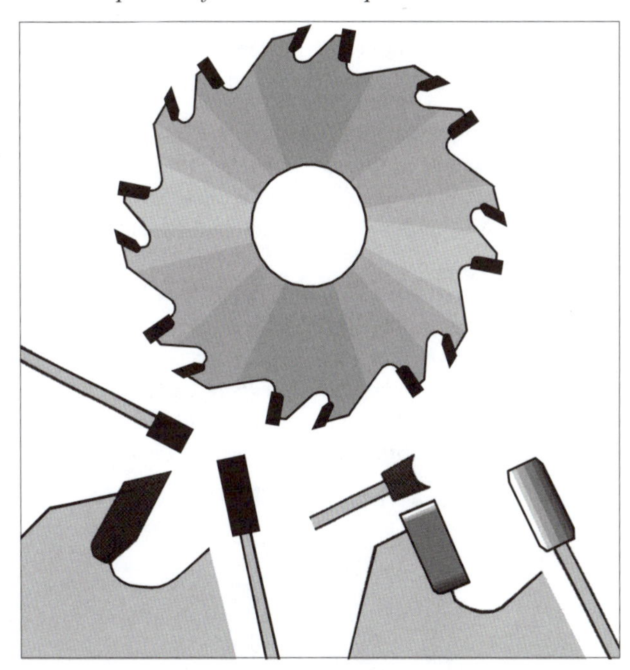

Fig. 13.20
This shows a possible version of the Woodsaw - now practical through modern grinding technology. The rip teeth are square-topped - but could have a top vee to scribe ahead of the main cut and more positively prevent side break-out - if necessary. The negative hook-angle planer teeth are fractionally wider than the rip teeth, have a hollow front face and are continuous-path side-ground to round-over the outer and inner corners of the tip.

Hogging Units.

Hogging units are intended for edge-trimming boards and similar, and simultaneously hogging the waste into small pieces that can easily be handled by the regular exhaust or blower system.

Two main types are used. The expandable type with individual sawblades or cutters, and the fixed-width type with inner sections fitted either with saw segments or cutters. In many cases the fixed-width type can be built-up by adding more sections.

Fig. 13.21
Two regular types of hoggers with individual saw sections to expand the width in specific increments (left) and a lightweight body type with skewed saw segments (right).
Source - Leitz

All hogging units have an outer sawblade with either flat-topped or top-bevelled teeth. Top-bevelled sawblades can have either alternate bevelled teeth or single bevelled teeth with high points facing out.

The outer sawblade, in many cases, contains gaps to allow inner teeth or cutters to project into the saw's kerf width - and so provide an unbroken level surface with the teeth.

Expandable hogging units. This type can have any number of inside cutters added to expand the hogging width in precise steps. Usually the inside cutters are staggered to reduce cutting vibration, and may be arranged in different diameters to step the cut.

In the latter case the cutters nearest to the outer sawblade are slightly smaller in diameter so as not to break-out in advance of the sawblade (which forms the clean edge of course) but away from this they increase in diameter to reduce the remaining waste to small enough sections not to break away and tear-off material at the trailing corner.

Fig. 13.22
This shows a hogger body carrying skewed saw segments - but separated from the outer sawblade.
Source - Leuco

Fixed-width hogging units -saw segment type. These have a lightweight body with skewed slots to hold two or more saw segments. The number of teeth in the segments, and the number of actual segments, varies with the cutting diameter of the set and its application. Where natural timber is to be hogged, the hogger unit should normally have the greatest number of segments. Tooth heights are often arranged to give a stepped cut, as per the expandable type and for the same reason.

Fixed-width hogging units - cutter segment type. Cutter inserts, each consisting of a removable body with two or more teeth separated by a gap, are fitted in a lightweight body section.

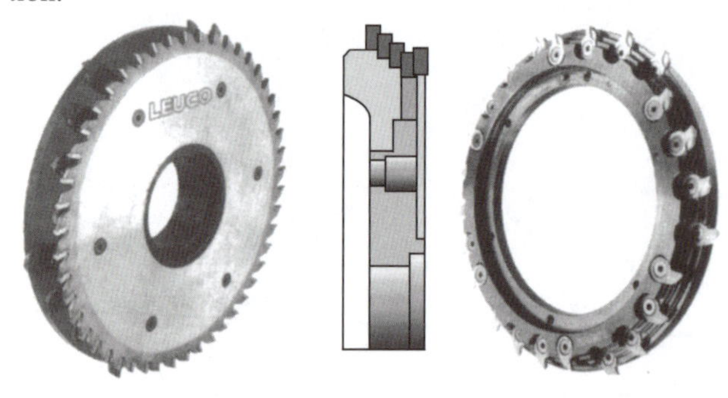

Fig. 13.23
This Twin Tec hogger set has stepped hogger teeth which, away from the sawblade, cut slightly ahead. With this arrangement the size of waste remaining near the end of the cut is so small that it does not break-off prematurely.
Source - Leuco

The inserts are staggered to alternate the cutting action between following inserts - and so form a perfectly flat cut. Additional sections can usually be added to increase the hogging width. An alternative, lower-cost, type uses turnblade cutters - but is only really suitable for hogging particle board.

Fig. 13.24
This shows the cutter-segment type of hogger unit (left) and the carbide cutter insert (turnblade cutters) type hogger unit (right). The latter has a hand-grip type quick-release fitting.
Source - AKE

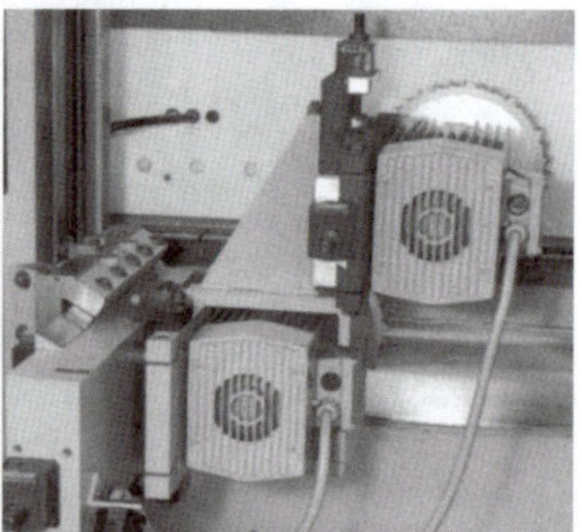

Fig. 13.25
This Universal Sizing Machine FL10/20, designed on the general lines of a double-end tenoner, is intended for splinter-free square or angular sizing, grooving, rebating, and sanding of all board material. The machine is double-sided to process both board edges at a single, high-speed pass. The infeed is shown on the left, and a double hogging unit on the right.
Source - Homag

Fig. 13.26
Left: A combination of a hogger and a trim saw gives a perfectly clean trimmed edge. **Source- Lueco**
Right: A stepped hogger unit. Source - **North American.**

Hoggers tend to splinter on the underside at their exit point. Where this is unacceptable this section can be supported on the underside by either a metal plate carefully set close to the hogger, or by a plastic or hardwood plate into which the hogger can be cut to give all-round underneath support.

Hogging units are also used to edge-trim board material on double-end tenoners, edge banders and similar. They are often used in pairs above and below the board, or with a hogger unit and a preceding trim saw. In both instances the hoggers and saw climb-cut *'in'* to form, between them, perfectly clean edges topside and underside.

Hoggers for vee-fold systems. These can be of the cutter or saw segment type and are intended for use on vee-fold machinery. The hogger is tilted to an angle of 45 degrees in use, so that the sawblade and cutter segments form a 45/45 vee cutout in plastic-coated boards - the cutters or segments being always arranged to give a perfectly square cut, .

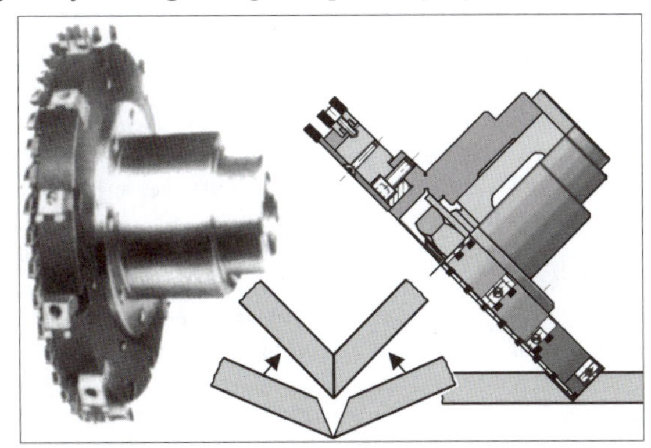

Fig. 13.27
Hoggers for vee-folds are mounted at 45 degrees as shown. The vee-cut board is then folded on its outer skin to produce a mitred effect.
Source - AKE

The hogger cuts through the inner skin and the structural material, usually chipboard or mdf, but not the outer skin - as this serves as the fold. When folded into a right angle the appearance is that of a mitred joint with the outer skin wrapped around the square corner as continuous surface.

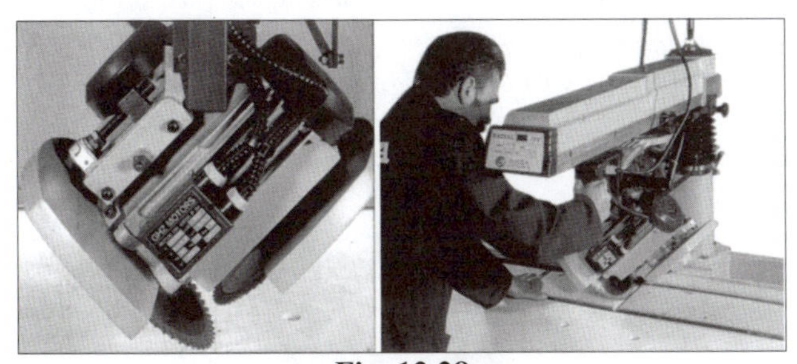

Fig. 13.28
If is also possible to form single vee-folds using two sawblades mounted on a double-head cross-cut saw, as show here. The sawblades are staggered and set to overlap precisely inside the outer skin where the external corner is formed.
Source - Omga

A pair of saw units mounted on a specially-modified pull-out cross-cut machine can also be used for vee-folding. In this case the sawblades are mounted at 45° to the face of the material and at right angles to one another. Adjustment between the two saw units allows the two sawblades to be aligned precisely at the point of the vee regardless of their diameter.

Surface hogging sets. This type consists of a hardened steel body with profiled and bevelled tct tips. Two hoggers of this type are regularly used as a pair to hog and clean-up board material to a perfectly square edge. In this case both are set to climb-cut, one hogging from the top and the other from the underside.

Another use is in preparing stock for edge-butted facings on soft-forming machines. The same type of head can also be used for forming shallow recesses.

Fig. 13.29
This compact, diamond-tipped hogger is for use on double-end tenoners and edge-banding machines for sizing panels. A similar diamond-tipped V-groove hogger is also made by Guhdo for folded panels.
Source - Guhdo

Finger-joint hoggers. These are for use in trimming mini finger joints - and are design to give a clean and quiet cutting action. They consist of a single outer sawblade with a large top bevel angle to produce a clean cutting action, and a straight-topped inner hogging cutter of slightly greater diameter. The number, type and style of teeth vary according to diameter and type of finger-joint.

Grooving and Trenching sets. These are made as a one-piece tool either as a grooving or trenching set with a width which is fixed or expandable in specific steps, or as an infinitely variable-width groover.

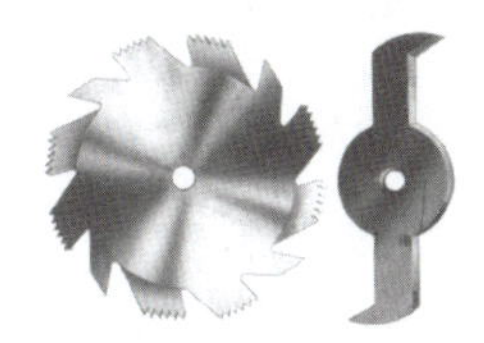

Fig. 13.30

The original Huther grooving or dado sets, consisting of two outside saws (left) and one or more inside cutters (right) can be used on many machines for grooving or trenching with or across the grain.
The outside sawblades are rather thick and of unusual design, with rip teeth followed by two sets of scriber teeth. The scriber teeth are of the triangular type, and the body is ground away to leave narrow teeth at one side only - which are filed to leave high points to the outside.
The following sets of scriber teeth alternate to one side and then the other, so each sawblade can be used on either the right or left-hand side of the set.
The rip teeth cut slightly below the scriber teeth to give a narrow groove between two shallow scribe lines.
The inside cutters are of a very plain design and available in different cutting widths. By using multiple cutters a set can be made up to almost any width required.

In all cases the outer cutting elements, either sawblades or scoring cutters, give perfectly clean edges when climb-cutting, and the inner sawblades or cutters give a perfectly flat and continuous surface between them.

They are used on double-ended tenoners to cut trenches square-across natural timber or panels, on cross-cut saws to cut trenches square or at an angle to the direction of grain*, on some dimension or variety saws and on cnc routers to form grooves or trenches in any direction on natural timber or panels.

**When used on cross-cut machines, trenching sets are usually of a smaller diameter than the sawblades regularly used. For this reason they run at a lower peripheral speed and - because of this and the cutting action of this type of tool - they have a greatly increased risk of snatch and run-out.*

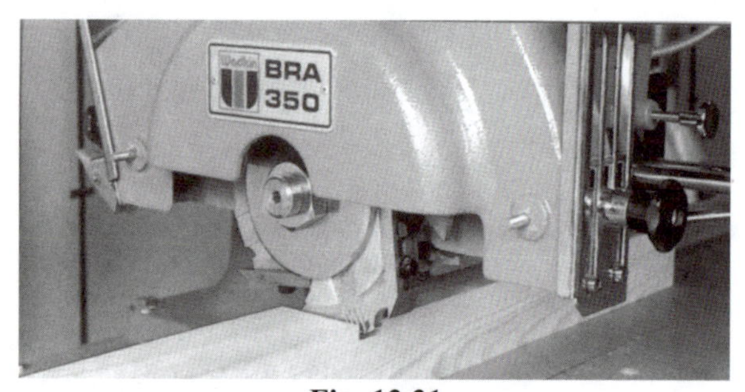

Fig. 13.31
Trenching heads can be used on pull-out cross-cut saws, but must be used with the workpiece securely clamped.
Source - Wadkin

For safe working it is essential to take precautions when using trenching heads on a manual pull-out cross-cut saw to avoid setting-up vibration - which can lead to snatch. Snatch can violently lift the workpiece and accelerate the outward movement of the cross-cut head.

Securely hold the workpiece down on the worktable and against the fence, preferably on both side of the trenching head, by using some form of quick-acting clamp.

Hand pressure alone will not hold material securely enough for trenching - it is dangerous to attempt trenching without clamping the workpiece.

Secondly, take particular care when using trenching sets on a manual machine, make slow and carefully-controlled outward and return movements - never let the head run forward out of control..

On automatic cross-cuts the traverse control speed can be varied and slowed enough to suit this operation - and should remain consistent once set - but here again use some form of clamp or clamps to secure the workpiece.

Because of the risk of run-out with any type of tool on these machines, but especially with trenching saws, always keep hands well clear of the operating area. The operating area, incidentally, should be clearly marked as dangerous for all cross-cutting work - not just trenching. Where cross-cut saws have a straight-line movement the operating area is a narrow strip, but with tilting and swivelling cross-cut saws the operating area can be a greatly extended quadrant.

Most other machines using trenching and grooving heads have either a mechanically-controlled feed system, or a manually-operated resistance feed action *(with the head rotating against the feed)* - so run-out is not a problem. Nevertheless they are dangerous tools and should be used with the greatest caution, fencing-in the operation as much as practical on manually-fed machines.

Also keep all tools sharp - dull tools cause accidents.

Fig. 12.32
Shown here are typical, conventional grooving saws (main and inset view) and a moulding cutterhead (main view), both tools that can be fitted to some fixed or sliding table saw benches in place of the sawblade.
The regular steel or aluminium table pocket is normally removed and a wooden or plywood insert fixed in its place. The grooving saw or cutterblock is then raised under power to break through this to give all-round support to the cut. The side and top pressures shown are used to control the cut when grooving along the grain.
The trenching and grooving saw shown is the Huther type consisting of two outer saws between which inner cutters are added or removed to vary the cutting width as required.
Source - Wadkin

Adjustable grooving sets One type once widely used on spindle moulders (shapers) comprised of a sawblade mounted between special collars that could tilt it to various angles - and so alter the groove width.

Unfortunately this type of tool starts quality-destroying vibration in use because, as the cutting width is increased, the tool also becomes increasingly out of balance - technically termed a couple.

In spite of this the wobble saw, as it was known, remained popular for many years. Because of the high speeds now used on modern machines where the effect of vibration is much greater (it increases fourfold when the rotary speed is doubled) the wobble saw is no longer acceptable.

Fig. 13.33
A typical adjustable grooving set, with adjustment made by means the handwheel.

Newer adjustable grooving sets are not designed to tilt a single sawblade to vary the cutting width. Instead they consist of two interlocking, overlapping and properly balanced halves which can be progressively separated or closed-in to vary the cutting width infinitely. This is possible via a handwheel with the actual cutting width indicated on a scale.

As with most tools of this type the teeth are of two types, rip teeth for removing the bulk of the kerf, and scriber or scoring teeth to prevent break-out beyond the groove width. The scriber teeth are formed with outer points which protrude slightly more than the rip teeth, and are angled to the outside, have a negative hook angle and large front and top bevel angles.

Saw and head mounting systems

Most sawblades fit on a straight arbor and are clamped between collars which hold them rigidly - a simple but effective method.

The only exception is on flitch breakdown machines where the saw bores may be splined to mount on a splined arbor - and no collars are provided. Instead the sawblades are steadied by extended and remotely-controlled guides which allows them to be shifted laterally between cuts - or even during the cut to follow any curvature within the flitch. The latter, known as guided curve sawing, is both to maximize the yield from flitches and to produce squared timber for structural timber in which the grain is maintained roughly parallel with the board edges.

On many machines using trenching and grooving heads there is a requirement for a better mounting system which allows them - and the associated scribe saws when both are mounted on detachable collar units - to be switched quickly and easily and with the least amount of trouble. Many of the tool manufacturers offer different means of doing this, for example using a twist-grip to lock the head when in place, or using an air line to release a mechanical locking system before loading or removing the head.

Hydraulic mounting systems

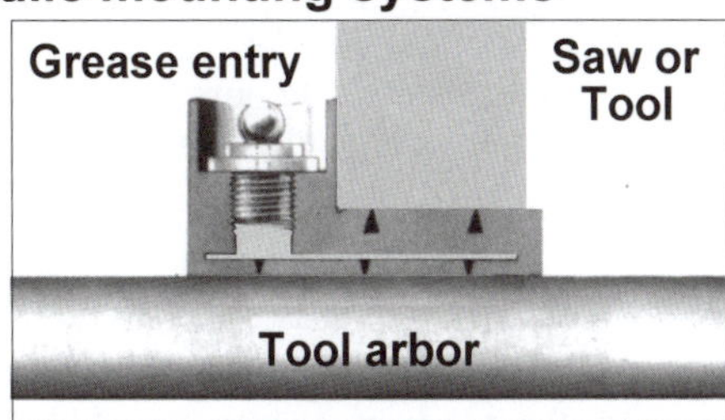

Fig. 13. 34
Grease within the lighter area of this hydraulic unit is pressurized manually by a grease pump via the entry nipple. This contracts the sleeve to grip the arbor, and simultaneously expands outwards to centralize and firmly clamp the tool - as shown by the pressure arrows. A built-in release valve immediately depressurizes the assembly for fast removal. Long sleeves can mount multiple tools.
Source - ETP

An alternative is to use an hydraulic mounting of the type made by ETP and sold under the trade name Hydro-grip. This type is available as a mounting for an individual tool, or as a sleeve which fits between one or more tools and their arbor.

Both types consist of an inner sleeve machined to be a sliding fit on a plain arbor. This is welded and sealed to an outer sleeve *(or the tool carrier itself)* leaving a narrow annular *(tubelike)* gap between the two which is filled with a special hydraulic grease.

Regular sleeves and some hydraulic units are pressurized with a manually-operated grease pump. This connects to a one-way valve in the sleeve or unit to pump more grease in - so increasing the internal pressure enough to secure both the unit and the tool without need of driving keys.

Before disconnecting the pump, the pressure within this and the connecting pipe must be released via the release valve provided. The tool itself can be removed easily and quickly after first releasing internal hydraulic pressure via a depressurizing valve on the unit.

Fig. 13.35
Left: A regular ETP hydraulic sleeve mounting a groover. Right: A quick-release type of hydraulic mounting for bossed multi splitter sawblades as used on a CNC router. Pressure is applied or released by a key (not shown),
Source - ETP

An alternative quick-fit type requires no grease pump to pressurize the grease, instead an externally-operated pressure-setting screw is used both to increase and to release internal pressure. If grease pressure is lost over time it can be quickly restored using a manual grease pump via a non-return valve.

Sleeves can be used with arbor and tool bores that are as little as 20mm. different in diameter. When a sleeve of these dimensions is pressurized, it contracts onto the arbor and expands into the tool bore by an equal amount.

When the tool is permanently or semi-permanently mounted on a hydraulic unit, the unit is designed to grip the arbor only when pressurized.

Fig. 13.36
The thin-kerf, bossed sawblades for louvre door slats are here mounted on a regular ETP sleeve.
Source - ETP/Super Thin Saws

With hydraulic units that are used with interchangeable tools that have a relatively small bore, then the unit acts in a similar way to a regular hydraulic sleeve. When the tool has much large bore than the diameter of the arbor, then two hydraulic annular gaps are provided in the hydraulic unit, an inner one to contract onto the arbor and an outer one to expand into the bore.

Hydraulic mountings are quick to use and give far more accurate self-centering than any mechanical system. They can be used both on the woodworking machine itself and on the grinder used to restore the tool - this is the way to ensure the best possible running accuracy of both sawblade teeth and cutting edges.

ETP makes a wide range of hydraulic sleeves and units, including the taper type of chucks now widely used on machining centres for automatic tool changers.

Wood-turning profile saw combinations. These combine two special wood-turning profile saws for use on CMS-Hit, Hempel and Zuckermann copy lathes. They provide a rotary tool which replaces the floating type of cutter used on the original copy lathes.

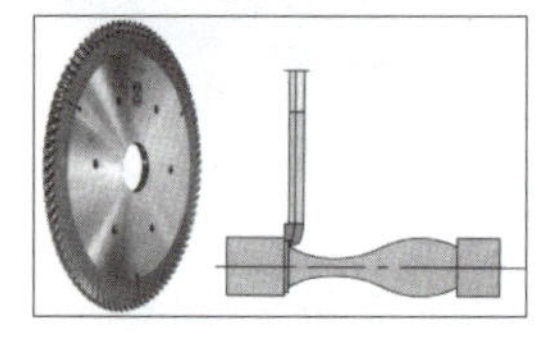

Fig. 13.37
Left: A purpose-made rotating sawblade as used on some copy lathes in place of a regular fixed knife or cutter. Right: A similar tool is offered for roughing-out square, hexagonal and octagonal balusters and other parts with similar-sections.
Source - Left, Leuco-Oertli; Right Simo-Bezombes

With regular copy lathes, a hard plastic or metal template of the required profile is fixed close to the blank workpiece. As the workpiece rotates at high speed a rounding tool first traverses the length of the workpiece to reduce it to a uniform size, and this is followed immediately by the floating knife which, in following the template, moves inwards and outwards to repeat the template profile on the workpiece itself.

By using a rotating profile-saw combination instead of a knife, much faster processing is possible - and it is then also practical to 'turn' sections other than round.

Maintaining Hoggers and Groovers

The sawblades associated with hoggers and groovers are sharpened in the regular way and as previously described. The time to sharpen these tools and most woodworking tools, is when the finish is no longer acceptable, when the cutting edges are broken or round-over by more than 0.2mm, or when the power consumption becomes excessive.

If face grinding is necessary at some stage, undertake this in the normal way - ignoring any difference between different types of teeth. When subsequently top grinding, carefully measure and record the finish-ground diameter of the sawblade using a dial indicator, then grind the inside cutters or saw segments to precisely the same diameter.

This is particularly important with certain tools, for example a vee-grooving hogger - the inside cutters of which must precisely align with the sawblade in order to form a perfect mitre. This also applies to grooving and trenching sets - though to a lesser extent. Most hoggers, other than vee-groove hoggers, are much less critical as regards working diameter - but still need to be well within the ball-park.

As previously mentioned, some hoggers are intended to produce a staggered cut, in which case follow makers instructions as to the height on inside cutters or sawblades.

Fig. 13.38
Sharpening a hogger unit complete on a TCT/2 attachment. The dial indicator provides the instant means of checking the relative diameters of the outer sawblades and the inside sawblades or cutters.
Source - Autool

Fig. 13.39
Some automatic machines are capable of dealing with hoggers, such as this CHC grinder. The left-hand view shows hogger saw segments being ground on a carrier, while, on the right, the same machine grinds a gap-tooth saw of the type used with hogger sets.
Source - Vollmer

Solid grooving and trenching heads usually have scribing or scoring cutters. It is important to repeat their original face and top angles for them to work as intended. Also these cutters must protrude fractionally beyond the rip teeth to sever the cross-fibres - otherwise the rip teeth will tear-out badly. A projection of 0.3mm is usually enough, but check against a new tool for confirmation - or contact the makers.

Where hogger saws are used in combination with other cutting tools, for example the inside sawblades or cutter segments, then these must be kept in step with one another - diameter-wise. To do this the amount ground-off the tops of both saws and inside tools must be precisely the same.

It is, of course, necessary to use the correct grinding wheel to match the tool being ground, see previous notes in Chapter 7. Due to the difficult nature of the grind many operators use manual grinding machines for hoggers - often a universal type of machine is the most suitable as it readily adjustable to a wide range of settings.

Saw segments from hoggers are normally removed and fitted to a grinding carrier for the purpose of grinding - but a few machines can cater for these units as a whole entity.

Chapter 14

Table and cross-cut saws

Safe use of circular saws

All woodworking machines are dangerous and must be used with the greatest care. In the United Kingdom they account for roughly 3/4 of all industrial accidents, most of these in sawmilling and joinery - and with a disproportionate number in the smaller premises.

Of all woodworking machines, circular saws account for roughly a third of the accidents, most of which result in amputation of fingers or thumbs. Many accidents occur when ripping or cross-cutting because of a missing or improperly-adjusted saw guard, or because the operator failed to use a push-stick or push block.

Fig. 14.1
This is the way it used to be done - at least with portable sawmills - but not any more! Yes it was a purely a demonstration at a steam fair, but the dangers are so obvious - no bottom enclosure for the sawblade, a badly-set riving knife (splitter), an ineffective top guard, no driving belt guard and, what is not shown, splitting round logs without using a push stick - and without dogging them to the table!
Source - the writer.

Modern machinery from major manufacturers incorporates the current and most effective guards and safety devices, and so give the greatest protection to machine operators. This includes fail-safe systems for the more sophisticated machines, and instantly accessible emergency stops and brakes - which on some types of machine is fully automatic following machine or material malfunction.

It is essential for management to apply new and improved safety equipment and measures as they are brought in - older machines may not have the current safeguards - so check and update them on a regular basis. Just as important is to ensure that any approved safety equipment is not only provided for all machines, but that operators know about it - **and use it!**

Operators become complacent about safety when they run the same machines day in day out, and familiarity leads to dangerous practices becoming commonplace - particularly in a small workshop where operators are often left to their own devices. This is probably the reason for the much higher percentage of accidents in small workshops than in larger concerns. In the latter safety regulations require the appointment of safety personnel - and it is the duty of safety personal to ensure that complacency in machine operation does not take over from sound and safe working practices.

Most of the recommendations to prevent accidents are just plain common sense. For those new to the trade, however, common dangers may not always be obvious.

The following comments, which are based on good practice, should be taken as a general guide only. Safety regulations vary throughout the world - so local regulations should be followed strictly as regards approved guards and safe working practices - not just to avoid liability on the employers part, but to ensure that trained staff remain uninjured and continue to serve the company. Although local regulations may go beyond the following recommendations, they do at least provide a reasonable basis for safe working practices.

Operating conditions

General cleanliness and tidyness.

The importance of having a clean working environment free of clutter, finished workpieces and non-essential machine parts cannot be overemphasized.

There should be sufficient working space around machines both to operate them safety and allow ready access for routine maintenance. There should also be space to comfortably house workpieces in progress, finished workpieces and waste - without interfering with machine operation in any way.

Fig. 14.2
This a fine example of how not to operate a woodworking machine, a cross-cut saw in this case. Rubbish, off-cuts and finished pieces are haphazardly strewn about - several accidents just waiting to happen.
Source - Health & Safety Executive.

Machines should be placed where the operator cannot be pushed, bumped into or easily distracted, and the floor must be level, free of any sawdust, off-cuts, general waste, liquids of any sort, and floor irregularities that could cause someone to trip or slip.

Floors that have become polished and slippy over time should be treated with a non-slip surface, or covered with a non-slip mat. Non-slip mats, where provided, should be regularly checked to ensure that they have neither worn through nor curled at the edges - either could in itself be a hazard.

Any machines that are hazardous to passing staff should be fenced-off. Examples of these are machines equipped with automatic transfer systems that start up without warning, such as on large panel saws and multi-machine grinding installations.

Stacks near to machines should be kept low and stable, and present no fouling points for work in progress. Stacks of finished parts should be removed when completed rather than being left lying around. Escape routes for use in emergencies - and all passageways - should be clearly marked, well lit and kept clear of all obstacles.

Clothing and protective gear

Proper protective gear is essential in the woodworking trade. These include industrial protective boots to avoid foot damage from dropped timber, materials or tools, and industrial gloves for handling both timber and tools.

Workwear should have no hanging ends that could be caught up in machinery - such as loose ties, belts, cuffs or hanging clothing of any sort.

Ear protection is essential in most mills, see following notes. Likewise protective eye glasses must be worn when operating grinding machines, and for general use when there is danger of flying particles of any type. Welding masks are necessary when undertaking manual welding repairs, etc., and certain other operations - see previous notes.

Dust extraction and ventilation

Dust in the air was long accepted as no more than a minor nuisance in woodworking mills, but fine dust from most types of timber is now recognized as hazardous to health. For this reason all sawing machines, in common with other woodworking machines, should have adequate dust extraction by connection either to the works system or individual dust collectors - and these must be routinely maintained to keep them operating efficiently.

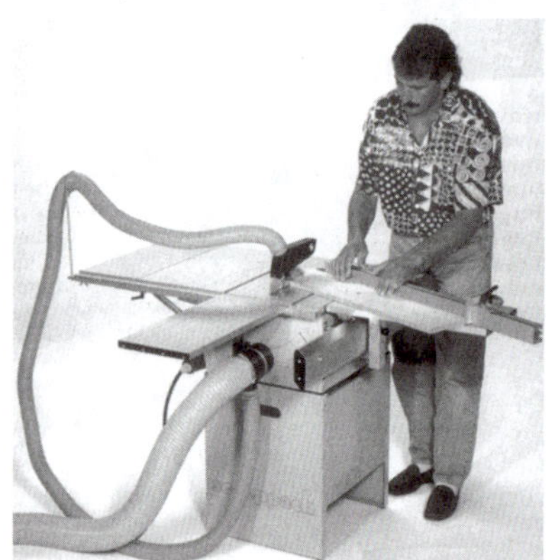

Fig. 14.3

To keep dust to a minimum the exhaust system should connect both above and below the sawblade, as on this table saw. The operation is angle-cutting using the sliding table.

Source -Scheppach

Works systems may simply vent extracted air to the outside, in which case adequate ventilation is needed to supply the same volume fresh air. Doing this looses air within the plant which has been conditioned - heated, cooled or dehumidified as needed - and this can create high ongoing costs.

One alternative is to vent individual machines from the roof or through the walls - but this is only practical on automatic machines that have been fully enclosed to reduce noise.

Where extreme outside conditions exist, extraction systems could recirculate the air to reduce air reconditioning costs - but these must completely clean the air or the dust hazard will get progressively worse.

Hazardous dust or fumes which cannot be treated effectively must always be vented to the outside, in which case a split system, part-vented and part-fully-extracted, is needed.

The health hazard from dust has actually worsened over the years due to the increasing use of wood and plastic-based manufactured materials - dust from the basic material and the bonding agent are both a risk. Asbestos, still worked in some specialized plants, is a particular hazard - but in this case the danger of asbestos dust is well documented.

Grinding machines using a coolant system present a particular problem. Wherever possible use fully-enclosed machines to prevent fine over-spray dissipating into the atmosphere - see previous notes.

In conditions where fine wood-based dust is created which the extraction system cannot handle, then operators should wear dust masks or respirators. Where a particular hazard exists, for example when working materials that create fumes, operators should wear a respirator or a full face visor with power ventilation. In any event such workplaces also need adequate ventilation.

Check with local health and safety agencies on what is hazardous, they will give an excellent guide as to what are the risks and what is needed to protect workers.

Lighting

Good natural or artificial lighting is essential in preventing accidents - hazards are more clearly seen in good light. Where overall lighting is not practical, then adequate lighting must be provided in or around individual machines, and for all gangways, passages and safety equipment. Particular care must be taken to avoid glare from either the sun or badly positioned lights.

Heating

Although cold working conditions are common in northern climates and accepted as routine during the winter, they can lead to lack of concentration - and cold hands reduce the ability of an operator to control the workpiece. Where the entire workplace cannot be heated, then radiant heaters should be placed near machines. Hot and humid working conditions should also be avoided as this slows down both the work rate and operators reaction time.

In any event, extreme working conditions that make workers uncomfortable could also badly effect both the natural and man-made materials being processed - certainly in finishing lines - so good working conditions are needed for this reason alone.

Noise

Many woodworking machines produce noise at well above the recommended tolerance level, much from manually-operated saws that cannot be totally enclosed. Use sound-dampened saw bodies, see the final pages of Chapter 9, and enclose saws as much as practical with sound-proof material.

Where enclosure is impractical, operators should wear earplugs, ear muffs, or some other type of ear-protective. Damage to hearing through over-exposure to loud noise has not always been recognized as a risk. The simple reason is that damage does not occur immediately - it builds-up over a period of years - with a result that those who have long been in the industry have invariably some hearing loss.

Warning notices

Notices should be displayed in areas where dust masks, safety glasses, ear protectors, hard hats or protective gear of other types should be used.

However, it is up to individuals to use them. If they do not, due to indifference or complacency, they may regret it - but not only to their loss, but also to the company they work for - the absence of works staff for any reason creates problems. For this reason safety officers should insist that safety notices are followed to the letter.

Sawing machine types

General operation.

All woodworking machines should have readily-accessible switches to isolate the machine electrically. These should be used routinely to prevent accidently start-up while the machine is being serviced, re-set, or the sawblades adjusted or changed.

Before start-up check that the sawblade(s) are free to rotate and that nothing can foul them when the machine is in use - note the comments in first few pages of Chapter 10. Start the sawblade(s) and allow them to reach full speed before starting the cut. When stopping the machine make sure that the sawblades are stationary before leaving it - and then isolate the machine electrically.

Circular saw benches

This general type includes fixed and sliding-table saws used for sawing both along the length and across the width of solid timber and panels.

Table saws

Fixed-table saws, usually termed plain or table sawbenches, have a one-piece fixed top and were originally designed for ripsawing solid timber only.

Fig. 14.4

The 66 is a typical example of a compact but weighty 10in table saw - with all the attributes of much larger models. It features a left-tilting sawblade to reduce kickbacks and to produce fuzz-free top-edge mitres.

Source - Powermatic

This application has been greatly extended by most makers adding milled slots for a cross-cut and mitre gauge for use when square and angled cross-cutting. Many also offer the options of an extended right-hand table for sawing wide boards to width, or an extended left-hand table for supporting long workpieces when cross-cutting. Certain makers can supply the left-hand table as a sliding extension table instead of a static one.

Fig. 14.5

A typical small table sawbench is the Delta 10" tilting arbor saw which, in standard form, has a long rip fence and a sliding mitre fence. When used with an extended table (right) large panels can be sawn to width. The saw guard is mounted on the riving knife.

Source - Delta

Sliding table saws

Fig. 14. 6

This sliding table saw has left and right extension tables and carries the top guard and dust extractor on the riving knife, but with extra support from a rear pillar.

Source - Startrite

Sliding table saws have a fixed table to the right of the sawblade, and a sliding table to the left of the sawblade for use when cross and mitre-cutting. Although generally termed 'sliding', they actually roll on a series of steel balls, rollers or ball bearings to give a smooth and precise movement.

The increasing use of lightweight materials for the sliding tables has allowed makers to combine exceptionally long traverse movements with absolutely rigid support even for large panels. Often the sliding tables have a cantilevered extension to the left for extra support. As with fixed-table saws, sliding table saws also can have extension tables to the right-hand side.

In old terms sliding table saws were called dimension or variety saws, but now, with far larger tables and longer traverse, they are generally called sliding-table panel saws.

When cross-cutting and mitring on the sliding table the workpiece can be clamped, so the operator can control the cut while gripping guarded control handles - well clear of the sawblade. Clamping the timber or board to the sliding table eliminates slippage and spoilage and makes the operation more precise and considerably safer.

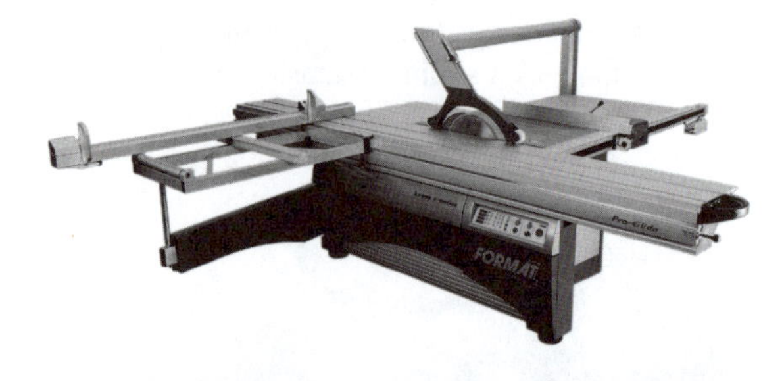

Fig. 14.7
Modern sliding table panel saws are high-tech machines suitable for rip and cross-cutting a variety of materials rapidly, precisely and cleanly. The Format-4- shown here has a Kapa X-motion with a CNC centralised command (below) for adjustment of sawblade height and tilt, rip fence setting and sawblade speed between 3000-6000 rpm. Also included is automatic fence adjustment to compensate for tilt setting of the sawblade, and cutting programme storage.
Source - Format-4

General safety requirements

Sawbenches should be enclosed below table level to fully protect the sawblade, and be provided both with a riving knife or splitter, and an adjustable top guard. Dust extraction points should be provided both on the machine base and on the top saw guard for connection to a dust extraction system.

As with all woodworking tools, the sawblades used on these machines must be of the correct specification for the work undertaken - and of the anti-kick-back type when ripsawing manually (see previous notes).

They must also be maintained properly and kept in a sharp and clean condition - dull and gummed-up sawblades cause accidents. Never attempt to clean gum from a rotating sawblade - remove and return a gummed-up sawblade to the saw shop, or isolate the sawbench electrically and treat in situ.

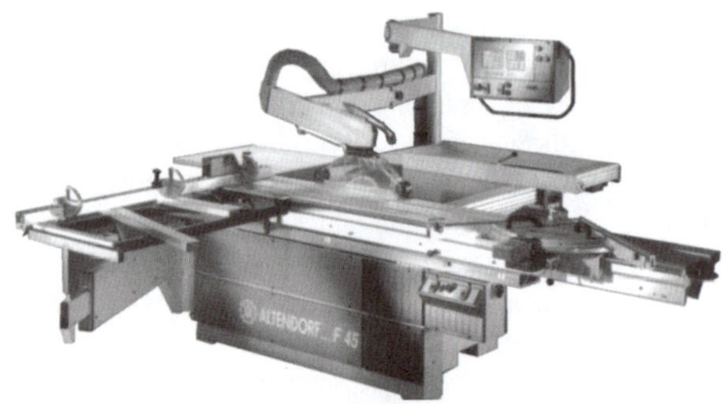

Fig. 14.8
The F 45 is typical of the upmarket sliding table saws - with four cnc controlled axes, an interface for PC and connector for a barcode reader.
Source - Altendorf

Table level supports, either wooden benches or roller supports, should be provided at both infeed and outfeed of saw benches for dealing with long material. The rear extension table should be long enough to provide a gap between the upcutting sawblade and the rear edge of the extension table of at least 1200mm (almost 4ft.) - to protect those removing cut pieces at the rear - who should always remain at the end of the extension table.

Sawblade adjustments

Most modern sawbenches have fixed-height table top and a rise and fall saw arbor which allows permanent fixed-height infeed and outfeed tables to be used. Most, but not all saw benches, also have a saw arbor that tilts on the table line to 45° to the right.

This contrasts with the early rise, fall and tilting table dimension saws which were once common. These were difficult and even dangerous to use - a correct-height and level table is essential for safe operation.

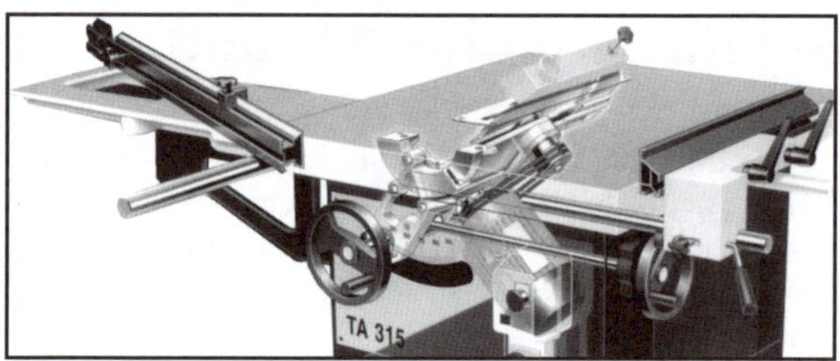

Fig. 14.9
The rise, fall and tilt mechanism is clearly shown ghosted-in on this TA 315 sawbench. The machine is basically a fixed-table sawbench with a sliding left-hand extension.
Source - Sedgwick

Riving knife or splitter

This is mounted behind and in line with the sawblade. It serves the dual purpose of preventing material closing-in on the sawblade, which could otherwise cause a kickback, and guards the up-cutting saw teeth to reduce the chance of accidental hand contact from the rear.

The riving knife must be securely mounted on the saw carriage, so that it rises, falls and tilts with the sawblade. The inside edge should be slightly tapered, curved to closely follow the saw teeth, and set to leave a gap between the knife and the sawblade at table level of no more than 12mm (1/2in).

Ideally several different curvatures of riving knives should be provided to suit the regular diameters of sawblades used.

The top of the riving knife should be no more than 25mm (1in) below the top of the sawblade, and its thickness should be, say, 10% more than the saw plate thickness - but must be less than the saw kerf.

Machines fitted with a rear-mounted top guard only require a narrow riving knife. A much wider knife is needed to give sufficient rigidity when the top guard is fastened to it.

Top guard

The regulations for top guards vary throughout the world, so check with local authority what is acceptable. The following, though, are general observations that apply to most.

The top guard prevents top contact with the sawblade, and should be set vertically to leave a gap between the guard and the workpiece top of no more than 12mm (1/2in).

Some guards adjust horizontally to allow the operator to more clearly see the cut, or are fixed horizontally and provided in clear plastic for the same reason. Most guards have instantly accessible adjustments for position, and many automatically lock in place once set.

Fig. 14.10
This rear-mounted top guard is spring counterbalanced for ease of adjustment. The guard rises automatically when timber is fed in and lowers on exit - to keep the sawblade fully guarded at all times - front and rear rollers aiding this movement. It is made from transparent plastic so the sawcut is clearly visible even with the sawblade fully protected. Dust is extracted via the top guard and from below table level. The riving knife is wide enough carry an alternative top saw guard used only when converting large sheets that would otherwise foul the regular saw guard pillar.
Source - Format-4

Some machines are supplied with automatic, counterbalanced guards which initially rest on the table. When material is moved towards the sawblade the guard automatically lifts and rides on the material, then re-sets back onto the table when it has passed through. Wheels at the front and rear of some guards assist this action, on others a sloping front beak is provided instead.

The top guard must fully guard the sawblade in whatever position it can be set. When the saw arbor has only rise and fall, then a narrow guard is suitable regardless of how it is mounted.

When the saw arbor has both a tilting action and a rear-mounted guard, then this must be wide enough to guard the sawblade when raised and tilted to its full extent. A narrow type of saw guard is suitable for all vertical and angular sawblade settings when this is mounted on the riving knife .

The saw pocket

The area immediately around the sawblade should have working clearance only for the sawblade and riving knife. Most have some provision for fitting replaceable wooden strips to narrow this gap and so prevent slivers being trapped.

Fixed-top saw benches have a wide saw-access pocket. This is often fitted with lips to mount wooden strips either at the front and rear, or on either side of the sawblade.

When intending to use a trenching or rebating tool, the pocket itself can be replaced complete by a hardwood infill. This should be made level with the table top, securely fixed in place and then broken through by carefully raising the rotating tool to give the smallest possible table gap. Sliding table saws may have a similar lips to support wood infills, but only at the right-hand side.

Wooden infills should leave the smallest possible gap around the sawblade and be maintained in good condition. Replace them routinely, and when they become even slightly worn or damaged in any way. Take particular care that the top of the infill is perfectly level with the surrounding table.

Ripsawing

Ripsawing is splitting timber along the grain. On these sawbenches timber is fed against an adjustable fence to gauge it to width. Timber and boards are conventionally fed manually for routine custom sawing, but for production runs a power feed should be used instead - these are safer to use and far more productive.

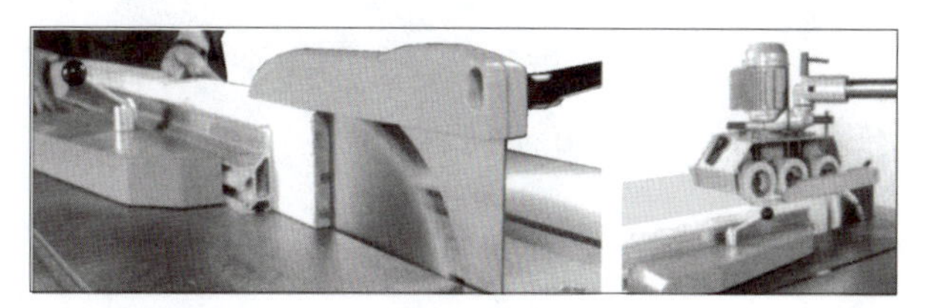

Fig. 14.11
The most common operation is ripsawing, as shown here on this 400mm AGS . Shown right is a power feed attachment.
Source - Wadkin

Fig. 14.12
Fence-mounted pressure rollers, such as these, keep thin material under control and prevent unwanted vibration when feeding manually.
Source - Scheppach

Push sticks and push blocks

Push sticks must be used both when ripsawing short lengths of material, and when feeding the last 300mm (12in) of longer pieces.

As push sticks are essential when ripsawing, each machine should be provided with one. Make sure that each push stick remains with the machine for which it is intended - a missing push-stick can lead to the type of accident it is intended to prevent.

The most common way to keep a push stick with its machine is to tie them together. Use a suitable length of twine or similar to tie the push stick either to the fence, or to any other point that can easily be reached when operating the machine. Make sure that the twine is long enough to use the push stick properly, and is so arranged to avoid becoming tangled with any machine part or the workpiece - both when in use and when lying idle.

Another way is to fit a hook to the push stick for hanging on the fence bar of those machines that have them, or purchase one with built-in magnets that stick to any convenient metal surface. An alternative is to have a pocket or mounting hooks on the saw guard for the push stick.

When using loose push sticks make sure that all operators are trained to routinely replace them immediately after use.

The operators left hand is commonly used to assist the feed - often at the same time or instead of using a push stick in the right hand - which is why so many older users lost thumbs. If doing this - but always in conjunction with a push-stick - take care that the operator's left hand always stops well short of the sawblade.

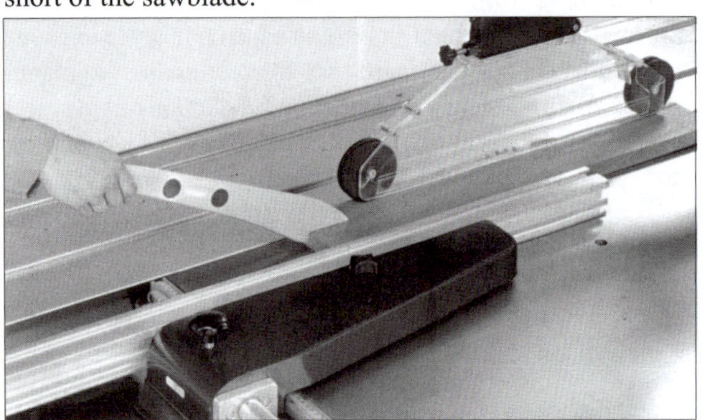

Fig. 14.13
This push-stick has embedded magnets which allow it is be magnetically clamped to any metal surface when not in use.
Source - Aigner

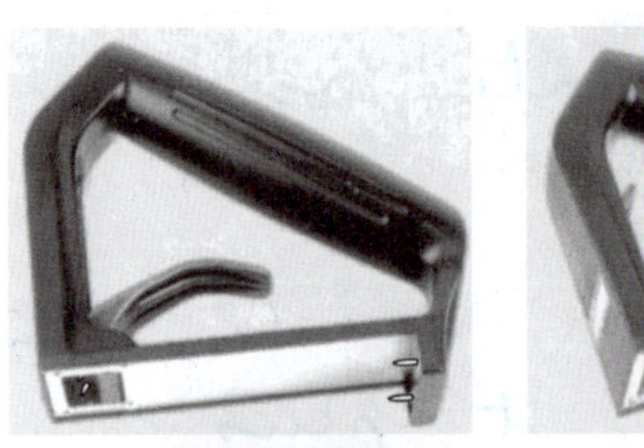

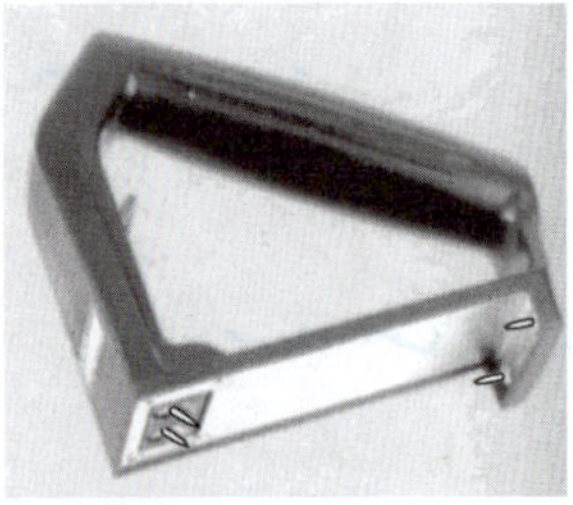

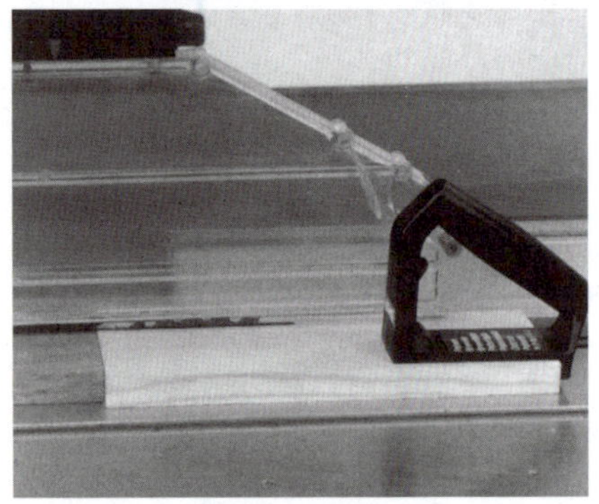

Fig. 14.14
An alternative to the push-stick is the push block. The Quicker shown here has a comfortable hand grip and a lever which engages top and rear spikes - so giving a firm grip for both pushing forward and drawing back.
Source - Aigner

Rather than using direct physical contact with a hand, a better way is to use a weighted push stick with a vertical spike - this will always keep the left hand well clear of the sawblade. With this type the spike is driven into thc material to be sawn using a quick, hammer-like blow. It then acts as a regular push stick when sawing, but also quickly and safely draws material back for a second and succeeding cuts when sawing a series of strips.

A manufactured version of the spike stick is the Quicker device which has a comfortable handle and two sets of spikes. By raising a lever the top spikes drive into the material and pull it back against rear spikes in the end stop - so gripping it securely.

The rip fence

This is a fence or guide on the right-hand side of the sawblade set parallel to the sawing line. Its purpose is to accurately gauge the width of material being ripsawn - when this already has a straight edge to run against the fence, of course.

It has both fine and coarse adjustment for setting purposes, and usually can be removed or turned over backwards to leave a clear table top - as needed when cross-cutting long material.

The fence plates on machines intended for ripsawing natural timber are shaped at the sawblade end to roughly follow the curvature of the sawblade. This type is adjustable laterally, and is traditionally set so that the sawblade end of the fence plate is roughly level with the base of the gullets.

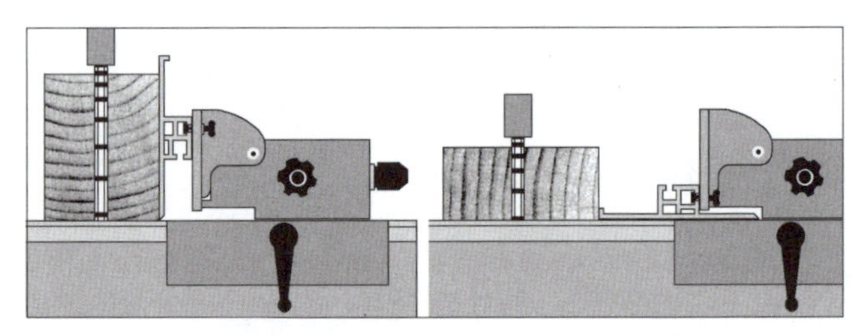

Fig. 14.15
This shows a typical fence which has two settings, one for deep sawing and the second for shallow sawing.

Modern sawbenches are intended for much wider use, for example in converting sheet materials, etc., and so are normally provided with the long fences needed for this purpose. Some machines may have a conventional rip fence as an extra an add-on.

On many machines the regular fence has two positions - for deep and shallow cuts - with provision for switching quickly between them. The regular fence is a deep one - to keep narrow boards stable and upright when making deep cuts. When making shallow cuts on thin material to produce narrow pieces, though, a deep fence makes it impractical to use a push stick in the conventional way - so use the shallow fence.

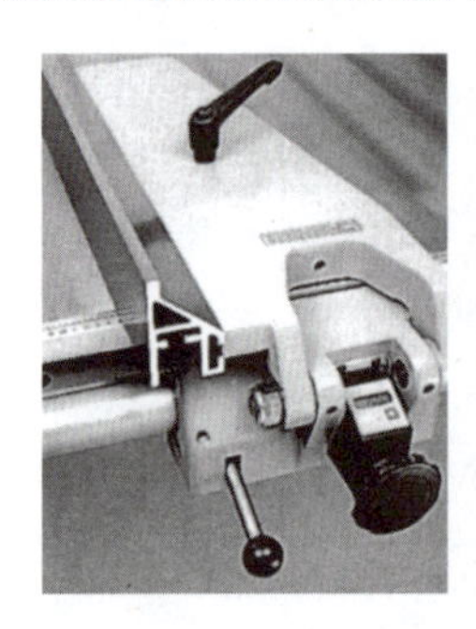

Fig. 14.16
The fence shown here is quickly switched between the deep and shallow positions. The fine adjustment has an electronic display, and shows the correct sawblade-to-fence dimension regardless of which fence setting is in use.
Source - Kolle

Angle ripping

The safest way to saw at an angle is with the sawblade tilted, and the material flat on the table and squarely against the fence. Use the shallow fence to avoid the chance of the sawblade

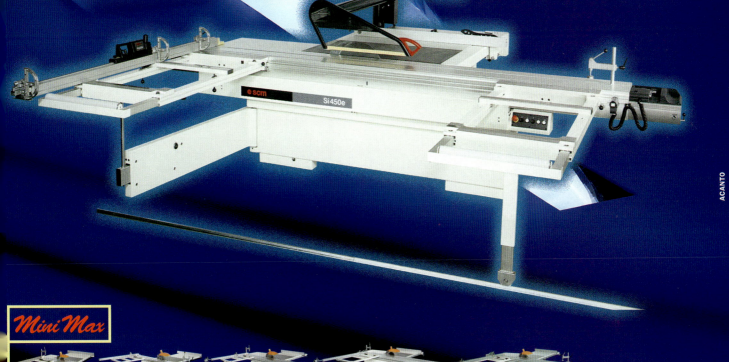

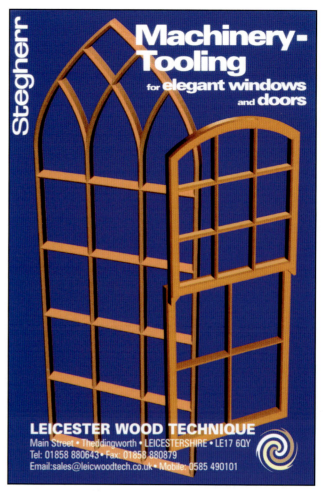

fouling - and only attempt angle-cutting on material that has the widest face down on the table top.

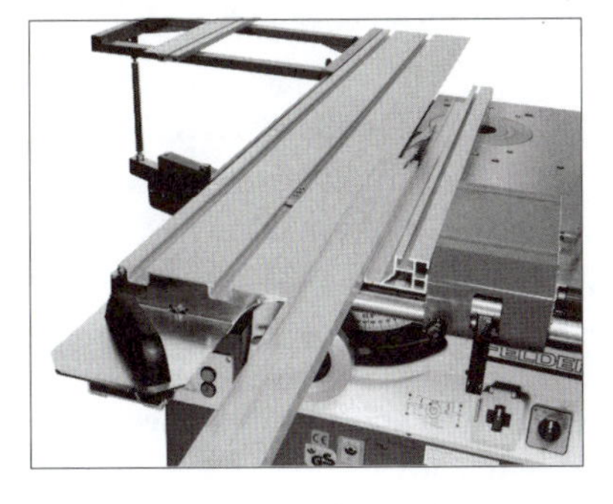

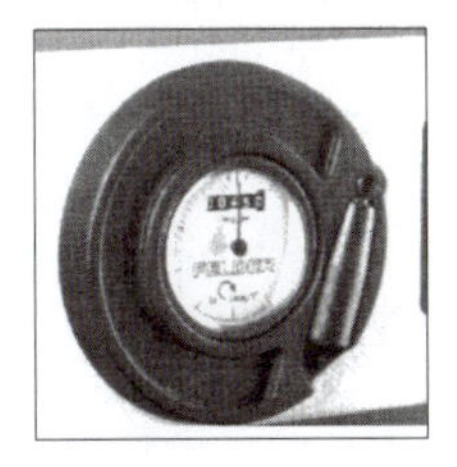

Fig. 14.17
The safest way to angle rip is to tilt the sawblade. Continuous height and tilt setting is clearly shown on the indicator(right). Digital readout is optional
Source - Felder

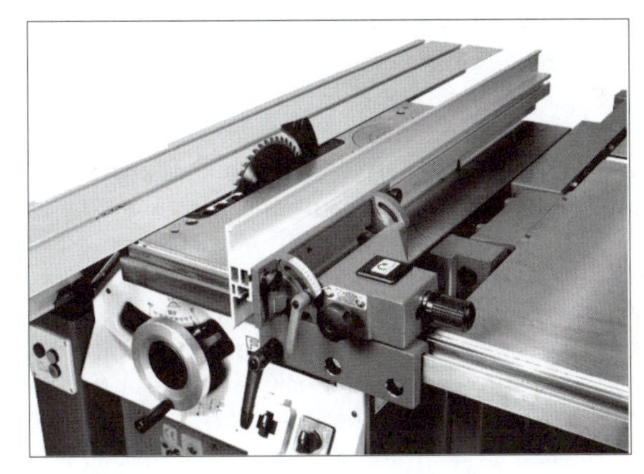

Fig. 14.18
The deep fence on this series 7 saw tilts for angle sawing.
Source - Felder

On non-tilting machines the traditional way was to tilt the fence to the required angle and manually keep timber against this during the cut - but this is not a safe way to operate. As only the under-corner rides on the table top the operation is unstable - there is always a chance of the workpiece slipping and being thrown back.

If there is no other choice, then the workpiece should be supported by a wedge-shaped wooden bed fastened to the table top. The wooden bed should be angled to form a right-angled trough with the fence to give proper support to the under-face and make the operation safe. Use a wooden 45 degree trough when diagonally splitting.

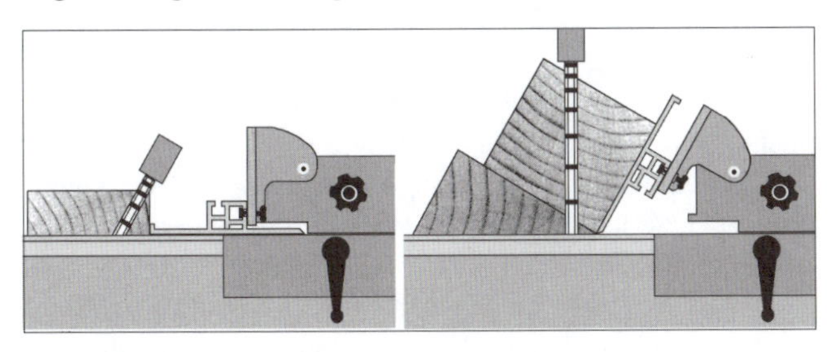

Fig. 14.19
Where the saw arbor tilts, use this feature for shallow angle sawing (left). For non-tilting sawblades tilt the fence instead, but support the underside of the timber with a wedge-shaped bed (right).

Rebate-sawing

Deep rebates *(rabbets)* can be sawn-out on a regular sawbench by forming two sawcuts to meet at right angles to one another. *(Deep rebates are sometimes needed for rough work where a sawn finish is acceptable, or to relieve-cut material prior to machining on a moulder.)*

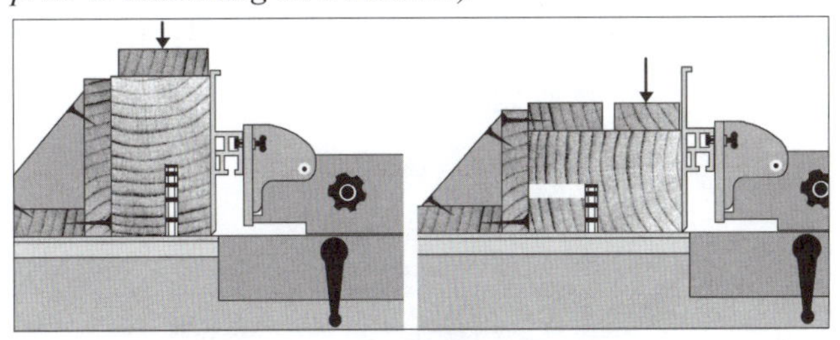

Fig. 14.20
Box-in the sawblade when making rebate cuts, both to prevent contact with it and to make the operation more stable and less likely to kickback. Apply pressure as shown by the arrows. Make the box long enough to prevent hand contact of the sawblade by anyone foolish enough to push the last piece through without using a push stick.
For both operations follow each piece of timber with another - in effect a continuous feed. Use a block of the same section for feeding through the final piece, and withdraw this only when the workpiece has been fed clear. This will prevent the off-cut shooting back at the operator.

The regular top saw guard cannot be used for rebating cuts so, in order to make the operation safe, it is essential to use overhead wooden pressures and side guides - in effect a complete boxed enclosure.

Where the under-face of the material is less than its depth - as it will be when in position for the one of the cuts - then an extra-deep fence and side guide is absolutely essential.

Grooving cuts can be made in a similar way. Use the regular fence for cuts along the grain, together with a long overhead side guard/pressure to make the operation safe. The cross-cut and mitre fence can be used for cuts across the grain, but in this case a long overhead guard only can be used.

NOTE: Guards used to protect the operator when rebate cutting and grooving should by wide enough and long enough to prevent anyone reaching the sawblade from any position when the timber being worked is not in place.

When working thick material this requires exceptionally long guards

Power-feed units

Regular ripsawing is safer and more productive when using a power-feed.

These are provided as detachable units which can be switched between machines - but usually they need holes drilled and tapped into the table top of any machines they are used with. (In some cases an extension plate can be clamped to the table rather than having to drill and tap it.)

They have several spring-pressured rollers and a choice of two or more feed speeds. Non-marking rubber feed rollers are fitted for use with planed material. In some cases these can be switched for serrated metal rollers to feed rough timber. The power feed supporting pillar must be well clear of the sawblade to avoid being fouled during the cut.

The regular setting for feed units is with the rollers contacting the top surface of the material being fed. They adjust readily for position, height and feed alignment - but take care that the feed direction is absolutely parallel to the sawblade. When feeding wide timber the feed unit can be alongside the top guard, or actually replace it, but for narrower timber it may have to be set close to the fence and short of the sawblade.

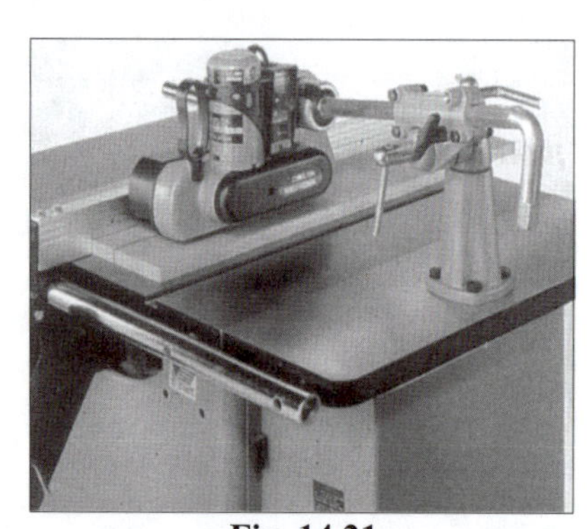

Fig. 14.21
This power feed unit can be quickly transferred between machines - and makes for safer ripsawing.
Source - Delta

For deep angle ripping using a vertical sawblade and a deep, tilted fence, the feed unit should also be tilted so that the roller axes are parallel with the angled fence.

This arrangement still requires a false angled bed, however, and particular care is needed to ensure that the timber feeds parallel to the table top or a with a ***slight*** *downward feed.*

To keep the timber running true and prevent it riding up the fence, fix either a top pressure or a top guide to the fence - just clear of the timber.

When using a feed unit to pressure timber against the fence, the nearest roller must be clear of the sawblade itself. Preferably follow each piece of timber with another to give a continuous feed. Use a push-stick to clear the final one.

Cross-cutting and mitering

Cross-cut is cutting squarely across the grain of timber, mitering is cutting across the grain at a specific angle, both use a cross-cut or mitre gauge.

Caution: *It is dangerous to attempt to cross-cut ripsawn lengths in a bundle simply by gripping them and moving them into the sawblade manually.*

This, unfortunately, is a common way of reducing off-cuts to more manageable size - don't do it.

Cross-cut and mitre gauges can be set square to the line of the sawblade, at specific mitre angles for mitre cutting, or at any other angle required.

Most have a graduated scale that clearly shows the set angle, and some have positive click stops at 90 degrees and some common mitre angles to give precise and repeatable settings.

If two pieces are mitre-cut at 45 degrees they form a right angle when fitted together. Two pairs of double-end mitred pieces then make up to a square or rectangle frame.

Mitre cuts for similar frames, but with a different number of parts, can also be mitre-cut using angles of 36^0 for 5 pieces, 30^0 for six pieces, 22.5^0 for 8 pieces and 15^0 for 12 pieces, etc. (Formula 180^0 divided by the number of pieces.)

Fixed-top sawbenches - Table saws

For use with a sliding cross-cut and mitre gauge, fixed-top sawbenches have at least one slot milled in the table top, parallel with and usually to the left of the sawblade - or slots at both sides of the sawblade. The gauge consists of a fence and graduated quadrant pivoting on and fastened to a long slide. This is a sliding fit in the slot(s), and the quadrant, fence and workpiece slide over the fixed table.

In regular use the workpiece is placed between the sawblade and the mitre fence, and is held firmly back against the fence during the cut.

Although traditionally hand-held, a mechanical clamp could be used to clamp the workpiece either to the mitre fence or to the slide, but this is only suitable for small timber sections - for boards a sliding table saw is recommended.

Regular sliding-type mitre gauge fences are always well short of the sawblade when set at an angle to the line of cut. To give better back support, and to make their operation safer and more stable, fit a thin hardwood fence onto the gauge fence long enough to extend slightly beyond the sawline. The false fence should be only fractionally deeper than the workpiece *(at the saw guard)* to pass under it.

Make an initial sawcut through the false fence and use this as a guide to accurate set length stops - or to position workpieces when these are pencil-marked for length.

The bulk of the work is on the gauge side, so it is not essential to extend the fence much beyond the sawline - and in normal use the end will be trimmed in any event by the sawblade.

However, extending the wooden fence by some 25mm (1in) beyond the sawline gives certain advantages:- support is given to both sides of the sawcut to make the final cut more stable, it helps to prevent break-out at the rear, and the initial sawcut provides precise left and right-hand sawlines.

Of course, this needs a rear bridge piece fitting to span the sawline and so support the extension. Care should be taken to stop the traverse movement well short of the bridge when cross and mitre cutting.

Using a false wooden fence may interfere with the length-stops provided with most mitre gauges, but a wooden stop can instead be clamped to the wooden fence.

Sliding table sawbenches

With sliding table saws, separate cross-cut and mitre gauges are often supplied, and both can be fixed in a wide variety of positions to the sliding table.

Whichever is used, though, the gauge and workpiece both traverse with the sliding table in a very precisely-controlled movement.

The fences of mitre gauges on sliding table saws are longer to give more stability, and usually adjusts lengthways to shorten the gap between the fence end and the sawblade regardless of the set angle.

In some cases a recess is provided at the sawblade end of these fences into which a wooden block can be fitted. A sawcut can be made into this for precisely positioning pencil-marked workpieces or setting the length stops. For the same reasons given above for table saw-mitre gauges, a wooden fence and rear bridge can be fitted to extended the back support to slightly beyond the sawline. Sliding table saws allow workpieces to be securely clamped to the sliding table itself - and for this reason are better for dealing with large cross-sections of timber and wide boards.

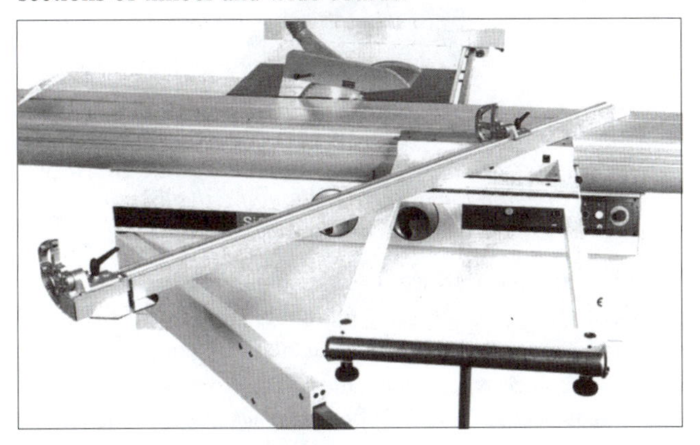

Fig. 14.22
The cross-cut and mitre fences on sliding table saws, such as the SCM SI 350, are adjustable both for angle and distance from the sawline.
Source - SCM

Using crosscut and mitre gauges

Double-end mitering. A common practice when double-end mitring large frames is to pre-cut the moulded sections square-across and slightly overlong before actually mitering them.

With small frames the parts can be mitre-cut from long lengths - without prior length cutting - provided the mitre fence gives enough support on the far side of the sawblade. This saves one extra step, but the way it is done and the mitre gauge used varies with the length and section being mitred.

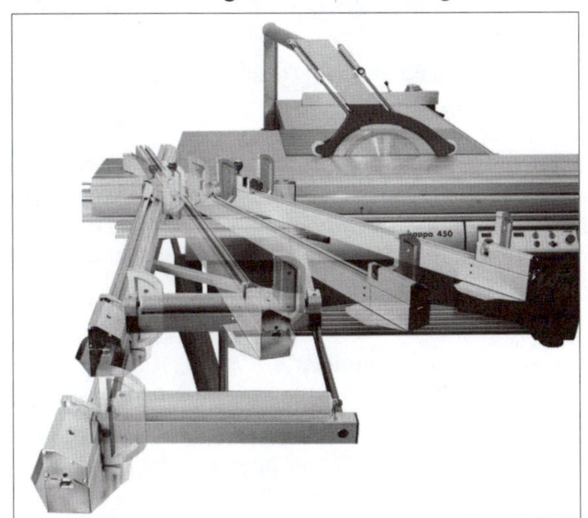

Fig. 14.23
The standard outrigger support on this Kappa 450 sliding table sawbench carries a telescopic mitre fence adjustable between 90 and 45 degrees, and loading rollers.
Source - Format-4

Using a single-mitre gauge. Mitre cuts can be made at both ends of each workpiece at following passes when converting square and rectangular sections having no specific face side and edge, .

Form the first mitre cut close to the end. Then turn the workpiece edge-for-edge and move it beyond the sawline by the length required to make the second mitre-cut at the opposite end. This cut is also the first cut of the following piece - so the only loss is a single sawcut. Repeat this for the remaining cuts.

For any final mitre-cut that otherwise leaves only a short off-cut, turn the workpiece end for end so that this, and not the shorter off-cut, is to the gauge side of the sawblade to be given maximum support.

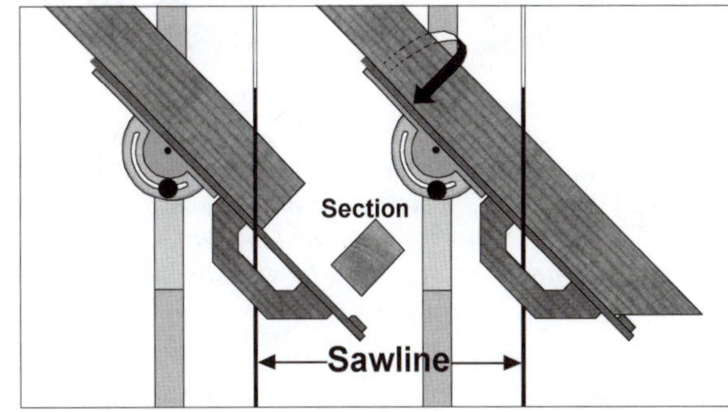

Fig. 14.24
Repeat mitering of rectangular sections for small frames at a single setting - turning edge for edge between cuts. This and the following sketches show the type of wooden bridge and fence extensions needed for the regular sliding-type of mitre gauges as used on table saws. Mitre gauges for sliding-table saws adjust to close the gap between the gauge end and the sawblade and need much less overhang.

With moulded architraves, picture frames and similar it is best to start the cut at the moulded edge to give a clean, spelch-free mitre - but this is not always possible.

The normal practice with small frames is to make a mitre cut at one end, then move the moulding beyond the sawline as above to make an second cut at the same angle. This cut is also the first cut of the next piece and is repeated as necessary. The final mitre cut to the opposite end of each pre-cut section is made with the mitre gauge angle reversed.

When using a small table saw with two milled slots make the first mitre cut using the slot to the left of the sawblade. For the final cut at the opposite end reverse the mitre gauge angle and transfer it to the right-hand slot.

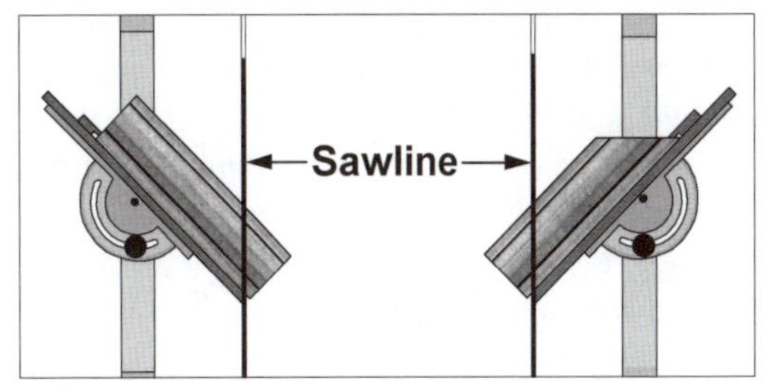

Fig. 14.25
On small table saws with two mitre slots use the mitre gauge to the left of the sawblade for one cut and to the right for the second. In this way the mould faces the sawblade in both cases, and so cuts cleanly. The stock is pre-cut to length in this sketch.
The end stops in this and the following sketches have a gap at the fence line to prevent waste lodging at the corner.

When using a sliding table saw, mitre-cut one end of all the parts, then reverse the angle of the mitre gauge to mitre-cut the opposite end. In one case the mould faces the sawblade and in the second it faces away.

NOTE: The above is for frame stock that has the mould

to the inside. The opposite applies when mitre-cutting frames with the mould to the outside.

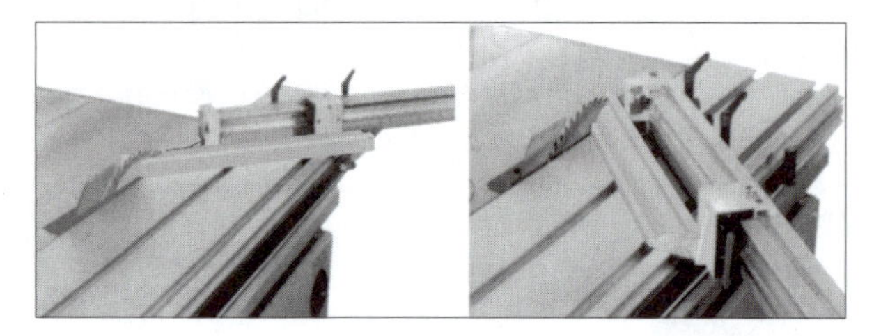

Fig. 14.26
The best practice is to work face up when mitring mouldings such as this architrave. On a sliding table saw two set-ups are needed when using this type of single mitre gauge. The mould faces towards the sawblade for the first cut (right) but faces away from it for the second cut (left).
Source - Felder

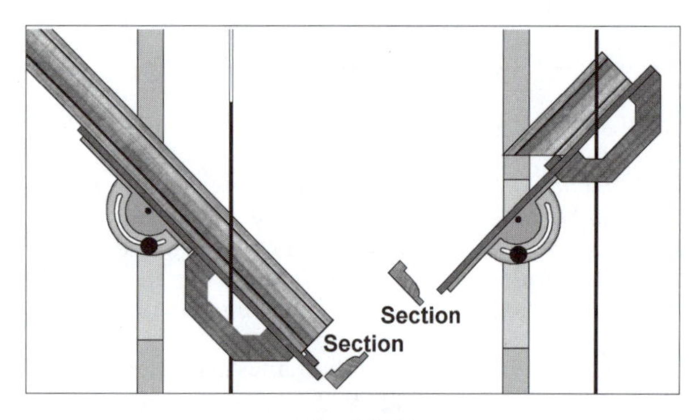

Fig. 14.27
For moulded stock the angle of a single mitre gauge needs reversing between the two cuts. In this case the parts are for a small frame, so long lengths can be converted without need of pre-cutting to shorter lengths.

Compound angle mitering. A compound mitre is needed if the two pieces being mitred together do not lie flat on their under-face when assembled.

Workpieces traditionally lie flat on the sliding table when being mitred, so the regular setting angle for the mitre gauge (relative to the number of pieces making up the frame) is not correct. The correct angle has to be developed-out geometrically, and the sawblade has to be tilted at some angle other than 90 degrees to make a perfect joint.

A simpler way of forming such compound mitres, is to use a jig to hold the workpieces at the assembled angle - then simply make a regular 45 degree cut with the sawblade vertical. As when mitering a moulding, a double mitre gauge is needed for moulded stock.

Fig. 14.28
Compound angle mitring with the sawblade tilted and using a regular mitre gauge.
Source - Sedgwick

Gauging the length of cross-cut and mitred parts
For limited production or custom work where parts are marked-out, use the initial sawcut in a wooden fence or block as a exact guide for the sawline.

Fig. 14.29
Angle-cutting a board on a sliding-table panel saw. Note the flip stop for accurately gauging the length.
Source - Felder

For a production runs use the end stops provided with most cross-cut and mitre gauges - which usually are of the turn-over or disappearing type. They accurately gauge the length of the workpiece when this is moved up towards the stop.

Turn-over stops are flipped clear when not needed. Disappearing stops pivot into the stop body when the workpiece is placed across it and pushed towards the fence. Using either allows two or more stops to be preset to different lengths and selected as required.

Stops may be manually set while cutting a marked-up workpiece.

Alternatively use the built-in scale and fine fence adjustment or, if provided, an electronic measuring system. All types of setting devices need correcting to a new left-hand sawline after changing the sawblade.

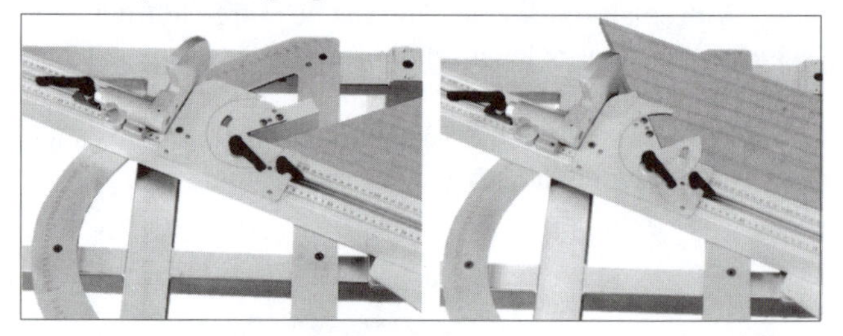

Fig. 14.30
Regular flip stops can be used to gauge the length of mitre-cut parts, but contact only a fine edge. In contrast, the adjustable-angle stops shown here on this Panhams type 690, can be set to give positive and full-width end contact for any angle.
Source - Panhans

Using a double mitre gauge.
An alternative on a siding-table saw is to use a double mitre gauge. This has two mitre fences at right angles to one another and at alternate 45 degree angles to the sawline.

One mitre-cut is made in the regular way against the leading mitre fence, but the opposite mitre cut is made with the workpiece against the rear of the trailing mitre fence.

Because the thrust of the sawblade is to separate the workpiece from the trailing mitre fence, make this the second cut and preferably clamp the workpiece down to the table.

Using a deflector
Cross-cutting on a sawbench produces numerous short off-cuts which the upcutting section of the sawblade can flick backwards towards the user with considerably speed - and these have caused eye damage in the past.

Although the operator may periodically clear-away these off-cuts to reduce the risk, doing this does not eliminate it altogether. Also, clearing the off-cuts - even when using a push-stick - is in itself hazardous.

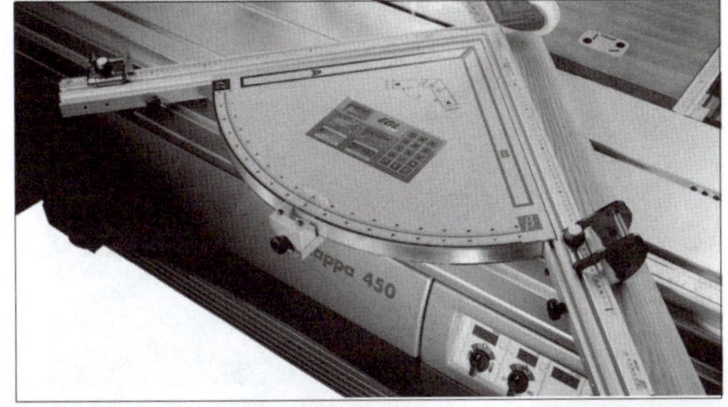

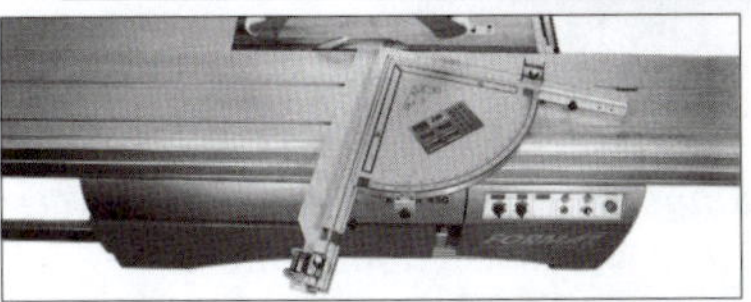

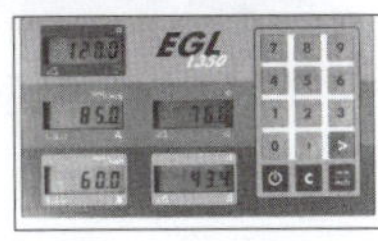

Fig. 14.31
The double mitre gauge on this Format-4 sliding table saw allows parts to be double-mitred at following passes. When setting the mitre angles for unequal section widths, left, below (or for a different frame angle) simply punch-in the two widths (or the required frame angle) to display the precise angle settings needed (right). A vernier scale provides length accuracy regardless of the angle setting.
Source - Format-4

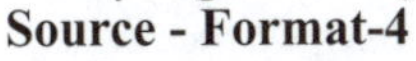

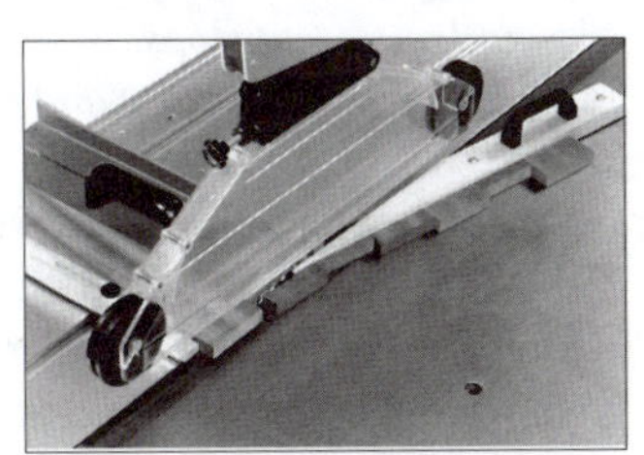

Fig. 14.32
A deflector prevents waste pieces being flicked back at the operator. This one clamps magnetically to the table top.
Source - Aigner

A far better option is to deflect off-cuts from the sawblade as they form, perhaps using a deflector which clamps magnetically to the table top. Using this type is simpler than attempting to somehow fix a wedge-shaped piece of wood to the table.

Panel cutting

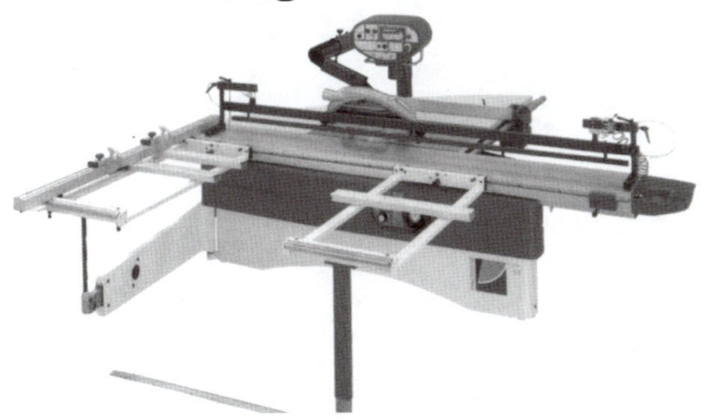

Fig. 14.33
Sliding table panel saws have a sliding table with a long, free-running movement. There is a wide choice of styles, types and sizes of these excellent machines. This particular machine, the P30N, has a pneumatics top clamp, a jump scoring saw, digital stops, and programmable setting for the rip fence and saw angle.
Source - Paoloni

Panel cutting is easy on the modern sliding table saws. The most upmarket machines have a large width and length-cutting capacity, and can be equipped with a wide range of devices that make setting fast, accurate and consistent.

Most sliding table panel saws are fitted with a scoring saw immediately in front of the main saw *(see section on scoring saws in Chapter 9)*. In some cases very precise alignment and projection of the scoring saw is possible from an external control.

Fig. 14.34
This shows the main and scoring saws (top left) and the scoring saw unit (top right) The scoring saw is aligned and set for width and height externally using the three controls shown in detail in the lower illustration.
Source - Felder

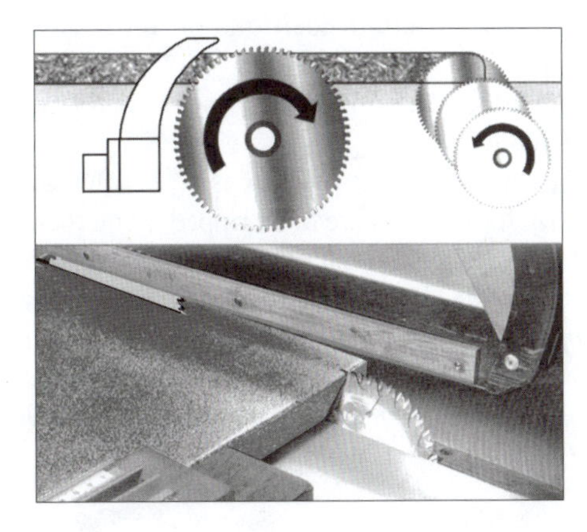

Fig. 14.35
The Synchroform system shown here ensures that the jump action (a mechanically-controlled rising movement of the scoring saw) coincides with the trailing end of post and soft-formed panels - to guarantee perfect cuts. Cross-cuts can be made square or to any angle, mitre cuts can be made as a single or compound angle.
Source - SCM

Soft and post-formed panel cutting

For cutting post and soft-formed panels *(see Fig. 15.17)* some sawbenches are equipped with a jump scoring saw that automatically rises as the panel end is reached. This completes its cut - in the trailing edge only - prior to the main sawblade reaching this point.

By doing this a perfectly clean edge-cut is formed. If a regular scoring saw is fitted, the main saw can break-out the trailing edge of this type of panel.

Cross-cutting large panels

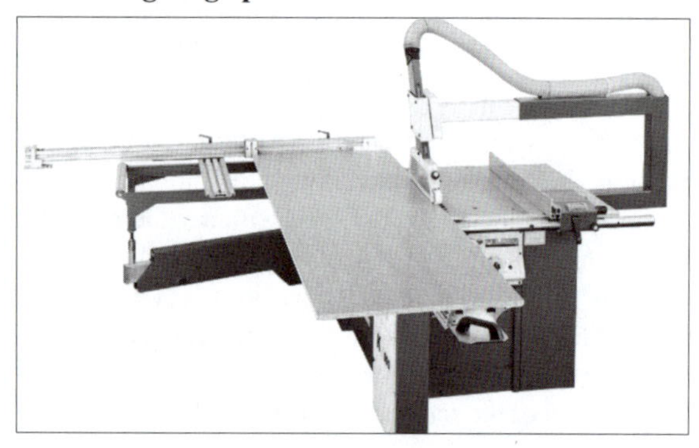

Fig. 14.36
A long traversing table allows wide panels to be accurately sized, as shown here, by manually holding the panel against a leading cross-cut fence. The cutting width of the panel is gauged either from a length stop on the front fence or by a length stop on the rear outrigger fence.
The outrigger supports have heavy-duty frames and telescopic arms with ball bearing rollers and dust wipers. Loading large and heavy pieces and sawing them accurately is easy on such modern machines.
Source - Felder

One regular practice when cutting large panels across their width is to use a leading cross-cut fence against which the panel is positioned and manually held during the cut. Alternatively a trailing cross-cut fence can be used, with the panel secured to the sliding table with a clamp.

Cross-cut and mitre fences and their clamps are readily adjustable for different positions on most sliding-table sawbenches.

Sometimes the same fence is used both for crosscutting and mitre cutting.

A twin or double table extension is often available if particularly long panels are to be ripped along their length.

With most machines a left-hand extension table is normally fitted to the sliding table remote from the operator, usually supported with a swinging extension arm. Many manufacturers offer a second extension table nearer the operator. This type is readily attached too, or removed from, the sliding table. It can be fitted at various points to suit the operation.

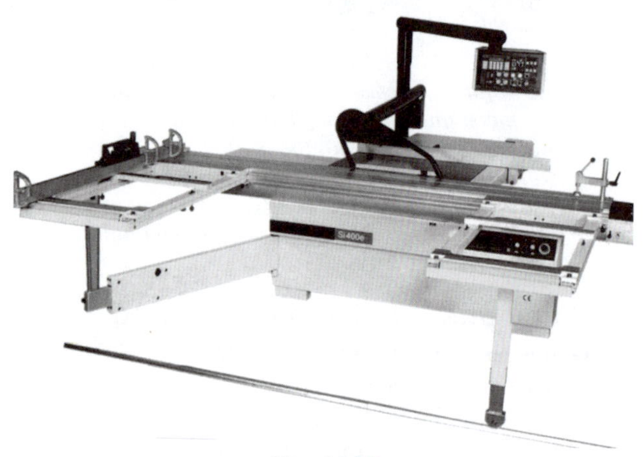

Fig. 14.37
This twin table can be quickly fitted and removed. It allows long and heavy panels to be sawn without additional support
Source - SCM

Some makers offer a servo-assisted movement for the sliding table for use when handling large and heavy panels.

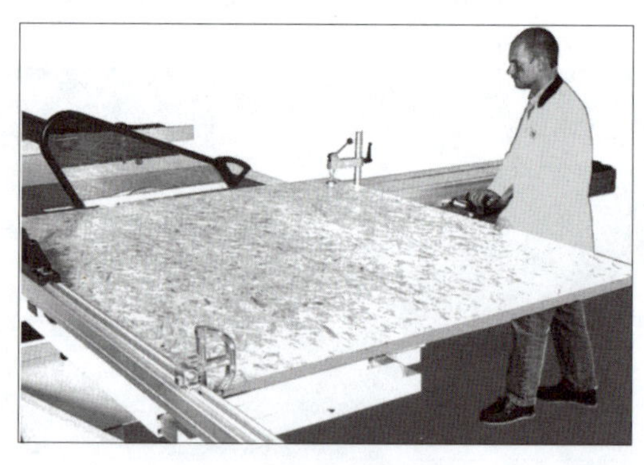

Fig. 14.38
On this SI 450 model, the board is clamped to the sliding table while forward against a leading fence. The operator grips a power-control handle with his right hand to traverse the table using a servo-assisted movement.
Source - SCM

Electronic and digital control

Many of the sliding-table panel saws offer highly sophisticated control systems for the regular machine operations. Often these include digital read-outs for precise setting.

Amongst the regular adjustments offered with this form of control are positioning of the rip fence, tilt setting of the saw arbor, angle setting of the mitre fence, and length-stop setting of the cross and mitre-cut stops.

In some cases a memory can store and precisely repeat regular settings.

Another feature is fast retraction of the sawblade below table level. This allows return of panels to the infeed without the danger of fouling the sawblade - and is coupled with instant re-setting to the original sawblade height when the panel is clear.

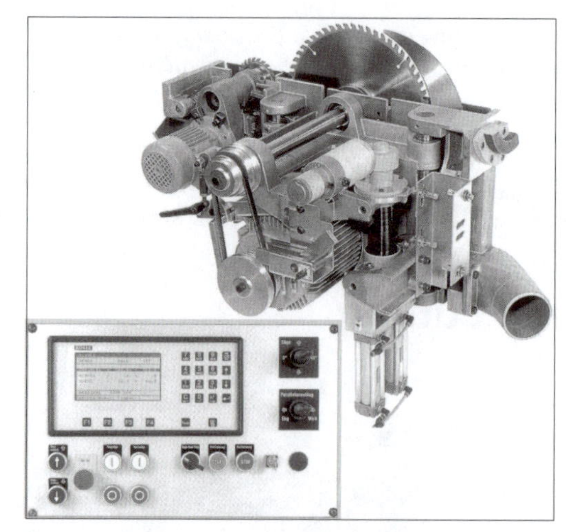

Fig. 14.39
This illustration shows the saw unit of a Formatic sliding table saw.
Clearly shown is both the drive for the digitally-controlled rise and fall movement of the saw arbor and the pneumatic, quick rise and fall movement.
The control panel (inset) shows the saw arbor speed, and incorporates push buttons to digitally preset the cutting width, also the height and angle settings of the sawblade.
Source - Kolle

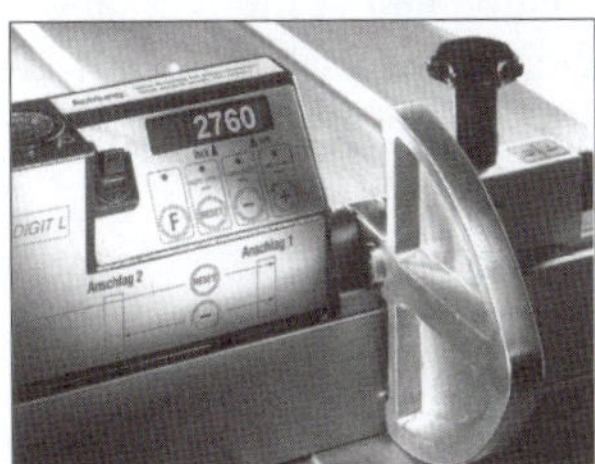

Fig. 14.40
This digitally-set rip fence on the F45 range of sliding-table saws (above) is accurate to within 0.1mm, and the large display is always clearly visible to the operator. The unit (below) digitally controls the settings of up to three adjustable cross-cut flip stops to an accuracy of $^{1}/_{10}$mm.
Source - Altendorf

Fig. 14.41
The Vanguard control panel fitted on the SI3 200 NS sliding-table panel saw enables the fence and sawblade to be quickly and precisely pre-positioned to any one of a number of stored combinations.
Source - SCM

Edging timber

Fig. 14.42
A leading-end angled clamp and a trailing-end 'X'-guide spiked bar securely holds timber when edging.
Source - Aigner

Trimming irregular edges of boards or flitches in a perfectly straight line is an ideal application for the sliding-table saw - provided, of course, that their length is less than the table traverse.

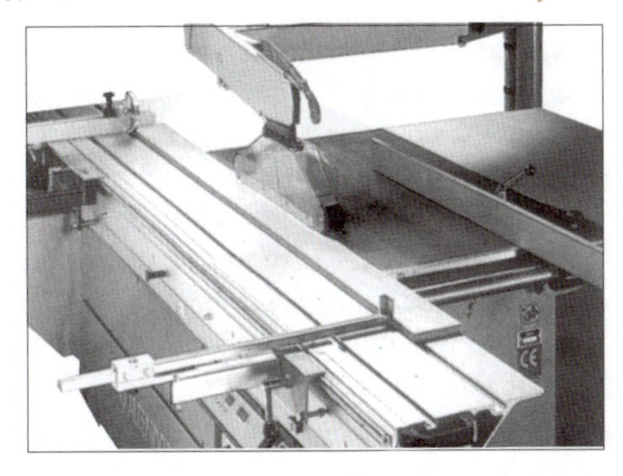

Fig. 14.43
An another way to position timber for edging is to use adjustable stops against their outer edge.
Source - Altendorf

It is essential, as with all non-guided work, to securely clamp the waney-edge timber or other irregular-edged material to the sliding table.

At the leading end a narrow, angled plate will hold the timber securely when this is forced under it, wedge-like. At the trailing end the operator could simply grip it, but a more secure way is to use a clamp or a spiked end-stop.

An alternative way is to use outer edge-stops near the leading and trailing ends of the flitch.

Lazer guide lines

A lazer projector can be provided for some sawbenches to project a fine red line along the sawline. This allows the flitch to be precisely-positioned so that the sawcut merely clips the waney edge.

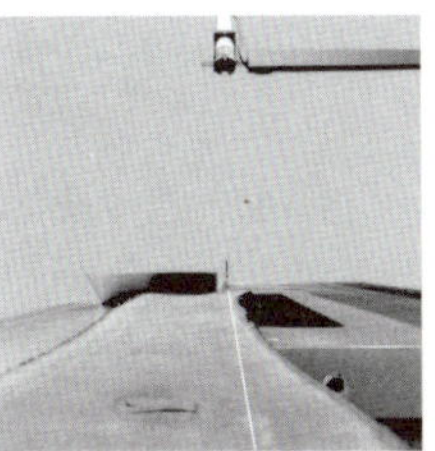

Fig. 14.44
Using a laser guide line to pre-set the timber relative to the sawing line.
Source - Casolin

Cross-cut & Mitre Saws

Cross-cut and mitre saws are used to cut timber across the grain to an exact length, or to mitre it to form part of a frame. These machines are mainly manually-operated, but some are automatic and certain types can use a grooving or trenching head in place of the sawblade.

Note: *The machines described here are the secondary machines used in building component and furniture manufacturing - cut-off saws for optimizing timber are dealt with in Chapter 16*

Safe use of cross-cut saws

UK factory inspectors report that most accidents on these machines result from the absence or poor adjustment of physical safeguards. This is often due to lack of training, because dust and off-cuts are brushed aside during rundown, or because the guard is held up to see the cutting line.

Manually-operated cross-cut and mitre saws. The sawblade should be enclosed as much as practical and for as long as practical during the cutting stroke. Machines should have a fixed hood to guard the non-cutting part of the sawblade and an enclosure at the rear to guard the sawblade in the rest position. Essential side guards may be adjustable or automatic.

All cross-cut and mitre saws should have fences high enough to support timber at either side of the cutting line - and have the means to close the fence gap at the saw gap when the sawblade is tilted and/or swivelled for angle cutting. Long workpieces need additional support in the form of roller stands or extension tables.

Preferably a no-hands area should be marked on the table. Operators should be clearly instructed to rigidly observe this - also never to cross the sawing line. The no-hands area should be marked roughly 300mm (12in.) on either side of the sawing line - bearing in mind the extreme swivel and tilt positions these sawblades can adopt.

Make sure that workpiece to be cross-cut or mitred is firmly down on the machine table, back against the fence and securely held manually or by using clamps - before starting the cut. When cross-cutting bowed timber set the bow against the fence at the sawblade gap - otherwise there is a risk of kickback. Waste removal should only take place when the sawblade is stationary, and then by using a push-stick.

Chop and mitre saws should be guarded as completely as practicable. They should by started only by a trigger on the operating handle. Because workpieces are inserted under the sawblade on these machines it is dangerous to leave them running - so never lock the trigger 'on' to run the sawblade continuously.

Chop and slide saws should be mechanically interlocked to slide only when fully down - and be fitted with a cross-cut type of sawblade (see previous notes in Chapter 5).

Radial and ram-type saws. A nose guard is essential on all such horizontal-stroke machines, and should be set to as close as practical to the workpiece top. An extension to the table may be necessary so that the nose guard never projects beyond. To give protection at any angular setting the extension should be a rounded one centered on the pivot point when the sawblade swivels, see Fig. 14.61.

Cross-cut sawblades only should be used on these machines. Also essential is an automatic brake or automatic return to the rest position when released - plus an anti-bounce-back device.

Care must be taken when using all radial and ram-type cross-cut saws to avoid run-out. This can happen if the wrong type of sawblade is used, if the sawblade is dull, or if the workpiece is not properly seated to twist during the cut.

Workpieces may be manually steadied, but when using grooving and trenching tools the workpiece **must** be mechanically clamped - and the forward traverse **must** be carefully controlled.

Trenching heads substantially increase the risk of kickback because of their width of cut and because their smaller diameter gives a lower peripheral speed.

Ripsawing on a radial arm cross-cut. Although some radial arm cross-cut saws can be used for ripsawing, they should only be used for this purpose when a regular sawbench is not available - and only then when the correct safeguards are fitted and used.

The saw unit should be turned and securely locked in the ripping position parallel to the rip fence. Fit a rip-sawblade, and set the unit the required distance from the fence. The sawblade must rotate against the feed and be provided both with a riving knife at the rear and an anti-kickback device at the front. The anti-kick device must be set so the fingers trail on the top surface of the timber.

This type of operation is particularly prone to kickback, so the greatest care must be taken. Use push sticks in both hands to clear the workpiece completely through the sawblade.

Automatic cross-cut and mitre saws. These additional remarks apply both to automatic cross-cut saws where the movement is triggered automatically, and semi-automatic saws where the cutting action is triggered manually.

Semi-automatic machines should be clearly marked with a no-hands area and should either be triggered only from a position remote from the cutting area with a two-handed control, or enclosed by guards so that contact with the sawblade is impossible. The latter is particularly important if a second person is removing cut pieces.

Automatic machines should have a tunnel guard that completely encloses the cutting area - including the clamping arrangement. Guards which protect the sawing area should be electrically interlocked to allow start-up only when they are correctly set. The design and operation of semi and fully automatic cross-cut saws is so varied that expert advice should be sought as to what are suitable operations and guarding arrangements for each of the different types.

Types of cross-cut and mitre saws

Fig. 14.45
A modern swivelling pendulum-type cross-cut saw. Note the nose guard and the flip stop.
Source - Bauerle

The pendulum saw. The original type of workshop cross-cut saw was the pendulum or swing saw. This was pivoted from an overhead wall bracket and swung in a shallow arc to cut across timber laid on a sloping table. It was capable only of straight cuts, and was used both to defect-cut* - take out knots and shakes - and used adjustable flip or pivot stops to cut timber to specified lengths. **In this respect up-cutting optimizing saws undertake this in the larger factories.*

A form of pendulum saw, but which *'swings'* in a straight line through a linked action, is still made. It can be manually or power-operated and supplied for cross-cutting at right angles only, or with a height-adjustable base and a swivelling movement for mitring up to 45 degrees.

The sliding table, double cross-cut saw was once widely used - and is still manufactured in this form by Bauerle.

The top model has a fixed left-hand saw-unit and a moveable right-hand saw-unit with an electronic traverse movement. The main sawblades have variable speed, automatic brakes and, together with the scoring saws, tilt up to 45° for mitre cutting.

The sliding table is actually two independent cantilever slides, one fixed to the left-hand unit and the other fixed to, and moving with, the right-hand unit - but linked together to move in step when crosscutting. A centre support and material clamps are provided. Automatic off-set of the main sawblades and automatic lowering of the scoring saws on the return movement of the slides produces clean-cut edges suitable for edge-banding without further treatment.

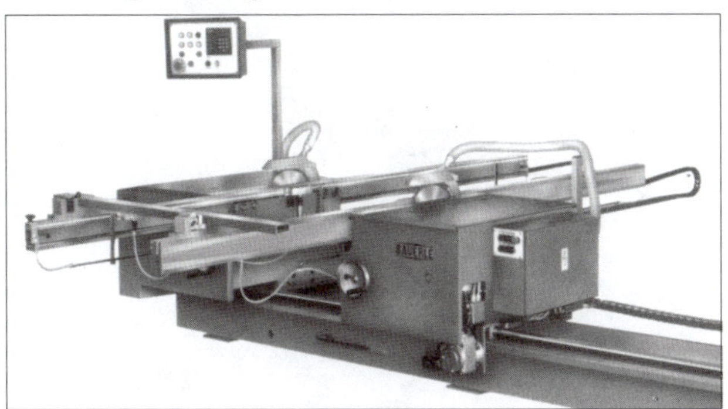

Fig. 14.46
This sliding-table double cross-cut saw has saw units carrying both main and scoring saws which rise, fall and tilt up to 45°.
Source - Bauerle

Most cross-cut saws used in manufacturing, though, are of the chop, chop and slide, radial or ram type.

Chop and mitre saws

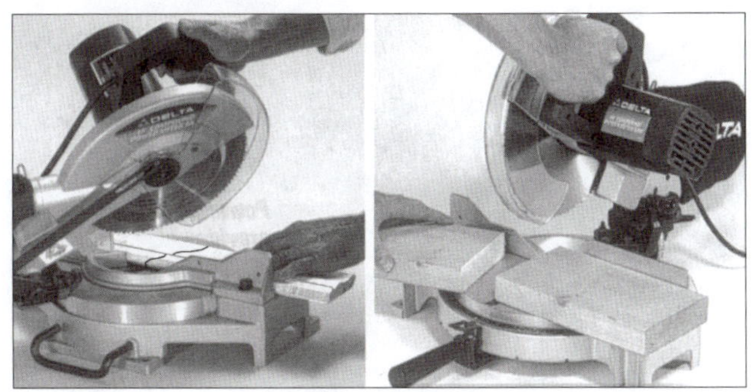

Fig. 14.47
The bench-mounted chop and mitre saw is capable of square cross-cutting (right) or compound angle cutting (left). Note the side guards which retract during the cut. Shown here is the 10in. 36-075 model.
Source - Delta

Fig. 14.48
A chop and mitre saw can also be floor mounted, and may be fitted with side clamps as shown on this T55 300.
Source - Omga

These machines are very compact and are simple in design and operation. They consist of a counterbalanced motor fitted with a direct-mounted sawblade.

The saw unit is pivoted at the rear, just above table level, and operates in a downward cutting action via a handle at the front. The cutting capacity of the sawblade is restricted by its diameter to smaller sections than can be cut on radial arm and ram-type saws.

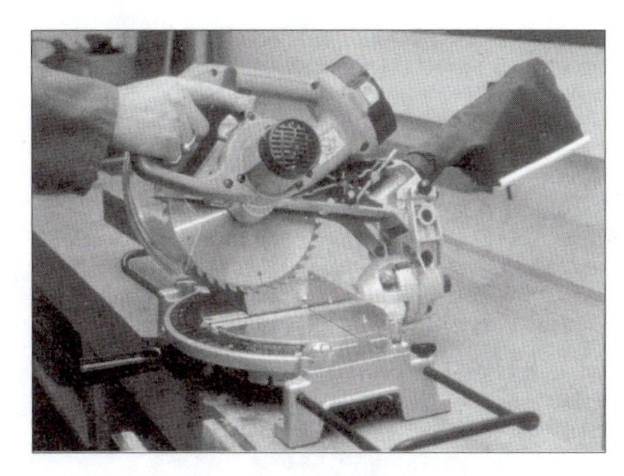

Fig. 14.49
Small chop and mitre saws are now made truly portable. This particular model, the LS800DWB-EX is a cordless machine. It makes precision cross and mitre-cutting possible for on-site carpentry on new buildings yet to have a mains connection.
Source - Makita

In its simplest form the chop saw cuts at right angles only. Other variants, classed as mitre saws, have a base which swivels on the fence line to allow setting to any angle up to 45 degrees in either direction. They are usually provided with a large and accurate mitre scale, and most have positive locks at 90° and the main mitre angles. Some mitre saws also tilt to the left for compound angle mitring.

With all types of chop and mitre saws an automatic saw guard is fitted which covers the sawblade in the raised position. Some models have a dust collector bag fitted, others can be connected to the factory exhaust system or to an internal dust collector. Clamps are recommended to securely hold the workpiece when dealing with short lengths and, as the table is generally short with all single machines of this type, extra table-level supports are needed on both sides.

The cutting capacity of these machines, in common with all cross-cuts, is affected by the cutting arc of the sawblade. It is usually quoted as two sets of figures - relative to width and thickness when set to cut at 90°- one when at maximum width and the second when at maximum thickness. Capacity is less when the sawblade is swivelled to cut a simple mitre, and further reduced when also tilted to cut a compound mitre.

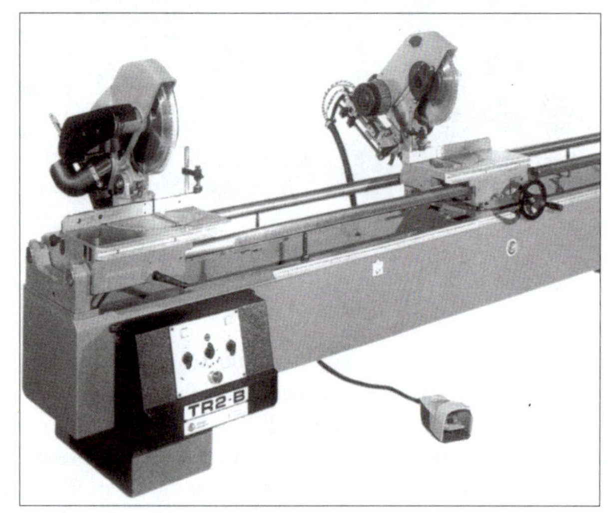

Fig. 14.50
This double mitre saw has power operation triggered remotely by a foot pedal.
Source - Omga

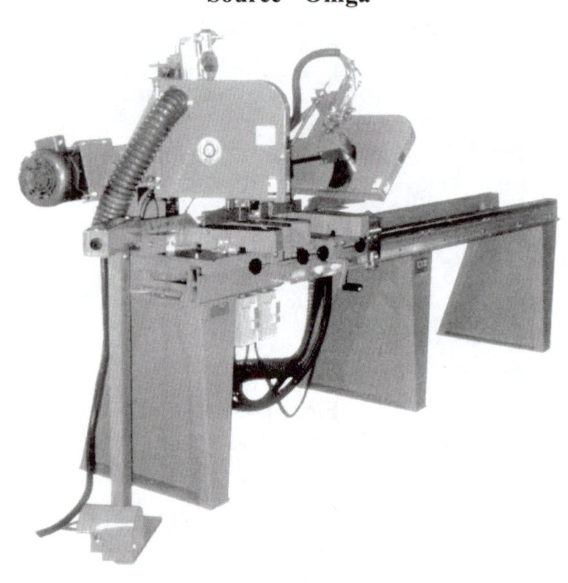

Fig. 14.51
The CDM60 16in Double end trim and mitre saw has digital length display and automatic clamping and operation.
Source - CTD

Chop and mitre machines are available with saws from 250mm (10in.) diameter and a sawing capacity of 146 x 60mm ($5\frac{3}{4}$ x $2\frac{3}{8}$in) or 86 x 94 ($3\frac{3}{8}$ x $3\frac{5}{8}$in), to saws of 500mm (20in) diameter and a sawing capacity of 280 x 130mm (11 x 5in) or 205 x 140mm (8 x $5\frac{1}{2}$in).

Small machines are manually operated and usually supplied for bench mounting. Large machines are usually floor-mounted either as a single saw, or as a double-saw combination - often with top and/or side clamps.

With single saws the saw motor is usually started by a trigger in the operating handle, and may incorporate a brake to automatically stop it within seconds when the trigger is released.

With twin-saw combinations a two-handed control or a foot switch is commonly used. This starts the cutting sequence for semi-automatic operation. Both should be electrically interlocked to trigger only after automatic clamps have already secured the timber. When mitre-cutting short pieces using a double-mitre saw the sawblades can be arranged to cut in sequence to avoid fouling one another.

These machines are good value for their money, have a surprisingly large cutting capacity for their size, take up little workspace and are simple to use. The smaller machines are popular for occasional work in large factories of all sizes, as portable saws for the building trade, and for DIY enthusiast. Some chop saws even double as a regular table ripsaw by reversing the table - but the main use for these is DIY.

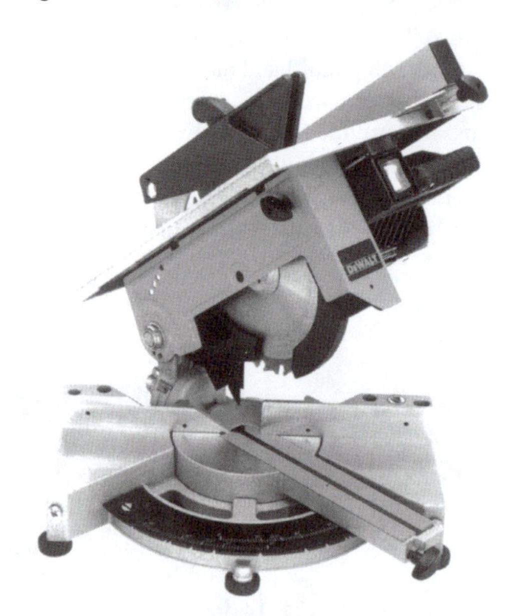

Fig. 14.52
Ideal for on-site building carpentry, the DW711 quickly switches from a chop and mitre saw to a table saw without use of tools. It can be supplied with legs and extended table for regular cross and mitre-cutting, and a sliding table for cross-cutting panels when used as a table saw.
Source - DeWalt

Chop and slide mitre saws

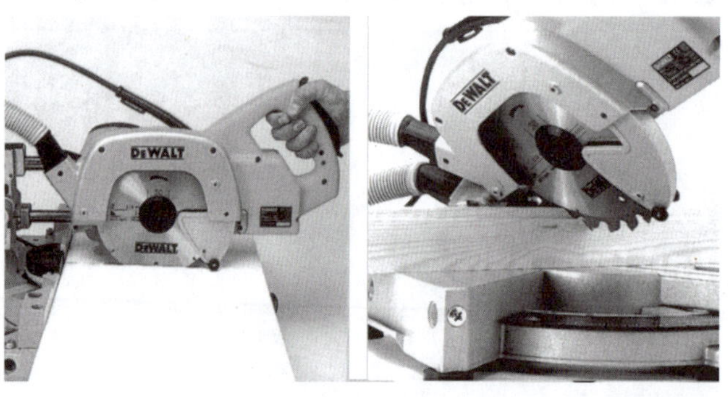

Fig. 14.53
This shows the wide range of work that chop and slide mitre saws are capable of - for wide cross-cutting (left) and for compound mitre-cutting (right).
Source - DeWalt

These are basically a chop saw but with an added telescopic

slide mounting to greatly increase their cutting capacity.

In use the sawblade is first lowered into the workpiece as a regular chop saw, and then drawn out to complete the cut (the two movements must be interlocked to allow downward movement only when in the back position). Like regular chop and mitre saws, this type is also capable of single and compound angle mitring.

Radial-arm saws

The radial-arm saw is the most popular and most copied type of cross-cut and mitre saw in the world. It is widely used in industry from the smallest one-man business to the largest mass-production factory.

The originator was the American DeWalt, a name by which this type is still commonly known, although the generic name is a radial-arm saw.

Fig. 14.54
The DW721 is typical of the modern range of DeWalt radial-arm saws. These machines are made with different sawblade diameters and cutting capacities, and are capable of straight, single and compound angle mitering, also ripsawing vertically or at an angle. Grooving tools can be fitted in place of the regular sawblade.
Source - DeWalt

This type of machine is usually floor mounted. The saw-unit freely traverses along a horizontal radial arm cantilevered from a rear, vertical column.

The arm can be raised and lowered on or with the column, and swivels to allow the sawblade to cut at 90° degrees to the fence or up to 45° on either side. The saw motor may also tilt to the left to cut compound mitres.

As the column is swivelled to different angles, the rest position of the sawblade also alters. As a result the fence-line may have to be shifted to retain the maximum width-cutting capacity - and to ensure that in the rest position the sawblade does not project beyond the fence-line.

Also the sawblade breaks through the fence at a different points as the angle is changed, so any previously-set length-stop needs adjusting to compensate.

When tilting for compound angles the sawblade both cuts through a different point on the fence and alters in height - so again adjustments are necessary.

Fig. 14.55
This 'Best' model is typical of the large-capacity radial-arm saws, having maximum cut sections of 20 x 915mm (3/4 x 36in) or 110 x 830mm (4 1/4 x 32 1/2) when fitted with a 400mm (15 3/4in) diameter sawblade. Note the ample top guard and automatically-retracting side guards.
Source - Omga

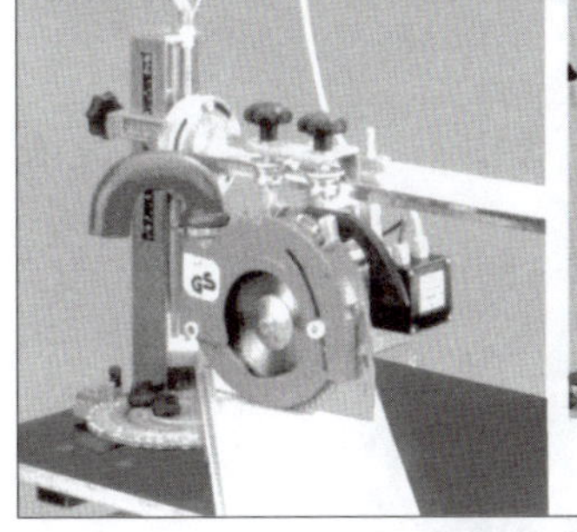
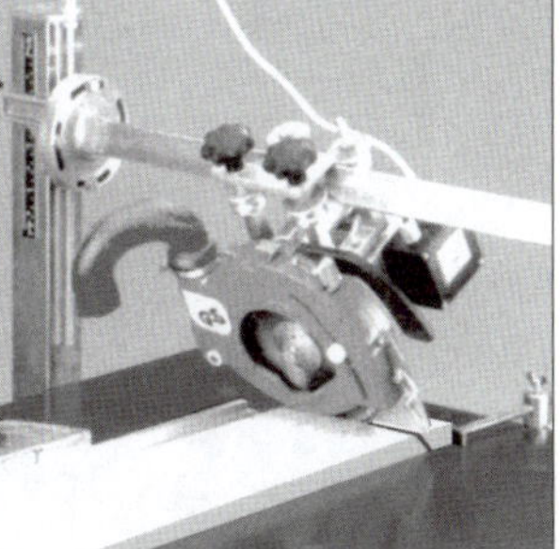

Fig. 14.56
These illustrations clearly show the typical settings of a radial-arm saw. This one is unusual in that the radial arm is a square bar and is angled to tilt the sawblade. Also the fence rather than the main column swivels for angle cuts. The illustrations show square cross-cutting (top left), ripsawing (top right), single angle cutting with the fence swivelled (lower left) and square angle cutting with the sawblade tilted (lower right).
Source - Eudora

Most machines are fitted with a wooden fence which can be quickly replaced, shifted across the machine, or repositioned front to back. When cross or mitre-cutting marked-up workpieces, or when using a marked-up sample to set an end stop, the simplest way is to position the workpiece relative to an accurate sawcut in the fence.

First set the sawblade to the correct angle(s) and height. Re-position the fence front to back, if necessary, so the sawblade lies behind the fence line when in the rest position.

Shift the wooden fence along so that the sawblade cuts through a clean section and use this, following the first cut, to position workpieces or stops. The fence should preferably project slightly above the workpiece so the sawline is clearly visible with the workpiece in place.

The saw-unit on some radial-arm saws rotates on a vertical axis and can be locked square to the radial arm at any point. At this setting it is used - with a ripsaw - for sawing timber or panels to width. Take particular note of the safety requirements previous stated on ripping with a radial-arm saw.

Radial-arm saw variants

Fig. 14.57
On the Metra-cut saw the radial arm pivots on the cutting line, so square and angle cuts always pass through the same point on the fence. It has a short-centre drive saw arbor to give a larger-than-normal depth of cut of 200mm (8in) from a 560mm (22in) sawblade.
Source - Alpine

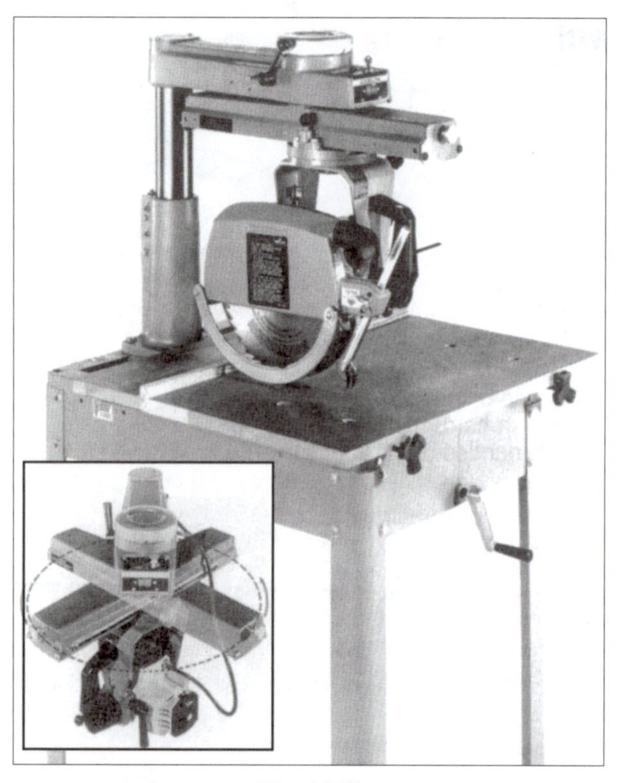

Fig. 14.58
One of the 33 series, this model has a unique turret action, as shown in the inset, which allows the sawblade to rotate 360° above the worktable. This maximizes the cutting capacity and makes mitre and angle cuts easier to set.
Source - Delta

To allow the sawblade to cut through the fence at a consistent point when cutting single angles, some machines have the pivot centre for the radial arm directly above the fence line instead of it pivoting on the support column itself.

With this arrangement the fence line never needs adjusting front to back- and the machine retains full mitring capacity at angle settings well beyond the normal 45°.

Radial-arm saws are also made with a range of non-standard features such as; exceptionally wide cross-cutting capacities; manual or power traverse; workpiece clamps - and can be supplied as single unit or twin-saw units.

Fig. 15.59
This Supercut 3558 has an electrically-powered movement. The close-up, below, clearly shows the electrical controls.
Source - OSC

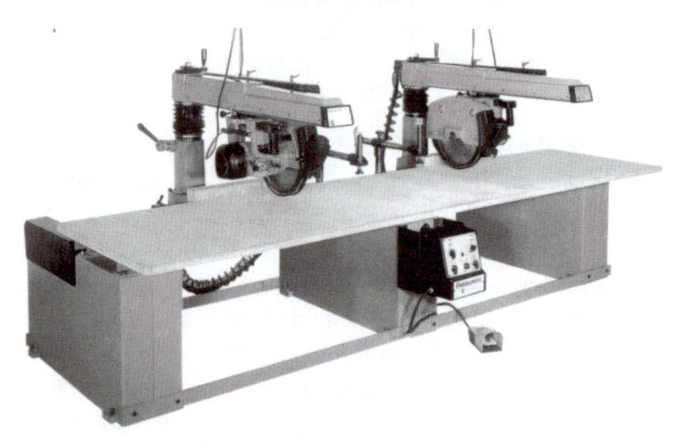

Fig. 14.60
Both saw-units on this two-unit machine independently adjust to any swivel or tilt angle within their range. The cutting cycle is triggered by a shrouded foot pedal.
Source - Omga

Twin-saw versions are made left and right-hand to face one another, and are semi-automatic in operation with power traverse, automatic clamps and centralized controls on a remote console. They may have one fixed and one moving saw-unit, or both saw-units may adjust manually or under power.

In spite of the quirks mentioned earlier, the radial-arm saw in its many forms has replaced practically all other industrial cross-cut saw types - a clear testament to its convenience and operator friendliness.

Ram-type cross-cut saws

These were once the favoured type of cross-cut saw for many UK mills, but the chop and radial-arm cross-cuts have virtually replaced them - although there are hundreds still in use. The saw motor is mounted on the end of a ram which slides above or into a casting on the rear column.

The rear column swivels 45° either side and the sawblade tilts 45° to the left. As with the radial-arm saws, making adjustments to either alters the cut-through point on the fence - which adjusts lengthways to compensate - although some models were designed specifically to avoid these particular problems. They were made both for manual and semi-automatic operation.

Fig. 14.61
A typical ram-type cross-cut set with the sawblade tilted to give an angle cut. With this particular machine the sawblade tilts on the sawline. Note the large curved table extension to prevent operator-contact with the sawblade - whatever its setting.
Source - Robinson/Wadkin

Twin mitre saws

Fig. 14.62
The Mod. C16 twin mitre saw shown here is the fastest and most successful machine in their range with a production speed of up to 1000 cuts per hour and a cutting capacity of 100 x 75mm (4 x 3 1/2in).
Source - Brevetti

These are special machines for mitring wooden picture-frame mouldings and similar using twin saws. The sawblades cut vertically, either rising or lowering into the cut, and operate in a rapid, automatic cycle to cut opposing mitres simultaneously.

The sawblades have a variable cutting speed and a fast return. They are slightly staggered in height and mounted close together to part the stock virtually at the same point on the nearside edge. This arrangement reduces waste between cuts to the absolute minimum.

The precise length of the mitred parts for picture, mirror and similar frames is gauged from the rebate (rabbet) line, so that the finished frame is correctly sized regardless of the actual shape and profile of the stock from which it is cut.

The two sawblades are mounted to form a 90° angle between them (45° in opposite directions to the fence) to produce precise mitre cuts for four-component frames.

On some machines the basic angles can be altered to also form the correct mitres for six or eight-component mitred frames.

Stock feed, clamping and triggering of the automatic cutting cycle in the low-cost versions is manual, the latter often via two separated push buttons for reasons of safety.

Fig. 14.63
This is a fully automatic line controlled by a computer to cut random-length stock to pre-selected lengths through an optimizing program.
Source - Omga

Semi and fully automatic machines are also offered, with automatic stock feed and a fully automatic cutting cycles. Timber can be entered individually by hand, but more commonly via a magazine feed.

The most sophisticated and efficient installations have full computer control and an optimizing stock conversion program*.

With these a suitable cutting program can be input via floppy disc - or selected from one of several hundred stored programs.

The computer then determines the best cutting sequence and, after displaying this for the operator to approve or change, starts operation on a single command. Once started, the fully automatic optimized sequence continues as long as the supply lasts, or until the required number of different lengths have been completed.

**An optimized program checks stock requirements against the random lengths as they are fed in, and calculates how these are best converted to reduce wastage to an absolute minimum.*

A screen shows a graphic display of the cutting cycle and the progress of the cutting program. It is also used when setting, and to undertake machine diagnosis.

Building site saws

Many of the smaller machines described in this Chapter can be also used on a building site - away from the general workshop. Some machines, however, are specifically intended for this type of work and are built with wheels to allow easy transport. They must, naturally, be used with the same care and attention as machines in the workshop - but a building site poses additional hazards not normally encountered in a workshop situation.

Fig. 14.64
This BSH combination rip and crosscut sawbench is intended for the rough and tumble of building site work, yet is built to the highest standards of modern manufacture and meets current safety regulations, including an automatic brake that stops the sawblade within 10 seconds of being switched off. In this operation timber is being split diagonally on an adjustable bevel gauge.
Source - Scheppach

Particular attention needs to be made as to where a sawbench is placed.

The site needs to be level, free of the usual builders rubble and protected both from loose items regularly dropped from buildings in construction, dust and fumes from the building process and from passing workmen - and their loads.

Machines driven by individual petrol (gas) motors must have safe and dependable speed regulation, and electrically-driven machines must have cabling well protected from damage by passing traffic, and with a dependable electric breaker.

The machine and the operator must be protected from being bumped by a carelessly moved load or an inconsiderate workman, so careful siting is essential.. Building sites are also notorious for loud and unexpected noises that can so easily distract the saw operator, so the operator must be aware of this possibility - whilst still remaining alert to any warning shouts.

Sawing operations within most workshops are generally more closely regulated - with only designated workmen allowed to operate machines. On a building site there is great temptation for other building workers to use an idle rip or cross-cut saw for occasional work - without bothering to employ an experienced operator. This is one of the most common causes of accidents - inexperienced users operating a machine in a dangerous way, leaving a machine incorrectly set or with a saw damaged by attempting to saw material for which it is not intended. In all these cases the next user could have an accident as a direct result of any of these malpractices.

Fig. 14.65
This shows the TKG 250E portable cross and ripsaw being used within a building - for finishing interior cladding in this particular case.
Source - Scheppach

Another common cause is to failure to electrically isolate a machine when left - so that it can be started accidently. Another is leaving a machine still running under power - a highly dangerous practice - particularly on a building site where there are so many distractions and where a running saw goes unnoticed. Electrical supply on building sites that already have them is often interrupted due to the nature of the work. For this reason machines should have electrical controls that prevent automatic start-up following a break in the power supply. Also the machines must be properly earthed (grounded) - building sites are often prone to cable damage and, with loose water commonly around, this is a lethal combination.

However, with a suitable machine properly used and supervised, building site saws can provide an excellent on-the-spot service that is convenient, and saves time, energy and money.

Chapter 15
Panel saws

Large-capacity sliding-table sawbenches are often called panel saws, but this same term is really more appropriate for travelling-head panel saws. These, also known as beam saws because the saw unit traverses on a rigid beam, are ideal for converting even the largest panels of virtually any type.

With a sliding-table panel saw *the sawblade remains in a fixed position and the panel is moved in sizing it. So, although the machine itself takes up little floor space, a large working area is needed - nominally, double the length and width of the largest panel handled.*

With a travelling-head panel saw *the panel remains fixed in position, either horizontal or vertical, and the sawblade is traversed to size it. Horizontal panel saws occupy more floor space than sliding table saws, but they need less working space around them. Vertical panel saws need the absolute minimum of both machine and working space.*

Simple and low-cost panel saws are intended for the smaller workshop where boards are loaded manually and finished panels off-loaded by hand.

At the other end of the scale are fully automated, computer-controlled panel lines that need virtually no manual effort. These may have two or more high-production panel saws, together with a transfer system to handle both stacked boards and cut panels.

All panel saws, however simple or complex, are fast, accurate and efficient - the accepted workhorses for large-scale panel conversion. They all operate in a similar way and have many common features.

Safe use of panel saws

The following are general guides that apply to most panel saws. These, however, vary widely in design and operation, have their own characteristics and differ in what is needed to operate them safely. More specific recommendations are available from the machine makers and local safety authorities.

There are quite specific hazards with these machines: Being trapped between the pressure beam and the panels or the table: being cut by the sawblade when in the end housings, as it rises into the cut or during its free travel before, between or after sawing the panels; being trapped by the automatic panel-feed mechanism.

Access to the travelling sawblade and clamp on undercutting horizontal panel saws should be prevented by a trip bar or similar at the infeed side *(and one at the rear if there is ready access)*. Also a screen of fingers should descend to guard the beam before this descends. Fixed guards are needed at both ends of the beam - with interlocked access to the saw unit to prevent opening until the sawblades are stationary.

On other types of automatic panel saws, the pressure beam and cutting cycle should be started by two-handed push buttons - and automatic side guards are essential.

Panel handling equipment on fully automatic lines must be effectively guarded as parts move quickly and start-up without warning. The area around handling equipment must be fenced-off and provided with interlocked doors. In certain instances photo-electric guarding is preferable to safeguard dangerous parts of panel handling equipment. Before entry into a potentially dangerous area is allowed the machine must be efficiently isolated electrically.

Safety equipment must be properly maintained. Machines should be checked at the start of each shift and at appropriate intervals - taking into account the use of the machines and manufacturers recommendations - and operators should be properly trained in a safe method of work.

Panel saw makers are fully aware of the potential dangers of their machinery, and safe operation is one of their prime considerations. One particular danger is with older machines built to comply with safety regulations at the time they were built, but which may no longer satisfy the current high standards of safe machine operation.

Types of panel saws

There are three basic types of beam-type panel saw:-

Horizontal overcutting beam saws.

Vertical panel or wall saws.

Horizontal undercutting beam saws.

Overcutting beam saws

A cost-effective way of converting panels in the smaller workshop is by using an overcutting beam saw. These have a horizontal table to support the panel, a side fence to square it up, and built-in rules to position it.

The single sawblade, on a beam above the table, can be traversed across the width of the table either manually or automatically. In the powered version the sawblade usually has a variable traverse speed in both directions, and is controlled by push-buttons on a remote console.

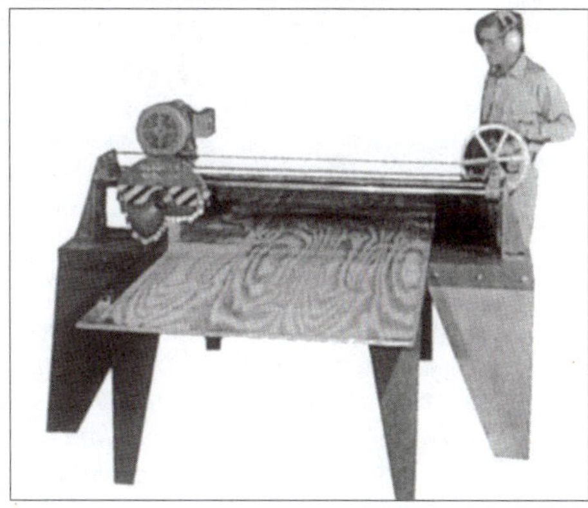

Fig. 15.1
This beam saw has a manual traverse movement via a handwheel, and link-type side guards for the sawblade.
Source - Hendrick

Fig. 15.2
With an automatic beam saw the sawblade traverse movement is controlled from a remote panel. The optional pneumatic hold-down allows cuts in both directions.
Source - Hendrick

Fig. 15.3
In this version the beam saw table has precision milled table slots at $22^{1}/_{2}{}^{\circ}$, 45° and 90° to the sawline. The leading moulded edge of kitchen counter tops is placed into these slots to make precise square or mitre cuts.
Source - Unique

Fig. 15.4
This large, custom-made beam saw has two saw-units at right angles to one another, and which are used alternately. One saw-unit traverses along the beam for cross-cuts. The second saw-unit is first pre-positioned on the beam, and then traverses with it along the length of the table for ripcuts.
Source - Elcon

The sawblade is arranged to climb-cut at a slow, controlled speed, then returns along the same path at a higher speed - so no hold-down is normally needed. To save operating time cuts could be made in both directions alternately by shifting the panel between cuts, but as the return sawcut tends to lift the panel, some form of hold-down is then essential.

Basic beam saws are manufactured in different forms:-
As a beam saw complete with table, fence and rules, plus an optional hold-down:
With an undercutting scoring saw for converting faced panels without danger of edge-splintering:
As a mitre-cutting saw for custom counter-top manufacture:
As a beam saw only - without tables - either for the user to fabricate his own table or for integration into a production line where the feed is provided by other equipment.

Vertical panel saws

Vertical panel or wall saws are ideal for the smaller workshop where space is at a premium. The machines, although normally placed against a wall, are usually self-supporting. They take up little space, and need only a small operating area compared to all other types of panel saw.

One person alone can usually operate these machines. However, panels need lifting onto a support fence just above floor level and placed against a rear support, so mechanical lifting gear or assistance may be needed to initially position panels when these are large and heavy. See Fig. 15.19

The rear support frame is tilted back slightly so that panels then rest safely back against the this and on bottom rollers or fence. In some cases fence rollers lift to allow the panel to be more easily positioned.

Vertical panel saws are made in two styles:
For both vertical and horizontal sawing:
For vertical sawing only.

Vertical and horizontal saws.

Fig. 15.5
Vertical & horizontal panel saws vary considerably in size and sophistication. This TRK 4164 is the smallest one in the Striebig range, yet is equipped with all the typical features of a high-performance saw. The operator is seen using the saw control lever.
Source - Striebig

The saw-unit on these machines is mounted on a horizontally-traversing vertical beam to the outside of the panel support frame. The unit has two, quickly-switched positions - with

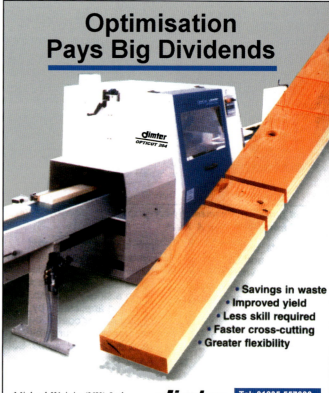

the saw parallel to the beam for vertical cuts, or swivelled through 90 degrees for horizontal cuts. The saw-unit and the beam are together traversed along the panel for horizontal cuts, but for vertical cuts the saw-unit is traversed up and down the beam - with this stationary.

When loading or positioning panels the sawblade is held in an outer rest position well clear of the panel. A lever first moves the sawblade inwards to start the cut, then out again after completing it. This allows the saw unit to be returned to the start position without the sawblade again contacting the panel.

For horizontal cuts

Fig. 15.6
This shows the strip-cutting gauge on the Economy 11 entry machine for making repeat horizontal cuts. The strip width is gauged from the previously-cut edge.
Source - Striebig

For horizontal cuts, i.e., when ripping to width, the saw-unit is set with the sawblade square to the vertical beam. After setting the saw-unit height to the required strip width *(gauged from the lower or intermediate fence on which the panel rests)* the unit is locked to the beam. The cut is then made by traversing the beam unit along the full length of the panel.

For repeat horizontal strip cuts a gauge can be supplied for reference against the panel top edge - set relative to a built-in rule on the saw unit. This is zeroed to the top sawline to give a direct read-out of the strip width it will produce.

For vertical cuts

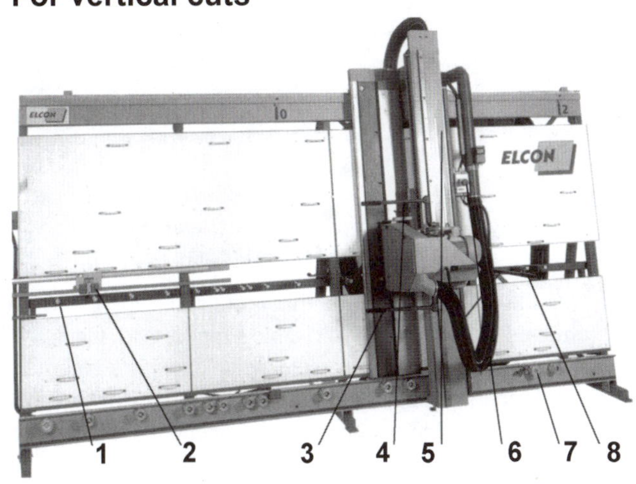

Fig. 15.7
In addition to all the regular wall-saw features, the RS Limpio models have a high-tech dust extraction system. The figures show; 1, Intermediate support fence; 2, Panel end stop for vertical cuts; 3, Saw unit; 4, Horizontal strip-cutting gauge; 5, Programming rail; 6, Motor switch; 7, Lower fence feed rolls; and 8, Locking lever.
Source - Elcon

For vertical cuts, i.e., for cross-cutting the previously rip-sawn panel strips to length, the counterbalanced saw unit is set with the sawblade parallel to the beam, and traverses vertically up and down the beam.

The beam unit can traverse the full length of the machine - and is locked for vertical cutting at any one of several positions along it - to correspond with vertical channels in the rear frame. These support the panel on each side of the cut and allow a clear cut through without damaging either the sawblade or the frame.

Wide panel strips rest on the lower fence support. An intermediate support fence is used when cross-cutting narrower panel strips - and which folds away when not required.

Fig. 15.8
This custom-built Holz-Her 1220 Super wall saw is an exceptional large machine. With a cutting length of 15.3M. (over 49ft.), it dwarfs the operator (centre).
Source - Holz-Her

The panel length is gauged by one or more end stops at the extreme left-hand section of most machines - which adjust horizontally over a distance of one metre. The fixed horizontal positions of the beam are at precise distances of one metre, so, simply by moving the beam along to other positions, the cross-cut length can be altered in one metre increments.

End stops are used in the conventional way, being set to cross-cut strips to standard lengths. Several stops can be fitted, with those not in use being turned clear.

Setting for width and length

On the economy range of machines the saw unit is positioned for horizontal cuts relative to a simple tape rule - the appropriate one of two or three rules zeroed to the base fence and intermediate support fences.

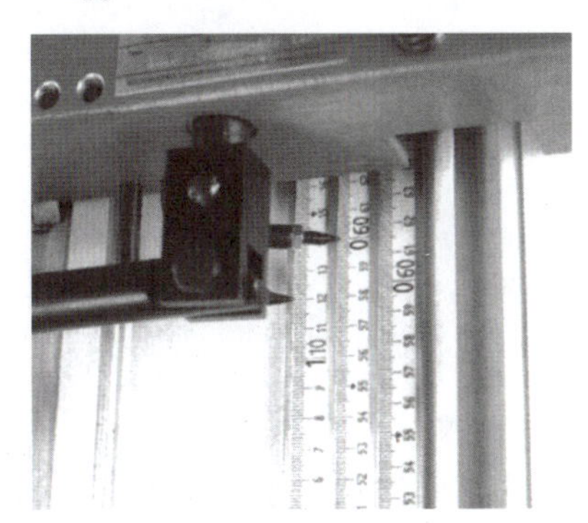

Fig. 15.9
Scales indicate the precise strip width when sawing horizontally.
Source - Holz-Her.

Panel strips are positioning for cross-cutting by butting them against stops set to a built-in rule - using these in conjunction with the fixed horizontal positions of the beam.

High specification machines can include digital indicators with fine-screw adjustment to give absolutely precise settings for both vertical and horizontal cuts.

An electronic, keyboard-controlled positioning system is also available for automatic cross-cut stop setting. This type can memorize and recall previous dimensions for precise repeat cuts.

Fig. 15.10
The DMS system is a digital measure-indicating system used for pre-setting stops for either vertical or horizontal cuts (left).
The EPS system is an electronic stop-positioning system for vertical cuts which can store up to 400 values (right).
Source - Striebig

Any variation in the kerf width when sawblades are changed can and should be taken into account with all types of setting rules and systems - so that they always give precise, first-time settings.

The saw unit.

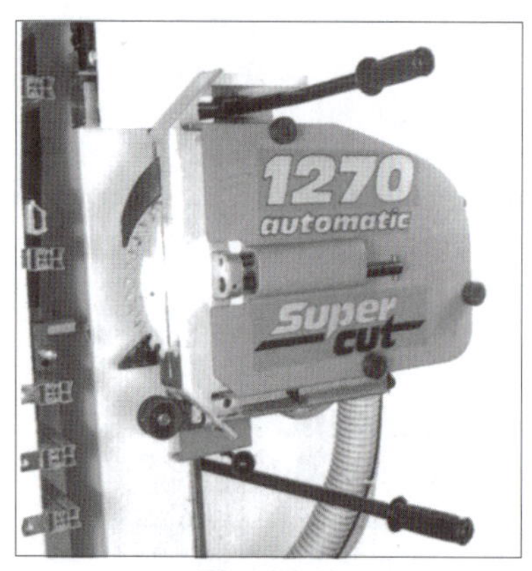

Fig. 15.11
This view of the underside of the 1270 Supercut saw-unit shows the riving knife, pressure pad, also the rollers and knife-scoring system to give tear-free cuts in coated panels.
Source - Holz-Her

The saw-unit has a low-friction, spring-loaded pressure pad which holds the panel firmly in position during the cut. To cater for different sawblade diameters, the inward movement of the sawblade is readily adjustable so that it cuts through the rear of the panel by no more than 10 or 15mm.

Fitted immediately behind the sawblade is a retractable riving knife which prevents both nipping and edge damage of the top board in the final stage of horizontal ripcuts.

Preventing tear and break-out. Many machines are fitted with a single sawblade only - which can break-out when sawing double-faced boards. To prevent this some makers offer an attachment for fitting immediately in front of the sawblade. This can be twin carbide scoring knives mounted between pressure rollers *(for tear-free cuts in coated facings)*, or a climb-cutting scoring saw (*for break-out-free cuts in all types of facings, soft or brittle)*.

Fig. 15.12
This VSA scoring unit gives perfectly clean cuts on hard, brittle and fibrous board surfaces.
Source - Striebig

Panel support frame

These vary according to the price bracket of the machine. Economy frames have series of fixed horizontal particleboard strips against which the panels rest. When ripsawing certain widths the sawblade will cut along and damage a strip, so they are easily user-replaced as they become worn or damaged.

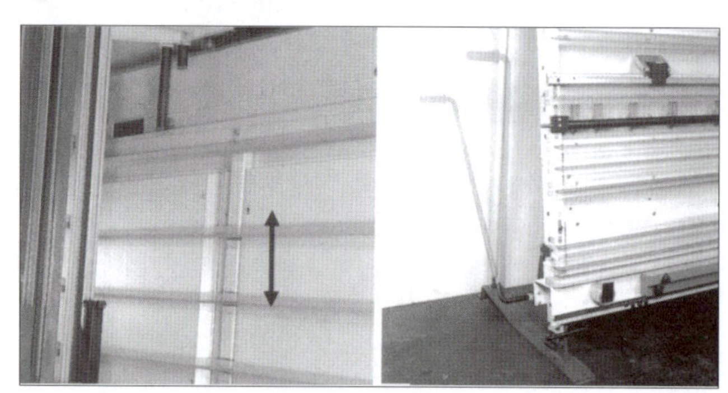

Fig. 15.13
Left, an automatic moving grid; Right a manually-moved grid, showing the lever in the two extreme positions.
Source - Holz-Her

Most regular machines have similar horizontal support strips, but mounted on a vertically-movable support grid. Once the sawblade is set for the required panel width, the frame is shifted to position the nearest strips close to both sides of the cut. This avoids strip damage and gives the essential back support to the panel.

On many machines the grid is moved manually by a lever, but on higher-specification machines it shifts automatically as the sawblade is positioned for width-cutting.

Dust extraction

A clean, dust-free working environment is becoming an increasing demand for all woodworking machines, especially for machines such as wall saws which often operate close to assembly lines and assembly workers.

Most makers are taking this very seriously, and offer machines which meet critical standards for clean air. An extraction duct on the saw-unit itself and rear extraction ducts can between them give efficient dust, chip and fume extraction.

The rear extraction arrangement on some machines is a combination of vertical non-return dust channels at the cross-cut settings of the beam, plus several horizontal dust channels for ripsawing which automatically align with the sawblade as this is positioned.

Another system has a wide but shallow, full-height dust collecting channel on the support-frame side. This is fastened to the beam and traverses with it when making rip cuts - to displace the rear panel support wheels as it passes across them. When cross-cutting, the sawblade traverses vertically along the dust collecting channel. In both cases the nearside faces of the collector give continuous support to the panel and seal the area so that the dust generated is efficiently extracted.

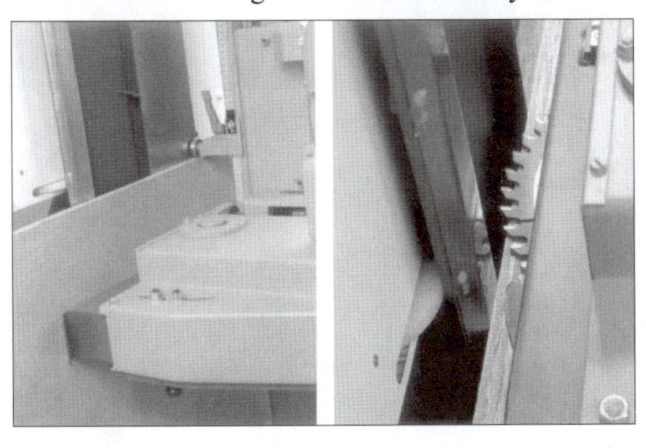

Fig. 15.14

This dust extraction system has a vertical dust-collector channel along which the sawblade traverses for vertical cuts, and which traverses with the sawblade for horizontal cuts. The sawblade is shown making a horizontal cut (left) and with the top panel strip removed to show the dust channel more clearly (right).

Source - Striebig

Automatic operation

Many machines operate quite satisfactorily with manual-operated vertical and horizontal traverse movements.

Automatic machines are better for large-scale production because of their higher productivity - and because their controlled cutting speed guarantees a consistently-high quality of finish.

On most automatic machines the plunge, traverse, retract and return cycle is started with a single push-button - with a sensor automatically triggering the return movement when the panel end is reached.

A choice of fixed or infinitely variable feed speeds in either direction is provided on some models - perhaps also with a powered rotary movement of the saw-unit, to instantly switch between rip and cross-cutting.

Both vertical and horizontal cuts can be automated on some machines. A series of following cuts is possible in a programmed sequence then, on completion, the saw-unit diagonally returns to the start position in readiness for the next panel.

Angle sawing fence

Some makers offer a precision angle fence for mounting on the lower regular fence.

Vee and square-grooving equipment

Machines of this type are also suitable both for square-grooving furniture panels and for vee-grooving sheets and panels for folded-type fabrication.

After the sawblade is replaced by a suitable square or vee grooving disc, the regular saw-unit movements are used to make any number of horizontal or vertical cuts.

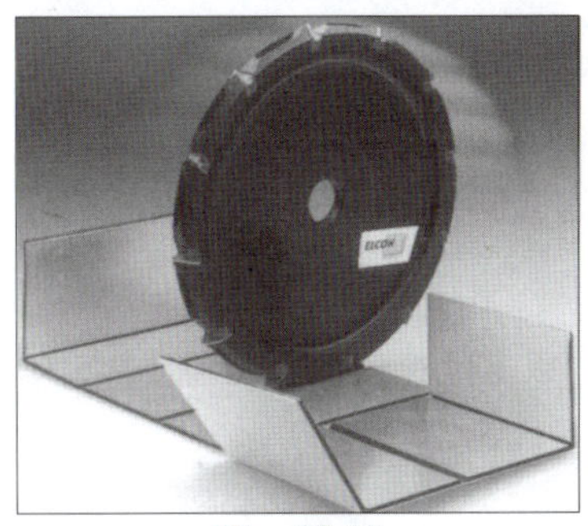

Fig. 15.15

For square grooving, or vee grooving as shown here, the sawblade is replaced by a vee or square cutterhead. In front are shown typical folded products.

Vertical saws

This type of panel saw is for vertical cutting only, and is intended for the more difficult-to-cut panels such as post and soft-formed types. The saw-unit has a fixed vertical traverse movement - behind the support frame - and the sawblade protrudes through a slot in it. On some a scoring saw is also provided. An outside full-length clamp holds the panel firmly against the support frame during the cut.

End stops on some machines can be set manually to a rule and provide an acceptable accuracy, but a digital readout is easier to use and gives more precise settings. Both types can be corrected to account for any saw-kerf variation.

Program cycle

The regular operating cycle begins with the panel being clamped. The sawblade then projects into the cut and immediately starts the downwards cutting stroke. When this is complete the sawblade retracts and returns to the rest position*, and the clamp releases. The cycle is usually started by push buttons on a central console, and the saw-unit movement is fully inter-linked with the pressure beam for safety reasons

Some machines allow for alternative cycle programs:-

For un-faced boards the cutting cycle is single downwards movement.

For faced boards the sawblade first scores the support side of the panel with a shallow cut, then projects the full depth on the return stroke to cut clean through. As the sawblade climb-cuts in both directions it gives clean, splinter-free cut on both front and rear faces.

For exceptionally thick boards some machines feature a reciprocating cut that gets progressively deeper, making several passes to completely cut through.

**NOTE: To give easy access when changing the sawblade, the rest position for the saw-unit is normally at the machine*

base. If this is the regular rest position, then the saw-unit has first to traverse to the top position (or may automatically reverse into the cut when the panel top edge is met) before the cut starts - but in either case this initial movement slightly lengthens the cutting cycle. Some machines allow the rest position of the saw unit to be changed - so that the cutting cycle always starts immediately after the saw-clamp engages.

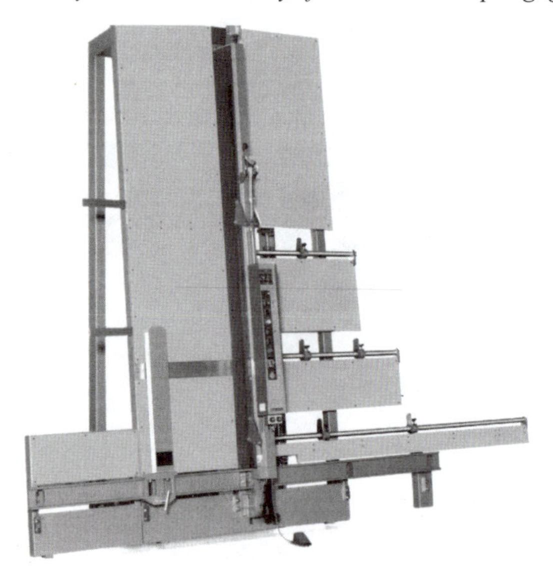

Fig. 15.16
This vertical-cutting panel saw, PRO-V, is fully automatic and has a dual-or infinitely-variable traverse speed in both directions. The machine can be programmed for a regular through-cut, for scoring prior to through-cutting, or for gradually-deepening reciprocating cuts for dealing with thick and dense materials. Several turnover length stops are provided on two or more horizontal bars. Pressure-sensitive tape on both sides of the clamp prevent the operator's hands from being trapped accidently.
Source - Hendrick

Cutting Soft and Postformed edges

Fig. 15.17
Three typical postformed edges are shown here (left) together with six soft-formed edges, (right).

An optional soft and postforming saw-unit can be fitted on some machines - separate and independent of the regular scoring saw on the main saw unit. The postforming unit is normally positioned near the fence line.

In the rest position the unit is held clear of the panel, but moves into the operating position as the primary scoring saw on the regular saw-unit begins its upwards movement. As this continues the postforming sawblade swings into the panel to cut through the moulded under-edge only, then retracts and returns to the rest position.

Meanwhile the regular saw unit reaches the panel top, reverses direction and, with the main saw engaged, traverses back down to complete the cutting cycle.

By carefully controlling the cut depth and degree of penetration of the soft and postform scoring sawblade, the under edge will be cut cleanly - regardless of its profile. It is essential that the main sawblade, scoring sawblade and postform sawblade are the same saw-kerf width - and precisely aligned. External adjustments may be provided for alignment purposes.

Fig. 15.18
This CVP vertical panel saw is specifically designed for cross-cutting post and soft-formed kitchen tops, but is also suitable for rip and cross-cutting regular panels. The separate saw-unit needed for scoring-through post and soft-formed edges can be seen in Fig. 15.19.
Source - Homag

Fig. 15.19
This shows the main saw, the regular scoring saw and the postforming scoring saw on the Homag CVP.
Source - Homag

Fig. 15.20
Vacuum lifting equipment makes loading of large and heavy panels onto a vertical panel saw- or onto a horizontal panel

saw - a safe and easy one-man operation.
Source - HSE

Undercutting beam saws

These machines, usually termed panel saws or automatic panel saws, have a travelling saw carriage that supports a main saw and a scoring saw. The carriage traverses parallel with and beneath a horizontal table - through which the sawblades project. Above the table a pressure beam holds panels securely during the cut.

When in the start position both sawblades are at one end and below the table line, and the pressure beam is raised in readiness to place a panel beneath.

Panels are initially positioned against a pusher fence at the rear. This moves them towards the sawline for a series parallel ripcuts to convert them into strips. The strips are then placed against a cross-cut fence, usually at the left-hand side, for cutting to length.

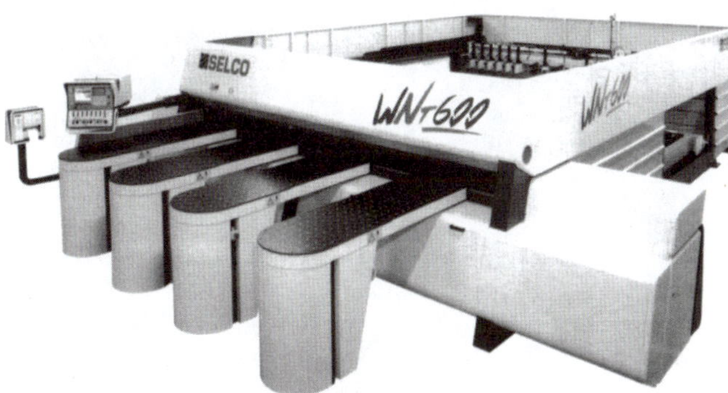

Fig. 15.21
Single line panel sizing centre WNT 600 with 145mm saw blade projection, cutting length 3800-5600mm, and with automatic panel loading from lift table
Source - Selco

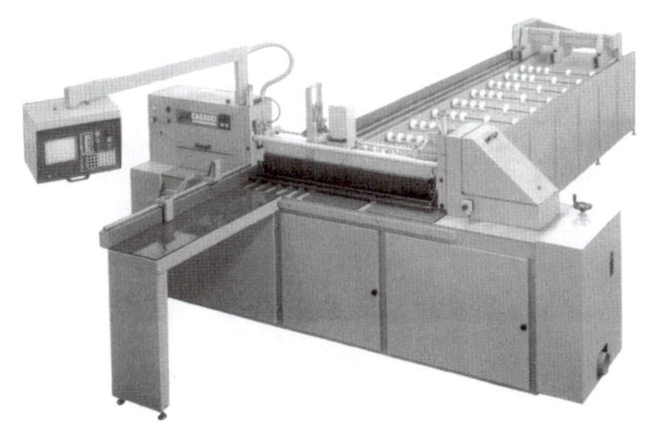

Fig. 15.22
This machine has a short traverse only and is intended for cross-cutting panel strips. It can be mechanically linked to a rip-cutting panel saw as part of an 'angle' production line, or used independently.
Source - Casedei

Most machines operate a fully automatic cycle. First the pressure beam lowers to securely clamp the panel. Both sawblades then rise into the operating position through the narrow table slot, and traverse along the length of the machine to completely the cut.

At the end of the cut the carriage stops, both sawblades are lowered to below table level and the carriage returns to the start position. As the carriage returns, the pressure beam rises for the sized parts to be removed and the panel or strips to be repositioned for a further cut, and so on.

The design and operation of most machines is much more complicated than this simple description implies. Many refinements make upmarket machines precise, efficient and surprisingly fast - yet still simple to operate. They may be computer-controlled to undertake complex cutting programs, and can be fully or partially equipped with automatic feed and take-off equipment.

Machines intended for short runs and custom work operate alone, with panels being manually moved around between the rip and cross-cut positions.

Semi-automatic stand-alone machines are fitted with panel handling equipment - to make rip and cross-cuts fast with a just single operator - even for large and heavy panels.

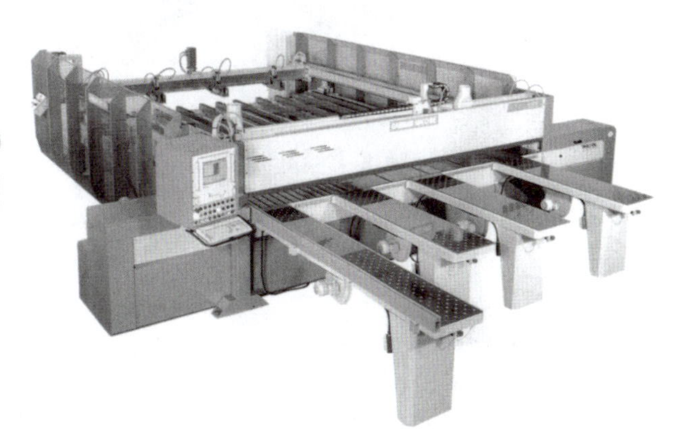

Fig. 15.23
Panel saws of the type shown here, in addition to working independently, can be modified at a later date to become part of an integrated system.
Source - Gabbiani

The real workhorses of the panel conversion industry, however, are the high production units. These consist of a long-stroke ripsaw and a short-stroke cross-cut saw, with semi or full-automatic mechanical transfer equipment linking the two together.

Whatever type of machine, though, there are many common elements:-

Traversing carriage

The carriage is a sturdy frame, supported by precision bearings running on accurate guide rails, and carrying the main sawblade and a scoring saw, plus their motors and control gear.

The cutting speed of the carriage is usually adjustable under the operators control to suit the nature of the cut and the condition and type of the sawblade.

The return speed is consistently fast to keep the work cycle times to an absolute minimum.

Some machines start with the sawblade at the cross-cut fence, which then traverses along the panel length or across the width of a strip before lowering and reversing direction. Early economy machines needed this point manually setting using an adjustable stop, but on most modern machines this is part of the operating sequence.

On other machines the saw carriage always traverses

the full length of the machine at each cutting cycle - but with a fast movement up to the panel and a slower traverse speed for the cut proper.

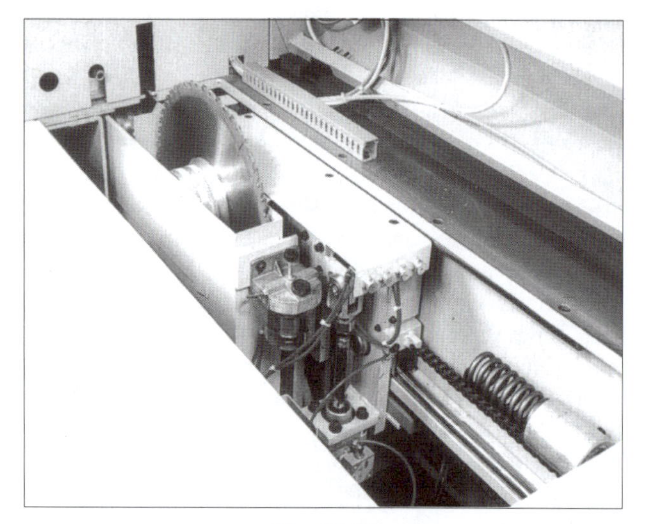

Fig. 15.24
The saw carriage on this Galaxy model is typical of its type, being sturdy and compact to ensure continuing high-quality service over many years.
Source - Gabbiani

The main saw

This has a choice of arbor speeds, which is selected according to the material being cut and the diameter of the sawblade.

The main saw diameter and power provided on most machines allow panels to be sawn individually or stacked several high - perhaps up to a maximum depth of 150mm (6in).

To give the best edge and top surface finish, the sawblade should break through the top surface of the top panel by no more than 15mm ($^5/_8$in).

Some machines have a choice of three or more fixed height settings, while others have sawblade height adjustment set automatically to the thickness of the panel or stack being sawn.

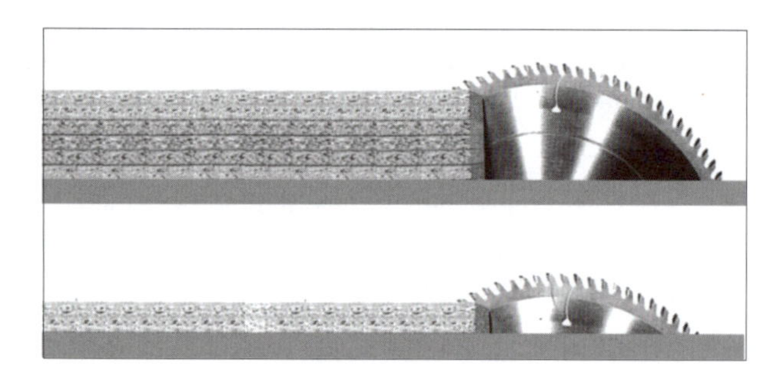

Fig. 15.25
Automatic sawblade height setting ensures that the quality of sawing is consistent regardless of panel thickness.
Source - Giben

Machines intended for custom and one-off work in the smaller workshops may have a sawblade that tilts up to 45 degrees for mitring worktops.

High production machines neither have nor require this feature. Invariably they operate with the sawblade always vertical.

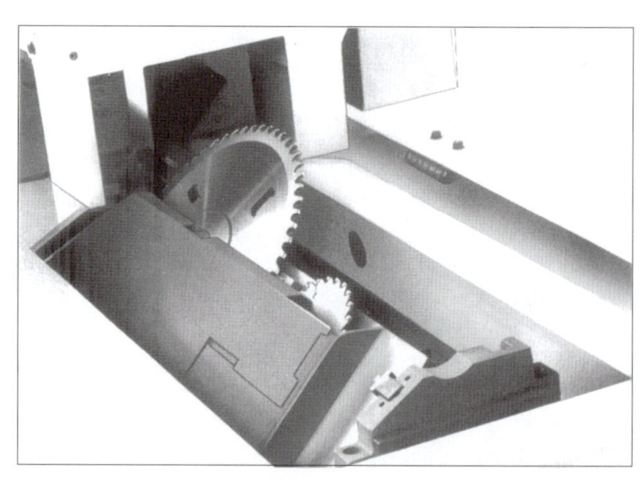

Fig. 15.26
This saw carriage has power tilt up to 45° - with electronic readout of the angle. Note the scoring saw
Source - Steton

Fig. 15.27
Horizontal and vertical adjustment can be made to this scoring saw - remotely and with digital readout on the Prismatic models - while actually in operation.
Source - Giben

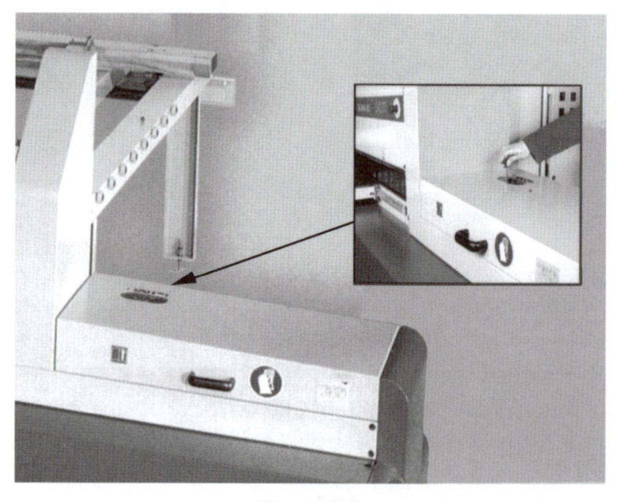

Fig. 15.28
Remote adjustment for the scoring saw height and alignment is well clear of the operating section on the Sigma range.
Source - SCM

The scoring saw

The scoring saw operates immediately in front of the main saw, and climb-cuts to avoid splintering or break-out on the underside of faced panels.

It must be carefully set to an appropriate kerf width relative to the main saw either prior to fitting or perhaps on the machine itself following a trial cut.

(On stacked panels the intermediate panels are clean-cut on the underside because each is fully supported by the panel beneath - so only the bottom panel needs scoring.)

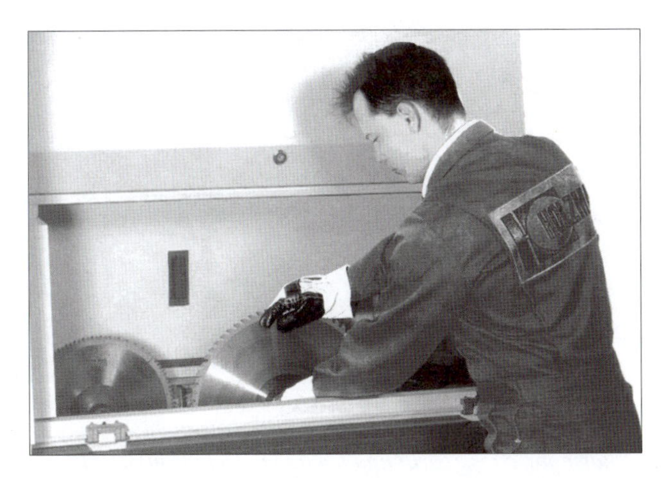

Fig. 15.29
Ready access to the sawblades is possible on all their panel saws - note the essential protective gloves.
Source - Holzma

The scoring saw must also be correctly aligned to the main saw - and set to the correct operating height - to produce quality sawing. On many machines these adjustments are manually or electronically accessible from the outside.

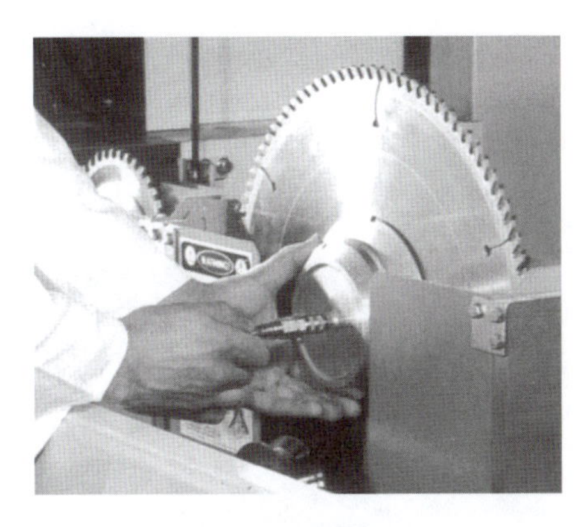

Fig. 15.30
On this Starmatic 850 a pneumatic sawblade release device allows rapid sawblade change.
Source - Giben

Both the main saw and the scoring saw may have a quick-change facility to reduce the time to change the sawblade - different systems are available from both sawblade and machine makers. Quick-change facility is more important on high-production machines than custom-production machines.

Pressure beam

The pressure beam clamps the panel under manual or automatic control - and with a parallel movement to ensure that panels are clamped evenly along their full length.

Some have a split beam which separately clamps on either side of the sawcut. When sawing very narrow off-cut strips the split clamp at that side lowers to table level. This then totally encloses the cut to make dust extraction more efficient, and improves machine guarding.

Fig. 15.31
A split pressure beam on the Prismatic range provides even pressure to the stock, better operator protection and better dust extraction.
Source - Giben

Fig. 15.32
Operator safety devices should include one or two trip bars running the length of the beam, together with protective fingers which rest on the panel top (or on the machine bed before the beam is lowered) as on this CV panel saw.

Source - Casadei

The pressure beam is a hazard, in particular when the machine operates fully automatically.

To prevent hands becoming trapped under the beam and in line of an advancing sawblade, the beam should have protective bars or a light beam. Either of these, if tripped, would instantly stop the cutting movement, lower the sawblades and raise the pressure beam.

Stand-alone machines intended for manual loading and operation by a single user may have a two-button cutting-sequence start - but a trip bar is still required for safety reasons.

The pressure beam should also be fitted with sectional fingers or flaps on both sides to rest either on the panel top, or on the table top when no panel is beneath.

This encloses the cutting zone both to give safer operation and improved dust extraction.

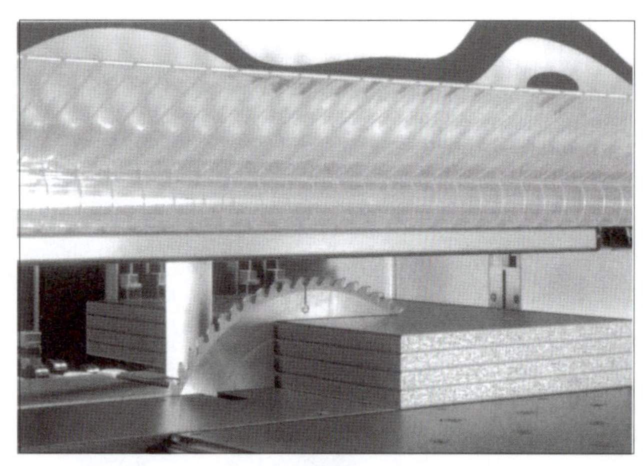

Fig. 15.33
This shows both the pressure beam and the sectional fingers raised to show the sawblade, panel and cross-cut fence. The machine would not be operated in this condition.
Source - Holzma

Pusher fence

The pusher fence is used both when rip-sawing panels to width and when cross-cutting strips to length.

Although called a fence, it is actually a traversing cross-beam carrying a series of in-line arms that contact the edges of the panels. The arms are more closely-spaced towards the cross-cut fence to give adequate end-location to all panels and strips regardless of their width.

Fig. 15.34
This general view of the centre section of their high-precision range of panel saws clearly shows the roller supports, also the pusher-fence cross-beam and several arms, each fitted with grippers.
Source - Schelling

The pusher-fence beam traverses absolutely parallel to the line of cut. The contact faces of the pusher fence arms are also both parallel to the line of cut and absolutely square to the cross-cut fence. So, whatever number of panels are stacked and handled together, they are guaranteed to be cut precisely to the same dimension.

Panels on manually-operated machines are loaded onto the infeed table, then passed under the raised beam until they are against the pusher fence. The cutting sequence is started with the fence at the furthest extent from the sawline. From this position, and between repeated cutting cycles, it advances towards the sawline in a series of pre-set steps.

The first cut can be an edge trim, and is followed by further cuts to produce strips, perhaps of different widths, until the panel is completely converted. Being under computer-control the cycle time is exceptionally fast.

The pusher fence movement has to be the strip width plus the saw-kerf width. So, after replacing sawblades, their precise details must be entered into the computer or controller to account for this and so maintain dimensional accuracy.

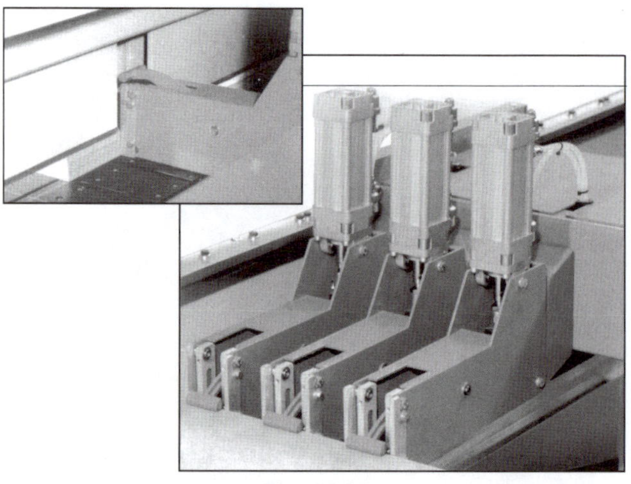

Fig. 15.35
The electronically-controlled pushers, shown on their Starmatic range, grip and keep the stack firmly in place during the full cutting cycle. The inset view shows how close the grippers can approach the sawline.
Source - Giben

Grippers in the form of a parallel scissor are often incorporated into the pusher fence arms. These grip and keep the panel *(or panel stack)* in full contact with the pusher arms during movement in either direction. They may also grip and move them from the front or rear loading position to the start position.

Some have a matching channel in the machine bed for the lower half of the gripper - so this can securely hold panels almost up to the cutting line.

Cross-cut fence

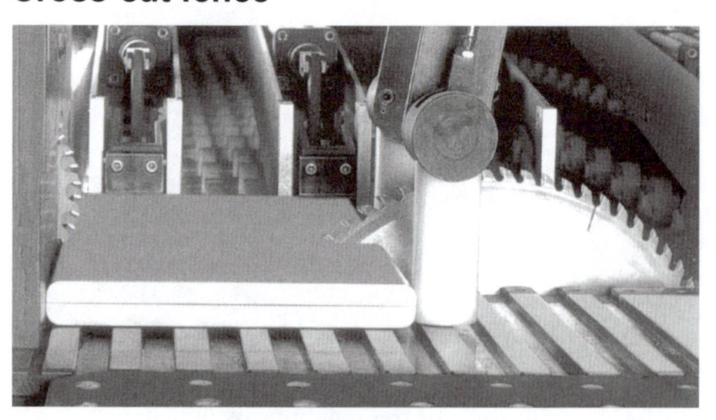

Fig. 15.36
Pressure rollers keep stock firmly against the cross-cut fence when cutting to length. Also shown is the additional scoring saw (left) for pre-scoring post and soft-formed board edges.
Source - Schelling

The cross-cut fence is normally at the left-hand side of the machine and square to the saw-unit traverse. Panels are edge-butted against this both when ripping them to width and when cross-cutting strips previously ripped to width.

Turn-over stops on some cross-cut fences can be used to gauge the cut length, or the pusher fence can position the strips under computer control.

Free-running spring-loaded rollers keep strips for cross-cutting in firm contact with the cross-cut fence during positional movement in either direction. With such positive and dependable control of this type, strips can be cross-cut edge-to edge, as single strips or stacked.

Main tables

Various types of main table are fitted. To provide easy panel movement they are provided with a low-friction finish, multiple ball bearings or rollers, or an air-cushion top.

The infeed table on manually-operated machines consists of several moveable supports - to meet differing operating modes - with space between for operator or loading access.

Both sides of the machine table proper are usually solid - except for recesses for the pusher fence grippers.

The rear table support consist of strips, between the pusher arms, provided with a low-friction finish, multiple ball bearings or rollers, etc.

Specialized equipment

Automatic scoring saw

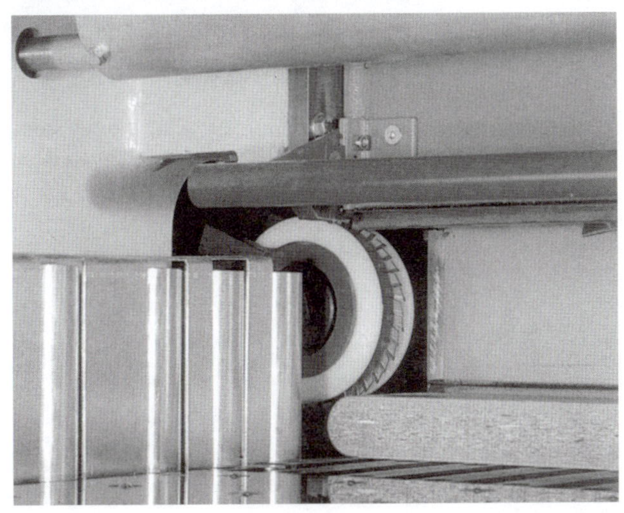

Fig. 15.37
A copy scoring attachment on their Prismatic range follows the panel edge shape to pre-score these and so guarantee chip-free cutting.
Source - Giben

An automatically-rising scoring saw is available for some machines to give a perfectly clean edge where otherwise the main saw would splinter or break out.

This can be the regular scoring saw arranged to rise as the panel edge is reached, or a separate unit at the cross-cut fence to clean-cut soft and postformed edges prior to the main saw reaching them.

Scoring saws for soft and postform edges operate in different ways according to the edge profile:-

Square edges can be pre-cut with a jump saw that rises clear of the edge, then moves in and finally downwards to complete the cut.

Topside moulded profiles are satisfactorily pre-cut using a scorer with a pendulum movement.

Rounded profiles are best pre-cut with a climb-cutting copy scorer which follows the profile.

Indexing veneered panels

Finger-type fences can be fitted to the pusher fence which correctly aligns the panels from the basic panel edge itself - regardless of the projection of the laminate or veneer on it.

Fig. 15.38
Special formatting stops can be fitted to the fence pushers to align veneered panels from the panel edges.
Source - Schelling

Mitre fences for angle cutting.

Fig. 15.39
Cutting panels at an angle on the Elcomat 320 H 'Hybrid' panel saw - it doubles as a dimension saw and a beam saw.
Source - Elcon

Angle and mitre-cutting machines

Special angle and mitre-cutting panel saws are the best choice for kitchen tops and similar. They have a swivelling infeed table that can be set to any angle required - and in either direction. *(See next page.)*

Laser light

A laser light can be provided when machines deal with panels and flitches lacking a true edge.

This projects a fine line to correspond exactly with the projected sawline, so allowing the board or panel to be precisely positioned for edging.

If necessary, flitches can then be steadied by contact with adjustable fingers against their outer edge prior to being clamped by the beam.

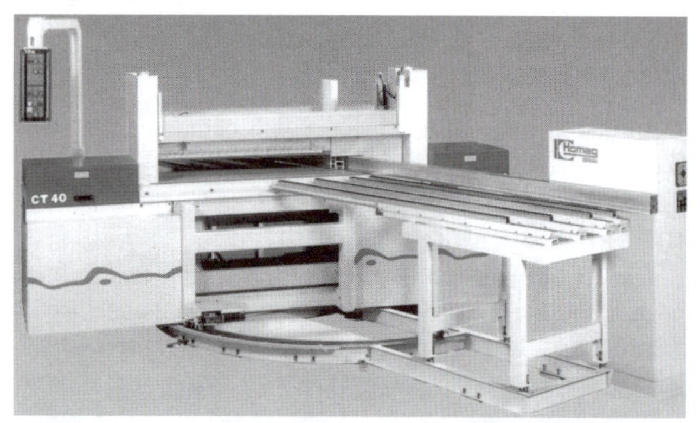

Fig. 15.40
This universal cross-cut panel saw, in addition to forming precise square cuts, can form precision angular cuts in either direction. The table has digital display of the swivel angle, and adjusts by power. The sawblade tilts, the length stop has electronic positioning, and a soft and postforming scoring saw-unit can be fitted.
Source - Homag

Fig. 15.41
The 430 H 'Hybrid' has a tilting saw and can size panels or rip and cross-cut timber. As shown here it can be used as a regular bench saw. It will also ripsaw timber at an angle, cut grooves or rebates and edge-trim waney-edge boards.
Source - Elcon

Handling thin flexible material

Flexible material such as laminates and veneers requires special handling equipment to ensure that they bed properly and square-up accurately before cutting. This handling equipment includes fingers which separate a predetermined number of thin sheets from a stack prior to grippers on the pusher fence securing them.

Stand-alone machine operation

The regular operation on stand-alone machines it to first fully ripsaw each panel into strips. These are then pushed over the infeed table as further cuts are made, allowing the strips to be removed and temporarily stacked.

The following operation is to cross-cut the strips to final size - by turning them through 90° and butting them against the cross-cut fence. They can be gauged for length using flip stops on the cross-cut fence, or the regular pusher fence could be used.

External turntables can be provided for many machines. These allow stacked strips to be quickly and easily rotated through 90° in readiness cross-cutting to length - so boards can be fully converted at a single handling and with little manual effort.

Fig. 15.42
Flexible sheets can be loaded several deep using equipment similar to this. The pack is delivered by a roller table and then raised in stages on a scissor lift as sheets are removed. The pusher fence has extra travel to traverse over the roller table and grip the sheets in readiness to move them into the machine for conversion.
Source - Gabbiani

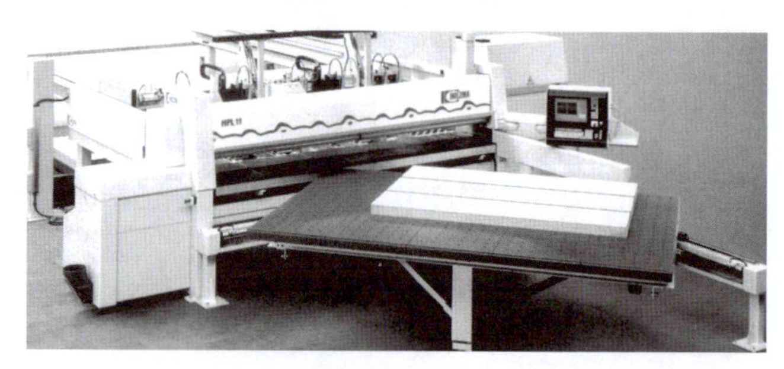

Fig. 15.43
On a stand-alone machines an automatic turning device can be used to rotate strips automatically for cross-cutting.
Holzma

Ripping and cross-cutting alternately

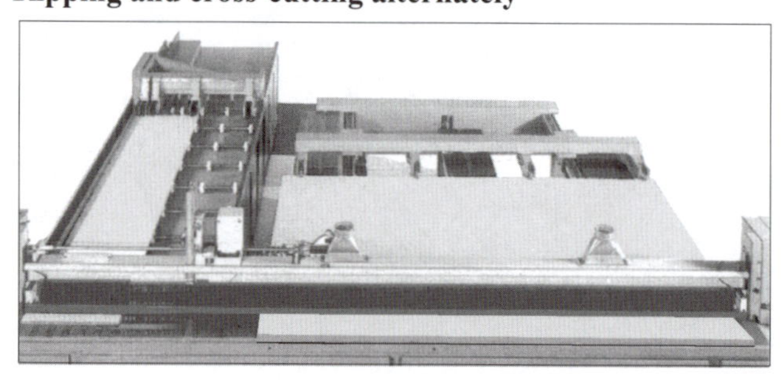

Fig. 15.44
This patented panel saw in the SV range has two independently-driven pusher fences controlled by a single electronic controller. This allows simultaneous ripping and cross-cutting at each pass of the sawblade.
Source - Casadei

It may be viable to use the pusher fence to ripsaw strips one at a time and then, with the main panel moved clear, use the cross-cut fence and stops to finally cross-cut each strip in turn. In this way strip stacking is unnecessary and finished parts are produced on a continuous basis.

However, this is only practical if the uncut panel width, plus the strip width, is less than the sawblade traverse. Also the panel must be positioned (when cross-cutting) remote from the cross-cut fence - and the machine must be operated under manual control only.

At least one machine is designed to rip and cross-cut at the same time. The pusher fence is split into a short section closest to the regular cross-cut fence for cross-cutting, and a longer section for rip cutting. The two sections operate totally independently of one another, to allow both rip and cross-cuts to be made at a single pass of the sawblade.

Automatic Panel Handling

The travelling-head panel saw is ideal for incorporating into an integrated, fully automatic panel handling system.

Automatic panel saws can be fed and off-loaded in different ways, for example, via a roller lift table from the rear or the side of the machine.

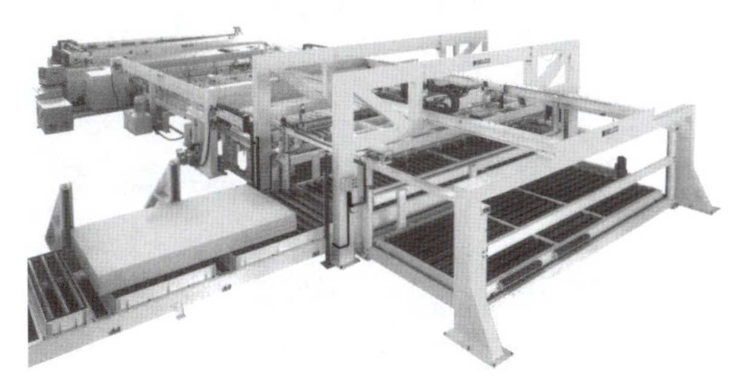

Fig. 15.45
Typical specialized panel handling system from Selco. Systems can be provided for a variety of production requirements, including vacuum loading and trim removal.
Source - Selco

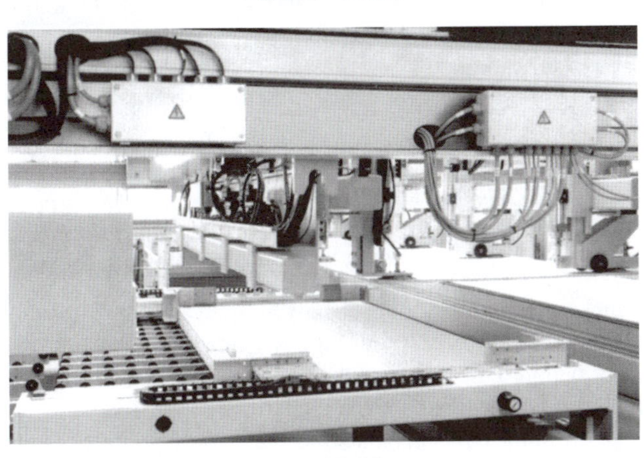

Fig. 15.46
A stacking table for separating and loading single boards.
Source - Holzma

Fig. 15.47
For thin, hard and flexible sheets a vacuum loader is used.
Source - Schelling

This places individual panels or stacks level with the rear table, from where it is passed under the pusher fence by handling equipment, or the pusher fence itself is used to grip and load panels direct from the stack.

Equipment can include roller-feed lift tables at both outfeed and infeed, automatic separators to feed a preset number of boards from an infeed table, and vacuum transfer equipment for handling delicate and thin panels.

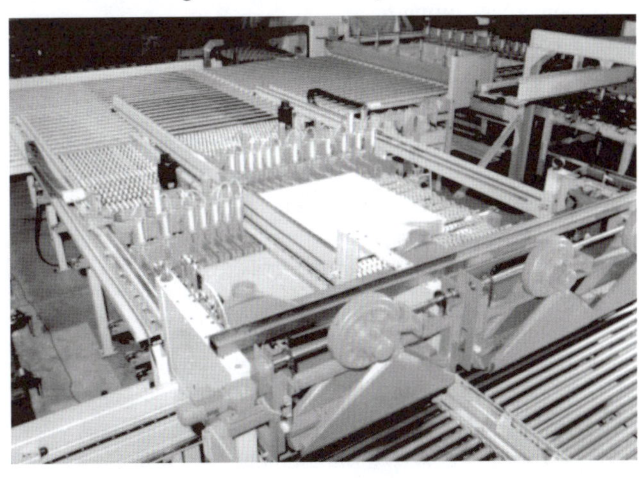

Fig. 15.48
Raw and finished panels are moved around on a series of conveyors of various sorts as shown on this angular saw installation.
Source - Anthon

Fig. 15.49
After cutting, panels can be sorted and stacked separately.
Source - Anthon

Feed systems can also transfer finished panels to an automatic sorting and stacking area. This is an essential process when using an optimization program to convert boards into a mixed range of sizes on a volume-production basis.

Many installations are purpose-designed to meet specific customer requirements, and often unusual and novel equipment can be offered by leading makers to customize complete installations.

Leading panel saw makers have considerable experience in this field, and can offer sound advice on this subject.

Full automation can be applied to an upmarket stand-alone machine. High-production plants, though, are usually based on a single machine for ripping only, and a second machine, for cross-cutting only, linked by transfer equipment and totally under computer control. These installation are often called angular saws.

Cut optimization

Modern and upmarket machines are capable of converting boards into a complex pattern of mixed sizes that cannot be reproduced by simply ripsawing a complete panel into strips and then cross-cutting them.

The reason for this complexity is to achieve true optimization - conversion of panels into the required sizes with the least amount of waste - combined with the greatest economy in machine time and operating costs.

Cut panel sizes, for example for kitchen furniture manufacture, vary considerably in size and numbers. If through rip cuts were to be used exclusively in converting all raw material, there could be considerable and costly wastage.

Deciding the various cuts needed for the mix of sizes common in cabinet manufacture requires careful calculations and planning to achieve efficient panel conversion.

This is something that could well baffle the average human brain. It has to take into account the varying numbers of each size required, keep a running total of finished and part-finished panels, work out and apply changing conversion patterns as numbers are met, while at the same time reducing to an absolute minimum the time factor, running cost, material cost and the amount of waste produced.

Optimization programs

Fortunately, this is something that modern technology can undertake in its stride. Most leading makers offer some form of optimization software that takes in account all these factors. It will then display suitable cutting programs as a series of screens showing the proposed cuts and cutting sequences.

The operator can accept or modify the cutting program*, and once this is approved the plant can then be started to operate fully automatically until all the parts required are produced, or the raw stock runs out.

**Changes can be made to the pattern where special circumstances demand this, but generally the computer is more efficient in doing this than the human brain. When stock runs out the machine remembers the balance numbers needed for when the supply is resumed.*

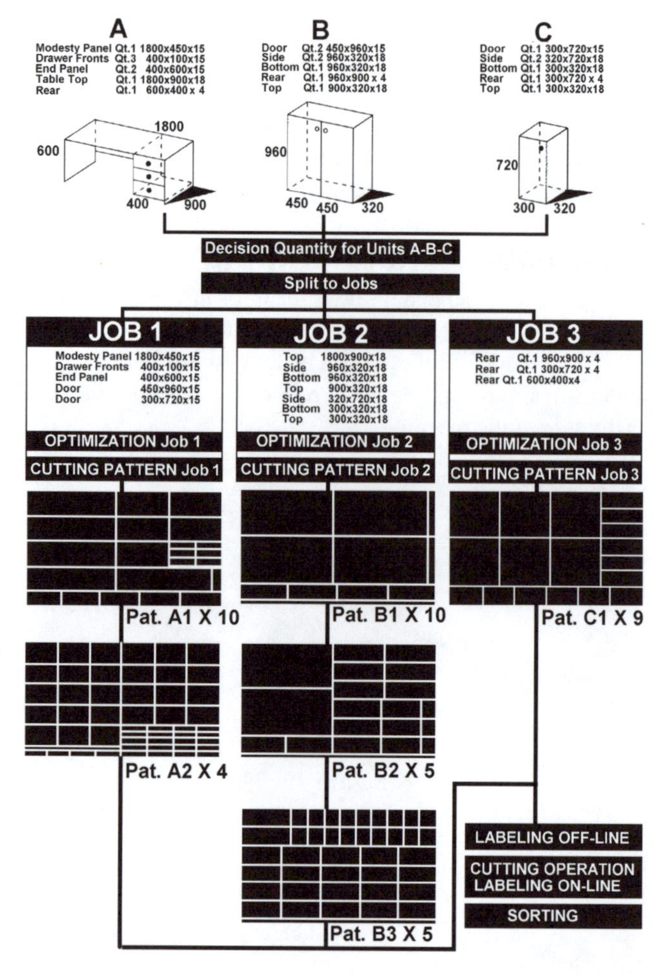

Fig. 15.50

This shows a typical optimization flow chart for producing the desk and cupboards detailed at the top. Different materials are used for doors, tops, sides, bottoms and rear panel., These are finally condensed into three combined cutting lists, each with different panel layouts.

Source - Giben

Panel conversion patterns displayed on the computer screen also include the date, order number, etc., and other relevant information. This information can be retained in the machines memory for recall whenever needed.

In some cases the machine can be on-line linked to the main office computer for batch processing and work progress, possibly also for label printing to identify parts and their pro-

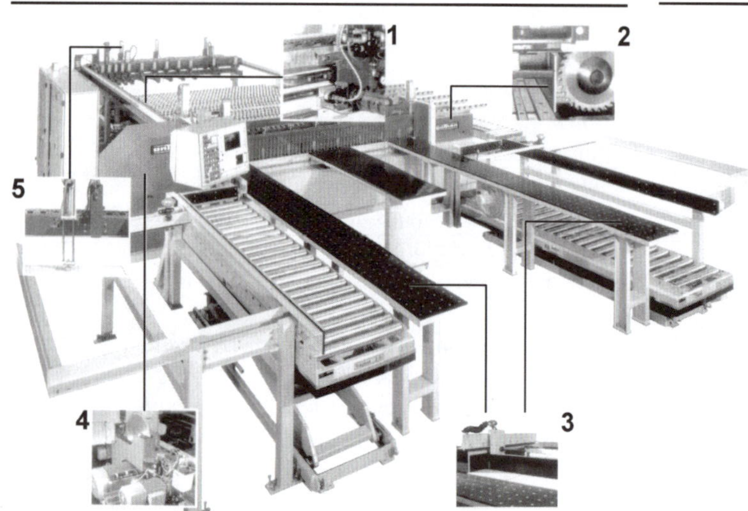

Fig. 15.51

Single panel saws can be equipped with mechanical handling equipment, such as the modular system shown here. This makes possible one-man operation. In this installation panels are fed from the rear and can be stacked at either side, using semi-automatic stacking boxes.

The numbered details are:-

1. *Servomotor for the pusher fence.*
2. *Vertical edge-scoring saw.*
3. *Semi-automatic stacking boxes.*
4. *Main and scoring saw unit.*
5. *Automatic stack-height measuring device.*

Source - Anthon

cesses. In other cases label printing can be included on the machine itself.

If producing only those parts immediately required for work in hand, the conversion factor from full-size boards may prove unduly wasteful. To make for greater efficiency in these circumstances the machine can also be programmed to cut stock strips or finished parts for later use - when these improve panel conversion efficiency.

It will then produce 'incomplete and finished-stock' labels to identify these parts until required - and keep track of their numbers in the computer's memory to discount them from subsequent production runs.

For true optimization so-called head cuts may be needed - in addition to the regular rip and cross-cuts.

Head cuts are cuts at right angles to the rip cuts, which are undertaken between the ripsawing sequences on whole or partially-converted panels.

Head-cutting requires newly-loaded panels, or panels which have been partially converted into strips, to be rotated through 90° for one or more head cuts, then rotated back for the further rip cuts - a process which may be repeated more than once in some cutting programs.

Turning panels during the conversion process is no problem with upmarket panel saws as a turntable can be incorporated within the handling system.

Panel saws, both the sliding-table type and those described in this Chapter are highly sophisticated machines capable of fast and efficient panel conversion.

They are also under continuing development, with companies continuously bringing out even newer and more efficient machines. Contact the makers for updating on current developments.

Fig. 15.53
This WNA combination unit is a compact, fully automatic angular panel sizing centre with three independent cross-cut stations and patented Shuttle System.
Source - Selco

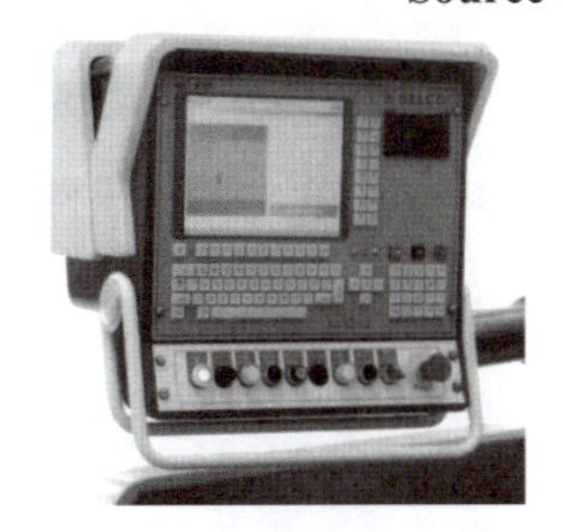

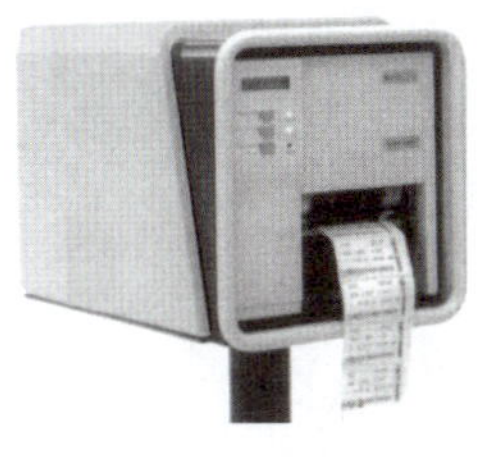

Fig. 15.54
Selco panel-sizing installations have powerful numeric control with user-friendly window-based screen displays. Software to facilitate the optimization of the cutting patterns, and other information, can be interchanged between the machine and the controlling mainframe The printer shown below produces bar-code labels to identify finished stock as it is produced.
Source - Selco

Stand-alone Systems

Optimization systems are not confined to the large panel-cutting systems. For example, Linkwood offer software systems that not only nest panels for maximum optimization and CNC programmes for sawing and drilling, but also produce complete planning systems in a modular and proven form.

These assist in product design and manufacture and produce cutting lists, bills of lading - in addition to quotations, invoices and delivery notes.

Fig. 15.52
The equipment shown below is a typical Holzma angular panel cutting set-up. Panels are brought in via a roller table, then onto a roller lift table at the left. Individual panels or panel stacks are transferred to the rip-cutting saw, centre, for conversion into strips. These are then fed to the cross-cutting panel saw, right.
To the far right is the control station and label printer. Finished panels are collected and stacked from the outfeed table. Panels can be turned on the ripsaw infeed table, when required, in readiness for any head cuts.
Source - Holzma

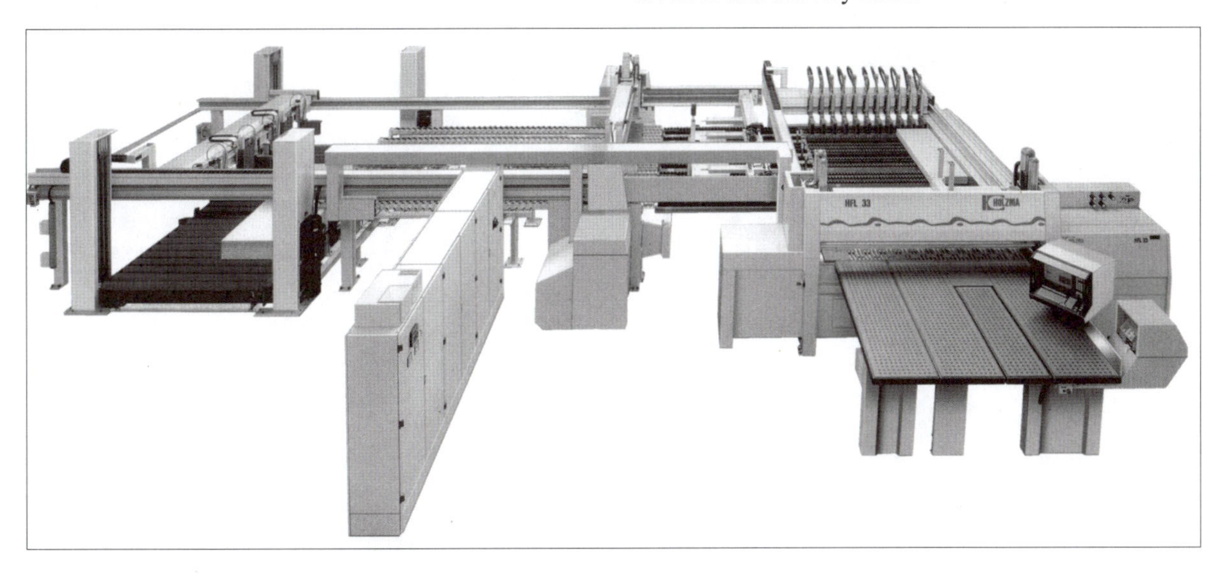

Fig. 15.55
This patented turning station on the WNA angular panel sizing centre shown on Fig. 15.53 rotates the panels for head cuts without any friction with the machine bed.
Source - Selco

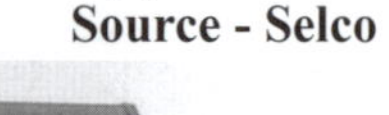

Fig. 15.56
At the delivery end of the Holzma angular panel cutting installation shown in Fig 15.52, the labels produced on the printer (right, with an enlarged example shown at the rear) are attached to finished panels to identify them for further processing and stocking.
Source - Holzma

Fig. 15.57
The illustrations directly above and below are all of the overhead type, high-production panel line shown in full below, centre.
The above illustration shows the two width-adjustable beam-mounted saw-units which traverse with the beam the full length of the panel in cutting it to width.

Fig. 15.58
The illustration below, centre, shows the complete installation, with the infeed section to the left, the panel saw centre, and the off-loading equipment to the right (fitted with suction cups for handling multiple parts).

Fig. 15.59
The illustration bottom right shows the cross-cut saw-unit which traverses along the beam. This also rotates to cut the panel to any angle required - without resetting the panel or needing any additional handling.
This unit, called a portal saw, is under two-axis computer control which allows it to square-cut across the panel, to cut along the length of the panel, or to cut diagonally across the panel at any angle.
In addition to cutting panels, the portal saw is also suitable for cutting, for example, roof truss members to length and to single or double angles.
Source - Anthon (Figs. 15.57, 15.58 & 15.59

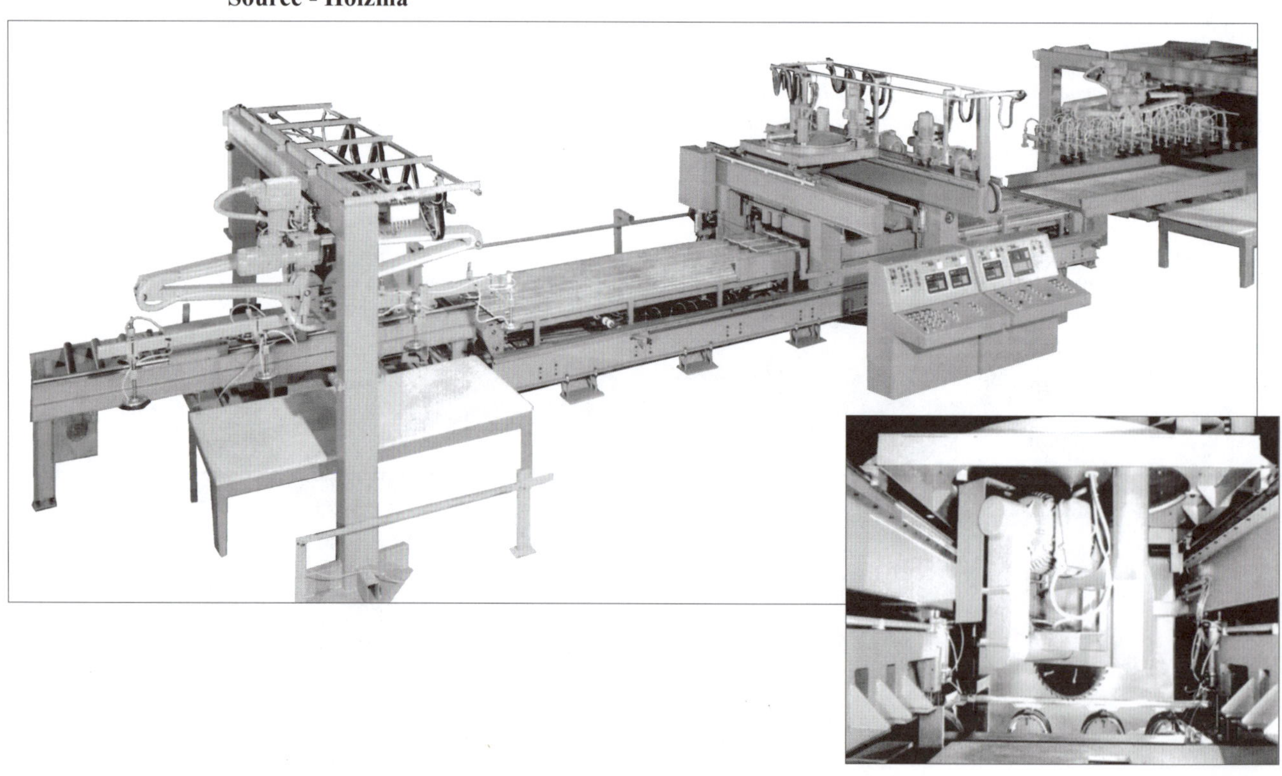

Chapter 16
Primary Conversion & Saw Data

Log breakdown

Frame or gang saws were once the primary means of log breakdown. Basically they are a development of the vertical pit saw, and consist of one or several straight saws stretched inside a frame. The frame is oscillated vertically via a crank, and the log is fed actually through the frame to cut it into two barkside slabs and several flitches at a single pass.

Usually the log is supported and clamped by travelling carriages at both ends, assisted perhaps by powered horizontal rollers.

One alternative type, a single-saw cantilevered type, allowed the log to be fastened to a carriage travelling alongside which gave a consistent and better support to the log - and a better cut. Machines were built with a single saw alone, but on others the single sawblade was cantilevered from a regular saw frame. Horizontal frame saws with a carriage travelling beneath the frame were also used for the same reason.

Fig. 16.1
The English circular logmill shown here was typical of this class of machine around the early 1900's.
Source - Robinson/Wadkin

In the American colonies there were huge and seemingly endless forest that were both useful for the timber they contained - and a hindrance to expansion. In fact, the attitude of the early Americans to these forests was reflected in their term lumber - meaning useless articles that hamper or obstruct. To clear land for cultivation and utilize this lumber, a fast means of log breakdown became an essential requirement in those early days.

The original wind, animal and water-powered frame saws were simply too slow and cumbersome, partly because there was insufficient power and technology then to drive the frames fast enough, and partly because they cut only on their downstroke.

Circular saws finally provided the means of continuous cutting around 1777 - and vastly increased the average feed speed.

When the first successful crucible-cast steel circular sawblades began to be imported from England to the Americas, their use in circular mills and rack benches quickly spread throughout both nations. They remained the regular method of log breakdown for many years.

The rack benches were so widely used that they passed into folklore. Even as a child I knew how these machines worked from the cliff-hanger serials favoured by the film-makers. They often depicted the heroine strapped to the bench in line with an advancing and wicked-looking sawblade while the hero and villain fought over the feed lever. It was years before I realized the real purpose of these machines.

Circular Mills and Rack Benches

Large logs were mainly broken down on twin-saw circular logmills, where the log was dogged *(securely fastened)* to a travelling carriage running alongside the sawblades.

Each log was first turned to get it into correct alignment, snugged against several adjustable head-blocks, and then securely gripped top and bottom by chisel-edged dogs mounted on them. By securing in this way the log and head-blocks could together be moved across the carriage - towards or from the saw-line - completely under the operator's control. By so doing the log was moved into a new position for following cuts and shifted clear of the sawblades for the return movements. Often the log was turned through 90 degrees after one or more cuts to give a better conversion factor or a more stable seating.

Smaller logs were converted on a single-saw logmill, or dogged instead on rack bench with a split travelling table. This straddled the sawblade and gave support to the log on both sides of the cut, perhaps aided by wooden wedges. The log was shifted across the table manually between cuts and re-wedged. For more stability a centre cut was made first, then the two halves were re-sawn individually, sawn-face down.

Originally logmills and rack benches used plate saws, but these mostly were replaced by insert-tooth saws from about 1860 - see Chapter 8. The latter was an American invention - the John Wayne of circular saws, rough, tough and ready for anything - but they have what is now considered a unduly wasteful saw-kerf width. When narrow-kerf bandsaws came a practical proposition from the late 1800's, they quickly replaced circular saws in both logmills and rack benches.

Band Mills and Band-Rack Benches

Vertical bandmills and rack benches are a direct descendant of first circular mills and rack benches. They were, and still are, used primarily for fast-feed through and through conversion of softwood logs - with turning of cants during conversion for certain cuts an established and regular practice.

Horizontal bandmills, originally converted from horizontal frame saws, are also used and these, with a carriage that traverses beneath the bandsaw, give better support to the log and ensure more precise cuts. These saws operate more slowly than vertical mills - because the cut flitch is best removed before a second pass is made - and so are mainly used to convert hardwood logs where quality matters more.

Following the change-over to bandsaws for log conversion, circular saws were relegated to relatively shallow cuts on very small logs, and for edging flitches where small diameter and relatively narrow-kerf sawblades could be used. With new and improved materials and techniques, however, the position has drastically changed.

Fig. 16.2
This SE gangsaw is a modern, fast-feed frame saw capable of converting logs and cants up to 760mm wide by 710mm deep (30 x 28in) at speeds of up to 24 M/min (78ft/min). The multiple saws are in two separate banks, and their spacing can be adjusted while the machine is running.
Source - Linck.

Modern, high-speed gangsaws, for example, are still used today for through and through log conversion in highly efficient production lines. In fact all the conventional means of log conversion, gang, circular and bandsaws are still used, each with their specific advantages and application.

Fig. 16.3
This modern log conversion installation uses cross chains to feed this high-capacity bandmill. The operation is controlled by the operator sitting alongside.
Source Primulti

Today insert-tooth and wide-kerf carbide circular saws are still used in the conventional way in some primary conversion mills. Their main use is to convert small round logs in low-cost installations, or as portable machines where their convenience and easy maintenance more than offsets their higher saw kerf wastage.

They have also been combined with vertical log bandmills *(with an extended carriage traverse)* in the form of efficient twin-arbor gangs of circular saws operating in front of the bandsaw. These bandmills then produce a barkside slab and several flitches at a single pass.

Another use is with a horizontal bandsaw where circular saws are fitted on a horizontal arbor to simultaneously edge and perhaps multi-saw the emerging flitches.

Fig. 16.4
This portable circular sawmill converts logs up to 1M22 (48in) diameter using a petrol-driven saw of 585mm (23in) diameter. The circular saw first makes a horizontal cut, then is flipped through 90º to make a vertical cut on the return stroke, cutting first one and then the opposite side of the log, finally reducing the mid-section in stages.
Source - Baker.

Fig. 16.5
This Tri-Scragg mill has a spiked drag feed chain for end to end feeding of scragg blocks up to 400mm (16in.) diameter and 19M (60ft). long. Two moveable 1M22 (48in.) diameter slabber circular saws, together with a horizontal bandsaw, produce barkside slabs and three-sided cants.
Source - Baker

Modern Circular Saw Machinery for Log Breakdown

Many conditions have conspired to make possible the present re-emergence of the circular saw for small and medium-diameter log breakdown.

These include:- improvement in circular saw design and manufacture; the introduction of advanced cutting technology in circular saw machinery for log breakdown; a steady decline in the diameter of logs now being harvested, and green conservation issues.

The main problem preventing the wider use of circular saws in earlier years was their excessive saw kerf width. This was because sawblades used singly needed to be large in diameter and this, in turn, needed a thick body and excessive side clearance of the tips to keep the sawblade stable. This particular problem was solved by using two-arbor machines with thin-kerf, paired and overlapping sawblades to share the cut.

Both sawblades each cut only part-way through the timber, so they can be smaller in diameter and consequently have a significantly reduced kerf width. Better quality saw bodies allow the use of thinner bodies than normal, and improvements in tips and in the way they are ground make these so efficient that they require less side projection. Strobe or other body slots now prevent burn-up on deep cuts.

Fig. 16.6
This close-up shows the twin-arbor circular saw section of a bandmill/multi-saw combination. The multi-saws cut a barkside slab and several flitches immediately prior to the final bandsawn cut. A round log is fully processed at four passes. See also Fig. 4.31.
Source -Primultini

Fig. 16.7
An extended carriage traverse movement can be provided for this SIB 4-head-block bandmill carriage to incorporate the circular gang saw unit shown above.
Source -Primultini

Circular saws are now regularly used in twin-arbor, multi-saw gangs for small and medium log breakdown. Normally they complete the final cuts, either as the cutting agent of a stand-alone machine, or as part of a log profiling line linked by transfer and feed equipment to preceding chippers and profilers.

Log Profiling Lines

Log profiling lines consist either of self-contained combination machines *(or several independent units such as chippers, profilers, multi gang-saws and possibly stationary or flying cross-cut saws)* all connected by infeed and outfeed units, log turners and various forms of transfer equipment. Such lines are capable of fully converting round logs into edged boards and squares without further processing.

Low-cost profiling lines have roundabout conveyors that allow partially processed logs *(cants)* to be returned to the infeed for further processing. High-capacity lines are based on a through-feed system, with duplicated machines and turners, where the log is fully converted at a single pass through.

Fig. 16.8
Above; general view of the outfeed section of a log profiling line. Note the operator, top left, controlling whole process.
Below, a view of the infeed to a small roundabout log line with circular saws. The primary feed of round logs is from the left, with two-sided cants returned from the right for a second pass.
Source - Linck

Chippers and Profilers

Chippers A chipper is a large-diameter disc, flat in the centre but formed as a flat cone towards the edges, around which it is fitted various types of knives. For a cleaner cut a flush-fitting circular saw can be fitted on some to the centre flat.

They are used regularly in opposing pairs to form two parallel flat faces on the log - to convert it into a two-sided cant - and simultaneously reduce the waste to chips that can easily be transported to storage bins*.

After the initial pass through, the log can be turned through 90 degrees to complete two more faces at right angles and so form a four-sided cant. This is done either by returning the two-sided cant to the chipper section of the machine, or by continuing to feed it forward, via a log turner, into a second, in-line chipper unit.

**With traditional log bandsaws the barkside slabs (the outer cut with one sawn face and the remainder bark) need separate handling equipment and hoggers to reduce them to saleable waste. All this is expensive and inconvenient - and the handling system often becomes clogged. By using chippers no such additional equipment is needed.*

Chippers do not necessarily completely square the log - this can be wasteful. Additional boards can be usually be extracted from part-waney squares by a following process which includes profilers and circular saws, see Fig. 16.12. This reclaims narrow boards that would otherwise simply be hogged to chips on chippers set to produce wane-free squares.

Fig. 16.9
Chippers (left) cleanly face the log and produce chips of suitable quality for the pulp and paper industry. These may be followed by profile units (right) which pre-edge the boards subsequently produced by the sawblades.
Source -Linck

Chipper knives can produce different qualities of chips.

As there are huge quantities of chips produced by log processing lines, the chips must be in a suitable form for whatever market exists for them. So the correct choice of knives has a significant effect on their value for resale - normally to pulp and paper plants.

Profiling units are similar to chippers - but with a square form. They can follow the chippers to form rebate-like cuts in the log corners so that the boards finally produced by the circular saws are already edged. With this system the additional process of edging is not needed.

Profiling units may be doubled or tripled-up to form staggered rebates - to produce pre-edged boards of different widths alongside one another.

Router units. These, in the form of a wide roughing cutterhead, precede the multiple circular saw units to remove projecting parts of the log that would otherwise foul the saw arbors. Primarily they reduce butt end flare where this is in line with the sawblade arbor.

Circular Saw Units.

The circular saw units following the chippers and profilers often fit a gang of saws, either on a single saw arbor, or on twin-arbor with matching saws.

The sawblades could be mounted on an extended one-piece arbor in both cases. For primary conversion, though, width-variation of the centre section is often needed when converting logs of different diameter. To allow for this the sawblades are regularly mounted on two in-line arbors - each of which is separately driven and individually adjustable towards and from the centreline of the log.

With machines having 'single-arbor' saw mounting there are in fact two in-line saw arbors. On 'twin-arbor' saws this arrangement is duplicated above and below the log as two pairs of in-line arbors. Each top saw arbor moves simultaneously with its corresponding lower arbor to keep the matching sawblades precisely in alignment.

One regular arrangement is for a single, outer sawblade on one of the two in-line arbors to split the centre square *(the centre heartwood section is of lower quality and is usually suitable only for construction timber)*. The gang of sawblades on either side simultaneously produces side boards of a higher quality from the outer sections. An alternative arrangement with larger logs is two centre sawblades to produce three sections from the heartwood, as shown below, or even several centre sawblades - see later diagrams.

Fig. 16.10
Twin-arbor matching sawblades in the saw unit of a log production line. The outer gang-saws have fixed spacings, but the in-line saw arbors themselves can be moved closer together or further apart - with both top and bottom arbors moving in step with one another. The space between the inner sawblades and the gang-sawblades can also be varied in this case, See Fig. 16.14. All these settings are controlled remotely by the operator to suit the size of log being processed.
Source - Linck

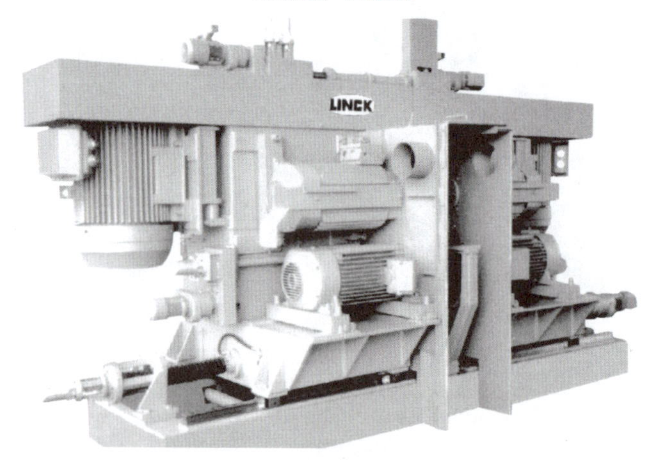

Fig. 16.11
This VPSN unit, one section of a log profiling line, combines both profilers and multi-saws. It produces centre lumber and sideboards from four-sided, waney-edge cants.
Source - Linck

Fig. 16.12
Stages in processing a round log, left to right. Chipping to produce a two-sided cant, turning and chipping to produce a four-sided waney-edge cant, profiling and double-sawing, turning, profiling, then multi-sawing.

The board thicknesses cut between the gangsaws is varied by switching their spacers, and remains fixed until the sawblades are re-spaced.

Size-changes between different log diameters are then made simply by moving the in-line gangs of saws apart or together. The width of gang-sawn boards and the size of the centre square both increase with log diameter - but board thickness remains constant.

In most cases the setting of the gang-saws is kept in step with those of the chippers. However there is an option with larger logs of increasing the outer board thickness to the next size up - if this produces a better yield.

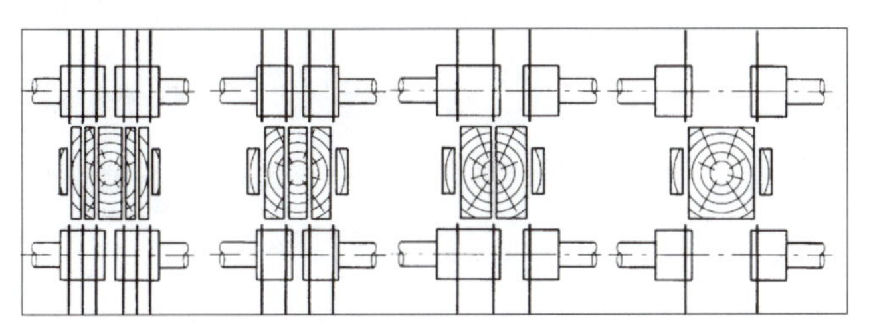

Fig. 16.13
The numbers and spacing of sawblades can be varied as needed. These diagrams show typical breakdown patterns for profiled cants.
Source - Linck

Variable-spacing multi-saws

For larger logs, and to give the facility to vary the centre widths, some machines also allow the setting of the outer sawblades on each in-line arbor *(the inner centre-splitting sawblades)* to be changed remotely.

Some machines have a splined arbor and two or more splined-centre sawblades which, without conventional collars, move freely on the arbor. They are positioned individually and held steady by guides at each side of the sawblade.

To change the sawn widths *(something that is easily possible between following logs or cants)* the guides are shifted manually or via electronic control to next pre-programmed position - and the sawblades move with them.

Another type uses telescopic arbors. With this arrangement each saw arbor carries a gang of saws, plus one or two inner telescopic arbors, each with a single, end-mounted sawblade.

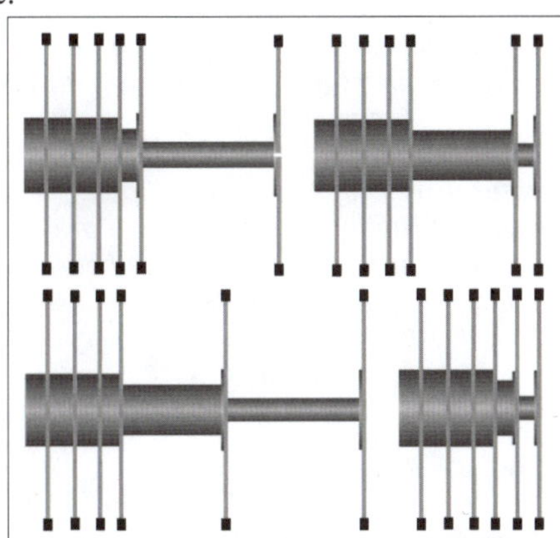

Fig. 16.14
This shows one type of remotely-adjustable, telescopic saw arbor.
The four inner saws have fixed spacings, but the two outer saws are independently adjustable to the extreme positions shown - or any setting between. Only the left-hand section of an in-line pair is shown, the right-hand section is identical but of the opposite-hand. This general arrangement is suitable for both 'single-arbor' and 'twin-arbor' gang saws.
Source - Linck

The gang saws are mounted between collars at pre-set spacings, while the outer gang saws and the adjustable saws are each fastened to an inner collar secured to its arbor. External hydraulic cylinders independently move both the gang of saws and the outer saws to the settings required.

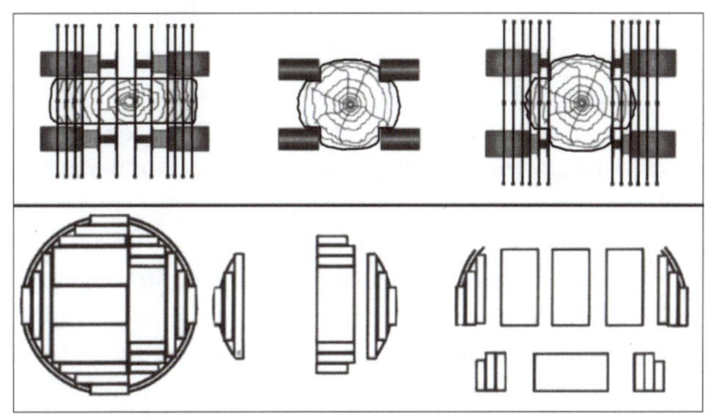

Fig. 16.15
Log conversion on twin-arbor multi-saw units equipped with router cutters (above, centre). These reduce butt flare where this is in line with the saw arbors. Typical cuts of a round log are shown below (left and centre). The final cuts in reducing the centre two-sided cants after turning are shown below (right). **Source - Linck**

Machine control

The spacing of the chippers, the setting of the profilers and the spacing and setting of the sawblades has to be changed according to the size and quality of the logs being sawn. This is controlled from a remote unit, usually an air-conditioned cabin overlooking the whole processes.

Although modern machine control allows fast changes, a more efficient way of working is to pre-select logs into batches of similar size and quality, then process these almost in a continuous stream with minor changes only needed between logs.

Optimizing Log lines

Fig. 16.16
From his cabin the operator of this Linck log processing line has a clear view of the entire operation, and can fully control it. Scanners view each log individually, then automatically turn and align it to give the optimum yield - if necessary to include curve-sawing capability. Screens show various sections of the plant, the scanner data and the proposed machine settings - which he can accept or override. See also Figs. 2.7, 2.8, 3.45 & 3.45
Source - Linck

In the most advanced lines of this type both the processing sequence and cutting unit settings are optimized by a computer *(or other form of control)* to obtain the maximum possible yield for whatever diameter and condition of log is being converted.

They incorporate a log scanner en route to the infeed section, which scans for automatic log rotation and alignment to give the best conversion factor and side board optimization. This takes into account the size-mix and dimensions of boards and squares required*. This information is displayed on a screen to the line operator who can either accept it or modify the settings *(if he judges this to be necessary)*.

**The controller or computer holds specific cutting programs in memory to meet these criteria. The operator can choose any suitable program from memory - or modify an existing program to suit a specific requirement. This permits rapid change-over and precise repeat settings.*

Control of this type facilitates rapid and informed decision-making, permits high feed speeds and produces high-quality sawn lumber with low production costs *(due to lower labour requirements, tools and energy)*. Put another way, it guarantees maximum yield at the lowest practical cost.

Fig. 16.17

The output of a saw unit converting round logs (above), and two-side cants (below). Splitters separate the centre and outer boards as they require different further processing. The infeed section has a log orientation and centering unit. This can be a manual operation, but on fully automatic lines the logs pass an electronic scanner connected to a processor - which then commands the equipment to automatically turn, centre and align the log to achieve the best possible yield.

Source - Linck

Fig. 16.18

This thin kerf, twin-saw headrig is combined with a chipping canter to process smaller logs requiring 1 or 2 side cuts. The main sawblades shown above are 840mm (33in) diameter and, in combination with the top sawblades, cut up to 430mm (17in) diameter logs at feed speeds of up to 140M/min (450ft/min). As a quad machine the main sawblades are 1M20mm (47in) diameter to cut logs up to 610mm (24in) diameter. The machine then includes routers or reducers to deal with flare butts. The main sawblades have a variable rim speed in both cases.

Source - Sawquip

Fig. 16.19

Shown below are the main units of an optimising production line for processing two-sided cants. The feed direction is left to right, first through the cant-canter, then onto the multi-edger. The infeed section, shown in Fig. 16.22, includes an optimiser - a 2 or 3 axis automatic scanning station - which determines the form and contour of each cant as it passes towards the canter infeed. With this information the computer pre-sets the infeed rolls of both the cant chippers and the multisaws to give the optimum recovery.

The line is also capable of curved and taper sawing. In dealing with tapered logs the cant chipping heads can partially open halfway into the cant - by a pre-determined board thickness - to recover short side boards via the multi-saw edger. The infeed can also off-set the feed to the cant canter for optimum recovery.

Source - Sawquip.

Logmill Units.

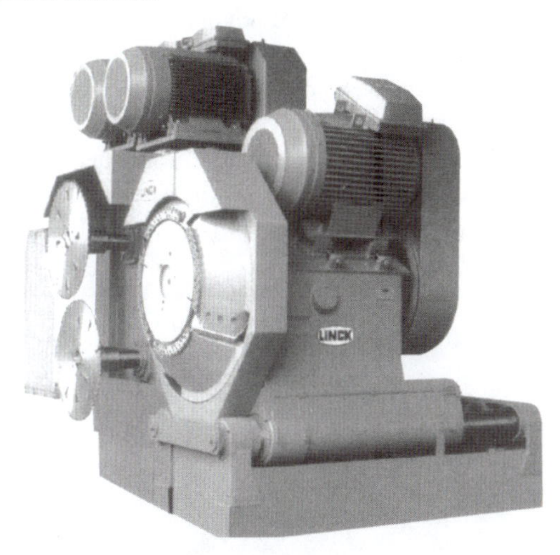

Fig. 16.20
This round log processing aggregate RBA (intended for use with a log carriage) consists of a combined chipper-canter and twin-arbor saw unit to carry single or multiple saws.
Source - Linck

In addition to their use in log production lines, between two and four cutting units can also be arranged process logs clamped to a log carriage - to face and saw one side only of the log at a single pass. After this the log is turned through 90 degrees for further processing. These units, or aggregates, can consist of chipper canters, profilers, vertical-arbor circular splitting saws or regular horizontal-arbor circular gang saws.

Fig. 16.21
This shows a vertical-arbor splitting saw unit, HK, in this case as part of a regular log production line. A similar unit is available as an aggregate for use with a log carriage.
Source - Linck

Curve and taper sawing

Although humans would wish otherwise, trees do not necessary grow straight, nor do they retain the same diameter along full length of the log. In the past these irregularities were simply ignored - logs were sawn exclusively on logmills and rack benches which made parallel, straight-line cuts irrespective of these factors.

The increasing concern with timber conservation lead to total a re-think on dealing with curved and varying-diameter logs. It was proved that regular straight-line conversion wastes valuable resources as this produces tapered outer timbers of limited use.

Sawing lines are now available that return the highest possible timber recovery by following, at least partially, the natural curve of the timber. Some equipment can also take taper into account - by stepping-out the chippers part-way along the log to the next board thickness.

Fig. 16.22
This shows the infeed section of the two-side cant-canter and multi-saw edger production line shown also in Fig. 16.19. The unit can process cants up to 250mm (10in) thick with thin kerf (3.2 - 3.6mm, 0.125 - 0.140in) climb-cutting guided sawblades mounted on a single arbor.
Source - Sawquip

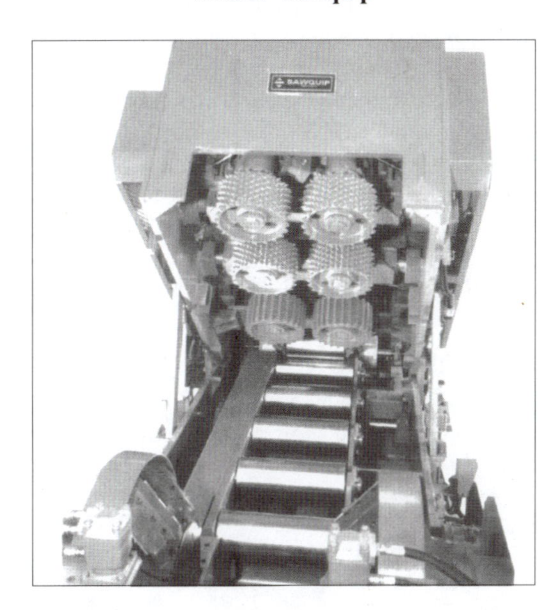

Fig. 16.23
The infeed section of the cant-canter shown in Fig. 16.19, with the feed section raised to show the vertical, powered feed rolls. They are automatically positioned to deal with curved sawing in a horizontal plane to 2.5% of cant length, and can process in two planes, i.e., "S" shape.
Source - Sawquip.

Curve sawing, widely practiced in the Northwest of the USA and Canada, gives far high conversion efficiency than conventional practice. No doubt this technique will eventually spread to other areas as the vast savings of valuable resources are more fully appreciated.

Due to the wide variation in cut depth and density of certain timbers, some machines of this general type incorporate an automatic system to vary the feed speed according to the load on the saw motors.

Double slabbers

Small diameter logs can also be broken down using double slabbers fitted with two, cluster-type saws - the spacing of which is instantly variable according to the log diameter.

Fig. 16.24
This is a traditional, semiautomatic double slabber capable of slabbing logs 100-300mm (4-8in) diameter using a dogged chain feed. For an electronically-controlled double slabber from the same company see Fig. 3.48
Source- Storti

Logs are fed via a chain to produce a two-sided cant and two barkside slabs. The two-sided cant can then be turned through 90 degrees and returned for sawing a second time *(or passed on to a second slabber)* to produce a square and two three-sided cants from the two-sided cant.

From the slabber the squares, cants and barkside slabs are usually passed onto a re-saw for further breakdown. The two and three-side cants are often passed onto a band or circular re-saw for further breakdown into flitches, then onto a double or multiple edger.

Barkside slabs are reclaimed by re-sawing and double-edging into smaller boards, with the final waste sent for pulping.

Single slabbers are also used, but usually with band-saws as the sawing medium because of their narrower and less wasteful kerf.

Double and Multiple Edgers

Following conventional log sawing, waney-edged boards are further processed by double or multiple edgers. These machines usually have fast, infinitely-variable feed speeds.

Fig. 16.25
Left: flitches are fed into an EDS edger equipped with two edging and a centre-splitting saws at feed speeds of up to 120M/min (390ft/min). Right: a CML Edger with one fixed and one moveable saw, and laser guide lines for both.
Source - Stenner

Double edgers are fast-feed, high production machines, usually with powered roller-feed system. They are often incorporated into a log conversion line, and are usually fitted with two sawblades to square-edge the flitch in a perfectly straight line and parallel - although some types follow the curve of a flitch.

To achieve the best possible conversion factor, the distance between the sawblades is quickly changed to match the flitch width. Usually they are re-spaced in specific steps to correspond with standard board widths. On some machines a centre sawblade can be used to split the centre board.

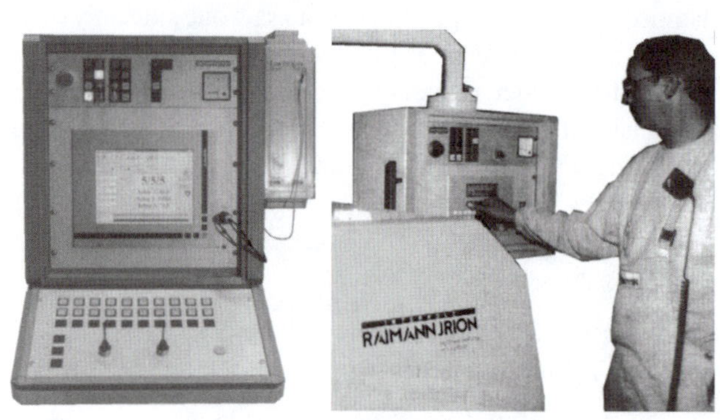

Fig. 16.26
Above, the KR 450 BV optimizing multi-ripsaw has up to three sawblades remotely-positioned from the control panel (right) or through a unique voice-control system (tuned-in to individual operators) via the alternative control panel, lower left. Laser guides above the machine move with the individual saws and project their precise cutting lines onto the timber.
The lower right illustration shows the Raimann system of sound-controlled ripping of timber. (See Fig. 16.33 for a full description.) This allows a narrower saw kerf to be used, so increasing the economic efficiency of the machine.
Source - Raimann (Weinig)

Fig. 16.27
This board edger CSM 60 is used in combination with a fully automatic alignment and feed system. The two bottom-cutting saw-units are individually driven and adjustable,

and can carry one or more sawblade(s) for cutting up to 60mm (2¼in) deep. A centre top ripsaw can also be fitted.

Source - Linck

Lasers on some manually-fed machines project the outer sawing lines onto the board so that the sawyer can position it accurately prior to feeding in. Older machines may simply have indicators showing the position of the sawblades - relying on the experience of the operator to correctly position and align the board or flitch.

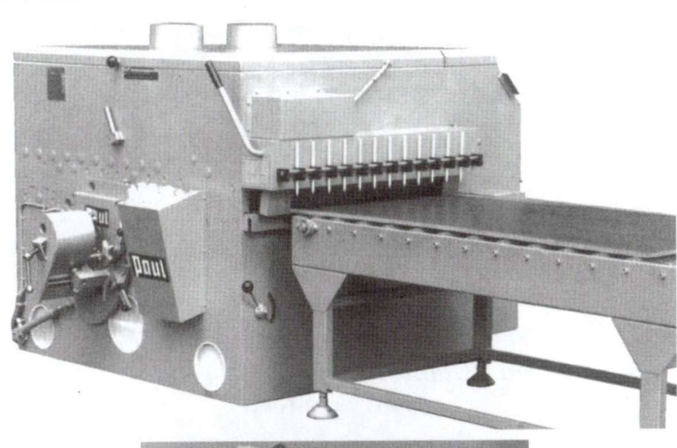

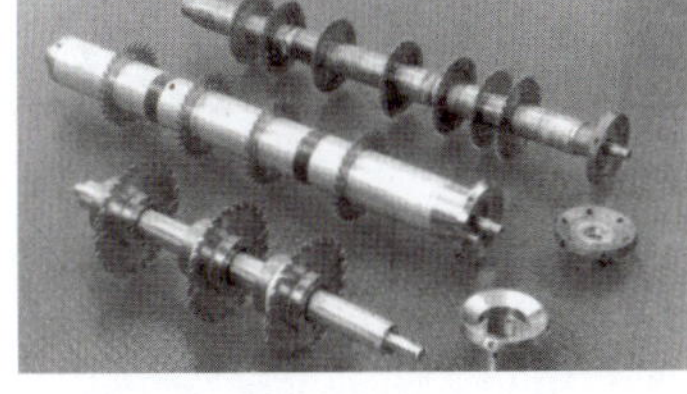

Fig. 16.28

The K 34 multi-edger shown above operates at feed speeds up to 48M/min (156ft/min) via several powered rolls, and has opening widths of up to 2200mm (7ft.) - according to the individual model. Sawblades can be spaced to give a wide choice of sawing combinations. The machine will also multi-saw veneer packs, thin foils and plastic sheets, and will groove panels.

Shown below are the different systems of sawblade mounting, including spacers, individual bushes and hydraulically-tensioned arbors.

Source - Paul

Multiple edgers have a several sawblades mounted on a common, large-diameter arbor, and a feed system provided by upper and lower powered rollers, or a combination of powered rollers and feed chains.

The saw arbor can accommodate sawblades in several groups, or in stepped spacings:-.

For group spacing. Several groups of sawblades are arranged to cut to standard board widths, with those in each group spaced to suit specific board-width combinations.

For stepped spacing the sawblades are arranged in rising widths from the drive end, See Fig. 16.29.

The irksome task of the sawyer with manually-fed machines is to feed the board through the group of sawblades which gives the best conversion factor. This he does by choosing and aligning the board at an appropriate width-position - but he has to do this quickly and decisively or production rates and productivity suffer.

Selection must take into account the straightness of the flitch, its edge-condition and the usable width along its full length. The conversion efficiency - and plant profitability - is highly dependent upon the skill of the machine operator in making rapid and correct decisions as he feeds each board or flitch in turn.

A perennial problem is in the actual spacing of the sawblades. It must be such that all boards likely to be processed can be converted in the most efficient way regardless of their width and edge condition.

Small wonder that many companies look to this area for computer control.

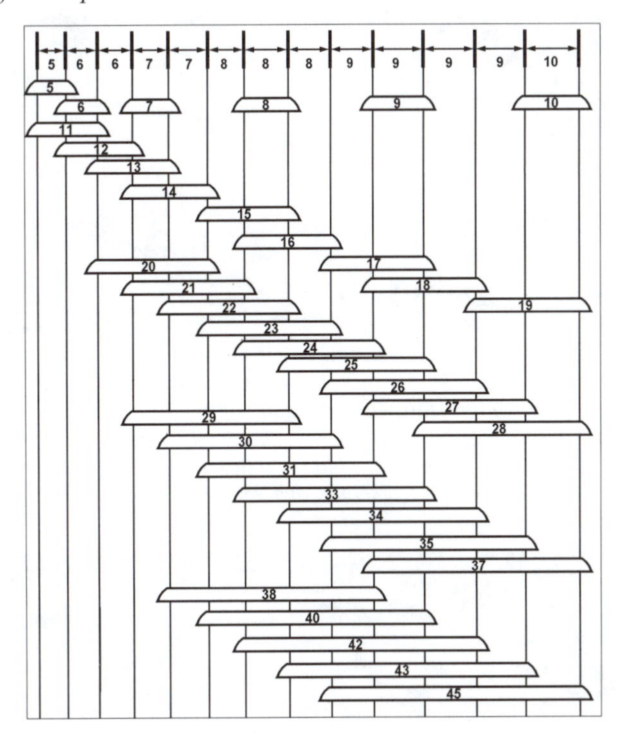

Fig. 16.29

Sawblades can be arranged in groups of different spacings, or in progressively increasing spacings as shown here. The wide multiple-choice of flitch positions illustrate how hard the operator's task can be. (Dimensions are in cm.).

Source - Paul

Both machines types can be fed manually. Where standard sawblade spacings apply to all boards within a pre-selected groups, however, a fully automatic self-centering feed system can be used instead.

Optimizing equipment.

Optimizing equipment is now widely used to make the conversion process faster and more efficient.

This incorporates the scanning of flitches for width and condition and then, based on this information, positioning them at the infeed of double and multiple edgers so as to give the best conversion factor. Such equipment produces the maximum possible yield - and is free of human operator shortcomings. Increases in yield through automatic feed systems over regular manual feed for edgers can be anything between 2% and 10% - an improvement that can make a huge difference to profits.

Pay-back time for the equipment - which can be used with both fixed and moveable sawblade types - can be as little as six months for their optimizing Compu-Rip installation (according to Barr-Mullin).

The most sophisticated systems of this type no longer need width pre-sorts. They also provide tight control of the breakdown process - and vastly improve the mill efficiency in terms of output, statistics and billing.

Fig. 16.30
For board thicknesses up to 80mm ($3^{1}/_{8}$in), this CSM 80 has value optimizing when used in conjunction with a semi or full automatic infeed system. The two in-line saw arbors each carry a gang of sawblades and two adjustable outer sawblades on telescopic arbors (See Fig. 16.14).
Source - Linck

Fig. 16.31
Viewed from above, the Compu-Rip gang-rip saw optimizer can be used with any regular fixed or movable-saw gang edger. The equipment projects laser lines (corresponding to the sawlines) onto the boards as they are passed to the optimizing station from the cross-chains.
The laser lines are quickly switched between the various sawblade groupings - using just a few buttons - so the operator can instantly choose the best conversion arrangement individually for each board. On passing to the edger the board is then automatically aligned for the corresponding sawblade grouping.
The ArborMaster program is included with the equipment. Once data covering standard widths and output-mix requirements is down-loaded, this program determines the most appropriate sawblade groupings.
The Compu-Rip works efficiently with the BMI Cut-off Saw Optimizer and their Rough Mill Command Centre. The latter controls scheduling through the plant and prints out cutting lists and customer data.
Source - Barr-Mullin

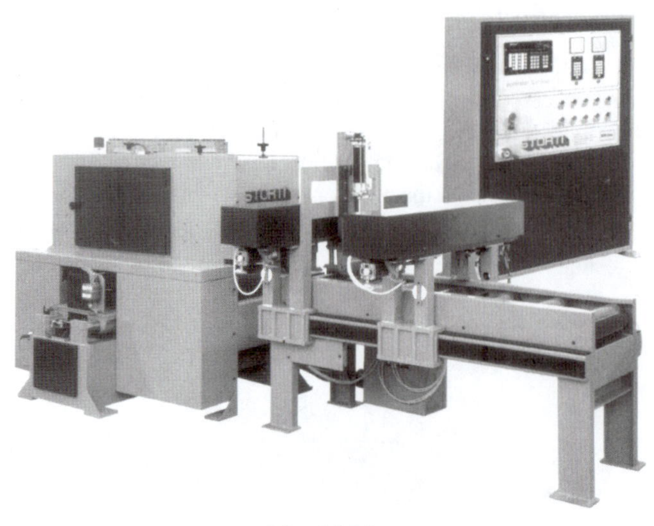

Fig. 16.32
The RM 500 is a compact optimizing double-edger for boards up to 170mm ($6^{5}/_{8}$in) thick.
Source - Storti

Secondary Conversion

Multi-rip saws.

These machines, although mainly used for secondary break-down in building and furniture mills, are also used in smaller primary conversion mills.

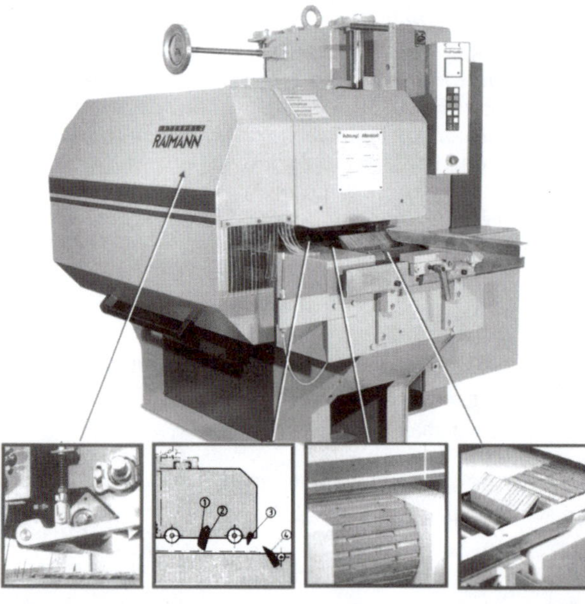

Fig. 16.33
The KR over-cutting single-arbor saw can be used with up to 5 group sawblades, or with one fixed sawblade for edging, and one sawblade which is moveable - with its position indicated on a digital display. Alternatively the Quickfix hydraulic locking system can be provided allowing each saw to be individually positioned manually (digital display optional). Any number of sawblades can be mounted in this way. Conventional bushes are not required and, as sawing-width changes can be made rapidly and without need of saw arbor removal.
The inset views below show the main features:- from the left, adjustable upper pressure rollers; four anti-kick-back devices; a wide feed chain; infeed anti-kick-back fingers.
This machine can be equipped with the Raimann sound control system (see Fig. 16.26). This monitors the sound of

cutting and alters the feed-speed accordingly (as determined from a program based on prior tests of various timbers). This provides optimum cutting conditions and is claimed to allow a narrower saw kerf to be used - giving a better yield, faster production, longer tool life, less down-time and lower tooling cost.
Source - Raimann (Weinig)

Two types are used, single-arbor gang saws for relatively shallow cuts, perhaps up to 120mm (4¾in.) thick, and twin-arbor gang saws for deeper cuts, up to perhaps 180mm (7in.).

These machines are particularly prone to kick-back, and should have efficient top and bottom anti-kick-back fingers, plus guards to prevent access to the sawblades from the side. The sawblades used are invariably strob-slot type due to the arduous nature of the cut.

Normally a fence guides the timber when this is square-edged, but a laser guide line can be provided on some machines for waney-edge boards.

Fig. 16.34
A laser projects the precise sawline on this M3 multi-ripsaw, and helps the operator to position a waney-edge board so that the wane is barely clipped-off.
Source - SCM

Machines intended for optimizing conversion of boards can have grouped sawblades working in conjunction with a multi-position fence - to precisely align square-edge boards according to the conversion pattern selected.

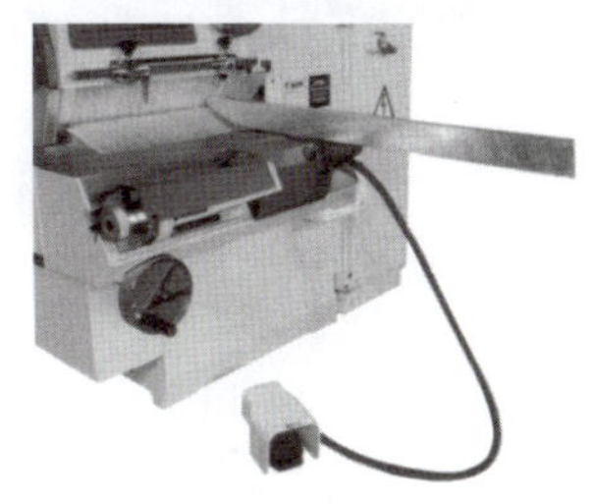

Fig. 16.35
The fence on this M3 multi-saw is controlled by a foot pedal - to instantly switch between any one of six pre-set positions corresponding to different saw groupings.
Source - SCM

Single-arbor over-cutting saws have one or several sawblades mounted on a single arbor above a single, wide feed chain. The chain is side-guided to dip immediately underneath the sawblades so they cut clean through the timber.

Multiple top feed rollers keep the timber moving smoothly, and a top guide cut into by the sawblades may be provided to control short pieces.

Fig. 16.36
This 18 series twin-arbor multi-rip saw cuts boards up to 180mm (7in) deep at infinitely variable feed speeds of between 6 and 25 M/min (20-80ft/min). The double feed chain, twin-arbor sawblades, feed rollers and anti-kick-back fingers are clearly seen with the hood raised.
Source - Storti

Twin-arbor saws have arbors which are slightly staggered in the direction of feed. Each arbor mounts one or several matching sawblades, which between them saw through the material with minimum kerf wastage for the depth of cut.

Feed is usually by two wide feed chains. One in precedes the lower gang of sawblades and the second immediately follows it - assisted by several top feed rollers.

A fill-in piece on the bed above the lower sawblades is sawn through to give continuous support, and a top wooden guide/pressure may also be provided.

Re-saws

Fig. 16.37
This twin-arbor gangsaw, part of a primary conversion installation, re-saws two and four-sided cants of up to 700mm (27in) wide by 350mm (14in) deep at feed speeds of up to 130M/min. (420ft/min).
Source - Linck

Re-saws are used both in primary and secondary conversion for dividing boards into thinner or narrower sections. They are regularly used within manufacturing units, both in bandsaw and circular-saw form. Heavy-duty circular re-saws are used in primary conversion for breaking down square cants, perhaps with a linebar or split taper feeding, and for straight or curve sawing.

Cut-off and cross-cut saws

Circular cut-off and cross-cut saws are general terms for machines that cut timber across the grain.

They are used throughout the industry - from log trimming in primary conversion - to factory use in trimming and frame mitring, etc. The latter have been dealt with in Chapter 14. This chapter deals with conversion and optimized cross-cutting of both rough and machined timber.

Log cross-cutting

Fig. 16.38
The shows part of a high-capacity round log processing plant for optimized log cross-cutting and grading.
Source - Linck

Cross-cut saws, along with fixed chain saws, are used in log lines for cross-cutting - after which the cut logs are sorted for length and grade before converting them. The sawblades used vary from insert-tooth cross-cut saws to carbide tipped types.

One of the original log cross-cutting machines was a below-floor-level circular sawblade which could be raised by a lever through a slot in the floor. The only evidence of the cross-cut sawblade was the floor slit, so it takes little imagination to appreciate how dangerous this arrangement was.

Another type had an unguarded floor-level sawblade which which moved forward on a carriage to cut a single log. Guards and fencxes were not supplied.

The ones that followed these were ceiling-mounted pendulum, ram, chop and under-table traversing saws.

Length and Defecting-Cutting Saws.

Saws of this type are used for cross-cutting timber to pre-set lengths and/or cutting-out defects. Although manual saws were once used for this, virtually all are now automatic - with either manual, semi, or fully automatic feed - often with electronic or computer control.

A few saws are of the chop type that cut downwards from above the timber, but most firmly clamp the timber from the top and then rise into the cut either from underneath or partially from the side.

Most are interlocked so that the sawblade cannot begin to cut until the timber is firmly clamped.

Fig. 16.39
This R600-700 cross-cut has a rising sawblade movement with a cutting capacity of 200x380 - 100x580 (8x15 - 4x23in). The cross-chain discharge and electronically-positioned end stop are optional extras.
On the larger models the vertical sawblade movement is further extended by a transverse movement to cross-cut wide boards or flitches of up to 100x1000mm (4x39in) .
Source - Bottene

Cross-cut saws that rise vertically have a cross-cutting capacity that varies with the diameter of sawblade used and the section being cut. Some makers include a diagram to show what alternative sections can be cut - or quote two width figures against maximum and minimum thickness.

In order to use the full cutting capacity, the rear fence has to be re-set between cutting wide, thin sections and near-square sections some of the lower-cost models.

Alternative diameters of sawblade may be fitted on the more up-market machines to cover alternative sizes-ranges, but in their case the fence remains in a fixed position. Some makers instead use an arcuate movement partially from the side. This increases the cutting capacity without need of switching sawblades - and also gives a cleaner cut when converting planed and moulded timber.

Fig. 16.40
The T20 has an inclined table to keep stock against the fence, and a top clamp for holding the timber. The sawblade is triggered by a foot control, and the pre-set stops are pneumatically operated. A wire mesh guard prevents access to the sawblade and clamp, and the machine can have an outfeed belt and a length selector.
Source - Omga

Originally flat tables were used - and still are for heavy sections and waney-edge boards and flitches - but most length and defecting saws handling edge-finished stock have inclined tables that automatically keep timber firmly against the rear fence.

To prevent operators trapping their fingers, the cutting and clamping area of all these machines must be well guarded by mesh or solid panels, as can be seen in the illustrations.

Manually-fed cross-cut saws In its simplest and lowest-cost form, the operator manually moves timber against the appropriate one of several pre-set end stops before cross-cutting.

Although many saws of this type are triggered by a foot pedal - or even by the timber contacting any stop - most safety agencies recommend a two-handed starting switch to ensure that the operator, or his assistant, cannot trap their hands.

The stops on horizontal-table cross-cut saws can be of the disappearing type. Those not used for the workpiece being cross-cut are de-activated by first drawing the timber away from the fence, and then pushing it *'in'* against the fence and towards the following stop.

Most machines, though, use either remotely controlled pre-set stops, or a flying stop that is positioned electronically. Both types are selected from a control box at the central operating position, and both allow timber to remain firmly against the fence during the cross-traverse movement.

The operators task is to crosscut lengths according to his cutting-list - in the most economic possible way - while also noting and also cutting-out defects. He must also memorize the number of lengths already cut and discount these as the work proceeds.

Profitability of this prime process depends heavily upon the speed and efficiency of the operator - and this is the weak link.

Manually-operated cross-cut saws are still used for this process in many mills, but because of the effect this process can have on the efficiency of the whole mill, there is a growing tendency to use specialized optimizing cross-cut saws - especially in large-scale production.

Optimizing Cross-Cut Saws

Optimizing cross-cut saws are controlled either by some form of simple processor or a full cnc system.

Both control systems allow the timber to be fed manually or automatically in an almost a continuous and fast-moving stream - with the machine stopping only at appropriate points to defect or length-cut - and with the processor or computer memorizing the numbers of cut lengths to discount these as conversion continues.

Computer programs provided with these machines can be arranged either to length or to defect-cut - or for both - with controllable priority between the two On the more up-market equipment other notable features can be included, as later described.

Using such a system is faster and vastly superior to manually-controlled cross-cuts lines. It can give a huge increase in the value of timber produced, a reduction in the processing time, and a considerable lowering of the labour costs.

Fig. 16.41
Here the operator is shown marking defects on the timber prior to feeding it into this R500 optimizing, push-feed cross-cut. After placing the timber against the back fence and towards the cross-cut, the operator starts the reader, in protected cage to his left, which traverses along the piece to measure it and note the marked defects. It then contacts the remote end of the timber and pushes it into the machine, merely stopping briefly at the points where cross-cutting takes place.
Source - Bottene

Typical Computer programs

Cutting to pre-set lengths When cutting only to pre-set lengths, the reader on the machine measures the length of each piece as it is fed through, but disregards any defects there may be. The computer then instantly determines the best possible yield from it - taking account the cutting list and the tally of prior-cut lengths - and controls the feed system accordingly.

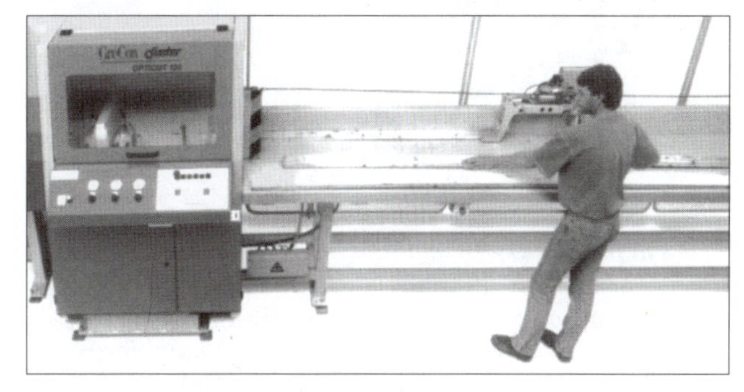

Fig. 16.42
On this Opticut 104 the operator holds a pre-marked length in readiness, while the read/pusher in front of him feeds the previously-marked length into and through the cross-cut.
Source - GreCon Dimter (Weinig)

Length and defect cutting Optimizing cross-cuts are also used for defecting - removing defects such as dead knots, shakes, edge wane or compressed grain. When also cutting to pre-set lengths the computer estimates what lengths can best be obtained after such defects are removed. It then cross-cuts them accordingly - trimming immediately after defects and sending the short length containing the defect to waste.

Fig. 16.43
The 11KE cross-cut is a through-feed type with one of up to three marking stations. The MKL version (Fig. 4.37), allows full optimization to alternative standards and can have automatic defect identification, width measurement, ink-jet printing and automatic sorting.
Source - Paul

Defect cutting only - defect removal prior to finger-jointing - ignoring length. *See the later section on finger jointing.*
Length, defect & short-stock cutting This is defect and length-cutting, plus defect cutting of short offcuts for finger-jointing.
Identifying defects
Defect recognition is possible either by the operator pre-marking either side of each defect, or via a scanning process.
Manual marking is by using a crayon which detectors at the measuring station can recognize.

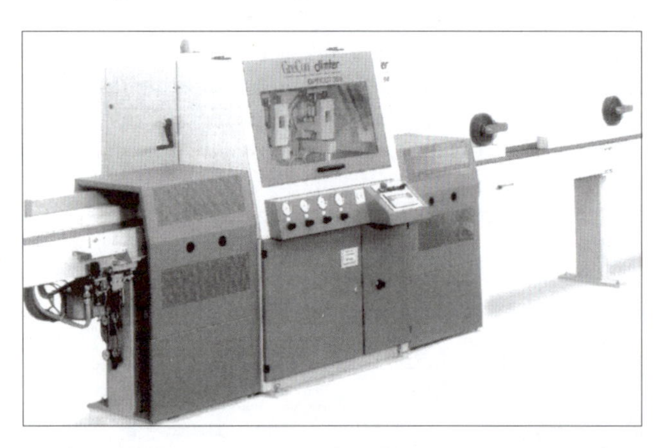

Fig. 16.44
One of the 300 series, this 304 optimizing cross-cut also defect-cuts. Of the through-feed type, its also offers options of width-measuring for billing - and is capable of cross-cutting extreme cross-sections, both small and large.
Source - GreCon Dimter (Weinig)

Automatic scanning equipment can be programmed to detect the specific defects that the user determines needs removal - and which has memorized programs customized to specific grades of timber.

Scanning takes place either on the cross-cut or via a separate installation. Scanning is possible using lights, sensors or lasers, possibly from the side, top and underside, and maybe coloured to identify defects which otherwise do not show. By collating this mass of information on length and defects, the computer can determine the cross-cutting program accordingly.

There are two basic types of optimizing cross-cut, semi-automatic types that have a reader/pusher, and the fully automatic, through-feed version.

Pusher optimizers have a reader/pusher that traverses along the infeed table under computer control. Starting at the saw end, the reader/pusher scans the full length of timber placed against the fence - to measure its length precisely and record any marked defects.

This information is automatically processed by the computer, which then reverses the direction of the reader/pusher to feed the timber into the machine for cross-cutting.

By using wide add-on pushers, boards can be handled as individual pieces, or as loose boards stacked in width, in depth, or as a pack.
Through-feed machines have a fixed reader station at a set distance from the actual crosscut saw. Individual pieces first pass thorough this to be read for length and marked defects. They are then fed directly into the cross-cut via a belt, chain or rollers, where a separate measuring wheel independently checks the traverse movement and so determines the stopping positions for cross-cutting. This type of machine is extremely fast in operation and processes timber fully automatically and in an almost continuous stream.

Fig. 16.46
The Turbo Wonder Saw is a through-feed cut-off optimizer which has a traverse speed of up to 200M/min (650ft/min), and a 1/4 second cross-cut speed. Like the Compu-Rip, the machine has a direct cellular phone and modem link to the makers to provide immediate upgrading and troubleshooting when needed.
Source - Barr-Mullin

Fig. 16.45

The Cell-Scan reads both sides of each board simultaneously and can identify whatever the user determines is a defect - which can include blue stain, knots, compression wood, twisted grain, crook, mineral stains and wane.
Source - Barr-Mullin

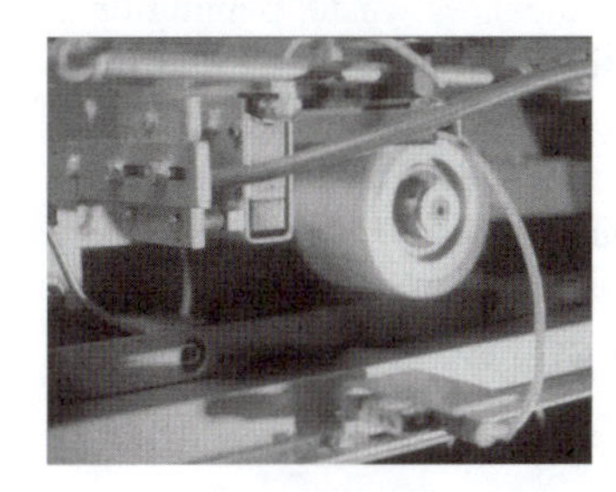

Fig. 16.47
*The defect and grade mark recognition section of the Turbo Wonder Saw. It can be programmed:- **for best yield** where size and quantity are unimportant (alternatively when sizes are important, or when both size and quantity are important); **for best value** (taking into account both lengths required and defect-removal); **for full defecting** (regardless of lengths and size-mix); **for different specifications** (with waste from a higher grade for use in a lower grade); **with finger joint added** (defect-cutting of all shorts for finger-jointing); **with glued re-rip added** (with edge-defective stock cut-out for ripping to clears); or **for furniture grade pine*** (leaving sound knots and cutting-out dead ones).*

**Although boards are normally defected before ripping, the defects can either be mainly central to be board or towards one or both edges.*

In these cases the Wonder Saw evaluates such defects and rejects those that give better return if ripped before cross-cutting - this produces longer lengths of clears.
Source - Barr-Mullin

Fig. 16.48
These composite views of the Turbo Wonder Saw show, left to right: 1/4 second cuts without tear-out; high torque digital AC servo motor drive; and a board-width sensor.
Source - Barr-Mullin

Optimizing cross-cuts can be equipped with automatic sorting stations, making timber handling fully automatic - even short off-cuts are mechanically handled.

Fig. 16.49 (Next column)
This HM-Z/S upstroke cross-cut saw is a dual-purpose machine suitable for straight, single and double angle cross-cutting, also notching, of components for roof trusses, timber-frame houses, decking, garden and household furniture, etc. The NC-controlled saw changes angle as boards are fed-in. Control is via an industrial computer with an LCD colour screen. A length and defect-scanning pusher provides optimization capability through length measurement and grade marking.

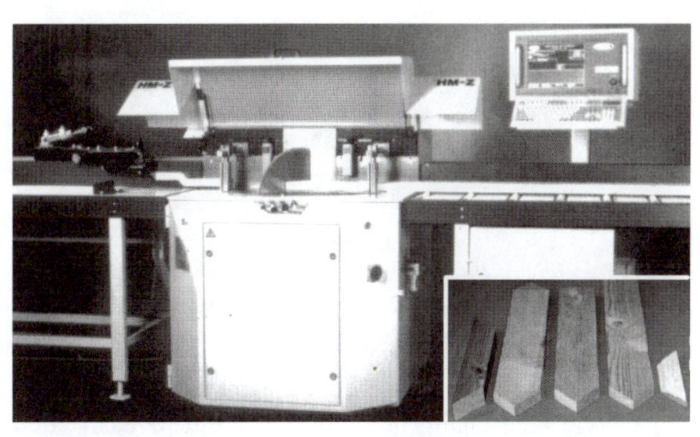

This view of the HM-Z/S shows the saw section, with the saw angled and the top guard both raised. The inset view shows typical single and double-angle roof truss parts.
Source - H&M Woodworking Machinery

Special Purpose Machines

Wood Block Machines

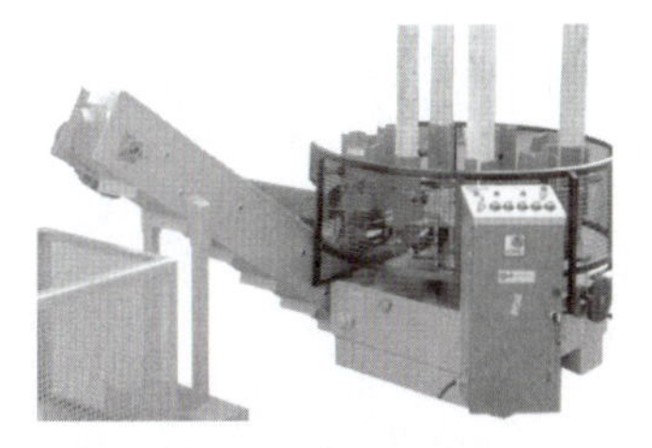

Fig. 16.50
These fully automatic machines cross-cut the separating wooden blocks used in pallet-making from random-length pieces. These are placed vertically in holders which rotate across a fixed-position sawblade. Finished blocks are extracted by conveyor to storage.
Source - Storti

Truss component saws

Fig. 16.51
In this general view of the Timber Mill 6-20 the component cross-feed is clearly seen - as are the indicators displaying the angle settings of the sawblades.
Source - Speed-Cut

These machines form multi-angle cross-cuts on both ends of timber parts intended for floor and roof truss construction.

Components are passed through logitudinally and edge-on, fed and supported by a parallel pair of dogged chains. These feed the components squarely past the multiple angle-cutting sawblades, where they are securely held down by overhead and off-set travelling pressure chains. Following manufacture these parts are gang-nailed to form roof trusses.

Fig. 16.52
This end view of the 6-20 shows short floor webs being cut using a single top pressure chain.
Source - Speed-cut

The normal machine arrangement is for one saw-unit, consisting of one feed chain, pressure chain and its associated sawblades, to be fixed, and the other saw-unit to be readily and easily moveable to whatever length of web or truss member is required - in some cases by power.

The sawblades range from 355mm (14in) to 760mm (30in) diameter. They are mounted flush-faced to the outside, so allowing the full saw diameter to be used - an essential requirement for long scarfing cuts (shallow angles cuts needed for low-angle roofs).

Three sawblades are provided for each unit. Two of these sawblades adjust to any angle - both pivoting on a common axis where their individual cuts intersect on the component end - so angle setting and corrections are be made without affecting the component length.

The pivot height relative to the bed line can be instantly altered, independently on both left and right-hand saw-units, to give centre-line or off-set angle cutting.

The third sawblade on the fixed unit can be adjusted to any angle needed for long scarf cuts, and is set independently of the two other sawblades.

The sixth, vertical-cutting sawblade on the adjustable saw-unit, square cross-cuts over-long components before they reach the following angle-cutting sawblades, so optimizing lumber recovery. Clear rotary scales and digital readouts make for rapid set-ups.

Edge-banding and post-forming machines

Fig. 16.53
The top illustration shows the rear of a typical multi-station edge-banding machine. The machine is one of the KL series, and is capable of applying the inclined, watertight and deep soft formed profiles shown below.
Source - Homag

Edge-banding machines are used to apply plastic edging to faced boards made from chipboard and MDF. The machine combines several different machining elements and a through-feed system consisting of a wide, travelling bed-chain and overhead pressure rollers. Panels can be edge-banded on one or both edges at a single pass through - depending upon the machine type. In line of feed the board pass a hogger to end-trim and remove excess waste, planing or moulding heads, glue and plastic edge applicators, end trim sawblades and finally corner clean-up stations.

Post-forming machines are similar, but wrap post-forming plastic around a shaped edge. Double-end trimmers *(universal sizing machines)* are sometimes used to edge boards prior to edge-banding.

Fig. 16.54
These end trim saws traverse diagonally to remove excess edging on the run - allowing a continuous panel feed.
Source - Brandt

Boring and Dowel Driving Machines

Fig. 16.55
The BD through-feed double-end dowel drilling and driving machine (right) is being fed from parts cut on the trim saw T-101-S (left). These machines constitute part of a drawer production line. The scribe and main trim sawblades of the T-101-S can be clearly seen on the left.
Source - Koch

These machines are widely used in the furniture industry for making dowelled joints between furniture parts and similar, or for drilling to suit knock-down assembly fittings.

As their name suggests, the part is end-trimmed and bored - with glue applied and dowels driven for the male assembly only. Single and double-end machines can be manually fed, but for production processing through-feed machines are used which have parallel bed feed chains and independent clamps at set work stations.

Fig. 16.56
This inset view shows the trimming sawblade and dowel drills at one side of a SBFD self-contained, double-end production centre for trimming, boring, shaping (for matching moulds) glue injecting and dowel driving.
Source - Koch

CNC Machining Centres

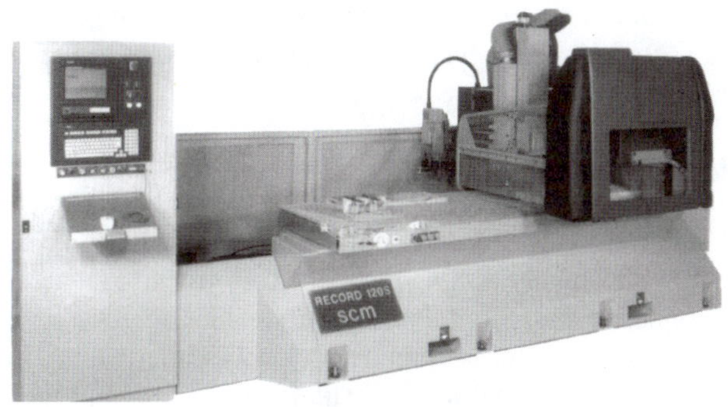

Fig. 16.57
This is typical of the modern machining centres, a Record 120. This has a fixed table and two-way traversing head. Finished parts can be removed and replaced by unmachined parts while the machine operates at a second station.
Source - SCM

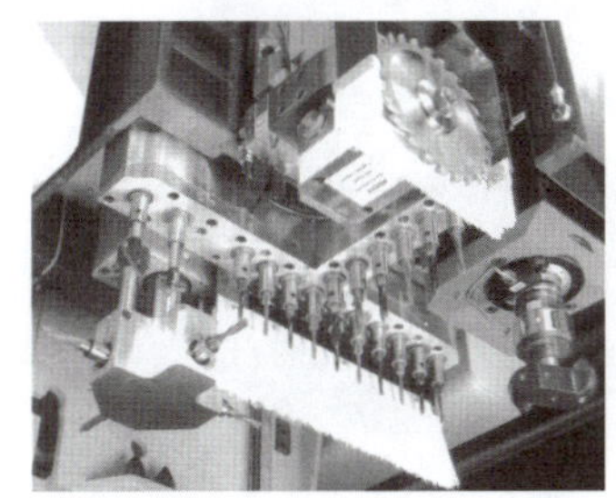

Fig. 16.58
This shows a cluster of several tools - including a trim sawblade - in the workhead of a machining centre. Any tool in the cluster can be brought into use as instructed by the computer, or quickly and automatically switched for any other tool held in readiness in a storage bin on the machine.
Source - Weeke/Homag

Machining centres evolved from earlier CNC routers - the term machining centre being adopted when the operations they became capable of far exceeded the original routing processes. There are several variations of the design, but all involve a workhead traversing around a workpiece lying flat on the table.

In early machines the table traversed in both directions. In later machines the table traversed lengthways and the head traversed width-ways. The more common arrangement now is for the cantilevered head to traverse in both directions leaving the table and workpiece stationary.

The movements of the head are computer-controlled and continuous-path, so that the head can traverse lengthways, width-ways, at any angle, or it can follow a complex multi-curved path.

The head also has a vertical cnc movement which is used to enter the tool into the cut, or to machine parts in three-dimensions. Some also have an angular movement of either the whole head, or only the section carrying the tool.

Fig. 16.59
Forming an angular groove on a machining centre using a tilting tool carrier left. Fitted with a cut-off saw for angle-cutting hip rafters (right). In both cases the workhead is simultaneously traversed across and along the length of the table to cut diagonally and at any required angle.
Source - Morbidelli (left), IMA (right)

Most have interchangeable tooling which can be drills or boring bits, routers or moulding heads, circular saws or groovers. The latter are used for length and cross-cuts, also for single and complex angle cuts or grooves in solid wood or panels. Being computer-controlled, setting and operation is extremely rapid and consistent.

Finger-Jointing Systems

Fig. 16.60
Opposed shaper units on this Combipact finger-jointing system trim packages of assembled pieces, then finger-joint and glue them at both ends before transfer to the assembly press.
Source - GreCon Dimter (Weinig)

Finger-jointing is the process of forming tapered fingers in the ends of timber off-cuts, then gluing the fingers and joining the off-cuts end to end to form a continuous length - which is then cut to the actual lengths required.

Once frowned upon as second-rate, modern finger-jointed timber has become an accepted quality product for a wide range of applications - from household trim to pallet boards.

Various finger arrangements are available. For example, joints can be made through the timber thickness when the edge is exposed, or across the width when the face is exposed (as when end-jointing outer boards for laminated beams).

Different types of finger joint can be used, regular or wide fingers, full-width fingers or fingers with butted outside edges.

The latter give a more acceptable face or edge than the original through finger-joint - which actually appears to be a butt joint from the two opposing faces or edges. (Full-width finger-joints often expose an unsightly glueline when machined parallel to the fingers).

Finger-jointing lines range from simple, single-ended equipment for manual or semiautomatic operation, to double-ended, fully automatic high-capacity production lines.

They all follow a similar process of end trimming using sawblades or hoggers, finger-forming, gluing, end-joining and length trimming.

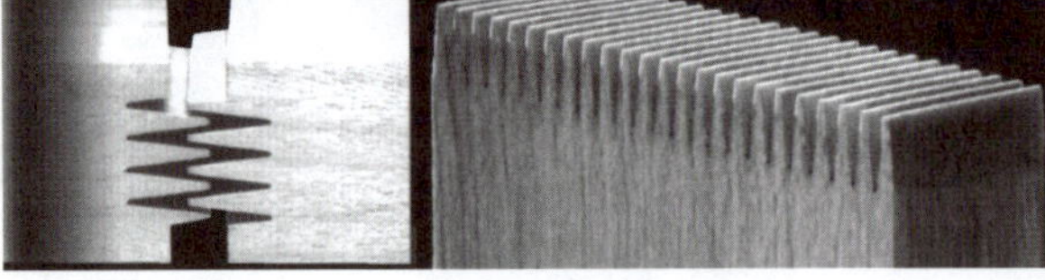

Fig. 16.61
These inset views of the Combipact show, left, a package of parts on the machining table ready for end-machining. The right illustration shows the trim sawblade and finger-joint cutters in close-up. The lower illustration shows butt-type finger joints (left), and through finger joints (right).
Source - GreCon (Weinig)

Finger jointing has become a fully automatic and very sophisticated process, now often a production line proceeding conventional woodworking processing.

The driving force has been the growing awareness of the need to conserve the valuable and shrinking supply of raw timber by using up the short ends that once were simply scrapped.

Originally only scrap pieces were finger jointed, but now there are production lines finger jointing raw, unused stock on a volume-production basis. These upgrade timber by removing all the defects and rejoining the pieces to any required length.

The reason for the development of these lines is that a finger-joint is stronger than the material it joins and finger-jointed sections are unlikely to fail through structural faults in the timber - these are all removed on a defecting cross-cut before reaching the finger-jointing line.

Finger-jointing is also used for producing beam-length, defect-free boards as used in laminated beams - where butt joints have previously been accepted. This obviates the weakness of previous construction occurring through natural faults in the timber and where boards are end-butted).

In fact, any application using a solid wood - such as lorry floors or wooden cores which are afterwards veneered - can use finger-jointed parts. Other examples are parts for household doors.

The very good reason for these applications is that not only are finger-jointed boards or strips available in any length and clear of structural and visual defects - they are also free of bend, warp and twist so often found in natural timber.

End users are now beginning to accept that finger-jointing is no detriment - even on exposed surfaces, so use of this process is likely to increase.

Single and Double-end Tenoners

Single-end tenoners form tenons on both ends of workpieces at two consecutive passes, producing the tenon at one end and then manually turning the piece end for end to machine the opposite end. Stops control the shoulder length.

Typical joints are produced either as interlocking fingers, or mortises and tenons. The tenon section and finger joints are produced on both single and double-ended tenoners when equipped with cut-off saws, tenoning heads and possibly scribe heads for matching moulded sections - this is the traditional machine layout - but some lower-cost tenoners have combined tenoning and scribing heads.

Fig. 16.62
The B5 single-ended tenoner, one of a series, has a tilting table and is a fully automatic machine with clamping and table movements controlled by a plc.
Source - Vertongen

Fig.16.63
This shows the tilting table of the B5 single-ended tenoner..
Source - Vertongen

Double end tenoners simultaneously form tenons at both ends

of workpieces by passing them through the machine logitudinally on dog-type bed chains, with pressures above to keep the stock running true.

Although originally designed for tenoning only, double-end tenoners are also capable of many other operations:- relishing tenons, edge trimming of panels using a jump scorer or of curved panels using a copy scorer, taper-finger jointing, slant tenoning and grooving, dovetail groove and edge machining, edging and sizing of windows and doors using contra-rotating and jump moulding heads, etc., etc.

Fig. 16.64
Double-ended tenoners are used for fast, through-feed production, tenoning both ends at a single pass. The machine has one fixed and one adjustable section, each complete with traversing bed and overhead pressure chains, cut-off saws, tenon, scribe and moulding heads
Source - SCM

Fig. 16.65
A typical scoring saw and a hogger assembly on a double-end tenoner for end trimming of solid timber or panels.
Source - SCM

Window Processing Equipment

Windows have traditionally been produced on classic machines for many years, and in most cases these machines are geared-up for quantity production.

Sash and frame sections are produced in quantity, then stored for cutting-off and end-joint machining as required. However, in some cases stock deteriorated, and there was always the problem of spelching when machining the end-joints.

A growing trend over the years has seen a gradual reduction in the market for mass-produced windows, and a rise in demand for custom-made windows and short-run production.

To meet this need new machine types were developed - ranging from single-arbor machines to duplex machine combinations. These completely machine squared-up stock, and so can produce fully machined window components in a matter of minutes.

They have rapid set-up facilities to produce short-runs and one-offs in the most economic manner possible. Often they have numerical control for precise initial and repeat machining setting - and possibly semi or fully automatic control.

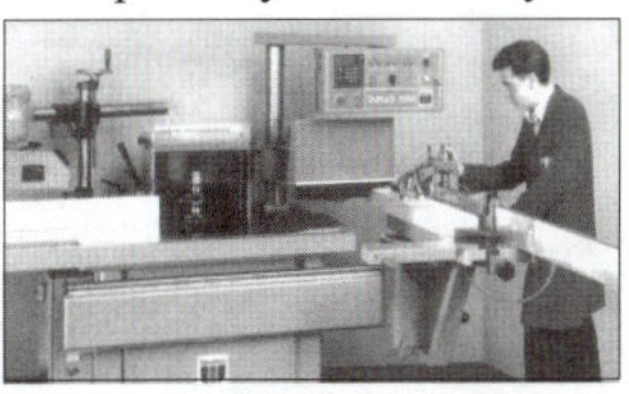

Fig. 16.66
Low-cost, manually-operated machines such as this Duplex 2000 are based on a spindle moulder equipped with stacked heads which simultaneously carry all the profiles required both for moulding and tenoning. They are quickly switched simply by raising or lowering the workhead via a programmable electric controller.
Source - Wadkin

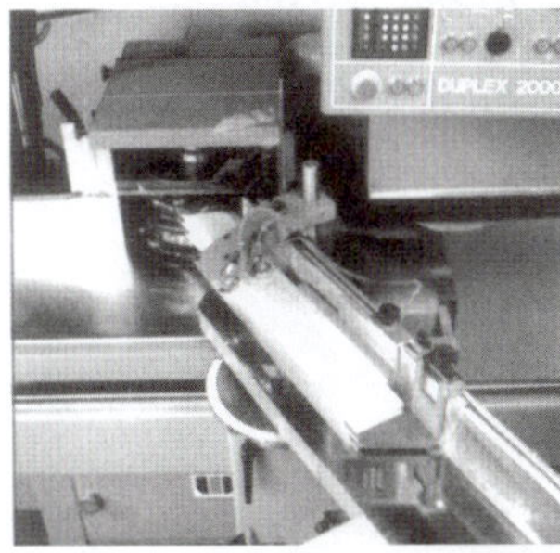

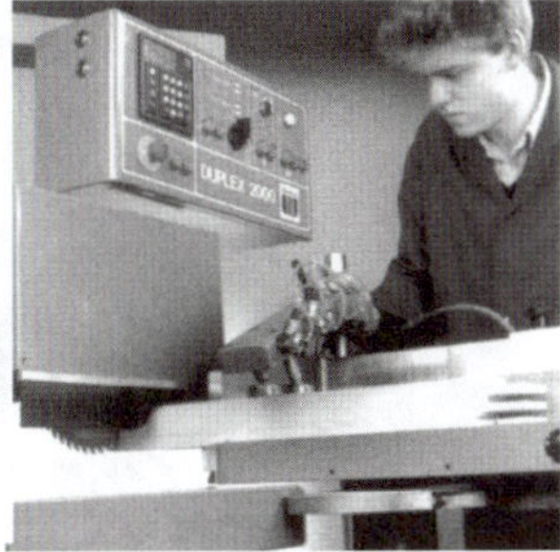

Fig. 16.67
A sliding tenoner table allows the squared workpiece to be cut to length and tenoned straight or on an angle, and then swings clear to allow the piece to be moulded on one or both edges using a power feed unit. When fully-assembled, the window unit can be outside edge-moulded - again using the power feed unit.
Source - Wadkin

Fig. 16.68
When used for moulding along the grain the machine is used as a regular spindle moulder, usually with a power feed (left). The moulding set-up may include a glass rebate bead removal unit (right).
Source - Wadkin

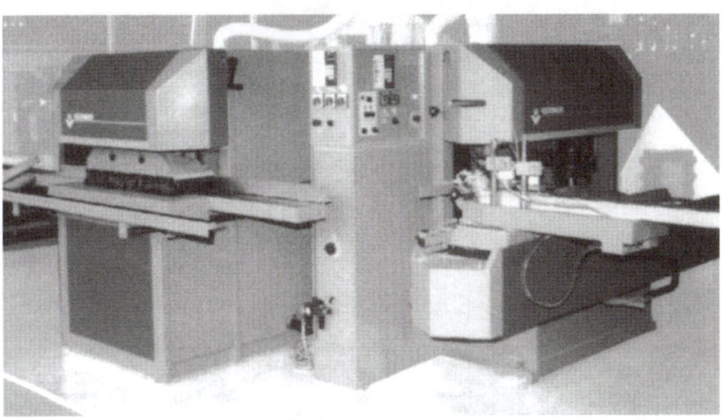

Fig. 16.69
The PenPro Standard 4 window production unit has separate tenoning and moulding sections, each with a vertical, multi-head arbor, plus a cut-off saw and glass bead saw.
Source - Vertongen

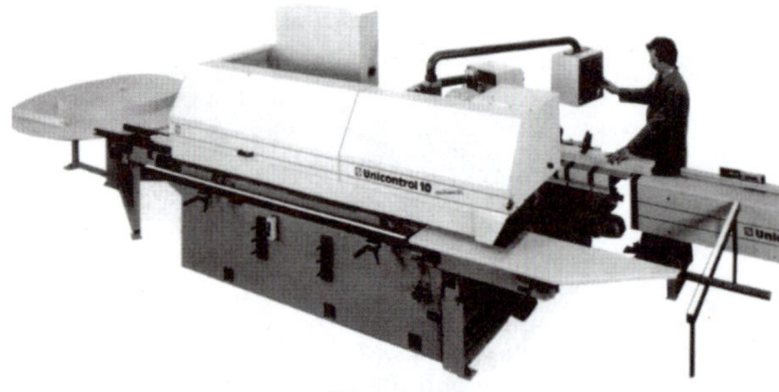

Fig. 16.70
The Unicontrol window machine operates on a continuous-production basis with tenoned workpieces being turned and automatically fed into a moulder alongside. Set-up is rapid with push-button control. The machine cuts timber to length, tenons, moulds along the grain and can finally stormproof assembled sashes.
Source - Weinig

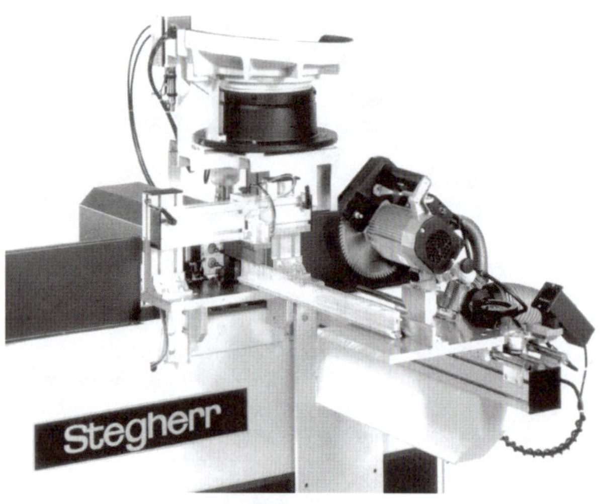

Fig. 16.71
On this NC-controlled KF-2 sash centre, the window section is shown in position for end drilling, gluing and dowel-driving. Also shown is the cut-off saw which is precisely positioned using a digital read-out. The same company also makes machines for closed halving joints for window production, and door processing machinery for dowel-type construction.
Source - Stegher.

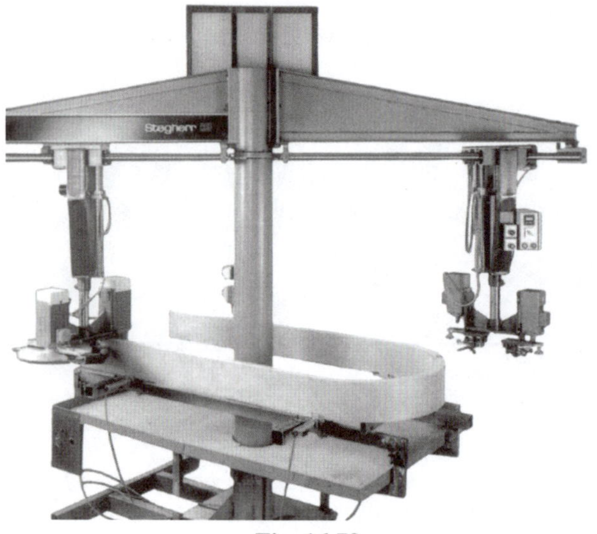

Fig. 16.72
Stegher also makes an Arc Cutting and Milling Machine for laminated arch windows. The saw unit, on the left, follows around the full profile of the window frame to edge-trim it, followed by the router unit, right, to edge-groove the frame for additional fittings.
Source - Stegher.

An alternative assembly system for windows is based on components being end trimmed and end-scribed for mould matching, then drilled, glued and one or two dowels driven at suitable points.

The matching frame is dowel-drilled as required, allowing the window to be assembled in a regular clamp. The cutting head carries stacked tooling to provide rapid switch between profiles.

Two counter-rotating profile heads are provided - and each cuts 'in' to just past centre - so fully profiled sections are processed without danger of tear-out.

Saw Calculations

The remaining section of this chapter deals with the calculations used to find a suitable feed speed for a specific sawblade, or the most suitable sawblade for a specific application. The following are examples shown in the charts and nomograms.

A nomogram provides one means of finding an unknown factor from two or more known factors - simply by drawing one or two lines across vertical scales. The examples show how.

It isn't necessary to actually draw lines, instead lay a rule to cut the scales at the appropriate points, then read-off the information on the third scale, or lay tracing paper on the nomogram and draw on this. An alternative way is to use a strip of clear plastic with a central line scribed on its underside.

Alternatively, on nomogram 16.67, write down the 'Belt speed' for the first part of the calculation, then use the same point on this scale for the second part of the calculation.

The same applies to nomograms 16.68 & 16.69, but in this case write down the 'Teeth per minute' figure.

Sawblade Details

Details	Saw X	Saw Y	
Saw diameter	300mm	350mm	
No. of Teeth	60	28	
Tooth Pitch	*16mm*	*39mm*	*(Fig 16.73)*
Hook angle	*10°*	*20°*	*(from text)*
Hook line off-set	*26mm.*	*60mm.*	*(Fig. 16.74)*
Motor Speed	3000 (50Hz)	3600 (60Hz)	
Motor Pulley Diam.	200mm	250mm	
Belt Speed	*31M/sec*	*44M/sec*	*(Fig.16.78)*
Saw Pulley Diam.	*135mm*	*300mm*	*(Fig.16.78)*
Arbor Speed	*4450rpm*	*3000rpm.*	*(Fig.16.77)*
Saw Rim Speed	*70*	*55*	*(Fig.16.77)*
Teeth/min. '000s	*267*	*84*	*(Fig.16.79/80)*
*Bite**	*0.035mm*	*0.6mm*	*(Fig.16.79/80)*
Feed Speed	*9.34M/min*	*50M/min*	*(Fig. 16.79/80)*
Depth of Cut	30mm	100mm	
Sawdust Area	*1.1sqmm/*	*60sqmm.*	*(Fig.16.81)*
Gullet area	*13sqmm*	*100sqmm.*	*(Fig 16.82)*
Material	Faced panels	Natural timber	

*Feed per tooth

NOTE: The details in italic are those obtained from the text, diagram or nomograms. All these are graduated in metric as this is the most convenient measure. For conversion to imperial measure check against any conversion chart.
Although the results given in the examples are calculated mathematically, absolute precision is not essential with woodworking applications - the scales are a guide.

Choosing the right sawblade

The variable nature of natural timber, and the variety of other wood-based materials regularly sawn makes it difficult to choose a suitable sawblade from the many different types and sizes offered. Although most woodworking sawblades work reasonably well within a wide range of operating conditions and materials - keeping to certain basic guide lines can make them more effective.

Hopefully the following data, when read with Chapters 2 & 3, will set basic guide lines out and - if nothing else - will at least get you into the ballpark.

As with other aids, these charts and nomograms can be used to find one factor from two or three known factors, and can be used in different ways; for example, taking Fig. 16.77:-

A - The rim speed for any sawblade can be found from its diameter and arbor speed, and from this its suitability or otherwise for the material to be sawn.

B - The correct arbor speed can be found for a given sawblade diameter by checking against the rim speed for the material to be sawn.

C - A suitable sawblade diameter can be determined from a given the arbor speed by checking this against the rim speed for the material to be sawn.

The regular order in which the calculators are used is in the sequence they appear in the following pages.

The examples show how to determine an acceptable feed speed range from certain basic details of the sawblade, the machine and the material to be sawn.

They can also be used in the reverse order - to find a sawblade specification from details of the material to be sawn and the feed speed preferred.

There are always certain factors which cannot be altered, perhaps the saw arbor speed or the types of sawblade available, etc. In these cases work out other details around these fixed factors.

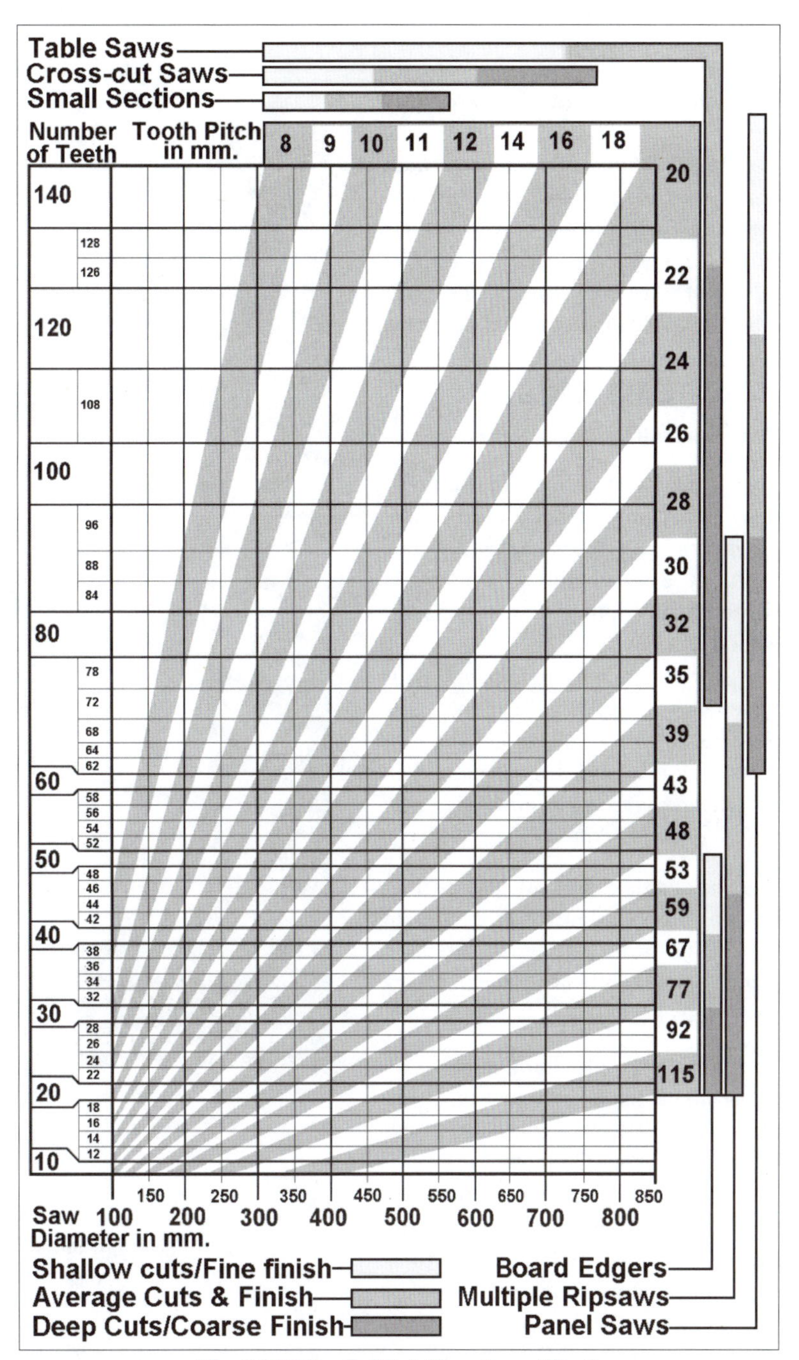

Fig. 16.73 Tooth Pitch/Number of Teeth

Tooth Pitch/Number of Teeth

This chart determines the tooth pitch from the diameter of sawblade to be used and the number of teeth in it - or the number of teeth for a given tooth pitch.

Saw X Locate **60** *(number of teeth)* at the left-hand scale and trace horizontally along to intersect with a vertical line from **300** *(sawblade diameter)* on the bottom scale, then follow the diagonal shaded band to indicate **16mm** *(the tooth pitch).*

Saw Y Locate **28** *(number of teeth)* at the left-hand scale and trace horizontally along to intersect with a vertical line from **350** *(sawblade diameter)* on the bottom scale, then follow the diagonal shaded band to indicate **39mm** *(the tooth pitch).*
NOTES: the tooth pitches shown are those commonly used for tungsten carbide tipped saws. In some instances manufacturers quote a given tooth pitch for a particular application - wide pitches for fast feeding of natural

timber and fine pitches for brittle plastic. Shown alongside is a guide for typical tooth pitches against machine types.:

Regulations in some countries require the use of a sawblade with chip-limiting shoulders for manual ripsawing - which in theory precludes the use of sawblades listed under 'Table Saws' having a pitch of less than 28mm.

Hook-Line Offset

(See Fig. 16.63)

This nomogram determines **(B)** the *hook-line offset* required for **(C)** the *sawblade diameter* and **(A)** the *hook angle* needed for the type and condition of the material to be sawn*, then set this measurement out as Fig. 16.76.

**Recommended hook angles for plate saws are given in Chapter 5, and on nomogram Fig. 16.63. Recommended hook angles for Stellite and tungsten carbide tipped saws are given in the following Notes.*

The hook angle can be calculated either by using the hook-line offset nomogram Fig. 16.74, or by setting-out the hook angle geometrically as shown in Chapter 6, Fig. 6.12.

To set-out a specific hook angle. Lay the sawblade flat on a wooden table and place a steel rule on the sawblade with its inside edge on the tooth point.

Keeping the rule edge on the tooth point, swivel it until it off-sets from the sawblade centre* by the amount indicated on the nomogram, measuring at right angles to the contact edge of the rule. See Fig. 16.76

To measure an existing hook angle. Lay a rule on the sawblade with its inside edge in line with the tooth front. Measure at right angles from the inside edge of the rule to the sawblade centre* and set-out this dimension, and that of the sawblade diameter, on the Nomogram Fig. 16.74 to determine the actual hook angle of the sawblade.

**To accurately pinpoint the sawblade centre fit a bush in the centre hole of the type used on a grinder, or measure to the nearest edge of the centre hole and add half its diameter.*

Use a square-ended engineers rule for measuring, or a square with outside edge markings - either will give a right-angle measurement.

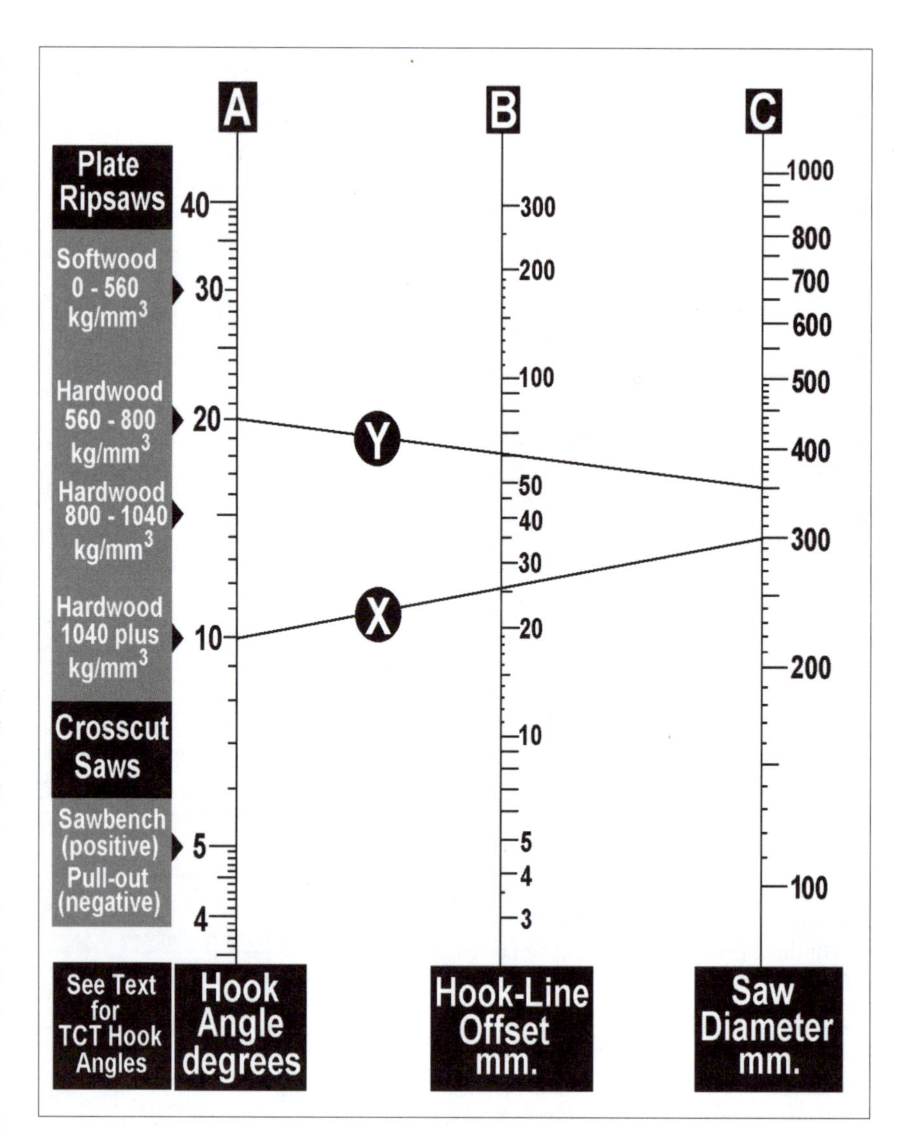

Fig. 16.74 (above) Hook-Line Offset

Hardness	10,000	9,500	16 - 17,000	65 - 80,000
Type	High Speed Steel	Stellite	Tungsten Carbide	Polycrystalline Diamond
Included Angle	40-45°	40-45°	50-55°	70-75°

Fig. 16.75 Minimum included angles for tips

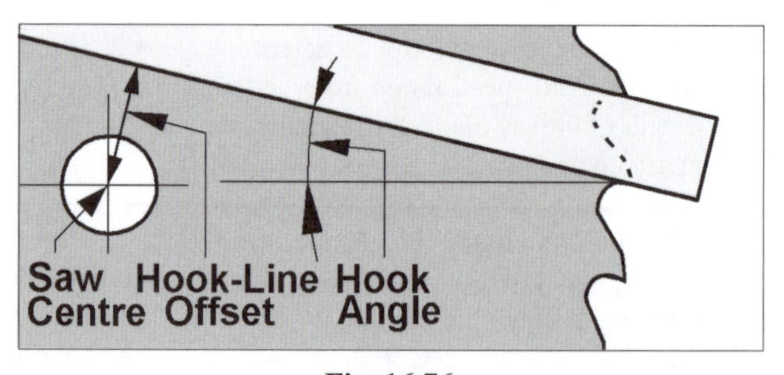

Fig. 16.76

This shows how to set out or check the hook angle of any sawblade by measuring the hook-line offset using a straight-edge and a graduated steel rule or outside-marked square.

Examples

Saw X. Draw a straight line from **300** *(sawblade diameter)* on scale **'C'**, to **10***(hook angle)* on scale **'A'**. Where this line crosses scale **'B'**, at **26mm**, indicates the required *off-set.*

Saw Y. Draw a straight line from **350** *(sawblade diameter)* on scale **'C'**, to **20***(hook angle)* on scale **'A'**. Where this line crosses scale **'B'**, at ***60mm***, indicates the required *off-set.*

**Read the following notes*

NOTES

Different types of tips require different minimum included angles, and this restricts the hook angle that can be used.

Recommended included angles for different sawblade tips are shown in Fig. 16.64.

The maximum practical hook angle is 90º less the minimum included and clearance angles combined. Based on Fig. 16.75, and assuming a regular clearance angle of 15º, the maximum practical hook angles are:-

HSS & Stellite 25º - 30º
Tungsten carbide 20º - 25º
Polycrystalline diamond 5º (This is based on diamond tip makers original recommendations, however, pcd saw makers have been regularly manufacturing sawblades with hook angles up to 15º.)

Regular hook angles for tungsten-carbide-tipped saws are as follows:-

For ripsawing natural timber.
Green timbers 25º - 30º
Seasoned timber 20º - 25º

For rip and Cross-cutting on table saws and panel saws
Sizing natural timber 20º
Sizing stacked panels 15º
Sizing single panels 10º
Sizing plastic materials and veneers 5º

For cross-cutting on radial-arm, ram, chop and mitre saws
Natural timber -5º
All panel materials -10º
Delicate mouldings (peg-tooth type) - 15º

Stellite-tipped saw hook angles are based on those for plate saws, see Chapter 5.

Insert-tooth hook angles are pre-determined by the maker, do not change.

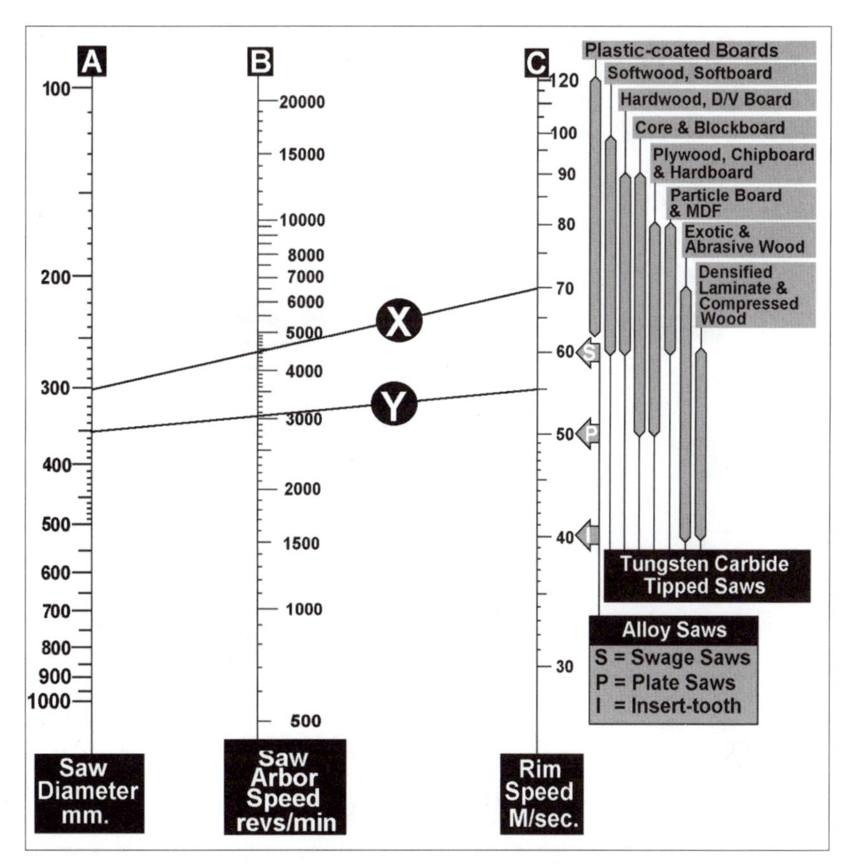

Fig. 16.77 Saw Arbor Speed

Saw Arbor Speed

This nomogram determines **(B)** the *saw arbor speed* needed for **(A)** the *diameter of the sawblade* to be used, to give the **(C)** the *rim speed* required for the material to be sawn.

It is used as shown to give the correct saw arbor speed for machines provided with an infinitely-variable speed saw arbor. On machines with fixed arbor speeds, instead check each speed against the sawblade diameter, and from this choose the one that gives a suitable rim speed for the material to be sawn.

Examples

Saw X Draw a straight line from **300** *(sawblade diameter)* on scale **'A'** to **70** *(low/mid-scale rim speed for plastic faced board)* on scale **'C'**, to indicate a *suitable saw arbor speed* of **4450** revs/min.

Saw Y Draw a straight line from **350** *(sawblade diameter)* on scale **'A'** to **55** *(suitable rim speed for Stellite tipped ripsaws for natural timber - based on alloy steel saws)* on scale **'C'**, to indicate a *suitable saw arbor speed* of **3000** revs/min.

NOTES:

Arrows **S, P** and **I** indicate the traditional rim speeds for alloy steel swage, plate and insert-tooth ripsaws, respectively. Rim speeds for these sawblades can, in practice, range between **60-70M/sec** for *Swage saws*, **50-70M/sec** for Plate saws and **40-60M/sec** for *Insert-tooth saws (the higher rim speeds in the latter case are for small diameter sawblades as used on edgers).*

The bands alongside give suitable rim speeds for tungsten carbide-tipped saws. They apply to various materials, e.g., 60 - 100M/sec for softwood and 40 -60M/sec for compressed wood.

Because a range of rim speeds is shown for the materials listed, a range of feed speeds can be also be used for both of the two sawblades, e.g., from **3700** to **4800** revs/min for saw **X**, and from **2800** to **3400** revs/min for saw **Y.**

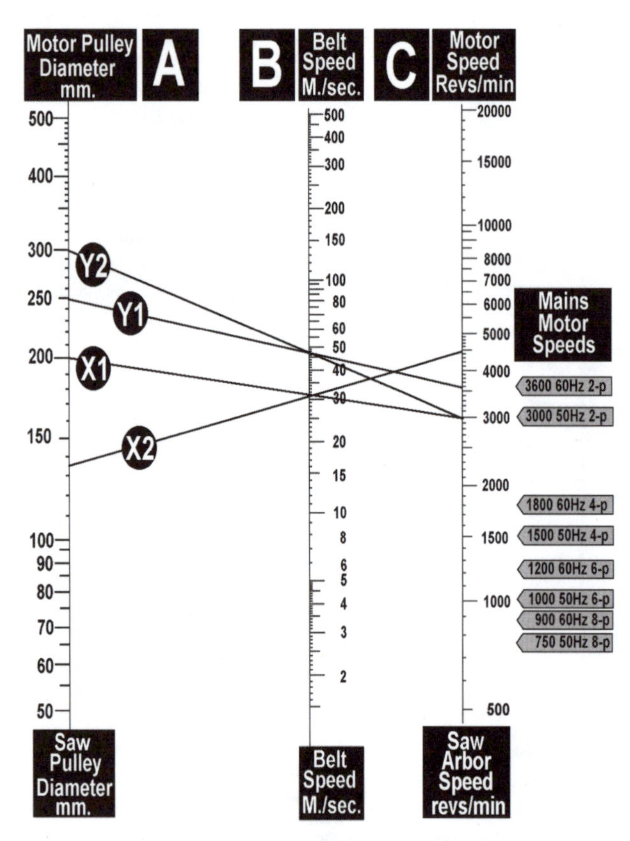

Fig. 16.78 Speeds & Pulley Sizes

Speeds & Pulley Sizes

This nomogram determines **(B)** the *belt speed* from **(C)** the *motor speed* and **(A)** the *motor pulley diameter - for this calculation use the top headings.*

From this can be determined **(A)** the *saw pulley diameter* from **(C)** the *saw arbor speed required (from nomogram 16.66)* by crossing at **(B)** the same *belt speed - for this calculation use the bottom headings.*

The same nomogram can be used to calculate pulley sizes for any belt drive.

Mains motor speeds are given to the right-hand side of scale 'C', and show motor speeds for 2, 4, 6 and 8-pole scr motors running off both 50Hz and 60Hz supplies. Saw motors normally are of the 2-pole type, drive motors usually have 4, 6 or 8 poles and run slower.

Examples

Saw X. Draw a straight line from **3000** *(50Hz, 2-pole motor speed)* on scale **'C'** to **200** *(motor pulley diameter)* on scale **'A'**, to indicate a *belt speed* of **31** M/sec on scale **'B' (X1).**
Then draw a straight line from **4450*** *(required arbor speed)* on scale **'C',** through the same point on scale **'B' (31)** to indicate **135mm** *(saw arbor pulley diameter)* on scale **'A' (X2).**

Saw Y. Draw a straight line from **3600** *(60Hz, 2-pole motor speed)* on scale **'C'** to **250** *(motor pulley diameter)* on scale **'A'**, to indicate a *belt speed* of **44** M/sec on scale **'B' (Y1).**
Then draw a straight line from **3000*** *(required arbor speed)* on scale **'C'**, through the same point on scale **'B' (44)** to indicate **300mm** *(saw arbor pulley diam)* on scale **'A' (Y2).**

* From nomogram 16.77.

Feed Speeds

See Fig. 16.79 for tungsten carbide tipped saws **& Fig. 16.80** or alloy steel, Stellite-tipped and insert-tooth saws. These nomograms determine **(C)** the *number of teeth* per minute from **(B)** the *number of teeth** in the sawblade and **(D)** the *saw arbor speed (from nomogram 16.66).*
From this can be determined **(A)** the *feed speed* from **(E)** the *tooth bite* by crossing **(C)** at the same *number of teeth* per minute.

**For flat-topped teeth use the actual number of teeth, for group teeth consisting of pairs (alternate bevel and flat/ triple-chip) use half the number of teeth. The calculations assume the use of flat-topped teeth in both cases. If paired group teeth are used the feed speed would be half that given to produce a comparable sawn-surface finish.*

On both nomograms the right-hand side of scale **'E'** shows a suitable range of bites (as the feed-per-tooth) for each of several different materials. As the centre band shows, smaller bites (shown lighter) give a better finish at a slower feed speed, and bigger bites (shown darker) give a poorer finish at a faster feed speed.

When converting natural timber a smaller bite is better for dry, hard and brittle timbers, and a bigger bite is better for wet, soft and stringy woods. The biggest bites should be used, where practical, with abrasive timbers. In all cases, whatever bite is used, this must be within the bands shown.

Examples

Saw X, Fig. 16.79. Draw a straight line **(X1)** from **4450** *(saw arbor speed)* on scale **'D'** to **60** *(number of teeth)* on scale **'B'** to cross scale **'C'** at **267**(000) *(number of teeth/min)***.**
Draw a second straight line **(X2)** from **0.035** *(mid-range for plastic laminated board)* to cross scale **'C'** at the same point **(267)** and indicate **9.34 M/min** on scale **'A'** as a suitable *feed speed* **(X2).**

Saw Y, Fig. 16.80. Draw a straight line **(Y1)** from **3000** *(saw arbor speed)* on scale **'D'** to **28** *(number of teeth)* on scale **'B'** to cross scale **'C'** at **84**(000) *(number of teeth/min)***.**
Draw a second straight line **(Y2)** from **0.60** *(mid-range for softwood)* to cross scale **'C'** at the same point **(84)** and indicate **50** M/min on scale **'A'** as a suitable *feed speed***.**

NOTE
As each of the materials listed shows a range of bites *(feed-per-tooth)*, a corresponding range of feed speeds could also be used.

For example for saw **X** the feed speed could range from **8M/min** for a *fine finish* to **14M/min** for a *coarse finish.*

For saw **Y**, the feed speed could range from **30M/min** for a *fine finish* to **70M/min** for a *coarse finish.*

If a different saw arbor speed is used, of course, this affects the feed speed and/or the quality of the sawn finish. There is not a free choice of arbor speeds, of course, this is limited by what is provided on the machine - check these with a speed calculator or use nomogram 16.67. Sawblades are prepared to run within a specific speed-range, so consult the makers if running outside normal limits.

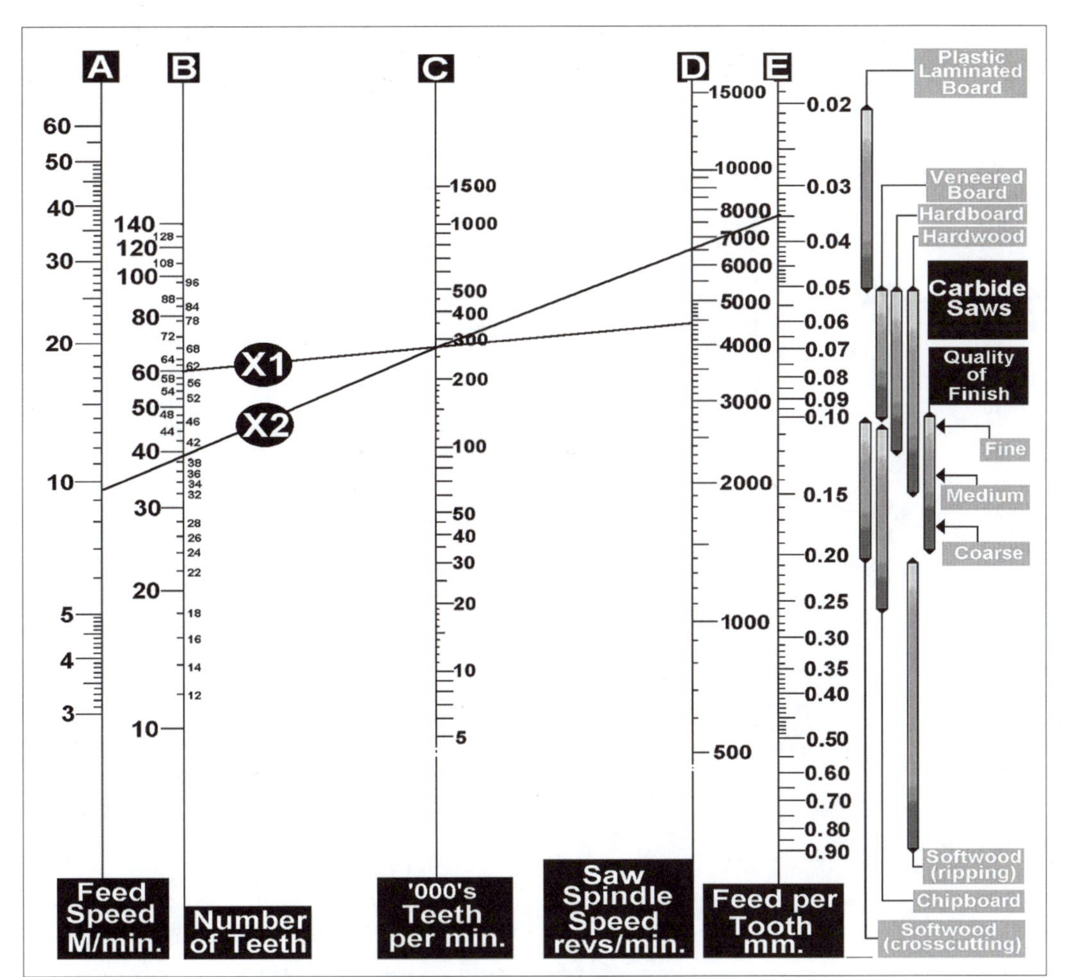

Fig. 16.79 (Left) Feed Speeds

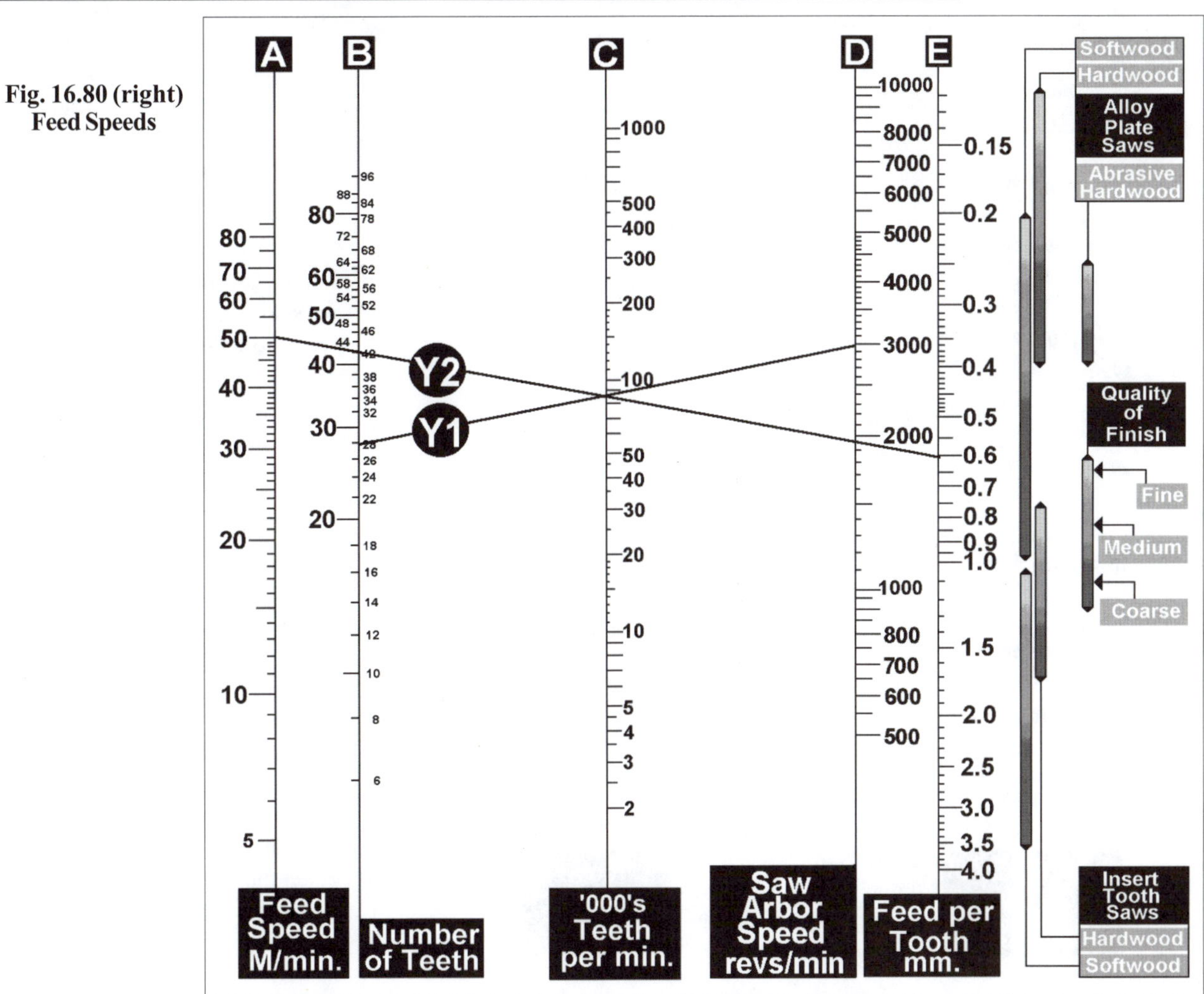

Fig. 16.80 (right) Feed Speeds

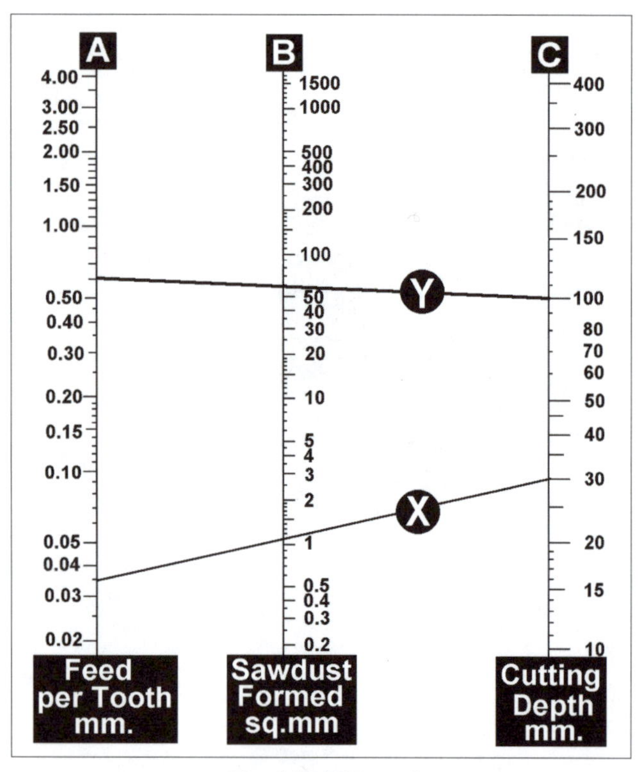

Fig. 16.81 Waste Space

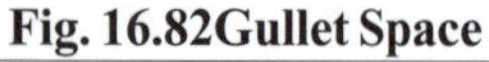

Fig. 16.82Gullet Space

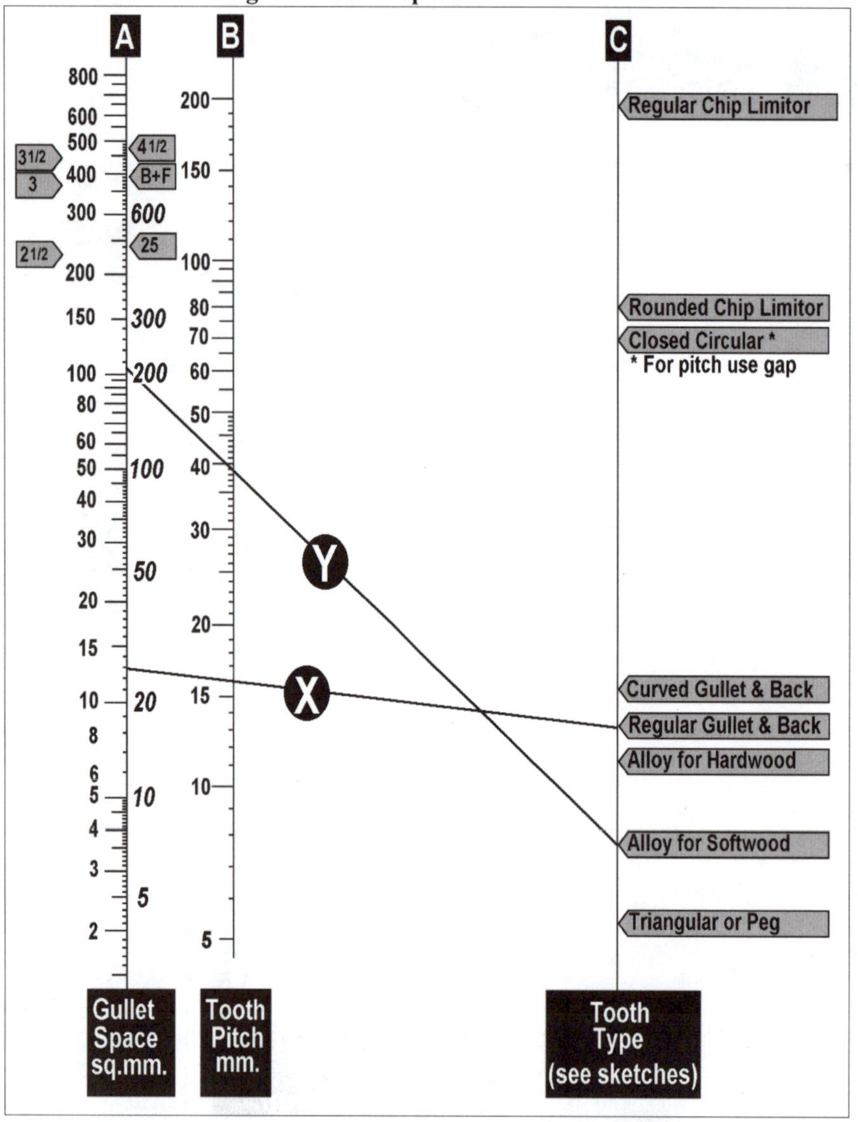

Waste & Gullet Space

Nomogram 16.81 determines **(B)** the *area of waste* from **(A)** the *feed-per-tooth (bite, from nomograms 16.68 or 16.69)* and **(C)** the *depth of cut.*
Nomogram 16.82 determines **(A)** the *gullet space* from **(C)** the *tooth type* and **(B)** the *tooth pitch.*

The two nomograms are used in sequence to determine whether or not the gullet is large enough to contain the waste generated during a single pass of each tooth, as follows:-

Examples

Saw X. Fig. 16.81. Draw a straight line from **30** *(depth of cut)* on scale **'C',** to **0.035** *(feed per tooth)* on scale **'A'**, and note where this crosses scale **'B'** to show **1.05** sq.mm of *waste* formed at a single pass.

Saw X. Fig. 16.82. Draw a straight line from **'Regular gullet and back'** *(type of tooth)* on scale **'C'** through **16mm** *(tooth pitch)* on scale **'A'** to indicate **13 sq.mm** of *gullet space.*
*In this case the waste of **1.1sq.mm** of waste generated is easily contained within the **13sqmm** of gullet space.*

Saw Y. Fig. 16.81. Draw a straight line from **100** *(depth of cut)* on scale **'C'**, to **0.60** *(feed per tooth)* on scale **'A'**, and note where this crosses scale **'B'** to show **60 sq.mm** of *waste* formed at a single pass.

Saw Y. Fig. 16.82. Draw a straight line from **'Alloy for softwood'** *(type of tooth)* on scale **'C'** through **39mm** *(tooth pitch)* to indicate *100 sq.mm* of *gullet space* on scale **'A'**.
*In this case the waste of **60sq.mm** of waste generated can be held comfortably by the **100sq.mm** of gullet space even when using the scale for fine dust when, in fact, the coarse dust scale would apply.*

NOTES
When sawing any material, the waste generated by each tooth at each pass is held in the gullet during the cut, then discharged when clear.

The waste expands when turned to dust and chips and so takes up more room than in its original solid state. The gullet must be large enough to contain all the expanded waste generated during a single pass, otherwise waste will pack into the gullet and can stall the saw.

For most sawing applications this is not a problem, but it can occur if sawing fast and deep while converting natural timbers, or if sawing stacked panels with a narrow-pitch saw.

The waste generated and the gullet space are both expressed as an area, not a volume. This is because variations in kerf or cutting widths do not count - the gullet space is increased or decreased in proportion to this factor.

The left-hand section of scale **'A'** of nomogram **16.81** is based on a waste expansion rate of **3 times**, the regular amount with fine tooth pitches on finished trimming.

When ripsawing natural timber at fast feed speeds larger pieces are broken off and, as these are not full broken down, the sawdust remains in a more compressed form.

Because of this the expansion rate of the waste can be anything between **1½** and **3** times that of the solid timber. - compared to a regular rate of **3** times for material converted with a smaller bite. As a result, the figures given for gullet space - **when fast-feeding natural timber only** - can be increased by up to **100%** - as indicated by the right-hand italic section scale **'A'**.

Because of this factor, and differences in the actual gullet area of individual sawblades, these nomograms should be taken only as a general guide - basically to warn if this factor needs more precise calculation.

The arrowed figures near the top of scale **'A',** Fig. 16.71 show the effective gullet areas of insert-tooth saws - which do not vary with pitch or numbers of teeth. As these are used mainly for fast-feed sawing of natural timbers, an expansion rate of **1½** times applies in these calculations - read off the left-hand section of scale **'A'**.

The following sketches show the tooth shapes annotated in Fig. 16.82

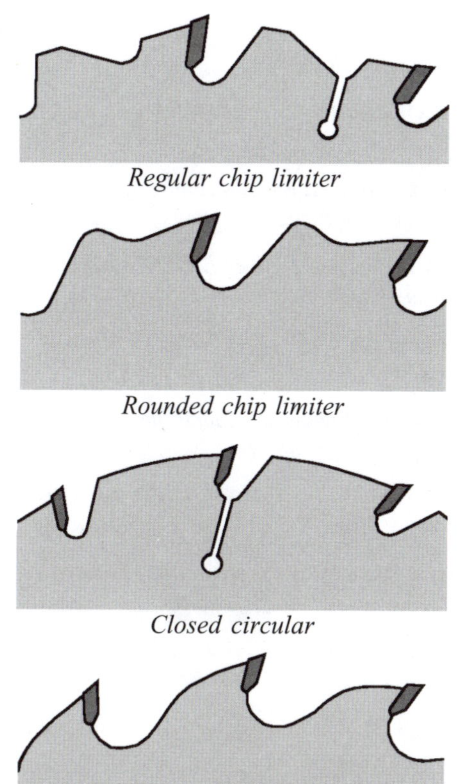

Regular chip limiter

Rounded chip limiter

Closed circular

Curved gullet and back

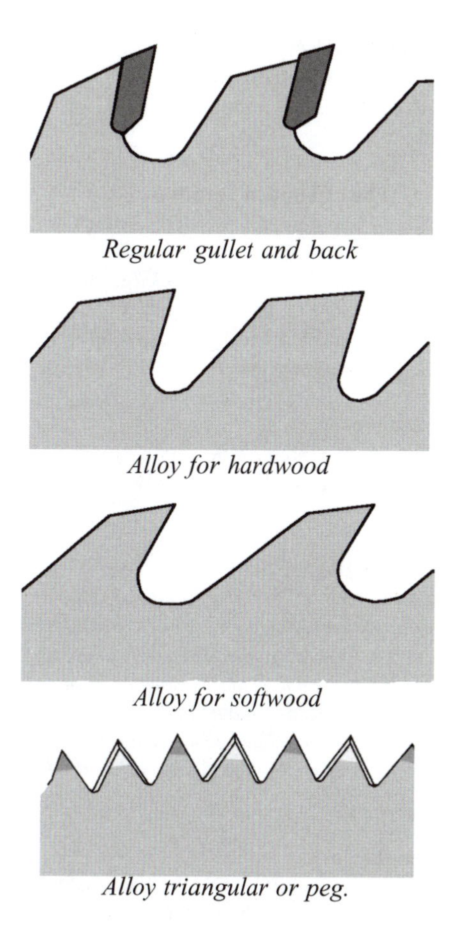

Regular gullet and back

Alloy for hardwood

Alloy for softwood

Alloy triangular or peg.

Calculating Non-standard Gullet Areas

For gullet shapes that do not correspond with any of those sketched above, the effective gullet area for fine dust can be found by using the graph below.

Draw the full-size tooth, tip-to-tip on a piece of tracing paper and place this over the graph. Count the number of squares enclosed by the gullet and multiply by ten to find the effective gullet space.

Use this figure instead of calculating the gullet space as shown in Fig. 16.82.

The sketch below the graph shows a typical gullet space within which the squares should be counted.

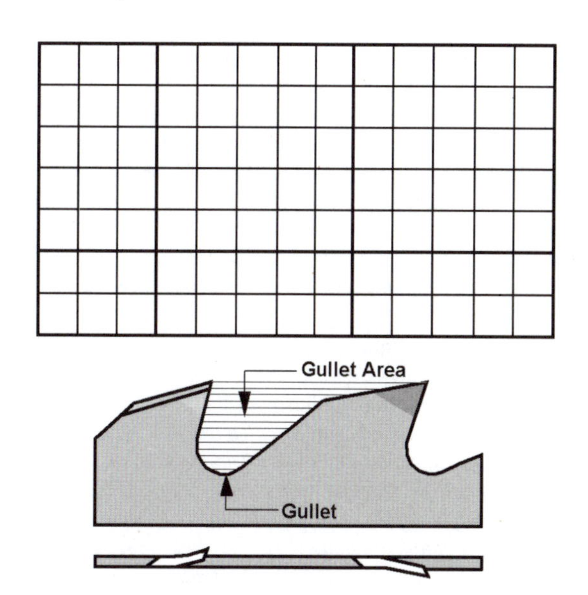

Woodworking Associations & Institutes

British Woodworking Federation,
56-64 Leonard Street, London EC2A 4JX, UK. E-mail: bwf@bwf.org.uk

The City & Guilds of London Institute
1 Giltspur Street, London EC1A 9DD. Contact: www.city-and-guilds.co.uk
City & Guilds has for many years provided a range of recognised vocational qualifications for the woodworking industry. The qualifications are hightly regarded, and accepted world-wide as indicating the level of competence of the individual.

FIRA International
(Furniture Industry Research Association)
Maxwell Road., Stevenage, Herts SG1 2EW, UK.

Forest Research
Sala St., Private Bag 3020, Rotorua New Zealand. E-mail www.forestresearch.co.nz. *Forest Research is a major provider of technology solutions and research services to the forest and wood products sectors in New Zealand and to a wide international cliental. Projects cover the full value chain in genetics, forest management, wood processing, product development, and market research. It conducts research and commercial consultancies with private and government agencies and it develops added-value products and processes.*

Forintek Canada Corp.
2665 East Mall, Vancouver BC, V6T 1W5 Canada. Phone (604) 224-3221. *Forintek Canada Corp. is a private, non-profit solid-wood products research and development corporation for the solid wood products industry manufacturing lumber, plywood, oriented strand board, MDF, particleboard, and value-added products. Research includes advancement in processing technology, and in-depth characterization studies to match market demands with Canadian forest resources.*

Forest Products Laboratory.
One Gifford Pinchot Drive, Madison WI 53705-2398 USA, Contact http;//www.tpi.ts.fed.us/. *FPL research concentrates on pulp & paper products, housing and structural uses of wood, wood preservation, wood and fungi indentification, and finishing and restoration of wood products. FPL is responding to environmental pressures on the forest resource by using cutting-edge techniques to study recycling, develop environmentally friendly technology, and understand ecosystem-based forest management.*

The Institute of Machine Woodworking Technology,
St. Keynes, Bowl Road., Charing, Kent TN27 0HB, UK. Website: http://www.imwood.org.uk. *This is an association of individuals who share an interest in machine woodworking. Meetings and works visits are held in the various divisions, and a Journal is issued twice a year together with regular Newsletters. The Institute retains a list of experts who are prepared to undertake training, is deeply involved in training programmes and health and safety aspects, and operates a stand at Woodmex making and selling wooden articles for charity.*

Saw Doctor Association,
Ken Wall, 23 Hill Farm Court, Chinnor, Oxon OX9 4NX, UK. E-mail: sawdoctors@westmids.fsbusiness.co.uk
This is an Association of companies with interests in metal and woodworking tool manufacture and repair. The Association holds Seminars and Technology Days on wide-ranging subjects allied to tooling, issues freely-available Information Sheets on their correct use, and produces regular Newsletters. The Association is in the process of preparing a Saw and Tool Technicians Training Course.

TRADA Technology Ltd.,
Stocking Lane, Hughenden Valley, High Wycombe, Bucks HP14 4ND, UK. E-mail: Jvert@trada.co.uk.
TRADA technology is a professional and independent provider of research, consultancy, information and training services for the timber and construction industries. It employs qualified architects, engineers, technologists, business analysts and marketing professionals with extensive industrial and commercial experience.

Wood Machinery Manufacturers of America,
1900 Arch Street, Philadelphia, PA 19103-1498 USA.
WMMA maintains a current listing of American woodworking machinery and cutting tool manufacturers, and constantly updates this to include their latest innovations. Information on what to buy and where to buy it is readily available by calling 1-800-Buy-WMMA (USA) or by faxing 001 215 963 9785 (overseas).

Woodworking Machinery Suppliers Association,
The Carriage House, Sir Richard Arkwright's Cromford Mill, Mill Lane, Matlock, Derbyshire DE4 3RQ, UK. Contact: www.wmsa.org.uk.
The WMSA is the UK trade organisation representing suppliers of woodworking machinery, both engaged in manufacture and import of products. WMSA aims to benefit the commmercial interests of members in an international and competative market by the delivery of world class services and representation.

Saw Filers Associations.
Reproduced by kind permission of Armstrong Mfg, Co., from their web site http://www.armstrongblue.com. Listing shows areas and secretaries addresses & E-mail (in the US unless otherwise stated).

British Columbia: Allen Gallant, *Box 57006, Scott Town RPO, Surrey, BC V3V 7V7, Canada. E-mail: sawfilers@telus.net.*

Cal-Western: Ed Branscomb, *21700 John Hymen Rd., Fort Bragg, CA 95437.*

Intermountain: Dale Sparber, *Box 786, Omak, WA 98840.* E-mail: bigd@northcascades.net.

Lake Erie & Ontario: Ed Chapman *8929 Adamsville Rd., Hartstown, PA 16131.*

Lake States & Midwest: Ralph Manting, *623 Lilac, Big Rapids, MI 49307. E-mail: rmanting@hanchett.com.*

Maritime: Tim Parsons, *28 Back Lake Rd., Tilley, NB E7H 1B9, Canada. E-mail: timlorpa@nbnet.nb.ca*

New Zealand: Keith Wrigley, *PO Box 2242, Rotoroa, New Zealand*

Northeast: Woody Higgins, *243 Maple St., Bangor, ME 04401. E-mail: Woody Higgins@yahoo.com*

South Africa: Don Priest, *C/O Salma, PO Box 189, Sabie 1260, Johannesburg, S. Africa. E-mail: salmastc@iafrica.com.*

Southeastern: Sharon Mason Seymour, *333 Fleming Rd., Birmingham AL 35217. Phone:205-841-2299.*

Southern: Lee Stockton, *862 Wells Road., Vicksburg, MS 39180. E-mail: lees@vicksburg.com.*

Western: Bill Saily, Box 939, *Weaver Rd., Myrtle Creek, OR 97457. E-mail: saily@pioneer-net.com.*

Zimbabwe: A. Chituwi, Forest Industries Training Center, *Mutare City, Zimbabwe. E-mail: sgopo@border.co.zw*

INDEX

Motors

Other Tools

Power supplies

Sawblade Types

Alloy

Insert-tooth Saws

Tungsten Carbide Tipped Saws

Saw Body Maintenance

Sawing Machinery

Sawing Techniques

Saw Tips

Saw Tooth Geometry

Saw Tooth Profiles

REFERENCES

The following companies kindly and willing supplied information on their products, or those of the companies they represent, and gave permission to reproduce their photographs and data - without which it would have been impossible to complete this publication. I am very grateful to them for their cooperation, and thank them for the invaluable help they gave.
The listing shows the type of equipment these companies manufacture or supply, grouped vertically. Agents & Representatives are shown in italics. The table is a guide only to manufacturers of the tools and machines detailed in the book, not a complete, world-wide listing - although every care has been taken to ensure accuracy of the details at the time of going to press.

SAWS & TOOLS
- Circular saws cv, hss, carbide & diamond .... A
- Hard tips and tipping materials .... B
- Saw brazing and tipping equipment C
- Ancillary equipment (mounting devices, etc.) .... D

SAW SERVICING EQUIPMENT
- Grinders for cv, Stellite & carbide saws .... E
- Eroding and grinding machines for diamond .... F
- Rolling, tensioning, testing & cleaning equipment .... G
- Grinding wheels, Grit, Diamond & CBN .... H

PRIMARY & SECONDARY LOG BREAKDOWN MACHINERY
- Band and circular log saws I
- Log profiling lines .... J
- Slabbers, Circular resaws, Edgers, Multi-ripsaws L
- Cut-off Saws .... M

TABLE & CROSS-CUT SAWS
- Table saws, fixed and portable .... O
- Sliding-table panel-cutting saws .... P
- Cross-cut and Mitre saws .... Q
- Safety & Ancillary Equipment .... R

TRAVELLING-HEAD PANEL SAWS
- Vertical Panel Saws .... S
- Horizontal Panel Saws .... T
- Angle Plants, Special Panel Saws & High-production, multi-saw machines .... U

OTHER MACHINES USING SAWS, & ANCILLARY EQUIPMENT
- Optimizing Equipment for any of the above V
- Machining Centres .... W
- Window & Door Lines X
- Edge Banders .... Y
- Other Machines using saws, listed by page number ....

Aigner, Thannenmais, D-94419, Reisbach, Germany .... (Safety Equipment) ....

Almac (Adrian Law Machinery,) Ltd, 7 Marshall Court, Marshall Street, Leeds LS11 9YJ, UK .... IJL ....

AKE Knebel GmbH & Co, Holzlestr 14 - 16 D-72336 Balingen-Engstlatt-Sud, Germany .... ABD ....

Allenwest Electrical Ltd., Eastergate Rd. Brighton BN2 4QE UK .... R ....

Allied Machinery Ltd, Star Works, Lupton's Buildings, Tong Road, Leeds LS12 3BG, UK .... A .... LM .... T .... VWX

Allied Tooling Ltd., Unit 2, 19 Willis Way, Poole BH15 3SS .... ABC ....

Allwood Machinery Essex Ltd, Unit 3, Galliford Rd, Maldon, Essex CM9 4XD, UK .... LM .... OPQ X

Alpine Engineered Products Inc, 2820 N Great SW Parkway, Grand Prairie, TX 75050, USA. .... M ....

Wilhelm Altendorf , GmbH & Co KG, Wettiner Allee 43/45, D-32429 Minden, Germany .... OP ....

AnThon GmbH & Co, Schaferweg 5, D-24901, Flensburg, Germany .... TU V

Armstrong Manufacturing Co, 2135 NW 21st Ave, Portland, Oregon 97208, USA .... CD .... EGH ....

Atkinson-Walker (Saws) Ltd, Bower Street, Sheffield S3 8RU UK .... AD ....

Autool Grinders. Ltd, Whalley Road, Sabden, Clitheroe, Lancs. BB7 9DZ, UK .... BC .... EG ....

Baker Products, PO Box 128, Ellington, MO 63638, USA .... ILM ....

Barr-Mullin Inc, 2506 Yonkers Rd, Raleigh, NC 27604-2241, USA .... LM V

Bauerle Maschinenfabrik GmbH & Co, D-73560 Bobingen/Rems, Bahnhofstrasse 25, Germany .... Q ....

Ernest Bennett (Sheffield) Ltd, Main St. N Anston, Sheffield, S Yorks S25 4BD, UK .... A ....

Biele SA, Bo: Urestilla, Apartador 80, E-20730, Azpeitia, Guipuzkoa, Spain .... TU V

Biele (UK) Ltd, Systems House, 24 Wilton Court, Newton Aycliffe, Co Durham DL5 7PU, UK .... (Automation Specialists) ....

Biesse Group, Via Della Meccanica, 16, 61100 Pesaro, Italy .... *(Automation Specialists) ....*

Biesse Group UK Ltd., Lamport Drive, Daventry, Northants NN11 5YZ, UK .... (Automation Specialists) ....

Birmingham Sawblades Ltd., Old Hill, Cradley Heath, Warley, W. Mid. B64 6PL, UK .... A ....

University of Birmingham , Edgebaston, Birmingham B15 21T, UK, ... Metal Cutting/Material Removal Research

Circular Saws

Heinrich Brandt , Maschinenbau GmbH, PO Box 6108, D-32647 Lemgo, Germany ... Y

MA Brevetti SrL, Via S Antonio 33, I-33080 Cecchini (PN), Italy ... Q ... V

Brook-Hansen - See Electrodrives

Bottene, Via Paraiso, 36/38 - I-36015 Schio (VI) Italy ... LM ..... Q ... V

Calder Wilkinson Aim Ltd, Station Road, Sowerby Bridge, W Yorks HX6 3LA, UK ... OPQR . T ... V

Carborundum Abrasives GB Ltd, Trafford Park Road, Trafford Park, Manchester M17 1HP, UK ... H ...

Gianfranco Casadei SpA, I-47040, Villa Verucchio (Fo), Italy ... OP ...... T ... V

Casolin , I-36036 Torrebelvicino (Vicenza) Via Marchioro 9 , Italy ... OP ...... T ...

J Chaland & Fils SA, 36 Ave de Bobigny, BP/PO Box 54, 93135 Noisy le Sec Cedex, France D ... G ... 188

Conway Saw Ltd, New Swan Lane, West Bromwich, W Midlands B70 0NS, UK ... ABCD ... LM ...... OPQ ... T ... Y

Cosmec, 53036 Poggibonsi SI - Loc. Fosci 28, Italy ... L ... W

A Costa SpA, Via Monte Pasubio 150, I-360 10 Zane (Vicenza), Italy ... L ...

CTD Machines, Inc, 2300 East 11th St. Los Angeles, CA 90021, USA ... M ...... Q ...

Cutler-Hammer , 4201 North St, Milwaukee, WI 53216 - 1897, USA ... 50

A L Dalton Ltd, Crossgate Drive, Queen's Drive Ind Est Nottingham NG2 1LW, UK ... E ... OPQR . S ... VWY

De Beers Industrial Diamond Division, Shannon, Co. Clare, Ireland. www.debid.ie ... B ... (Diamond & CBN Products) ...

Deloro Stellite Inc, 471 Dundas St. East, Belleville, Ontario K8N 5C4, Canada ... B ... (Hard tipping materials) ...

Deloro Stellite Ltd, Cheney Manor Ind, Est, Swindon, Wiltshire SN2 2PW, UK ... B ... (Hard tipping materials) ...

Delta International Machinery Corp, 246 Alpha Drive, Pittsburgh, PA 15238, USA ... A ... OPQR ...

DeWalt Ind. Tool Co, 626 Hanover Pike, Hampstead, Maryland 21074, USA ... A ... OPQR ...

DeWalt Power Tools, 210 Bath Rd, Slough, Berks SL1 3YD, UK ... A ... OPQR ...

Didac Ltd, Kingswood House, South Rd, Kingswood, Bristol BS15 2JF ... (Training) ...

Diehl Machines, 981 S Wabash St, PO Box 465 Wabash, Indiana 46992, USA ... L ...

Dustraction Ltd., Box no 75 Mandervell Rd., Oadby, Leicester LE2 5ND ... (Dust extraction equipment) ...

Dymet Alloys Ltd, Station Rd West, Ash Vale, Hampshire GU12 5QT, UK ... B ... (Hard tipping materials) ...

Dynashape Ltd., 117 Station Rd., Old Hill, Cradley Heath, Warley, W. Mid.B64 6PL, UK ... A ...

Elcon , PO Box 72 - 2450 AB Leimuiden, Holland ... STU ...

Electrodrives Ltd, Cakemore Rd, Rowley Regis, Warley, W Midlands B65 0QT, UK ... (Motors & Control Gear) ...

Equipment Ltd, PO Box 3580, Hickory, NC 28603, USA ... ABCD .. EGH ...

ETP Transmission, Box 1120, S-581 11 Linkoping, Sweden ... (Hydraulic Sleeves) ...

Eudora GmbH, Gunskirchner Strasse 19, A-4600 Weis, Austria ... Q ...

Felder Maschinenbau, Heiligkreuzerfeld 18, 6060 Hall in Tirol, Austria ... AD ... OP ...

Felder UK Ltd, Unit 80-82, Tanners Dr., Blakelands Est, Milton Keynes, MK14 5BP, UK ... AD ... OP ...

Foley-Belsaw Institute, 6301 Equitable Rd, Kansas City, MO 64120, USA ... (Filers & Setters) ...

Format-4, Loreto 42 A-6060 HALL in Tirol, Austria ... AD ... OP ...

Format-4, *Unit 80-82, Tanners Dr., Blakelands Est, Milton Keynes, MK14 5BP, UK ... AD ... OP ...*

Frankland Grinding Products, 35 Woodfield Ave Accrington, Lancs BB5 2P ... (Filers & Setters) ...

Freud, Via Padova, 3 Zona Ind le Feletto Umberto 33010 Tavagnacco (Udine) Italy ... A ...

Freud Tooling UK Ltd, Unit 3, Emmanuel Trd Est, Springwell Rd., Leeds LS12 1AT ... A ...

Gabbiani Macchine, via Statale Marecchian 34, I-47827, Villa Verucchio (RN) Italy ... TU V

General Electric Company, Worthington, Ohio 43085, USA ... (Diamond & CBN Products) ...

Gerrymet Ltd, 5/6 Maybrook Ind Est Brownhills, Walsall, W Midlands WS8 7DG, UK ... AD ...

Giben Impianti SpA 40065 Pianoro, Bologna, Italy ... TU V

Giben UK Ltd, Unit 4, Central Court, Finch Close, Nottingham NG7 2NN, UK ... TU V

Gomex Tools Ltd, Finedon, Wellingborough, Northants NN9 5JF, UK ... A ...

Greenall's WW Machinery Ltd, Unit 2, Sovereign Business Park Bontoft Ave, Hull HU5 4HF, UK ... TU V

GreCon Dimter (See Michael Weinig) ... (Finger Jointers) Q ... V_

Guhdo-Werk, Herbert Dorken GmbH & Co KG, D-42905 Wermelskirchen, Germany ... A ...

Hallamshire Hardmetal Products Ltd, 315-317 Coleford Rd, Sheffield S9 5NF, UK ... (Hard tipping Materials) ...

Hanchett Manufacturing Inc, 906 North State St, Big Rapids, MI 49307 ... CD ...... EG ...

Hendrick , RWH Industries Inc, 32 Commercial Street, Salem MA 01970, USA ... ST ......

Holz-Her - Reich Spezialmaschinen GmbH, D-72622 Nurtingen, Plochinger Str., Germany ... S ... V

Holzma -Maschinenbau GmbH Holzmastrasse 3, D-75365 Calw-Holzbronn, Germany ... TU ...... VY

Holztechnik Machinery Ltd, Unit 8, Erivan Park, Wetherby, W Yorks LS22 4DN, UK ... L ... T ...

Homag Maschinenbau AG, Homagstrasse 3-5 D-72296 Schopfloch, Germany ... STU .... V Y

Homag Espana SA, PO Box 35 E-08480 L'Ametila del Valles, Barcelona, Spain ... STU ....

Homag UK Ltd, The Old Exchange, Holmer Green, High Wycombe, Bucks HP15 6SU, UK ... STU .... VY

Homag UK Ltd, Old Parkside, Estcourt Road., Darrington, Pontefract, W Yorks WF8 3AJ ... STU .... VY

Interwood Ltd, Stafford Ave, Hornchurch, Essex RM11 2ER, UK ... OPQ ... T ...

Iseli & Co AG Maschinenfabrik CH-6247, Luzernerstrasse 31, Schotz, Switzerland ... BC ... EG ...

JKO Ltd, Hughenden Ave High Wycombe, Bucks HP13 5SQ, UK ... ST ... W

Jordan Woodworking Machinery Ltd, Houghton-le-Spring, Tyne & Wear, DH4 4UG, UK ... OPQR ...

Jonsereds AB, WMD S-433 01 Partille, Sweden ... L ...

Gerhard Kock , GmbH & Co KG, Industriesgebiet Greste, 33818 Leopoldshohe, Germany ... (Borers & Dowel drivers) ...

Kock UK (Sales), Paul Tipler, 5 Hardrush Fold, Manchester M35 9GQ, UK ... (Borers & Dowel drivers) ...

Kolle Maschinenbau GmbH, Mettinger Strasse 103-105, D-73728 Esslingen, Germany ... OP ...

KWO-Werkseuge GmbH Aalener Strasse 44 , 73444 Oberkochen, Germany ... AD ...

KWO Tools (UK) Ltd. , 4 Strawberry Vale, Vale Rd, Tonbridge, Kent TN9 1SJ UK ... AD ...

Lancashire Saw Co. Ltd., Imperial Mill, Gorse St., Blackburn BB1 3EU UK ... ACD ... EGH . IJLM ... OPQR ...

Leicester Wood Technique Ltd., Main St. Theddingworth, Lutterworth, Leic. LE17 6QY ... A ... (Special Machines) ... X

Gebr Leitz, GmbH & Co, PO Box 1229, D-73443, Oberkochen, Germany ... AD ...

Leitz Tooling (UK) Ltd, Flex Meadow, The Pinnacles, Harlow, Essex CM19 5TN, UK ... AD ...

Hans Lenze Maschinenfabrik, KG Postfach 100, D-4923 Extertal 1, Germany ... (Variable-speed drives) ...

Leuco Oertli: Ledermann GmbH, Stadionstr 2, D-72160 Horb am Neckar, Germany ... AD ...

Leuco (Great Britain) Ltd, Twyford Bus. Park, Bishops Stortford, Herts CM23 3YT, UK ... AD ...

Licom Systems, Licom House, Davenport Road, Coventry CV5 6PY, UK ... V

Linck Holzverarbeitungstechnik GmbH, Appenweierer Strs. 46, D-77704 Oberkirch, Germany ... IJLM ... V

Maggi Engineering, I-50052 Certaldo (FI) - Via delle Regioni 299, Italy ... Q ...

Otto Martin , Maschinenbau GmbH & Co, Postfach 1160, D-87720, Ottobeuren, Germany ... P ...

Matsushita Choko Sangyo Co, Ltd, 7-27 Hikuma, 2-Chome, Hamamatsu City 430, Japan .... A ... EG ...

Metcalfe & Tattersall Ltd, Brookside Mill, New Lane, Oswaldtwistle, Lancs, UK ... AD ...

Morbidelli SpA, Strada Montefeltro 81/3, 61100 Pesaro, Italy ... W

Multico Ltd, Paragon House, Flex Meadows, Pinnacles, Harlow, Essex CM19 5TJ, UK ... AD ... OP ...

Ney Ltd, Falkland Close, Charter Ave Ind. Est., Tile Hill, Coventry CV4 8UA, UK ... A ... P ... TU ... VY

NLS Tools Station Approach, Eleanor Cross Rd.,Waltham Cross, Herts., EN8 7LZ, UK ... AD ...

NMA (Agencies) Birds Royd Lane, Brighouse, W Yorks. HD6 1LQ, UK ... Q ... T ...

North American Products Corp, 1180 Wernsing Rd, Jasper, Indiana 47546, USA ... AD ...

Northfield Foundry & Machine Co, 320 N. Water St, Northfield, MA 55057, USA ... AD ...

Norton Abrasives Ltd, Welwyn Garden City, Herts, UK ... H ...

Omga SpA Via Carpi Ravarino 146 41010 Limidi MO, Italy ... M ... Q ... V

Onci RBD sa ZT 36, Avenue de Bobigny-BP 90 93135 Noisy-le-Sec Cedex, France ... AD ...

Origin Framing Supplies, Ridges Yard, 107 Waddon New Road, Croydon, Surrey CR0 4JE, UK ... Q ...

The Original Saw Co, 1465 Third Ave SE, Britt, Iowa 50423 USA ... Q ...

Pacific Grinding Wheel Co, Inc, 13120 Highway 99, Marysville Washington 98270, USA ... H ...

Pacific Hoe , 2700 SE Tacoma St., Portland, Oregon 97202, USA ... A ...

Ant. Panhans , GmbH, Postfach 140, D-72488 Sigmaringen, Germany ... AD ... L ... R ... T ... V

Paoloni Macchine SrL Via F Meda, I-61032, Fano (PS) Italy ... P ...

Paul Maschinenfabrik GmbH & Co, D-88525 Dumentingen, Germany ... LM ... Q ... V

Paul Saws & Systems , Unit 7, New Mills Ind Est, Inkpen, Hungerford, Berks RG17 9PU ... LM ... Q ... V

John Penny WW Machinery Ltd, Abingdon Science Park, Oxfordshire OX14 3YT, UK ... AD ... OP ...

Walter Perske , GmbH, Friedrich-Ebert-Strasse 80-84, D-6800, Mannheim1, Germany ... (Motors) ...

E Petschauer , GmbH 72411 Bodelshausen, Erienbrunnenst 5, Germany ... BCD .... EG ...

PF Cutting , 2820 N Great SW Parkway, Grand Prairie TX 75050, USA ... Q ... V

Powermatic , 619 Morrison Street, McMinnville, Tennessee 37110, USA ... OPQ ...

Precision Sales Ltd, Derwent House, The Bridge, Milford, Derbyshire DE56 0RR UK ... OP ... T ...

Primulti Filli & Cie sas Viale Europa 70, I-36035 Marano Vic. (VI) Italy ... IL ...

Robinson - see Wadkin

Robland International BV, Groote Markt, NL-4524 Sluis CD, Holland ... OPQ ...

Robland International BV, Jockey's Hall, Combs, Stowmarket, Suffolk IP14 2NH ... OPQ ...

Rotherwood Machinery Ltd, Somersham Rd., St. Ives, Huntingdon, Cambs PE17 4WR ... L ... P ... TU ... V

Circular Saws

RW Woodmachines Ltd, Rowood House, Murdock Road, Bicester, Oxon OX26 4PP UK ... T ...

Safety Training (Woodworking), 18 Chelwood Road, Cherry Hinton, Cambs CB1 9LX, UK ... (Training) ...

Saturn - Sigrist & Muller AG, CH - 8197 Rafz, Switzerland ... ABCD .. EG ...

Savecrest Machines Ltd, Stepney Grove, Bridlington, E Yorks YO16 7PD, UK ... WX

Sawquip International Inc, 91 Boisjoly, Lavaltrie, Quebec, J0K 1H0, Canada ... IJL ...

Scheer & Cie GmbH & Co, 75446 Wiernscheim, Lindenstr 70 Stuttgart-Feuerbach, Germany ... T ...

Schelling Anlagenbau GmbH A-6858, Schwarzach, Austria ... P ... V

Schelling UK Ltd, Unit 2, Sandbeck Way, Wetherby W Yorks LS22 7DN, UK ... P ... TU ... V

Joseph Scheppach Maschinenfabrik GmbH & Co, D-89335 Ichenhausen, Germany ... AD ... M ... OPQ ...

J Schneeberger Mashinen AG, CH-4914, Roggwill, Switzerland ... EH ...

J Schneeberger UK, Unit 15 Hunt. Ct, Westminster In Est Measham DE12 7NQ ... EH ...

HO Schumacher & Sohn *(See K.W.O.)* ... AD ...

SCM SpA, Via Casale, 450 47827 Villa Verucchio (Fo), Italy ... AD ... L ... OPQR . TU ... WY

SCM GB Ltd, Dabell Ave, Blenheim Ind Est, Bulwell, Nottingham NG6 8WA, UK ... AD ... L ... OPQE . TU ... WY

M Sedgwick & Co Ltd, Stanningley Field Close, Swinnow Lane, Leeds LS13 4QG, UK ... AD ... OP ...

Selco (Biesse Group) Via Della Meccanica 16, I-61100 Pesaro, Italy ... TU ... V

Dave Sharp Woodworking, Featherstone, Pontefract, W. Yorks WF7 5LR ... AD ... L ... R ...

Siko GmbH, Post fach 1106, D-79195 Kirchzarten, Germany ... (Digital Indicators) ...

Siko Ltd, 45 Murdock Rd, Bedford MK41 7PQ, UK ... (Digital Indicators) ...

Simonds Industries Inc, 135 Intervale Rd, Flitchburg, MA 01420, USA ... (Insert-tooth Saws) ...

Skellingthorpe Saw Services Ltd., Old Wood, N. Skellingthorpe, Lincoln LN6 5UA, UK ... ABCD .. EGH ... V

J.J.Smith & Co, Ltd., Moorgate Rd, Knowsley Ind. Park, Kirkby, Liverpool L33 7DR, UK ... AD ... LM ... OPQR ... V

SMS Machinery, PO Box 5333, Mortimer RG7 2YP UK ... S ... XY

Spear & Jackson International Ltd, Atlas Way, Atlas North, Sheffield S4 7QQ, UK ... AD ...

SpeedCut 2820 N Great Southwest Parkway, Grand Prairie, TX 75050, USA ... 269

Stegherr Maschinenbau GmbH & Co KG, Fabriktrasse 2-4, D-8413 Regenstauf, Germany ... 274

Stehle GmbH & Co, Allgauer Strasse 51-53, Postfach 1865 D-8940 Memmingen 1, GermanyABD .... EH ...

Stenner Ltd, Lowman Works, Blundells Road, Tiverton, Devon EX16 4JX, UK ... IJL ...

Steton SpA, I-41012 Carpi MO, Italy ... T ...

Stewart Lumber, Block 6, Lomond Ind. Est, Alexandra, Dumbartonshire G83 0TL, Scotland ... IJLM ... V

Storti Srl, I-26045 Motta Baluffi, Via Bassa di Casalmaggiore 7, (Cremona), Italy ... ILM ... Q ... V

Striebig AG, Grossmatte 26a, CH-6014 Littau, Switzerland ... S ... V

Super Thin Saws Inc, PO Box 299, Moretown Com. Center, Waterbury , Vt 05676, USA ... A ...

Supreme Saws Ltd, Newton Ind. Est, Eastern Ave West. Romford, Essex RM6 5SD, UK ... A ... (Diamond Saws & Tools) ...

Tewkesbury Saw Co, Ltd, Newtown Ind Est, Tewkesbury, Glouc GL20 8JG, UK ... AD ... L ... OP ...

TM Machinery Sales Ltd, 76a Leicester Rd,Wigston, Leicester LE18 1DR, UK ... A ... P ... T ...

Transwave (Power Capacitors) Ltd, 30 Redfern Rd, Tyseley, Birmingham B11 2BH ... (1-3 phase converters) ...

Trymwood Mach Sales, Woodpecker House, Balaclava Rd, Fishponds, Bristol, BS16 3LJ,UK ... A ... OP ...

Universal Superabrasives Inc, 27588 Northline Road, Romulus, MI-48174-2867, USA ... H ...

Universal - Unicorn Abrasives, Doxey Rd., Stafford ST16 1EA, UK ... H ...

Unique Machine & Tool Co, 4232 E. Magnolia St, Phoenix, Arizona 85034, USA ... Q ... T ...

Vertongen , New, NV, Schipstraat 30, 2870 Puurs, Belgium ... (Tenoners) ... 272

Vickers - see Electrodrives

Vollmer Dornhan GmbH, Postfach 11, D-72173, Dornhan 1, Germany ... EF ...

Vollmer Werke Maschinenfabrik GmbH, Ehinger Str. 34, D-88400 Biberach/Riss, Germany BCD .... EFGH ... V

Vollmer UK Ltd, Town Street, Sandiacre, Nottingham NG10 5BP UK ... BCD .... EFGH ... V

Wadkin Ltd, Green Lane Road, Leicester LE5 4PF, UK ... ABCD ... OPQ ... 273

Walter AG, Derendinger Strasse 53, D-72072, Tubingen, Germany ... D ... EH ... W

Walter GB Ltd, Walkers Rd, N Moons Moat Ind Est., Reddich, Worcestershire B98 9HE, UK ... D ... EH ... V

Michael Weinig AG Weinigstrasse 2-4, D-97941 Tauberbischofsheim, Germany ... A ... LM ... 131, 271, 274 V

Michael Weinig (UK) Ltd, 5 Blacklands Way, Abingdon Business Park, Oxon OX14 1DY, UK ... A ... LM ... 131, 271, 274 V

VR Wesson , Waukegan, IL, USA ... B ... (Hard Tipping Materials) ...

Woodtech Mach, Ltd, Unit 7, New Mill Ind Est, Inkpen, Hungerford, Berks RG17 9PU, UK ... ILM ... Q ... V

The Woods Group (Bolton) Ltd., Jackson Street, Farnworth, GM BL4 9HB ... ABD ...

Winter Diamantwerkezeuge, Schutzenwall 13-16 D-22844 Norderstedt, Germany ... H ...

Yew Lodge Health & Safety Training , Old Vicarage Gardens, High Street, Henlow Beds, SG16 6AD. ... (Health & Safety Training)